GLOBAL POSITIONING SYSTEM

Papers

published in

NAVIGATION

VOLUME IV

reprinted by

THE INSTITUTE OF NAVIGATION

Alexandria, VA

International Standard Book Number: 0-936406-03-8

Published in the United States of America
The Institute of Navigation
1800 Diagonal Road
Suite 480
Alexandria, VA 22314

CONTENTS

Overview

RONALD R. HATCH
Magnavox Electronic Systems Company, West Coast Division

This is the fourth volume of papers published in *Navigation* on the subject of the Global Positioning System. Several years have passed since the last issue was published. During that time the implementation of Selective Availability, which limits the available accuracy, and the development of carrier phase techniques, which provides a new level of accuracy, have resulted in an increasing emphasis on differential GPS systems. That emphasis is reflected in this volume. Only those GPS papers which address differential GPS or the new kinematic GPS, which makes use of the carrier phase measurements, have been reproduced in this volume.

The selected papers are separated into three broad categories. These are:

Characterization of DGPS Error Sources
DGPS Reference Stations and Wide Area Differential GPS
Kinematic and Carrier Phase Aided DGPS

The last category has been addressed so extensively that it is further divided into sub-categories. These are:

General Implementations
Land Survey Applications
Hydrographic Survey
Aircraft Positioning and Landing
Attitude and Azimuth Determination

Within each category or sub-category the papers are arranged chronologically. This allows the reader to obtain a sense of the progress being made.

A severe selection criterion has been applied by limiting the papers to those which have appeared in *Navigation*. However, when one looks at the increasing size of the proceedings of the navigation and GPS symposia, it is clear that a severe limitation was required to prevent this volume from becoming much too large. While some significant papers may not be found within this volume, the reader will find papers discussing virtually all aspects of differential and kinematic GPS developments. In addition, if the reader wishes to pursue any particular topic in more detail, the references cited in the papers can be used to access a dramatically expanded library of information.

It seems a bit ironic that the DoD, in pursuing a policy of limiting the accuracy available to the commercial user, has stimulated the development of DGPS systems which, when augmented with carrier phase measurements, have demonstrated a new level of accuracy. Clearly, this "centimeter" level of accuracy will prove useful in a wide spectrum of applications. Precision aircraft landing, prescription farming and remotely piloted vehicles are only a few of the multiple applications which can be expected to develop in the future.

NAVIGATION: *Journal of The Institute of Navigation*
Vol. 33, No. 4, Winter 1986–87
Printed in U.S.A.

Measurement Errors in GPS Observables

RICHARD L. GREENSPAN, JAMES I. DONNA

The Charles Stark Draper Laboratory, Inc., Cambridge, MA
Received June 1986

ABSTRACT

Airborne users of GPS may benefit from differential (relative) navigation to the extent that the errors in their measurements of pseudorange and/or delta range to each satellite are correlated with observation errors for the same satellite signals received at a reference site. This paper describes some results of an experiment to characterize the measurement errors of one airborne GPS user equipment. The results suggest that pseudorange errors are dominated by slowly varying systematic processes that are likely to be highly correlated from user to user, whereas delta-range errors appear to be zero-mean processes that are dominated by user generated noises, especially clock jitter and tracking errors.

TECHNICAL APPROACH

A 5-channel GPS receiver manufactured by the Collins Radio Division of Rockwell International for Phase-2B GPS Development Test and Evaluation was loaned to Draper Laboratory by the United States Air Force for this inflight test program. The receiver was operated in the unaided mode (no altimeter or inertial sensor inputs). During the flight tests the receiver was mounted on a pallet installed in a de Havilland DH-6 (Twin Otter) aircraft that has been modified to carry the Aerial Profiling of Terrain System (APTS)[1] developed by the Draper Laboratory as shown in Figure 1.

The GPS receiver was delivered with provision to format and record extensive diagnostic files including observed pseudorange, delta range, current values of propagation corrections, and interchannel delay bias as well as all broadcast satellite data. GPS observables were recorded every second for all four satellites being tracked.

APTS was used as the navigation reference for post-flight reduction of the GPS data. It provided ground-truth values for the aircraft location and velocity. These were incorporated into predicted values of pseudorange and delta range to the GPS satellites that were tracked by the receiver. These predicted observables were adjusted to account for propagation effects, earth rotation, receiver interchannel biases, and the net drift and drift rate of satellite and user clocks. The residuals between measured and predicted values of GPS observables are presented and discussed in a later section on interpretation of results.

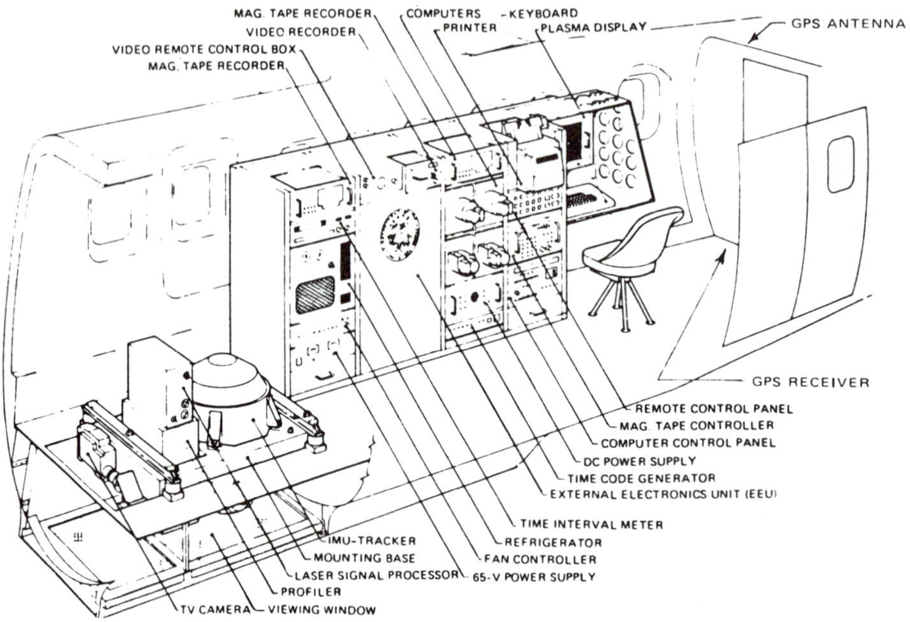

Fig. 1—Location of GPS Antenna and IMU in APTS Aircraft.

EXPERIMENTAL DESIGN ISSUES

Time Synchronization

The design goal in this test was to calibrate the APTS clock relative to GPS time accurately enough to keep the contribution of synchronization errors to the growth of position errors to less than 1 centimeter/second, and to maintain velocity error below 1 millimeter/second. With an aircraft speed of 50 meter/second and maximum acceleration on the order of 1 g, the corresponding time accuracy requirements are 200 microseconds and 100 microseconds, respectively.

A conservative goal was set to align time bases to 80 microseconds. This was achieved by exploiting the fact that the GPS receiver generates a precisely timed output pulse once every second. This pulse is synchronized to the internal receiver clock. The receiver also records a time mark data block every second, containing GPS time and data valid at the time of the pulse. Assuming good satellite geometry, the GPS time recorded in this block should be accurate to 100 nanoseconds or better.

APTS also has an internal clock. To calibrate APTS time against GPS time, a 12.8 kHz pulse train synchronized to the APTS clock was used to increment a counter. The GPS time mark pulse was used to reset the counter every second at a known GPS time. (In this experiment, this pulse was the only interface between the GPS receiver and APTS.) APTS sampled the count once every 2 seconds at a known APTS time. The count tells the time difference (in units of 1/12800 second) between the GPS receiver pulse and the recorded APTS time. The time offset between GPS and APTS is given (in seconds) by

$$\text{GPS time} - \text{APTS time} = \text{Recorded GPS time} - \text{recorded APTS time} + \text{count}/12800 - 0.0375$$

The extra 0.0375 second is present because the time of validity of APTS position and velocity data occurs 0.0375 second before the time that labels the APTS recording list.

The only possible remaining synchronization error is the ambiguity in determining which GPS pulse corresponds to each GPS record. This ambiguity introduces an error that is an integral multiple of 1 second. In practice it was easily resolved by inspection of the APTS and GPS data. A time mismatch of 1 second leads to position differences of 50 meters and velocity difference of up to 3 meters/second, both of which are easily detected.

Lever Arm

The position and velocity navigated by APTS are those of the center of the IMU stable member, which is mounted forward near the floor of the aircraft (Figure 1). GPS position and velocity are those of the phase center of the antenna, which is mounted toward the rear on the roof. For comparisons, it was necessary to add the displacements and velocities contributed by the IMU-to-antenna lever arm to the APTS navigated position and velocity before comparing with GPS. The IMU-to-antenna displacement was measured in airplane body coordinates (forward, right, down) to be (-5.288 ± 0.010, 0.000 ± 0.010, -1.360 ± 0.020) meters. The vertical uncertainty reflects the fact that the antenna is packaged in a box 4 centimeters high. Possible misalignments between the APTS navigation base and the aircraft increase the uncertainty by about another 1 centimeter in each coordinate. With maximum angular rates of about 0.1 radians/second, a 1 centimeter lever arm error contributes only 1 millimeter/second velocity error. Thus a lever arm uncertainty of a few centimeters does not seriously affect the comparison. The antenna will remain permanently mounted on the APTS aircraft. If it ever becomes necessary to measure the lever arm more accurately, it will be possible to do so.

Data Collection

Data were collected during a series of four flights from about 9:30 p.m. to 12:30 a.m. in May and June 1985 when four or more GPS satellites were simultaneously visible in Eastern Massachusetts. Data collection for the APTS reference navigation system and for GPS navigation were accomplished using independent equipment and later processed independently. The various flights involved the conventional operation of APTS; no special accommodations were made to the GPS mission except for the time of the flights, and the aircraft flight pattern.

A 5-channel Rockwell-Collins Global Positioning System (GPS) receiver was installed aboard the Aerial Profiling of Terrain System (APTS) aircraft during the week of May 20, 1985. This receiver was in the 5-channel DT&E configuration as modified for use by the Range Instrumentation Joint Program Office (RAJPO) at Eglin Air Force Base. The GPS receiver was functionally tested on the night of May 23 and joint APTS-GPS data collection flights were conducted on the nights of May 29, May 30, June 10, and June 11.

Figure 2 shows the flight path on May 30, with Hanscom Field in Bedford, Mass. at the origin. The flight altitude was approximately 2000 feet. The locations of the five retroreflectors are shown as diamonds in the plots; the X marks indicate 100-second time increments along the flight path. The sites labelled Mohawk, Natick, Acton, and Sudbury form a "racetrack" that allows APTS navigation to be updated using tracker data approximately every 200 seconds. The May 30 flight was the most successful in the sense that all retroreflectors were acquired on all passes and there were no APTS recording gaps.

These five sites had been used previously during APTS performance validation flights in January and February of 1984.[2] Their positions were surveyed by the U.S. Geological Survey (USGS).[3] Their relative horizontal positions can be expected to be accurate to about a part in 10^5 (corresponding to 10 centimeters at a maximum retroreflector spacing of 10 kilometers), and their relative elevations are expected to be accurate to within about 2 centimeters.

The coordinates of the retroreflectors are referenced to the North American datum of 1927 and the North American geodetic vertical datum of 1929. GPS navigates relative to the World Geodetic Survey 1972 (WGS 72) ellipsoid. Fortunately, USGS was able to supply the WGS 72 coordinates of several sites. In the limited region in which the APTS aircraft was flown, the offset of these retroreflector locations with respect to WGS 72 is given by a constant shift vector as follows:

WGS 72—North American Datum = (6.8, 27.6, 33.0) meters (north, east, down)

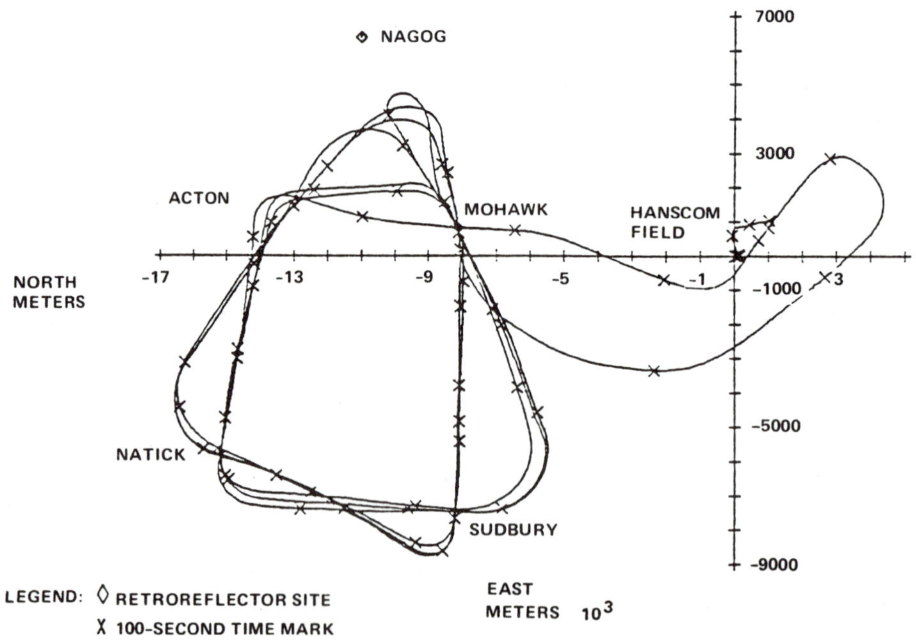

Fig. 2—Flight Path on May 30, 1985.

The difference can be used in adjusting the APTS navigated position when comparisons are made.

APTS Navigation Accuracy

The APTS aircraft incorporates an inertial navigation system which produces position data in ECEF coordinates (altitude, latitude, and longitude). Laser pointing and ranging to surveyed retroreflectors provides a companion navigation system which removes the long-term increase of position errors attributable to drift, misalignment, and gravitational anomalies in the inertial solution, and ties that solution to local geodetic coordinates (Figure 3).

The tracking laser is mounted on a gimballed carrier which provides a precise measurement of the direction from the aircraft towards any retroreflector that is illuminated. These angles, combined with the two-way pulse transit time uniquely locate the aircraft in 3-dimensions relative to that retroreflector. The measured transit times are corrected for propagation delay before their use in the navigation solution using values of pressure and temperature measured at the aircraft.

The peak errors of the APTS *real-time navigation* system are 60 meters and 20 centimeters/second in position and velocity, respectively. This is adequate for locating and acquiring retroreflectors on the ground. Recorded data is post-processed to yield much higher accuracies. One-sigma errors of *post-processed data* are typically 1.0 centimeters and 0.3 millimeters/second during lock-on to a retroreflector; errors increase to 50 centimeters and 5 millimeters/second 150 seconds after lock (assuming one retroreflector is acquired every 5 minutes).

The GPS specifications for navigation errors are 15 meters "spherical error probable" (SEP) and 0.1 meters/second rms velocity errors for each axis. We see that the corresponding APTS errors after post-flight processing are at least one order of magnitude smaller. Therefore, APTS is a suitable standard reference for measuring GPS navigation errors.

The APTS post-flight filter and smoother embody the APTS error models.

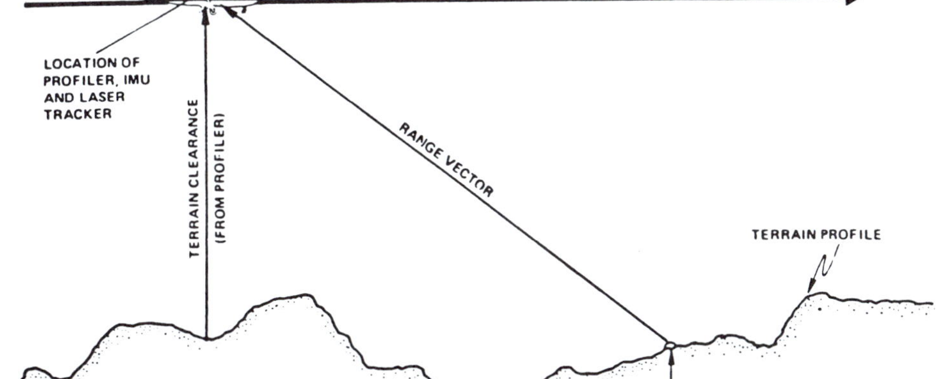

Fig. 3—APT System Concept.

The square root of the diagonal of the smoother's covariance matrix represents the probable error remaining in APTS navigation after postprocessing. The peak value calculated during the May 30 flight is 0.8 millimeters/second. Experience shows that this smoother estimate is too small by a factor of 2 or 3, possibly because of an over-optimistic choice of APTS measurement noise parameters. We estimate that APTS errors in velocity estimation during that flight are less than 2.4 millimeters/second.

As an additional check on APTS velocity measurement accuracy, the flight of May 30 was refiltered, omitting measurements from the Mohawk lock at 4966 seconds. The peak difference in velocity error between the two differently processed APTS outputs during the periods from 4794 seconds to 5178 seconds is 2.5 millimeters/second. We infer that APTS velocity measurement error is always less than 2.5 millimeters/second when retroreflector locks are no more than 400 seconds apart, and is likely to be much smaller for shorter intervals between lock because there is evidence that the error grows as a quadratic function of time.

DATA REDUCTION

In the absence of any errors except those caused by clocks, the two types of GPS observations of any satellite signal are related to the user position as follows:

$$\rho(t) = \text{Pseudorange} = |\mathbf{S}(t - \tau(t)) - \mathbf{u}| + c(t_u - t_s) \tag{1}$$

and

$$\Delta\rho(t) = \text{Delta range} = \int_{t-T}^{t} \frac{d}{dx} |\mathbf{S}(x - \tau(x)) - \mathbf{u}(x)| dx + \frac{cT}{f}(f_u - f_s) \tag{2}$$

$\mathbf{S}(t)$ is the satellite position at time t
$\mathbf{u}(t)$ is the user position at time t
t_u is the offset of the user clock from GPS time at t
t_s is the offset of the satellite clock from GPS time at t
$\tau(t)$ is the line-of-sight propagation time
f_u is the offset of the user clock from the GPS carrier frequency
f_s is the offset of the satellite clock from the GPS carrier frequency
T is the interval over which delta-range is counted
f is the GPS carrier frequency
c is the speed of light in free-space

Equation (1), where "free-space" propagation is assumed and effects of the propagation medium are treated as error sources, states that pseudorange is the line-of-sight range plus a clock error and (2) states that delta-range is the *change* in line-of-sight range over T seconds plus a clock drift (rate) error. When T is small enough and the relative dynamics of the satellite and user are small, then the delta-range is proportional to the average velocity of the user towards the satellite.

In the standard GPS navigation solution **u**(t), t_u, and f_u are estimated and t_s and f_s are reported via the broadcast data message, as is an approximate expression for **S**(t). The analysis reported here reverses this procedure as follows:

(1) Pseudorange and delta-range environments output by the GPS user equipment are recorded once per second. These recorded measurements include the effects of propagation delays and interchannel bias. The receiver's estimates of these error sources were individually recorded and included in a separate output block. Therefore they were available to incorporate in our predicted values of the observables. (See Step 7)

(2) **S**(t) is computed from the broadcast ephemeris.

(3) **u**(t) is calculated by transforming the APTS navigation solution to WGS-72 coordinates. In this transformation the lever arm from the APTS reference point to the GPS antenna phase center is incorporated.

(4) t_s, f_s are obtained from the satellite data message.

(5) t_u, f_u are obtained from a 17-state Kalman filter solution obtained by post-flight processing of GPS observables and APTS inertial sensor data. User clock errors determined by a high-order filter solution for a combined GPS/INS sensor suite are being used as truth for performance evaluation of the unaided GPS receiver. The APTS navigation reference solution is not used to estimate user clock errors.

(6) Pseudorange and pseudo-delta range are predicted according to (1) and (2) and the residuals

$$\rho(t) - \hat{\rho}(t); \Delta\rho(t) - \hat{\Delta}\rho(t) \tag{3}$$

are recorded for analysis. The satellite orbits are given in ECEF coordinates. Therefore, the rotation of the earth during the satellite to user propagation time must be included in the transformation from terrestrial coordinates to WGS-72.

(7) Step 6 is repeated including various refinements in the prediction of the GPS observables using data output by the GPS receiver. This data includes a) ionospheric delay corrections, b) tropospheric delay corrections, and c) receiver interchannel delay biases. Note, that no special corrections are available for the delta-range observations.

Ionospheric Delay

The test receiver uses dual-frequency measurements to estimate the ionospheric delay (advance relative to free space propagation) on each satellite to user path. This data was available to incorporate into the data reduction.

Tropospheric Delay Correction

The GPS receiver predicts the following values for excess delay attributable to propagation through the troposphere:

$$\Delta\text{range (meters)} = \frac{2.4224}{0.026 + \sin E} \exp[-0.13345\ h] \qquad (4)$$

where

E is the elevation, angle of the satellite above the horizon (degrees)
h is the user altitude (km)

Interchannel Delay

The experiment uses a GPS receiver that continuously monitors differential delay between the four channels that process satellite signals and a fifth (reference) channel used for ionospheric measurements. Any common channel delay would be "absorbed" in the navigation filter's estimate of receiver clock offset. These delay estimates were incorporated directly in the pseudorange predictions using the attribution of receiver hardware channels to specific satellites that is part of the data output by the receiver.

(8) The residuals remaining after Step 7 are the unmodeled or unavoidable errors in interpreting GPS observations for navigation.

INTERPRETATION OF RESULTS

The composite of data from the May and June flights suggests that one should consider two regimes of operation when discussing the quality of GPS observables. These are:

(1) Nonstandard conditions.

(2) Standard (nominal) conditions.

Nonstandard conditions are typically the result of poor satellite geometry, unusually high noise bursts, or sporadic user oscillator "glitches." The satellite geometry effects error modeling in two ways:

(1) Geometric dependence of propagation errors.

(2) Poor GDOP, leading to poor clock error estimates.

Pseudorange Residuals for the Flight of May 30, 1985

The flight of May 30, 1985 was selected for intensive analysis because there were no failures or gaps in the performance of APTS as the reference system.

Figure 4 shows the dilution of precision factors that apply during the data collection on May 30, They are significant because clock bias must be estimated from satellite data. Note that the DOPs are poor until about 4100 seconds after the APTS reference time. We can see the consequences of the initially poor geometry in Figure 5, which presents the pseudorange residuals and in Figure 6, which presents the delta-range residuals for all six satellites that were observed during the tests.

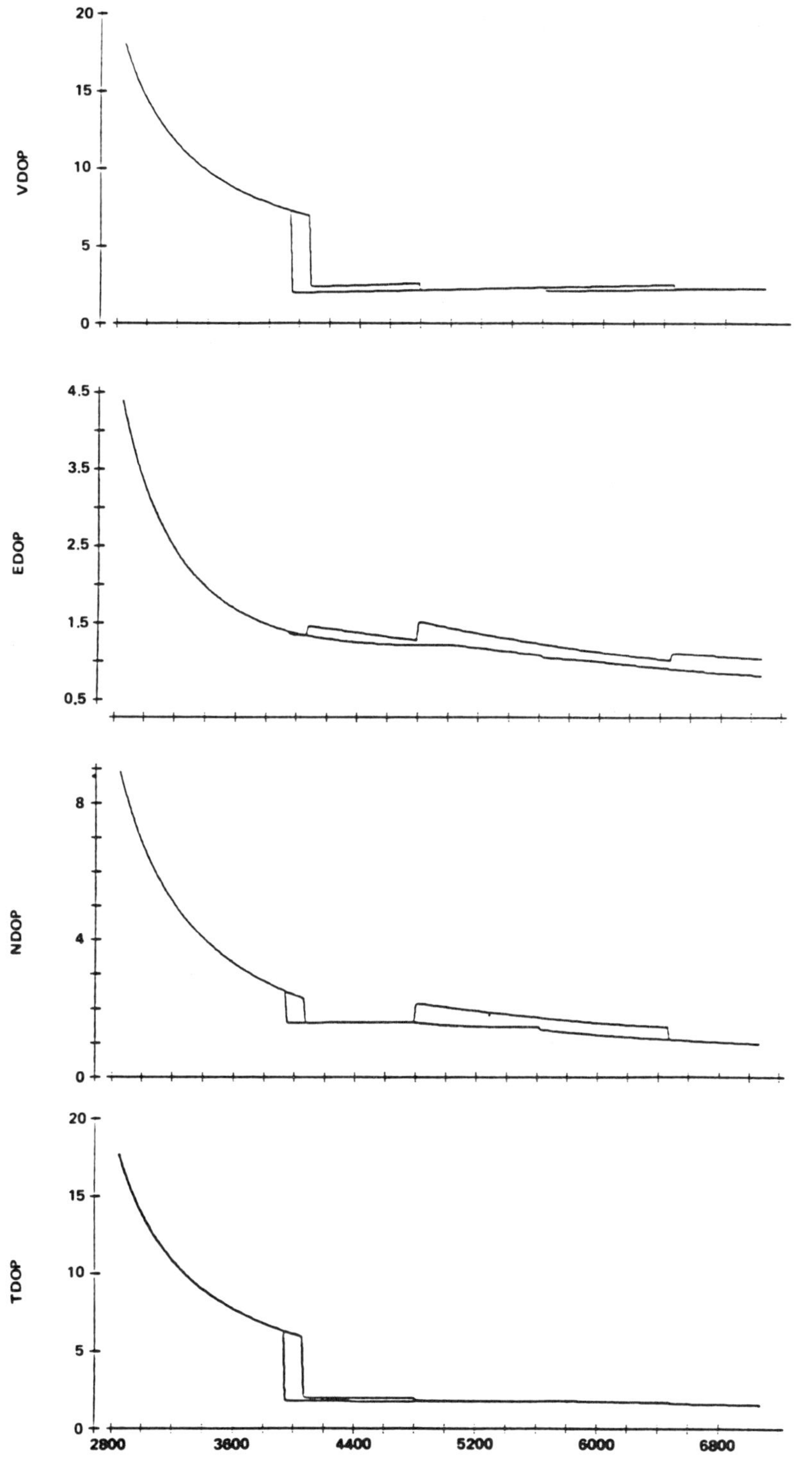

Fig. 4—NDOP, EDOP, VDOP, TDOP on May 30, 1985.

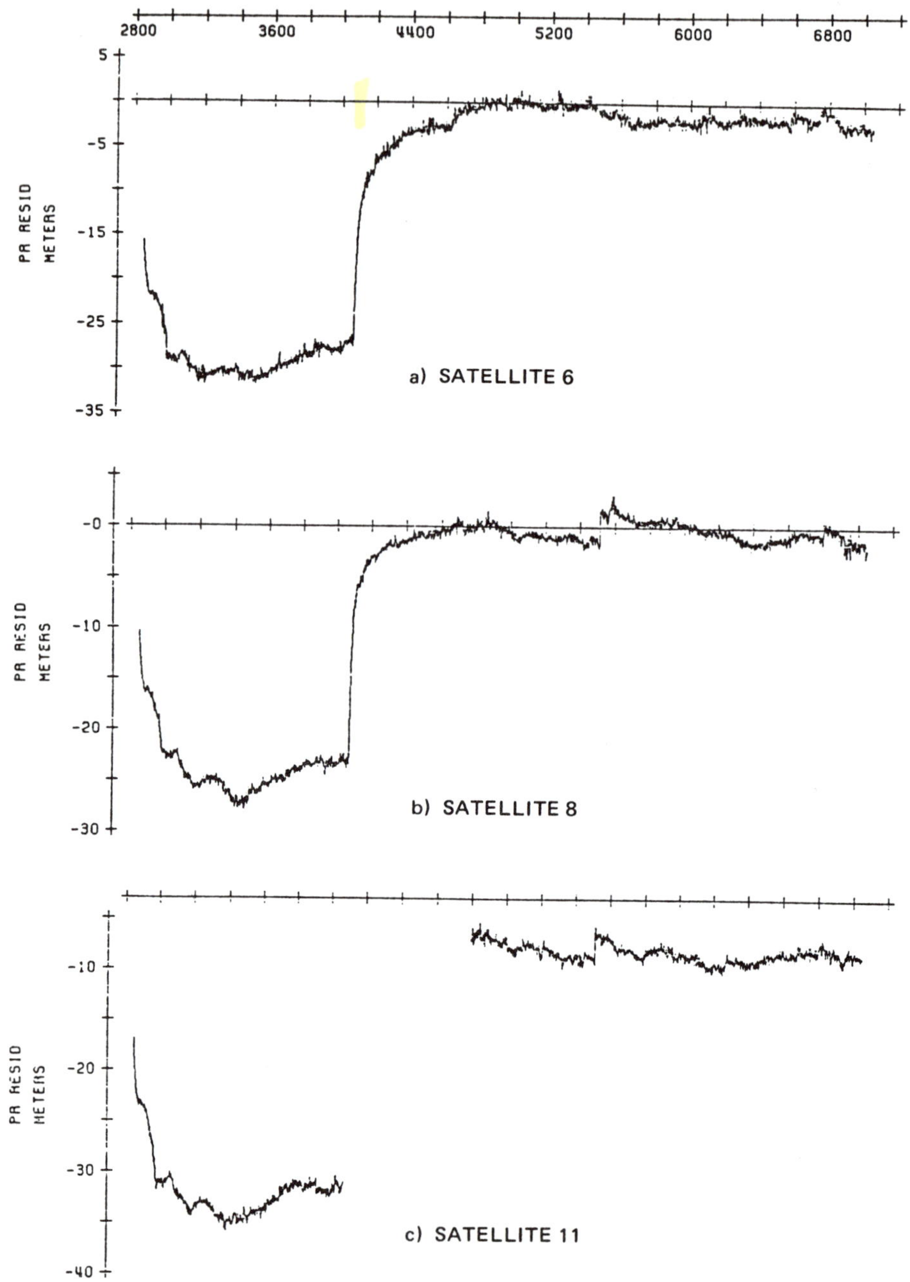

Fig. 5—Pseudo-Range Residuals with Corrections and Improved Clock Offset State (May 30, 1985).

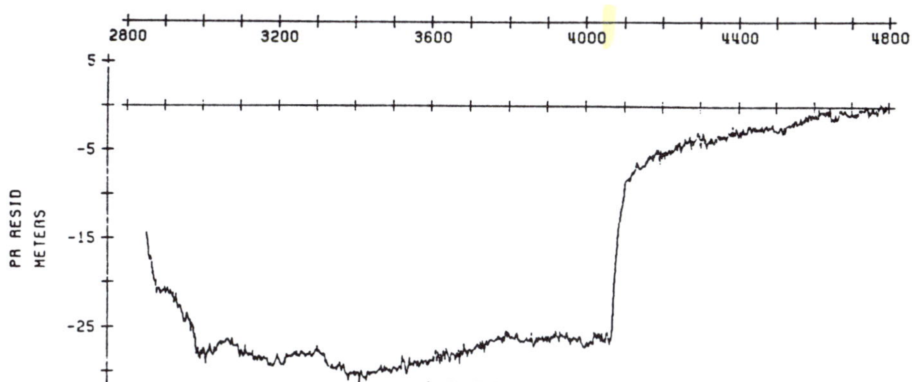

d) SATELLITE 9

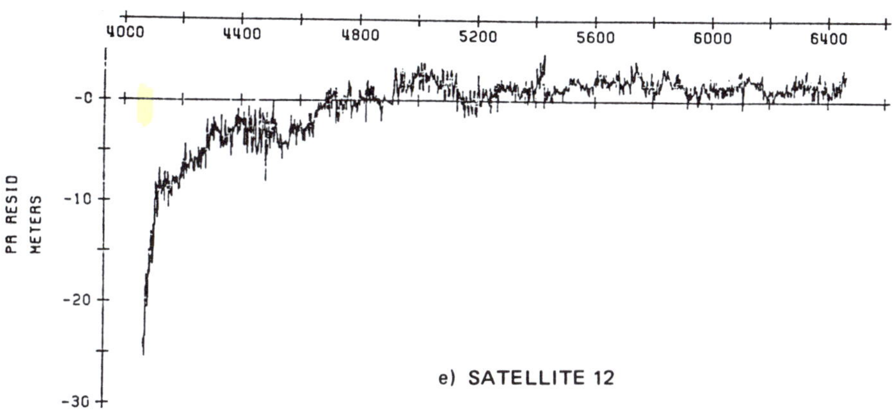

e) SATELLITE 12

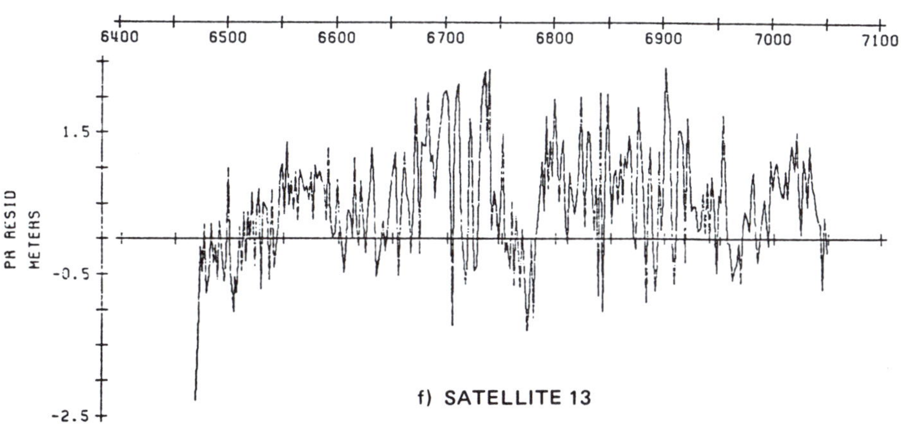

f) SATELLITE 13

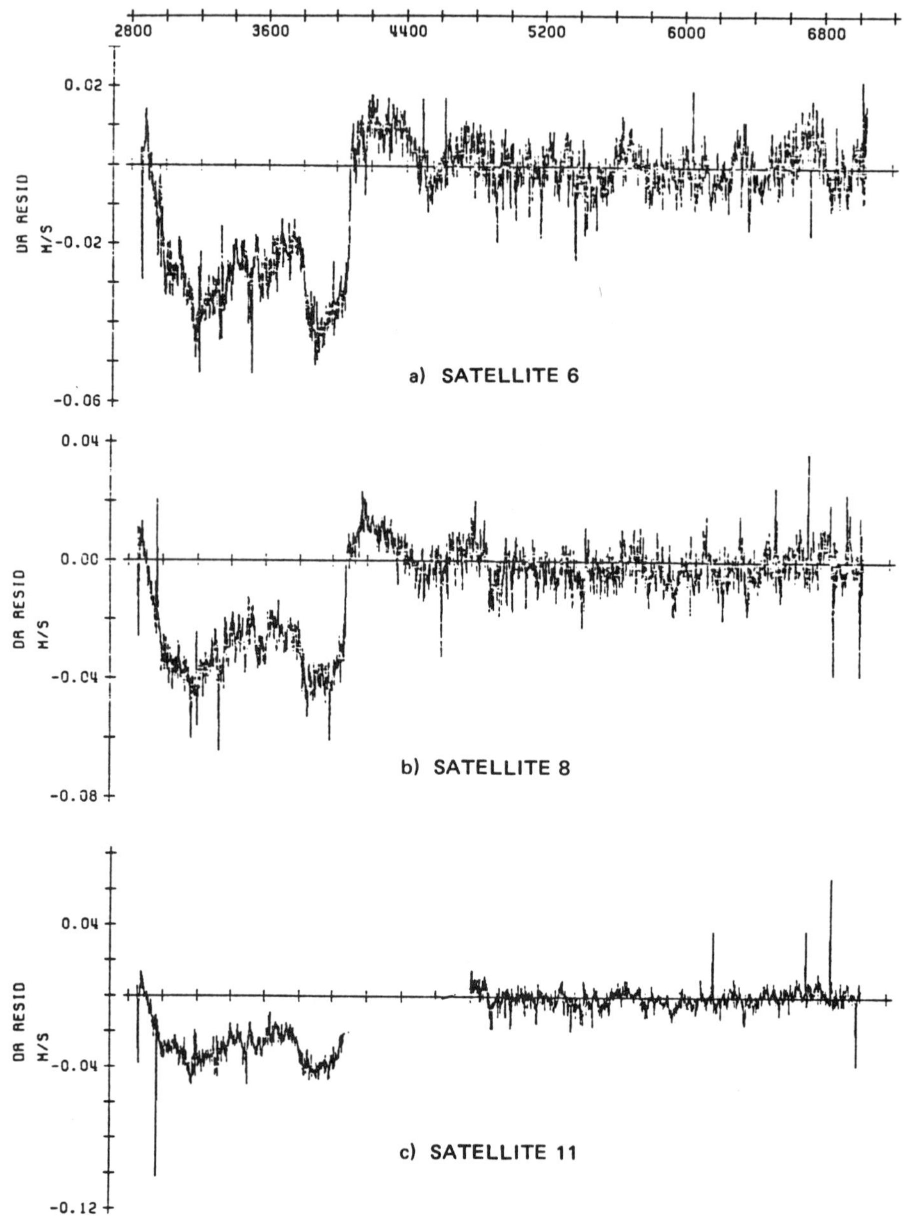

Fig. 6—Delta-Range Residuals with Corrections and Improved Clock Drift State (May 30, 1985).

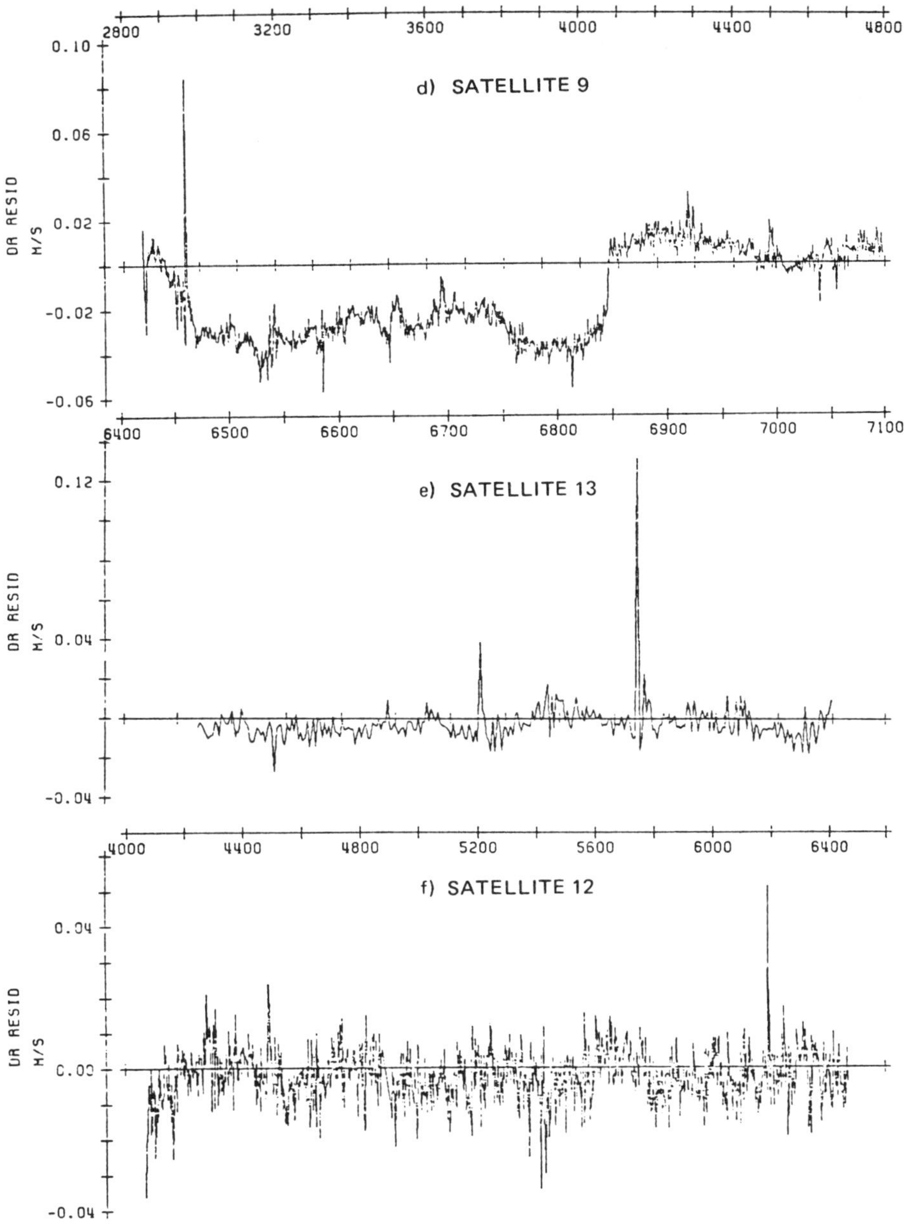
d) SATELLITE 9
DR RESID
M/S
0.10
0.06
0.02
-0.02
-0.06
2800
3200
3600
4000
4400
4800
e) SATELLITE 13
DR RESID
M/S
0.12
0.04
-0.04
6400
6500
6600
6700
6800
6900
7000
7100
f) SATELLITE 12
DR RESID
M/S
0.04
0.00
-0.04
4000
4400
4800
5200
5600
6000
6400

The residuals for Satellites 6, 8, 9, and 11 show a large negative term of about 30 meters from turn-on to t = 4100 seconds. This bias is caused by an error in the estimated clock offset, which in turn is large because of the poor satellite geometry. As soon as satellite 13 becomes visible at t = 4100 seconds the geometry improves, the clock errors become very small, and the pseudorange residuals become small.

During "standard" conditions (e.g. Figures 5 and 6) we observed that the GPS pseudorange residuals were typically on the order of three meters or less, and that they were dominated by a slowly varying component plus a zero-mean additive noise whose standard deviation is on the order of 0.5 to 1.0 meter. To this degree of approximation the pseudorange observable can be written as PR(t) where

$$PR(t) = R(t) + c\delta T + \delta r_p + n_r(t) + b(t) \tag{5}$$

and where

- $R(t)$ is the line-of-sight range from the user to the satellite
- δT is the composite of satellite and user clock error plus receiver interchannel biases
- δr_p is a propagation correction comprised of a predicted troposphere correction plus the ionospheric delay measurement
- $n_r(t)$ is a zero-mean noise process representing code-loop tracking errors
- $b(t)$ is a very slowly varying bias process

The model for pseudo-delta range observations is even simpler:

$$PD_r(t) = D_r(t) + cT\frac{\delta f}{f} + n_v(t) \tag{6}$$

where

- $PD_r(t)$ is the pseudo delta-range
- $D_r(t)$ is the delta-range from LOS geometry
- $\dfrac{cT\delta F}{f}$ is the error attributable to frequency offsets
- $n_v(t)$ is a zero mean noise process

We were not able to detect any consistent bias terms in the delta-range observations at the level of 0.01 meters/second or higher. The errors are dominated by a zero-mean process whose standard deviation is about 0.5 to 1.0 centimeters/second. In principle, the combined effect of refraction in the ionosphere and troposphere could produce ray-bending that would be observed as a wavelength (frequency) change. If these effects were present, they were too small to be detected.

The delta-range measurement residuals were too large to attribute to phase measurement errors caused by thermal noise in the carrier tracking loops. That noise produces an error whose standard deviation σ_ϕ is estimated by

$$\sigma_\phi = \sqrt{\frac{1}{\rho}} = \sqrt{\frac{B}{(C/n_o)}} \text{ (radians)} \tag{7}$$

where ρ is the measurement SNR and B is the tracking loop bandwidth. For the unaided 5-channel set B is 5.0 Hz and (C/n_o) exceeds 42 dB · Hz.

For these parameter values the rms effect of thermal noise on delta-range accumulation over a 0.78 second interval (the delta-range accumulation interval in the 5-channel Collins receiver) would be no more than

$$\delta DR = \frac{\sqrt{2}}{0.78} \frac{\sigma_\phi}{2\pi} \lambda = 1.0 \text{ mm/s}$$

where $\lambda = 19.3$ cm at GPS L_1 (1.575×10^9 Hz). As noted above the rms value of observed delta-range measurement noise is about ten times larger.

The most likely source of the observed "noise" is either receiver oscillator jitter, or quantization of the phase tracking error in the digital filter that mechanizes the receiver's carrier loop. If the oscillator is the noise source, the error is a "common mode" factor that will not affect navigation accuracy. If the noise is caused by quantization, then its effect is independent from satellite to satellite.

CONCLUSIONS

In conclusion, we see that a few mechanisms account for most of the observed residuals. It appears that for a dual-frequency receiver with a "good" model of tropospheric delay, and an accurate measurement of inter-channel biases the *pseudo-range errors* are dominated by the bias term b(t), and *delta-range errors* are dominated by random fluctuation noise.

The bias term is probably driven by ephemeris errors or by errors in the broadcast values of satellite clock corrections. It would be instructive to repeat the calculation of residuals using the most accurate non real-time ephemerides available at the GPS master control station to see if the bias term is significantly reduced.

As a last comment, we believe it is a tribute to the GPS Program Office that they have coordinated the action of all GPS segments (space, control, and user equipment) so that the dominant pseudorange error is in fact only a bias of 1 part in 5 million (or less) and that the dominant delta-range error arises within the user equipment.

ACKNOWLEDGMENTS

The authors would like to thank the following organizations that contributed to the success of this program.

The United States Geological Survey which sponsored the program, and especially Mr. William Chapman who has fostered the development and use of APTS aircraft and ground support facilities.

The United States Army Avionics Research and Development Activity, which contributed to funding the project, and especially Mr. Joseph McGowan who arranged for the loan of GPS user equipment and who contributed useful insights to the interpretation and presentation of our raw data.

The Rockwell-Collins Division of Rockwell-International (Cedar Rapids, Iowa); and, in particular, Mr. Merv Dosh who routinely answered our questions about the Data Blocks output by the GPS user equipment hardware, and the navigation software.

REFERENCES

1. Soltz, J. A., G. Mamon, W. Chapman. *Aerial Profiling of Terrain System Implementation,* Presented at Second International Symposium on Inertial Technology for Surveying and Geodesy, Baniff Alberta Canada, June 1981.
2. *APTS Applications Test Report: Sudbury Point Positioning Project Results,* CSDL Report R-1782, June 1985.
3. *APTS Calibration Range; Final Report,* USGS Document, June 1983.

NAVIGATION: *Journal of The Institute of Navigation*
Vol. 34, No. 2, Summer 1987
Printed in U.S.A.

Experimental Measurement and Characterization of Ionospheric and Multipath Errors in Differential GPS

J. W. SENNOTT

Bradley University, Peoria, IL

D. PIETRASZEWSKI

Coast Guard R&D Center, Avery Pt., Groton, CT
Received January 1987

ABSTRACT

Real-time differential GPS is fundamentally limited by a) signal estimation errors, b) satellite ephemeris errors, c) propagation decorrelation errors, and d) local multipath errors. In this paper we address the latter two error sources. An understanding of these is needed to establish an error budget for any monitor station network that the U.S. Coast Guard would consider for a nationwide system.

To estimate these error sources, extensive two-frequency P-code and integrated Doppler data have been collected over 150 kilometer and 1500 kilometer baselines. A processing scheme is developed which employs both code and carrier observations. Applicable to real-time differential systems, as well as to post-collection data analyses, the estimation scheme treats multipath errors and carrier-code offsets as state-variables. The resulting Kalman filter yields the very precise carrier observation, corrected for carrier-code offset.

Day-to-day repeatability of multipath errors is illustrated. The decorrelation of ionospheric errors across short and long baselines is evaluated by combining the integrated Doppler estimation scheme with the standard two-frequency ionospheric measurement technique. Also discussed are the impact of multipath and ionospheric error sources on Differential GPS.

INTRODUCTION

This research is sponsored by the U.S. Coast Guard in an effort to determine the limits of coverage provided by a GPS differential monitor station. The error sources that reduce the accuracy of GPS have a variety of natural and systematic origins. Broadcasting corrections determined at a fixed location is one method to compensate for some of the errors. The corrections themselves lose their effectiveness with the passage of time and separation in distance from the monitor station. This research studies the effect of separation on the level of correlated errors along 150 kilometer and 1500 kilometer baselines. The level of correlation is an important factor in establishing the error budget needed to design the type of monitor station network that the U.S. Coast Guard would consider for a nationwide system.

The goal of the overall research program is to develop a cost effective, accurate, and reliable radionavigation service for both harbor approach and harbor phases of marine navigation. The requirement for these phases is a predictable accuracy (2 drms) of 8 to 20 meters. At the present time, no federal radionavigation service satisfies this requirement. GPS augmented with a differential correction network, DGPS, possesses the potential to satisfy this requirement.

Unlike post-processed geodetic GPS, real-time vehicular DGPS is more susceptible to short-term drifts between vehicle and reference station error sources. While the effects of satellite ephemeris errors are a function only of relative geometry, there are significant effects whose magnitude and rate-of-change are less well understood. For single-frequency DGPS operation the degree of correlation of ionospheric errors is of particular interest. Care is required to separate this error contribution, and others such as multipath, from receiver noise. To accomplish this, the much more precise integrated Doppler observables, together with code loop data, have been integrated in a Kalman filter system.

Within the filtering scheme, the carrier-code offset and multipath components are treated as state variables, and are removed before the very precise L1 and L2 carrier observations are used to track ionospheric decorrelation across the baseline. This same technique appears generally applicable to DGPS reference stations and user filters. A related code/carrier measurement technique has been reported elsewhere.[1,2,3] An estimate of the multipath contribution to DGPS errors occurs as a byproduct of the filtering scheme.

In support of the study, the USCG sponsored GPS data collection activities over two baselines, Groton-Boston and Groton-Peoria. Data was collected during the spring and summer of 1986. The data collection systems deployed at each site were constructed and programmed at the U.S. Coast Guard Research and Development Center. Each system contains a Texas Instruments 4100 GPS receiver and a Hewlett Packard 200 series computer. The TI-4100 located at the USCG R&D Center was connected to a rubidium frequency standard. The HP-200 computer controls the TI-4100 through the "CDU" port and monitors the data at the "INSTRUMENTATION" port. The software is designed to give information to the operator during data collection, enabling the operator to conveniently monitor the performance of the TI-4100. The software also replaces many of the normal "CDU" operations, such as loading of almanacs, with single keystroke inputs.

Data are recorded in two files. The first of these, used for reviewing the quality of data immediately following a run, is generated by software resident on the HP-200. The second file contains a record of all "INSTRUMENTATION" port data, and is recorded on a Telebyte "PACKETAPE" in the TI-4100 binary format. The short baseline data were taken over a period of 18 days. A total of 31 data collection runs were recorded. Eight long baseline runs, spanning 16 days, were also collected. Using software developed at Bradley University, the data were decoded from the binary records stored on the "PACKETAPE" and processed as discussed below.

OBSERVATION MODEL

In a preliminary attempt at measuring the ionospheric delay, L1 and L2 P-code observables were employed. While simple low-pass filtering of L1-L2 differences reduced the code tracking loop contribution to the estimate to below 1 nanosecond, it soon became apparent that the remaining low-frequency variations were too large and too rapid to be ionospheric. Geodetic users have ascribed this low-frequency term to multipath, a conjecture strongly supported by the high degree of day-to-day correlation demonstrated later in this paper. This suggested a state variable approach, in which the multipath is treated as a colored noise source within a carrier bias resolution scheme. The corrected carrier observable, measured separately on L1 and L2, provides a measurement of sufficient accuracy to discern decorrelation of the DGPS ionospheric term over baselines in the 100 kilometer range.

Consider the signal tracking loop outputs of the coherent GPS processor. These consist of both code delay and integrated carrier Doppler, measured relative to the user clock. The differential equations describing code and integrated carrier Doppler observables are

$$\dot{z}_1(t) = \dot{\rho}(t) + L\dot{T}(t) + \dot{m}(t) \tag{1}$$

and

$$\dot{z}_2(t) = -\dot{\rho}(t) + L\dot{T}(t) - \dot{a}(t) \tag{2}$$

where

$\rho(t)$	=	user-to-satellite range, measured in code chips
$T(t)$	=	user-satellite clock offset, in seconds
L	=	a scale factor, seconds per chip
$m(t)$	=	multipath error, in code chips
$a(t)$	=	the bias between integrated Doppler counter and code delay counter, in code chips

The $\rho(t)$ term consists of free-space path delay, as well as delay contributions from ionosphere, troposphere and receiver. It is important to observe that the sense of range-rate is reversed between the code and Doppler channels. A positive range-rate (user-satellite range increasing) is seen by the code channel as an increase in code delay. A positive range rate is seen as a negative Doppler frequency, resulting in a decrease in the integrated carrier count.

Integrating the above equations, and expressing them in time-discrete form, we have

$$z_1(k) = \rho(k) + LT(k) + m(k) + w_1(k) \tag{3}$$

and

$$z_2(k) = -\rho(k) + LT(k) - a(k) + w_2(k) \tag{4}$$

where w_1 and w_2 model tracking loop noise on code and carrier tracking loops, respectively. In order to make use of the far greater measurement resolution of $z_2(k)$, it is necessary to estimate the bias parameter, $a(k)$. In the absence of receiver cycle slips, this variable need only be estimated each time a new satellite rises. However, in the filtering scheme described below this quantity is modeled as a slowly varying state, which enables recovery from cycle slips occuring due to temporary signal attenuation or severe multipath. Although

w_1 is far larger in variance than w_2, the code observation will be used together with the carrier observation to estimate the quantity a(k). With sufficient smoothing time, a(k) may be estimated to much better than a carrier cycle.

Note that adding (3) and (4) the user-to-satellite path delay term is eliminated. We will treate this sum as a new observation

$$z(k) = 2LT(k) + m(k) - a(k) + w(k) \tag{5}$$

A typical plot of this observable, collected at Groton on Julian day 156, is shown in Figure 1. Several features of the data are of interest. First, since user-to-satellite dynamics have been removed in forming the observable, it is possible to see quite small fluctuations in z(k). Over the 900 second duration of the plot, the total change is 0.2 chips, or 20 nanoseconds. The most obvious feature of the data is the linearly increasing trend, with a slope on the order of .02 nanoseconds/seconds, which results from the user-satellite clock drift term T(k).

Superimposed on the linear change in z(k) are medium-frequency terms, which fluctuate over intervals of 10–100 seconds. Local peak-to-peak variations are on the order of 5 nanoseconds. These terms, which are of much lower frequency content than tracking loop noise, are contributed by multipath reflections from objects within a code chip delay. The resulting error in code tracking is typically found to be 1–2% of the code chip length. Even though it averages out over intervals of several minutes, this delay error component is of some concern in real-time DGPS applications. In the estimation work to follow, m(k) will be treated as a colored noise source. It should be noted that the impact of multipath on the carrier observable, z_2, is much less than on code, and is seen

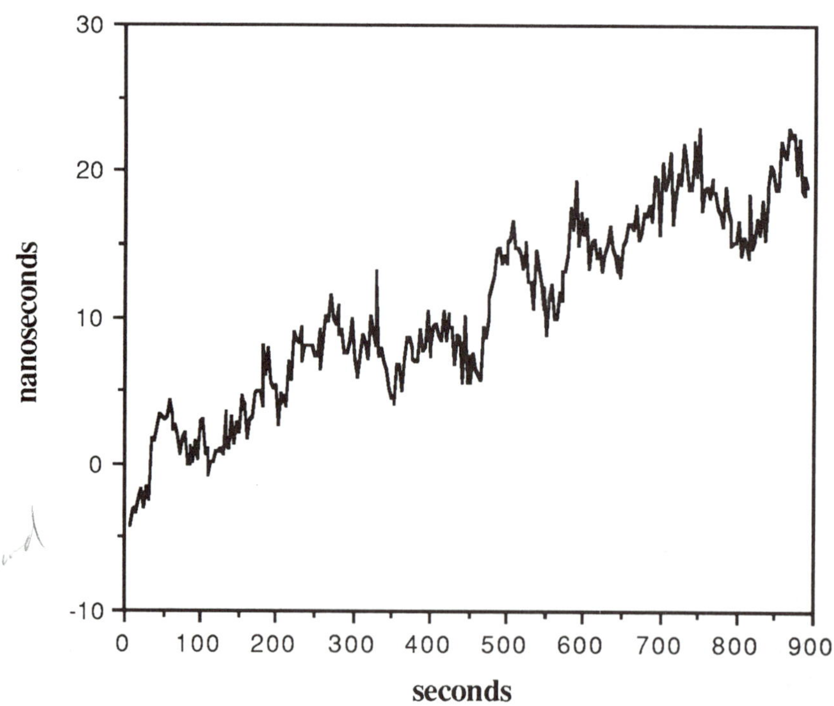

Fig. 1—Integrated Carrier Plus Code, Z(k).

as a small phase shift of the carrier. While severe multipath may momentarily capture the carrier tracking loop, causing cycle slips, these slips will ultimately be regained in tracking the bias state, a(k).

The term m(k) can be largely removed by filtering. The far more precise observable $z_2(k)$, corrected with the estimate of a(t), can then be employed in the DGPS reference station, or user position-fixing estimator, or in satellite tracking and ionospheric studies. The following sections explore this approach.

MULTIPATH ESTIMATOR

The state variable model for this estimator is shown below:

$$a(k) = c_1 a(k-1) + w_a(k) \tag{6}$$

$$m(k) = c_2 m(k-1) + w_m(k) \tag{7}$$

$$T_1(k) = T_1(k-1) + \Delta T\, T_2(k-1) \tag{8}$$

$$T_2(k) = T_2(k-1) + w_T(k)$$

The state variable $T_1(k)$ is defined as the drift between the user and satellite clock, from the start of the data filtering interval. The (common) offset at the beginning of the interval is absorbed in the user-to-satellite range variable $\rho(k)$.

A bank of eight filters has been implemented. Each filter is assigned to an L1 or L2 channel on one of the four satellites. This system is illustrated in Figure 2. Although the main purpose of the filtering system is to estimate and remove the carrier bias from $z_2(k)$, a useful byproduct is an estimate of the local multipath environment for code tracking receivers.

A measure of the effectiveness of the filter design is obtained from examining the power and "whiteness" of the filter residuals. Figure 3 is a plot of the original data and residuals on one of the eight channels. The standard deviation is about 2 nanoseconds, contributed mostly by code tracking loop noise. The ability of the filter to isolate the multipath and other lower frequency terms is evident here.

MULTIPATH EFFECTS

Figure 4 is a plot of the filter estimate of m(k) for the data set shown previously. With the clock rate term removed and the receiver noise contribution to the estimate of m(k) reduced to under 1 nanosecond, the magnitude and rate of change of the multipath term is now more clearly evident. The low-frequency character of these swings demonstrates the difficulty of obtaining real-time DGPS accuracy to the meter level with code data alone. The 4 nanosecond swings seen in the P-code data translate through GDOP into DGPS errors of several meters. The C/A-code results are not expected to be as good.

The ability to estimate and track the bias parameter, a(k), in the presence of strong multipath is of critical importance, as errors in this estimate will be passed on to the corrected carrier observable. Within the estimator there is an obvious trade between the ability to reject multipath and the speed of response to carrier cycle slips. Figure 5 shows the results of using process correlation times of 10 and 100 seconds, for m(k) and a(k) respectively. The estimate of

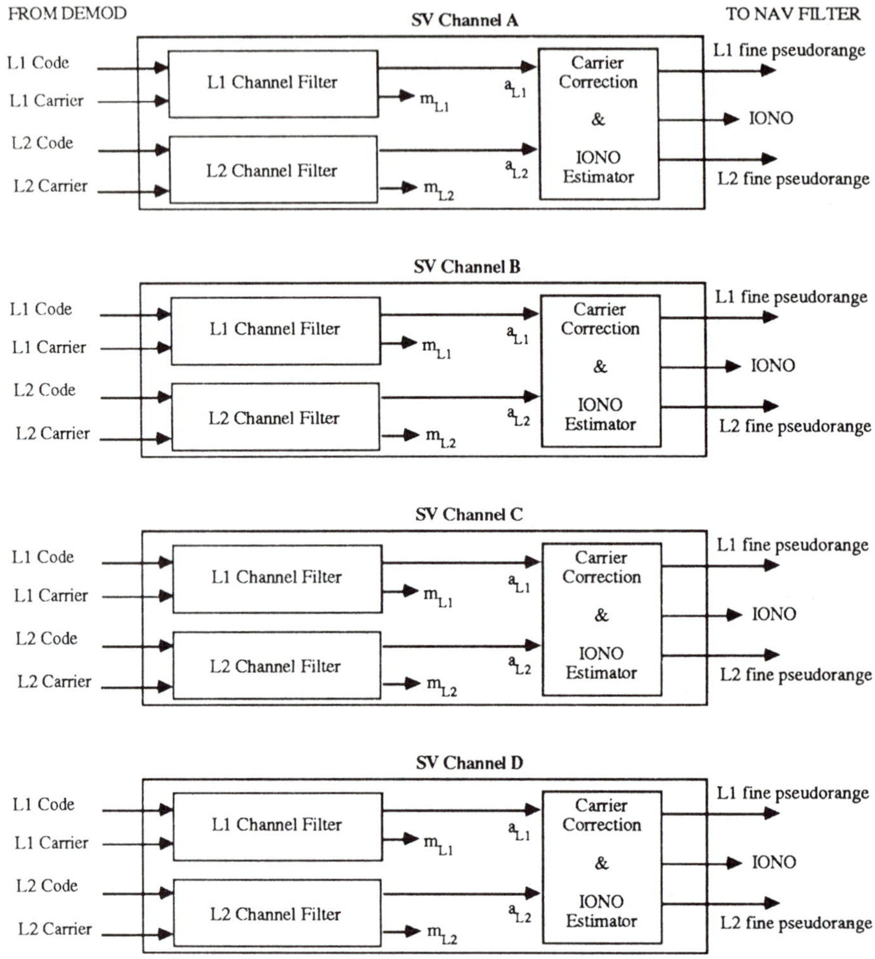

Fig. 2—Carrier and Code Estimator Structure.

a(k) holds nearly constant, with a variation of under 0.2 nanoseconds, even though the multipath has a variation of 4 nanoseconds.

Of the three sites, the Peoria site was judged to have the most severe multipath problems. To demonstrate the repeatability of the multipath component, data from the Peoria site were compared for Julian days 217 and 219. Figure 6 shows a segment of multipath estimator output for SV8, observed on days 217 and 219. The plots were obtained by recording the multipath output of the above estimator system. The time axes were then shifted to obtain approximately the same satellite elevation and azimuth angles on each day. While the plots are not identical there is a high degree of correlation. Careful attention to antenna location and design is one way to reduce the severity of this effect. Alternatively, one can employ the carrier tracking and estimation scheme developed above.

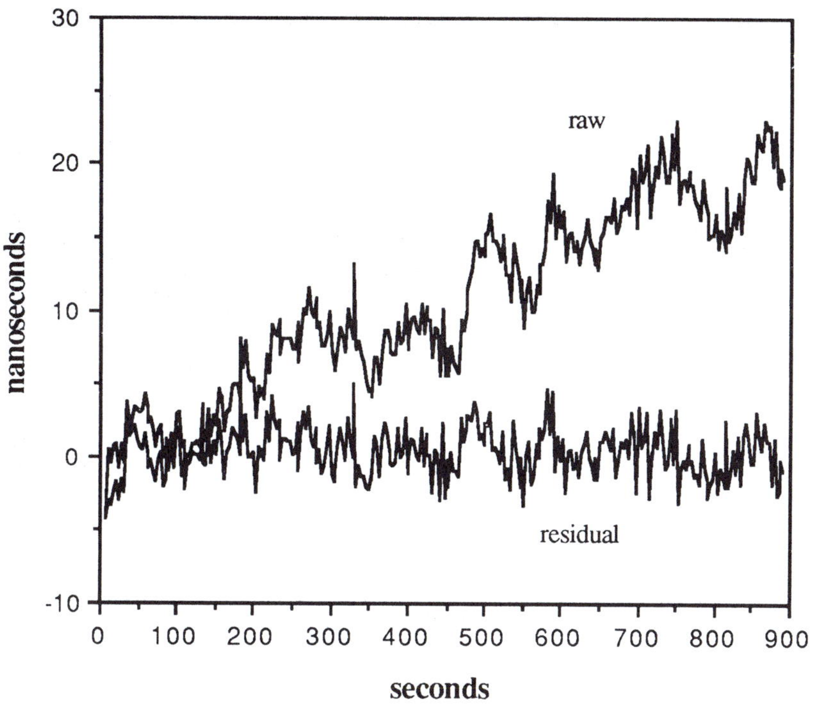

Fig. 3—Raw Data and Filter Residuals.

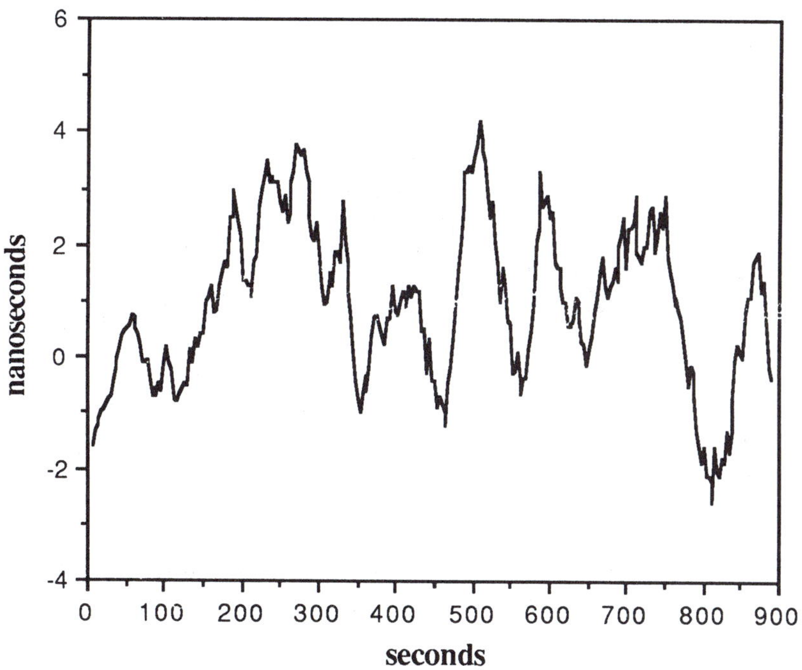

Fig. 4—Multipath Estimate.

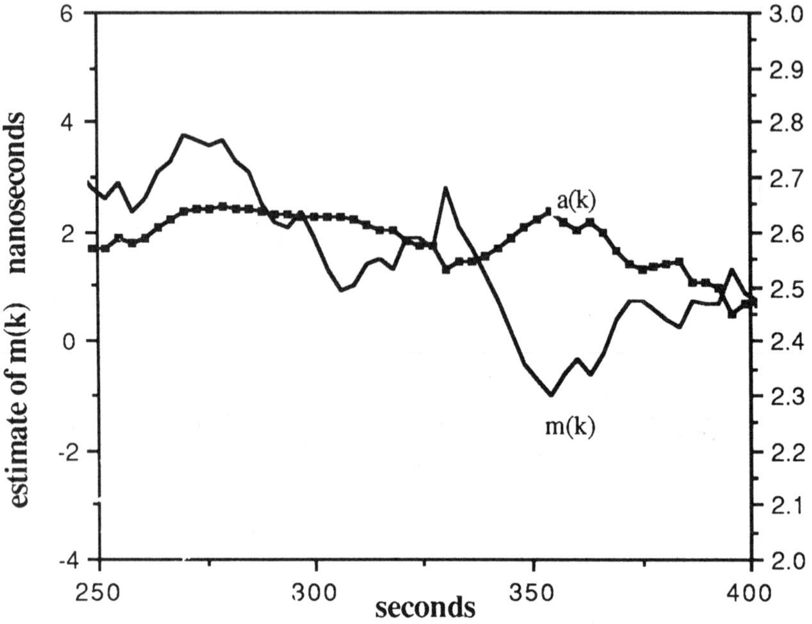

Fig. 5—Ambiguity and Multipath Estimates.

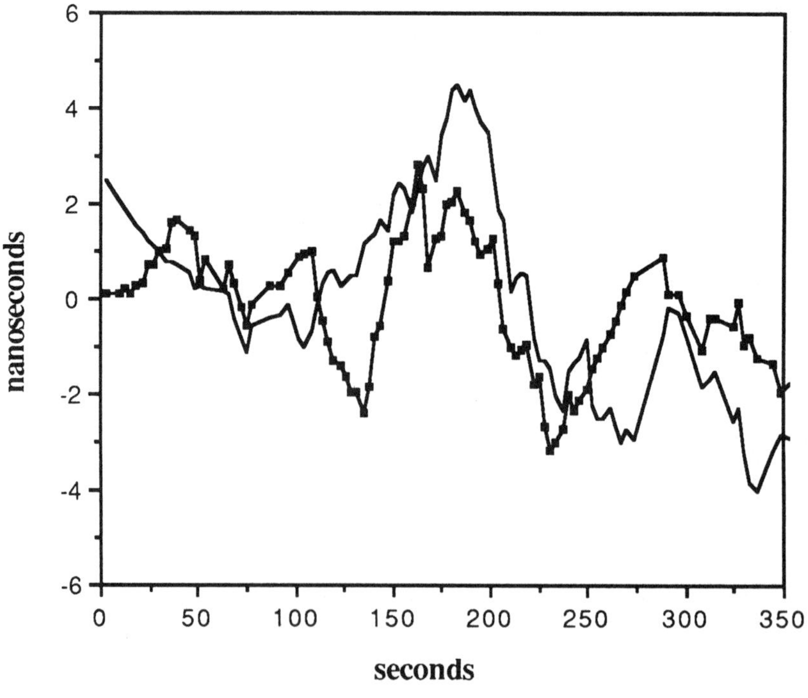

Fig. 6—SV Multipath Comparison Day 217 and 219.

IONOSPHERIC ESTIMATOR

In estimating the ionospheric delay term, we may take advantage of the common-mode user-satellite clock drift rate parameters on L1 and L2. This modification of the carrier technique yields a more rapid and accurate convergence of the carrier offset parameters on a given satellite. Consider L1 and L2 versions of (5):

$$z_{L1}(k) = 2LT(k) + m_{L1}(k) - a_{L1}(k) + w_{L1}(k) \tag{9}$$

$$z_{L2}(k) = 2LT(k) + m_{L2}(k) - a_{L2}(k) + w_{L2}(k) \tag{10}$$

Rather than process the above observables in separate filters, as done in estimating multipath on each satellite frequency and path, these quantities can first be differenced to produce a new observable

$$\begin{aligned} \Delta z(k) &= z_{L1}(k) - z_{L2}(k) \\ &= \Delta m(k) - \Delta a(k) + \Delta w(k) \end{aligned} \tag{11}$$

There are two filter states defined here, $\Delta m(k)$ and $\Delta a(k)$. This significantly reduces the computational burden, requiring only four two-state filters, one for each satellite. Now consider the carrier data seen on L1 and L2 channels

$$z_{2,L1}(k) = -\rho_{L1}(k) + LT(k) - a_{L1}(k) + w_{2,L1}(k) \tag{12}$$

$$z_{2,L2}(k) = -\rho_{L2}(k) + LT(k) - a_{L2}(k) + w_{2,L2}(k) \tag{13}$$

Differencing these uncorrected carrier counts gives:

$$z_{2,L1}(k) - z_{2,L2}(k) = -\Delta a(k) - [\rho_{L1}(k) - \rho_{L2}(k)] + \Delta w(k) \tag{14}$$

Adding the estimate of $\Delta a(k)$ to (14) yields the two-frequency difference of pseudoranges. Multiplying by 1.54 gives the desired estimate of ionospheric delay on frequency L1:

$$IONO = -1.54[z_{2,L1}(k) - z_{2,L2}(k) + \Delta\hat{a}(k)] \tag{15}$$

It should be noted that the $\rho(k)$ terms in (12) and (13) are offset by receiver group delays. Since these differ on L1 and L2 receiver channels, they remain in (15). However, when estimating the ionospheric decorrelation across the baseline, the difference in L1/L2 biases between receivers will be estimated.

IONOSPHERIC DECORRELATION

As discussed earlier, the separation of ionospheric errors from multipath and receiver noise terms posed a special challenge. While heavy smoothing of code data would reduce greatly the effect of receiver noise, some low-frequency multipath artifacts would remain. Additionally, any fast changes that might be present in the ionosphere would be masked by such smoothing. The carrier approach provides both fast response and greatly reduced multipath and receiver noise contributions to the ionospheric measurement.

As reported elsewhere,[4] the ionospheric decorrelation is expected to be in the range of one to ten parts per million of the baseline length. Over the 150 kilometer baseline we would anticipate ionospheric differences of the order 0.15

to 1.5 meters. To gain confidence in the measurement technique several sets of the short baseline data were first analyzed.

Figure 7 is typical of the many plots obtained on day 156. The two curves show nearly identical delay dynamics. Note that the plots contain the L1/L2 inter receiver group delay terms. This inter-receiver L1/L2 difference was found to be very stable for the two TI4100 receivers used in the experiment. Combining all short baseline data on day 156, the best estimate of this delay was 6.8 nanoseconds. The calibration was repeated for day 136, and the value was estimated to be 7.1 nanoseconds. In all subsequent data reduction a value of 7.0 nanoseconds is employed.

Differencing the curves in Figure 7, and correcting for the inter-receiver L1/L2 group delay, Figure 8 is obtained. The quantization of this plot is 0.1 nanosecond. The ionospheric difference across the baseline is seen to fluctuate between −1.5 and −1.0 nanoseconds. The data obtained for SV8 is typical of the other satellites observed, with the ionospheric difference fluctuating several tenths of nanoseconds about a magnitude of less than 2 nanoseconds. The baseline differences observed for days 136 and 156 are given in Table 1. Day 156 shows a somewhat larger range of decorrelation, with values ranging between +1.5 and −2.5 nanoseconds. While much data remains to be processed, these values suggest that the contribution of the ionospheric term, for reference baselines of 150 kilometers or less, is adequately small for single-frequency DGPS marine users.

In this preliminary analysis of the data, two days of long baseline observations were processed. Of particular interest during the observation window was

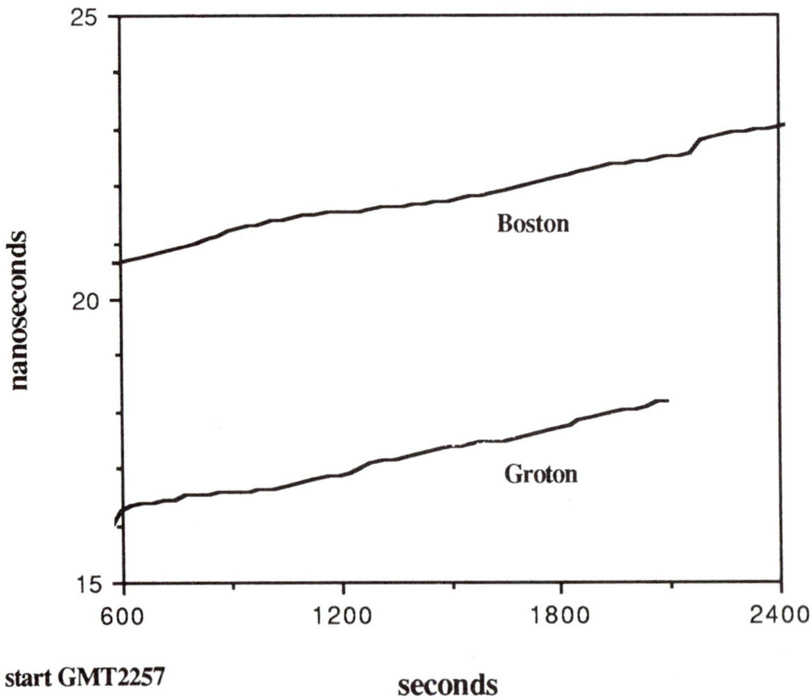

Fig. 7—SV6 Boston-Groton Ionosphere Plus Receiver Delay JD156.

150 Km separation

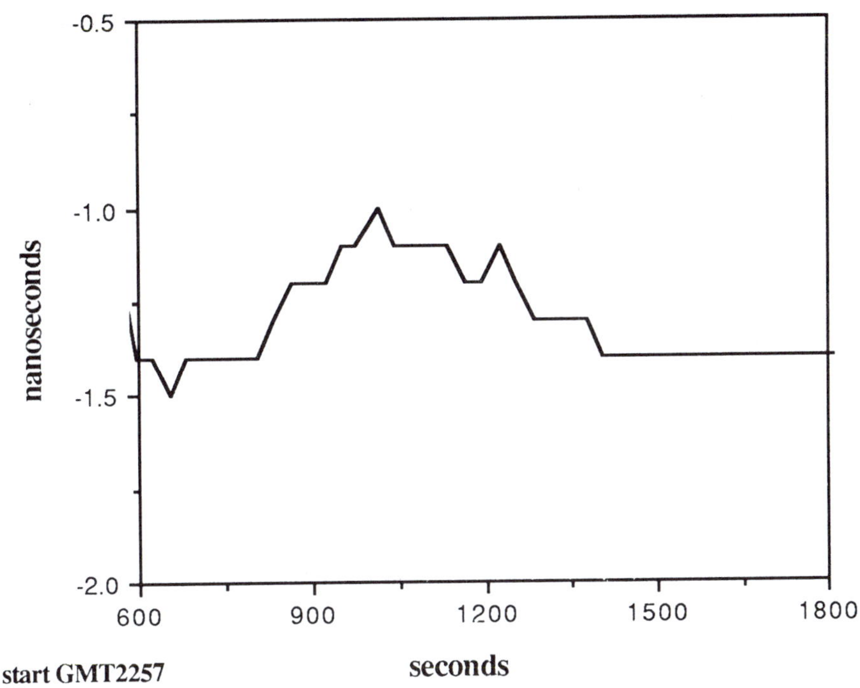

Fig. 8—SV6 Boston-Groton Ionosphere JD156.

Table 1.—Average Ionospheric Difference, Boston-Groton Baseline

Day 136	*SV6* −1.5 ns	*SV8* − .7 ns	*SV9* +0.5 ns	*SV13* +0.1 ns
Day 156	*SV3* +1.5 ns	*SV6* −2.5 ns	*SV8* −0.8 ns	*SV11* +0.2 ns

Above data includes 7.0 nanosecond inter-receiver L1/L2 channel offset correction.

the day-night transition, as seen first by the eastward station at Groton. It was anticipated that larger static and dynamic ionospheric differences would be observed, due both to differences in obliquity and offsets in day-night transition times.

Plots of the ionospheric estimates at each site showed very similar changes in delay. However, the differences in ionospheric delay were significantly larger than for the short baseline data. Unlike the short baseline, there were also some major differences in the delay dynamics observed. Figure 9 compares the delays seen for SV8. Note that the plots have not been compensated for inter receiver L1/L2 delay. Figure 10 is a plot of the compensated difference, showing a difference between sites varying from −4 to −8 nanoseconds. The magnitude and rate-of-change is much larger that observed for any of the short baseline data. It is conjectured that during the day-night transition, both the eastward satellite azimuth for SV8 and large east-west offset of the sites contributed to the delay difference.

Table 2 summarizes the ionospheric differences observed between Peoria and

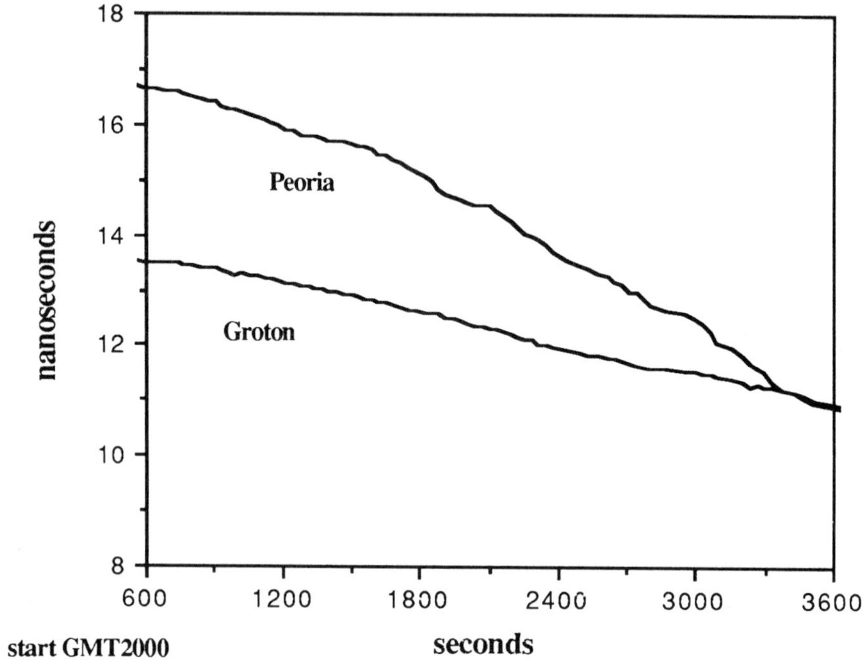

Fig. 9—SV8 Peoria-Groton Ionosphere Plus Receiver Delay JD219.

1500 km Separation

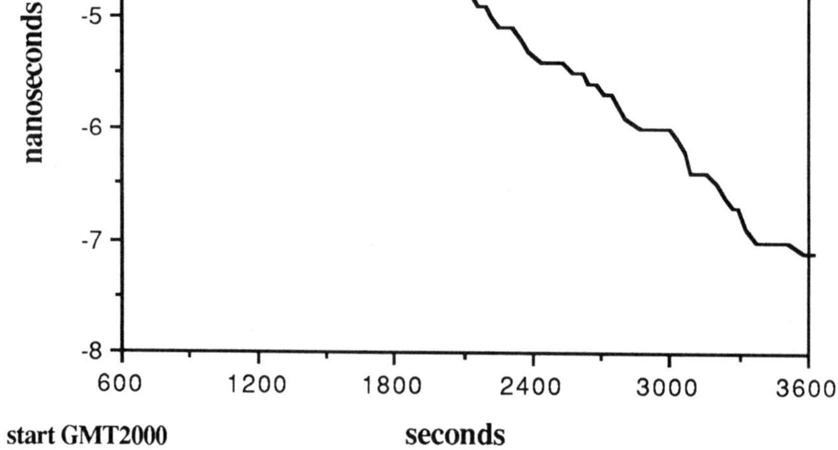

Fig. 10—SV8 Peoria-Groton Ionosphere JD219.

Table 2.—Average Ionospheric Difference, Peoria-Groton Baseline

Day 203	SV6	SV8	SV9	SV11
	−.75 ns	+.5 ns	−4.7 ns	+6.9 ns
Day 219	**SV6**	**SV8**	**SV9**	
	−2.7 ns	−5.5 ns	−6.8 ns	

Above data includes 7.0 nanosecond inter-receiver L1/L2 channel offset correction.

Groton. Day 203 values range between −4.7 nanoseconds and +6.9 nanoseconds, with Day 219 values ranging between −2.7 and −6.8 nanoseconds. These values are clearly larger than acceptable for single-frequency DGPS marine operation in harbor areas.

CONCLUSION

This study is aimed at determining the error sources which affect the C/A-code differential GPS user. In order to separate the various error sources, a combined code and carrier filtering scheme was developed. The scheme enables accurate separation and measurement of ionospheric errors and local multipath errors. Modeling the carrier-code offset and multipath terms as state variables allows the precise carrier observable to be used both in error analysis work, and in real-time DGPS applications.

At typical sites the multipath error was found to contribute a P-code range delay of 4 to 6 nanoseconds. Further study is warranted to determine the magnitude of this error on the C/A-code. A preliminary analysis of short and long baseline ionospheric decorrelation was carried out. Short baseline (150 kilometer) data suggest that the typical decorrelation is in the range of 1 to 2 nanoseconds. Over the 1500 kilometer baseline, larger rates of change and absolute offsets were observed, with decorrelations typically in the range of 4 to 6 nanoseconds.

At the present time the remaining data sets are being analyzed for multipath and ionospheric contributions. Using the same data sets and an expanded version of the data reduction filter, satellite ephemeris contributions are under study. These efforts are one facet of the ongoing U.S. Coast Guard effort to develop a cost effective, accurate, and reliable radionavigation service for both harbor approach and harbor phases of marine navigation.

ACKNOWLEDGEMENTS

The authors would like to thank Jack Kraemer of the Department of Transporation, Transportation Systems Center, Cambridge, MA for encouragement and helpful suggestions. Lee Luft of the Coast Guard R&D Center, Groton, CT developed much of the data collection software and hardware and provided field support during the experiment

REFERENCES

1. Hatch, R., *The Synergism of GPS Code and Carrier Measurements,* Proceedings of Third International Geodetic Symposium on Satellite Doppler Positioning, DMA/NOS, Washington, D.C. 1982.

2. Seeber, G., Schuchardt, A. and Wubbena, G., *Precise Positioning Results with TI4100 GPS Receiver on Moving Platforms,* International Geodetic Symposium on Satellite Positioning, Austin, TX 1986.
3. Myerhoff, S.L. and Evans, A.G., *Demonstration of the Combined Use of GPS Pseudorange and Doppler Measurements for Improved Dynamics Positioning,* International Symposium on Marine Positioning, USGS, Washington, D.C. 1986.
4. Henson, D.J. and Collier, E.A., *Effect of the Ionosphere on GPS Relative Geodesy,* Proceedings IEEE Plans 1986.

NAVIGATION: *Journal of The Institute of Navigation*
Vol. 35, No. 1, Spring 1988
Printed in U.S.A.

U.S. Coast Guard Differential GPS Navigation Field Test Findings

D. PIETRASZEWSKI, J. SPALDING, C. VIEHWEG, L. LUFT
U.S. Coast Guard R&D Center, Groton, Connecticut
Received September 1987
Revised January 1988

ABSTRACT

The U.S. Coast Guard has a research program to investigate the capability of C/A-code GPS and methods to improve its accuracy and integrity. A differential GPS demonstration system is being assembled and tested at the U.S. Coast Guard Research and Development Center in Groton, Connecticut. Analysis of static and dynamic testing is discussed. Details of instrumentation supporting the test program are also presented.

BACKGROUND

As a primary government agency responsible for providing aids to navigation to the civil community, the U.S. Coast Guard views Global Positioning System (GPS) as a potential replacement for existing systems. GPS is also considered to be a tool that can be used by the U.S. Coast Guard to meet present and future mission objectives. A differential GPS (DGPS) makes two significant improvements to the GPS. It can correct for significant natural and unnatural errors that may affect navigation accuracy. It also provides a mechanism to detect and warn users of potentially dangerous malfunctions during routine operation of GPS more quickly than the GPS Control Segment. A failure warning method is a necessary and valuable feature existing in other systems such as Loran-C.

The details of the differential GPS portion of the U.S. Coast Guard research program were outlined in a paper by Quill [1]. The development program has reached the first stages of static and dynamic testing. The system is called the differential GPS demonstration system. Four major elements make up the DGPS demonstration system: DGPS monitor station, DGPS user equipment, data link, and test instrumentation. The primary DGPS monitor station was developed by TAU Corporation and is described in [2]. A Magnavox monitor station also was used for a portion of the dynamic tests reported in this paper. The DGPS user equipment is the commercially available Magnavox T-set, modified for differential operation. (It is available from Magnavox with modifications.) A Radio Technical Commission For Maritime Services Special Committee-104 (RTCM SC-104) message format [3] is being used during early field testing. The radio beacon data link described by Quill has been constructed, but was undergoing separate testing at the time these tests were performed. The test instrumentation was developed at the U.S. Coast Guard R&D Center.

Analysis of differential GPS performance is being studied at two levels of user dynamics. The first level looks at the stability and accuracy of navigation information provided by stationary user equipment, but not using static navigation filter parameters. The navigation filter is set up for dynamic conditions in these tests, but the antenna remains stationary. Offsets between user sets, short and long term stability, bias conditions in the user equipment and differential monitor station operation, and differential versus non-differential performance are some of the areas that are being studied at this level. The second level attempts to study the same performance characteristics of user equipment undergoing dynamic maneuvers. The maneuvers were performed by a U.S. Coast Guard 42 ft patrol boat.

Prior to the static and dynamic test series, the compatibility of the equipment making up the demonstration system was tested under static conditions. A considerable amount of effort was expanded during this earlier testing to assure compatibility of the equipment. An important part of the U.S. Coast Guard development of a differential GPS demonstration system is the use of equipment from a variety of manufacturers. Most of the test results presented involve equipment from at least two manufacturers. The primary differential monitor station consists of Trimble 4000 A receivers and data processing software developed and implemented by TAU Corporation on an Indtech computer. The user equipment is the modified Magnavox 5-channel T-set. During static testing, the monitor antenna and user antenna are located within 10 m of each other. The results of earlier tests showed that a differential GPS can incorporate equipment built by different manufacturers. Some position bias differences still exist between T-sets. Similarly a small range bias seems to exist in the TAU corrections. Work to resolve these relatively minor problems will continue through the future phases of the U.S. Coast Guard research program.

STATIC DGPS TESTING

Many tests using stationary equipment were conducted during April and May of 1987. The satellites were observed during the primary constellations and the "secondary" constellation (PRN 6, 9, and 12 using fixed altitude). Analysis of the test results showed that the performance was within the 8 to 20 m target established at the start of the project. During the best of conditions, the results were better than 8 m (2 drms). The reintroduction of NAVSTAR 4 (PRN 8) into the GPS created a three-satellite constellation of marginal utility. The analysis presented here takes a look at the performance improvement differential GPS can provide during the use of a marginal constellation. The test also provided an extreme set of conditions for the differential GPS demonstration system.

A test was conducted on 15 May in such a way as to exaggerate the natural error conditions affecting C/A-code GPS. The objectives of these static tests were to determine if both the differential monitor and user equipment could track low satellites and what level of performance improvement could be realized under these conditions. The Magnavox 5-channel differential T-set tracked satellites PRNs 8, 9, and 12 with a known fixed altitude. During this period, PRN 8 was operating with a crystal clock and was 2 to 5 degrees (rising) above the horizon. PRN 9 was 8 to 1 degrees (setting) above the horizon, and PRN 12 was 53 to 46 degrees (setting) above the horizon. Three major error sources affected this test. They are uncertain crystal clock drift, high tropospheric delay, and high ionospheric delay. The horizontal dilution of precision (HDOP) was about 4.5 (latitude DOP of 4 and longitude DOP of 2).

Analyses of the test results are shown in Figures 1 through 4. These graphs are produced using a program written by the project team. The same program can process data from a moving platform. It was used to prepare the analysis contained in this paper. The discussion of the analysis in this section includes a description of the graphic presentation. Later in this paper, only the analysis will be discussed.

The geodetic WGS-84 location of the antenna used during the static tests is known to within 2 m. In Figures 1 through 4 the "zero" reference lines are the true location of the antenna. Position variations from the zero represent the apparent drift of the user equipment position solution away from the correct position solution. The distance between the "zero" latitude and "zero" longitude is the error in the calculated position. Because the U.S. Coast Guard research is aimed at the use of GPS for marine navigation, this paper deals only with errors in latitude and longitude at approximate sea level.

The radial error is the distance between the true coordinates of the antenna and the location calculated by the user equipment. It is always represented as a positive number. The radial error combines latitude and longitude information into one parameter. Figure 1 presents the horizontal (latitude and longitude) radial error as the receiver was switched between three operating modes. The three modes are full differential GPS, normal GPS with ionospheric and tropospheric models used, and normal GPS with ionospheric and tropospheric

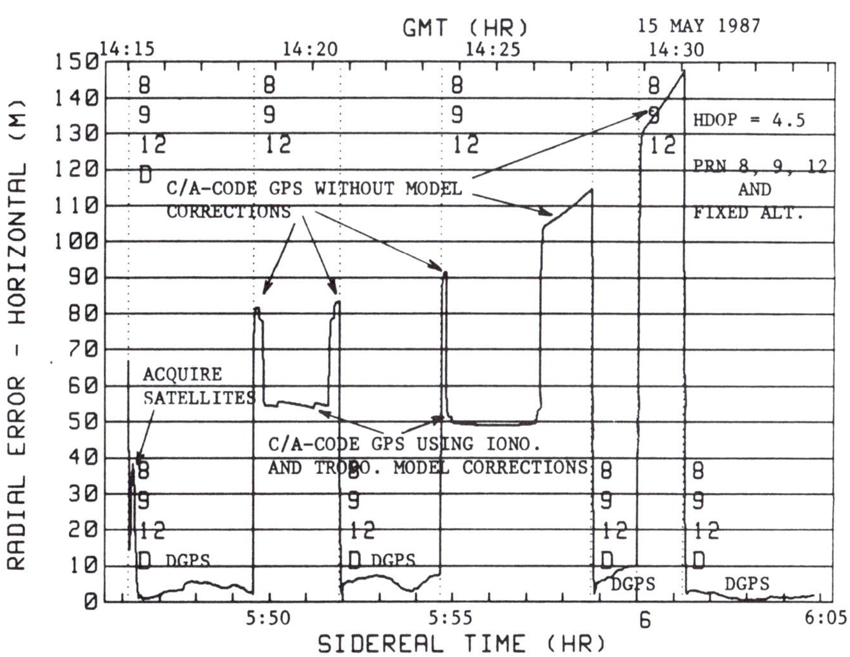

Fig. 1—Static Comparison of C/A Code GPS and Differential GPS in Presence of Natural Errors (Horizontal Radial Error)

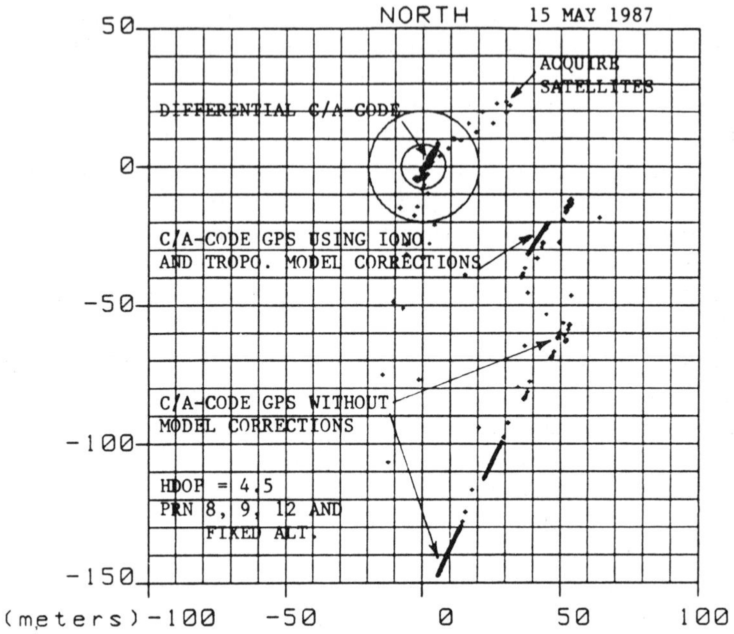

Fig. 2—Static Comparison of C/A Code GPS and Differential GPS in Presence of Natural Errors (Scatter Plot)

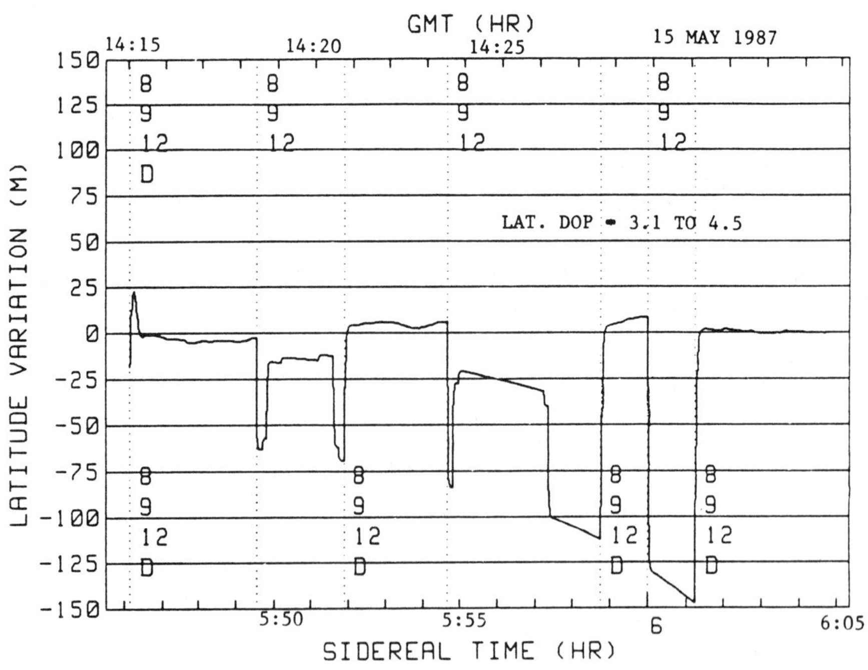

Fig. 3—Static Comparison of C/A Code GPS and Differential GPS in Presence of Natural Errors (Latitude)

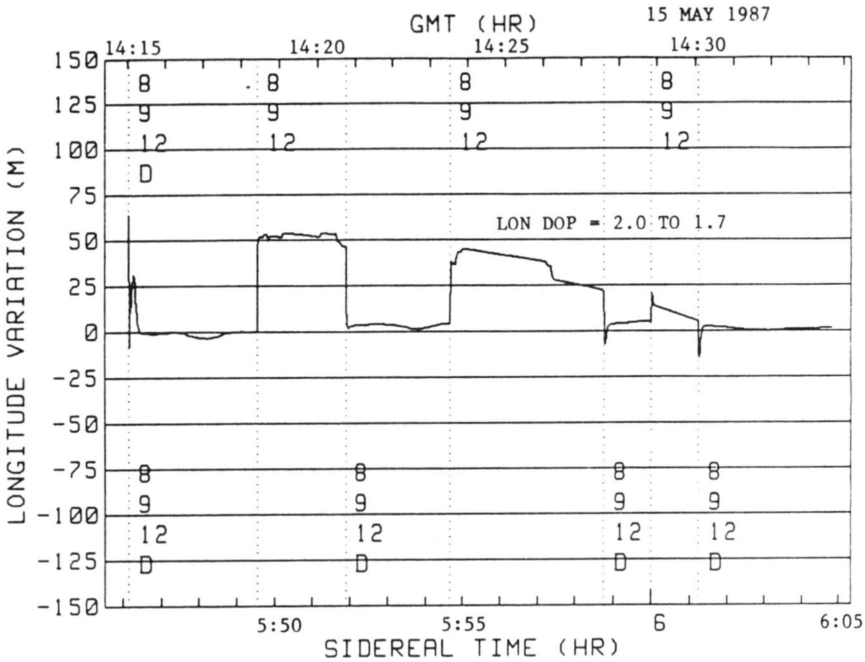

Fig. 4—Static Comparison of C/A Code GPS and Differential GPS in Presence of Natural Errors (Longitude)

models not used. Under the extraordinary conditions of this test, the three modes result in distinctly different position solutions.

Figure 2 presents the latitude and longitude in a two-dimensional format. It is used to study the direction and migration of the error. This representation is called a scatter plot. It shows the correlation of latitude and longitude errors. An ideal scatter plot would show a tight cluster of points around the "zero-zero" reference. The ratio of the north-south versus east-west variations often follows closely the ratio of the latitude and longitude DOPs. The individual latitude and longitude errors versus time are shown in Figures 3 and 4.

The location of the monitor station antenna provides a view of the sky with a low horizon. Using a TI-4100 antenna on the differential monitor and using a Magnavox marine antenna on the T-set, satellites are reliably tracked down to 2 degrees. Acquisition can usually be accomplished between 2 and 3 degrees. The ability to acquire and track low satellites was also demonstrated aboard the boat during the dynamic marine tests discussed below. These observations suggest that visibility calculations should review the use of satellites as low as 2.5 degrees when evaluating marine GPS coverage. While coverage can be maintained (acceptable DOPs) using low satellites, the non-differential user will suffer some position degradation caused by ionospheric and tropospheric errors that cannot be modeled.

Several conclusions can be drawn from looking at Figures 1 through 4. First, the position solution can be significantly improved using differential GPS. In this analysis the position solution is within 10 m of the known location. In order to verify this finding, similar tests were conducted on succeeding days.

In each test the differential corrections provided a position solution within 10 m of the reference position. Second, modeling of ionospheric and tropospheric delays can improve normal GPS performance. The ionospheric and tropospheric models used in the T-set during this test were designed to correct pseudoranges of satellites with an elevation of 5 degrees or higher above the horizon. Extending the modeling below 5 degrees could improve GPS performance over the analysis presented here. Third, after repeating this experiment several times, the direction of the offset error from the known location was found to vary significantly. This erratic behavior is attributed to the fact that PRN 8 is driven by a crystal clock. The only cure for this problem is differential GPS or frequent uploads by the Control Segment.

The results of the static test demonstrate that GPS equipment can track low elevation satellites. The resulting position solution is greatly affected by natural sources of error that can be partially removed using modeling algorithms. The use of differential corrections can restore full performance by eliminating most of the natural errors. In the case of unnatural error sources, such as the crystal clock modeling problem, differential GPS also can eliminate a significant part of the error. The results of this test gave us confidence that the differential GPS demonstration system being assembled was working, and that plans could be made to conduct dynamic tests.

DYNAMIC DGPS TESTING

A major part of the dynamic test and subsequent analysis is dependent on how well the position of the moving GPS antenna is known. It is easy to understand what is meant by the true position of a static antenna. It is the geodetic location of the antenna throughout the test period. In order to perform analysis similar to that done on static test measurements, it is necessary to know the exact location of the GPS antenna at the time of the dynamic measurements. The analysis of the dynamic data requires a position by position comparison of the GPS receiver solution with the true antenna position. During the dynamic analysis, the antenna reference position (the equivalent of the static "zero" positions in Figures 1 through 4) becomes the actual antenna location at the time of the GPS receivers calculated solution. Taking the difference between the GPS receiver's position solution and the true antenna location creates an ensemble of relative positions. This ensemble can be analyzed using the same techniques and software used during static analysis. The value of the results derived using this approach depends heavily on good antenna position truth and careful synchronization of time during the test.

A U.S. Coast Guard 42 ft patrol boat was selected to be the dynamic platform. Simple maneuvers that represent typical boat accelerations were established for a 3 mi^2 test area. A microwave ranging system was selected to provide the true location of the patrol boat during the dynamic maneuvers. A laser optical system was selected to provide periodic calibration of the microwave ranges. A Kalman filter was selected to convert the corrected microwave ranges into positions. A survey team from the Defense Mapping Agency (DMA) established the positions of the shore locations selected for the microwave reference stations. Also located on shore at the site of the differential TAU monitor were a Magnavox monitor receiver, a Magnavox T-set receiver, a TI-4100 receiver, a laser optical system data recorder, and equipment to monitor the RTCM SC-104 format broadcasts. All measurements were recorded using a combination of digital magnetic tapes and Hewlett Packard Series 200 computers.

The microwave ranging system is a Motorola Miniranger Falcon 484. Four shore-based reference stations were used. Range measurements to the four stations are converted to positions using a Kalman filter. The true GPS antenna positions are calculated for the times of the GPS position solutions as determined by the T-set or TI-4100 receivers.

Determining the true GPS antenna location after the test from four range measurements taken during the test, allowed the use of a "non-linear" technique to supplement the Kalman filter. Only two of the four range measurements are really needed to calculate a position. The "non-linear" technique takes advantage of the extra information available during each filter cycle. During each cycle of the Kalman filter, Falcon range measurements were compared to their expected values. If a single range measurement exceeded a predetermined difference threshold, that single measurement was excluded from that cycle of the Kalman filter. If the signal-to-noise ratio of a measurement fell below a predetermined level, the measurement would also be excluded. This elimination of potentially misleading measurements helped smooth the Kalman filter output without reducing its dynamic responsiveness.

The laser optical system is a Measurement Devices Limited Lasertrak L5000 range, bearing, and elevation positioning system accurate to better than 1 m. During the dynamic tests, it served as a mechanism for calibrating the Falcon. Between test maneuvers, the boat was allowed to drift for several minutes while laser optical and Falcon data were recorded. This calibration data is used to correct each Falcon range measurement. A Kalman filter processes the corrected range measurements to calculate the Falcon antenna position. For the truth data used in this paper, the laser optical system provided range corrections between 0.3 and 2.4 m.

Time correlation of the Falcon and the differential GPS positions is very important. GPS time was selected as the common time base to be used throughout the test program for the purpose of synchronizing measurements and controlling the analysis. A vessel traveling at 20 kn moves about 10 m/s. Because position bias and noise levels were expected to be on the order of meters, elimination of time skews during the tests and during the data correlation process was important. By running the Falcon in "range only" mode, each sequence of range measurements can be triggered by a 1 pulse-per-second signal. During the tests, this pulse was provided by a TI-4100.

The time of a TI-4100 pulse is known to within 20 ms of GPS time. After the pulse, the Falcon takes 35 ms to complete the range measurement sequence. The 20 ms uncertainty and 35 ms corrections are taken into account during analysis. TI-4100 receivers provided time synchronization signals aboard the boat and at the differential monitor station. In addition to time synchronization, the TI-4100 receiver on the boat was operated in normal C/A-code or P-code modes. This provided the ability to compare the differential GPS performance with normal GPS.

DGPS PERFORMANCE IN FIGURE EIGHT MANEUVER

Several figure eight maneuvers were done on August 14 at both 10 and 20 kn. Figure 5 shows a trackline plot of Falcon derived positions (line plot)

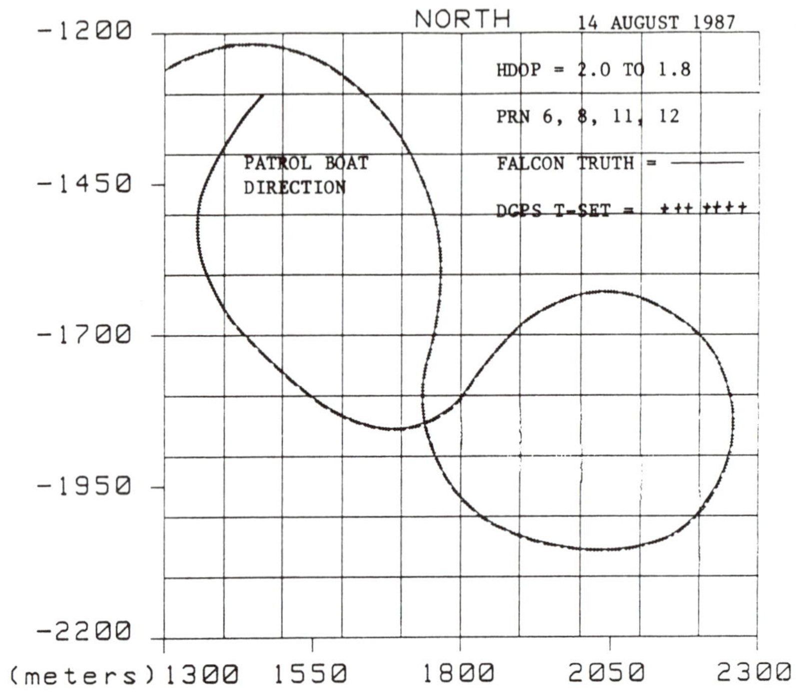

Fig. 5—Trackline of FALCON and T-set Data During 10 kn Figure Eight

and differential T-set positions (crosses) during the 10 kn maneuver. This presentation gives a good overview of the actual patrol boat maneuver but makes detailed performance evaluation difficult.

Figure 6 shows the position scatter plot created by the stationary shore-based differential T-set. This provides information about the general performance that could be expected on the boat if the boat were stationary. The stationary T-set used the same differential corrections being broadcast to the dynamic T-set. The acceleration parameters used in the navigation filters were the same in both receivers.

The differences between the dynamic differential GPS positions and the true antenna positions derived from the Falcon range measurements are shown in Figures 7 and 8. Comparing figures 6 and 7, it is obvious that Figure 7 contains more noise. Dynamic data contains three major sources of error not affecting static data. They are the dynamic response of the T-set navigation filter, the noise in the Falcon range measurements, and the dynamic response of the Kalman filter used to process the Falcon measurements.

During the 10 min it took to complete the 10 kn maneuver, the average latitude of the static T-set was −0.8 m and the longitude was −1.6 m from the known antenna location. The position calculated by the dynamic T-set was an average of −1.6 m in latitude and 0.1 m in longitude from the true antenna position. This is a radial bias offset of 1.7 m between the two T-sets. The standard deviation of the dynamic T-set positions was double that of the static T-set positions.

During the 20 kn figure eight maneuver, the radial bias offset between the

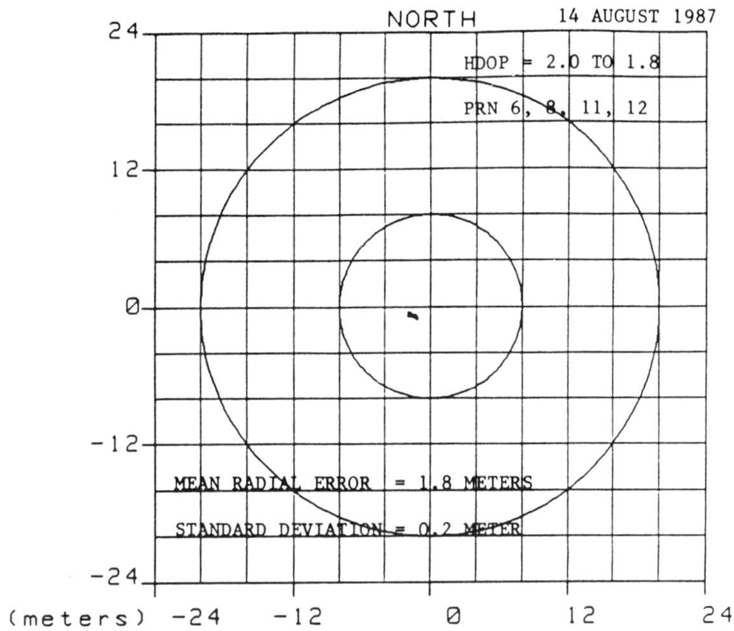

Fig. 6—Stationary Differential T-set Position Error During Execution of 10 kn Figure Eight

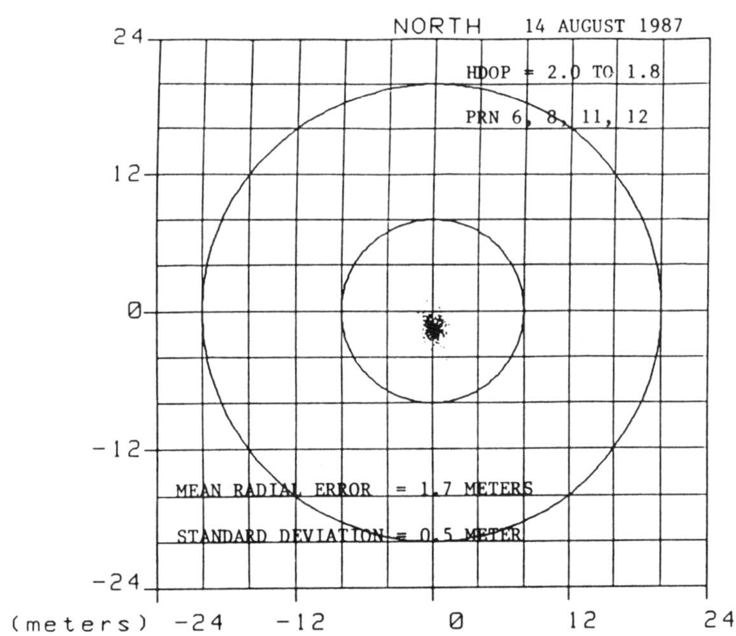

Fig. 7—Dynamic Differential T-set Positions Relative to FALCON Positions During 10 kn Figure Eight

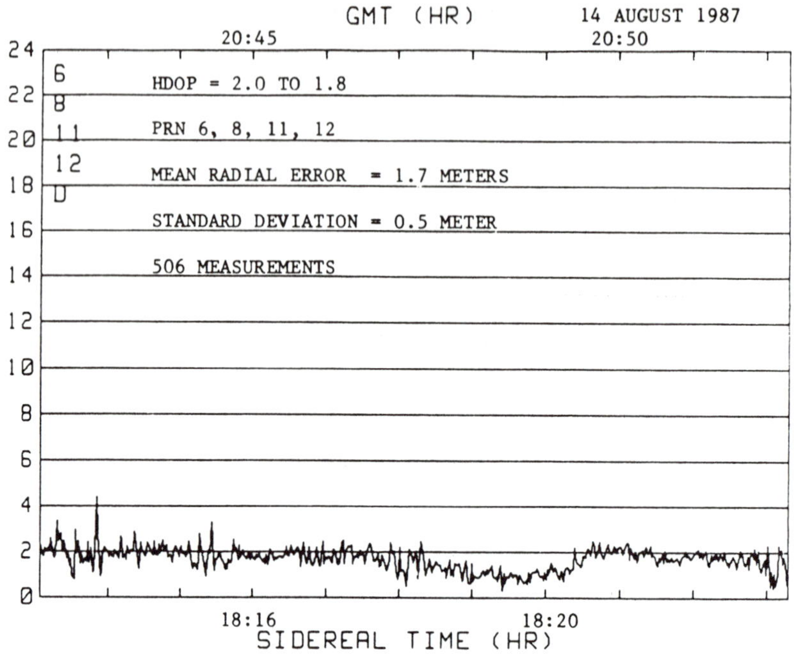

Fig. 8—Horizontal Radial Error of Dynamic Differential T-set Positions Relative to FALCON Positions During 10 kn Figure Eight

static and dynamic T-sets was 1.5 m. The standard deviation of the dynamic positions was four times that of the static positions. Most of the standard deviation is due to the data in the first minute of the test. The magnitude of the bias offsets is within the bounds observed in earlier static T-set performance comparisons. The increase in the standard deviations is attributed to the increased number of error sources in the dynamic data and analysis as discussed above.

During the first minute of the 20 kn figure eight, the T-set position error increased to 8 m for a short period of time. The increase can be seen in Figures 9 and 10. The reason for this short excursion is believed to be related to the dynamic response of the T-set Kalman filter and the fact that the maneuver began with the boat accelerating from 0 to 20 kn. The filter acceleration parameters are controlled by the operator. This excursion was probably caused by the acceleration parameters that had been selected for the T-set during this test. This response will be investigated in future tests.

With the exception of the first minute's data in Figure 10, the radial error in Figures 8 and 10 is generally less than 3 m. Although this analysis is a sampling of the many test measurements recorded, similarly low errors were obtained in the analysis of other maneuvers. After analyzing the DGPS performance in a cross section of maneuvers, it became clear that carrier-aided differential GPS is accurate enough to expose errors in the instrumentation or truth measurements. It did cause us to consider the possibility of instrumentation error in the truth system during the acceleration noted above. However, after more detailed analysis, the acceleration induced error is still believed to be caused by the T-set Kalman filter.

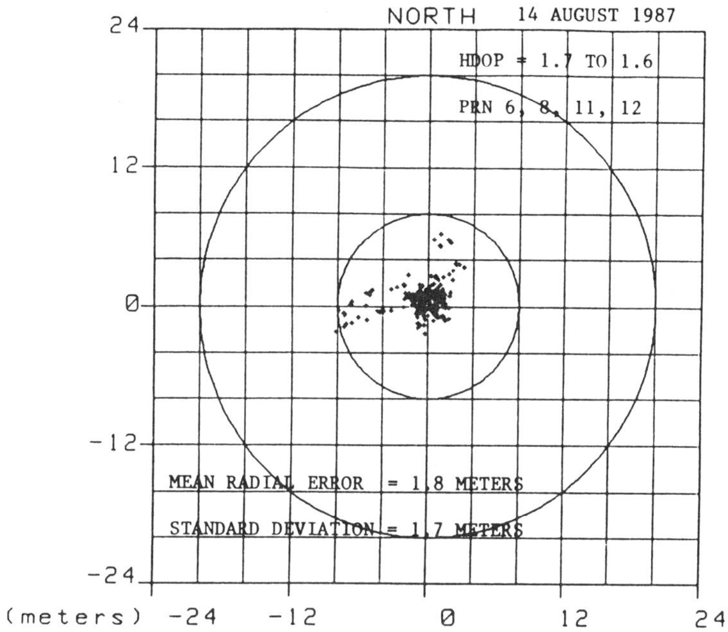

Fig. 9—Dynamic Differential T-set Positions Relative to FALCON Positions During 20 kn Figure Eight

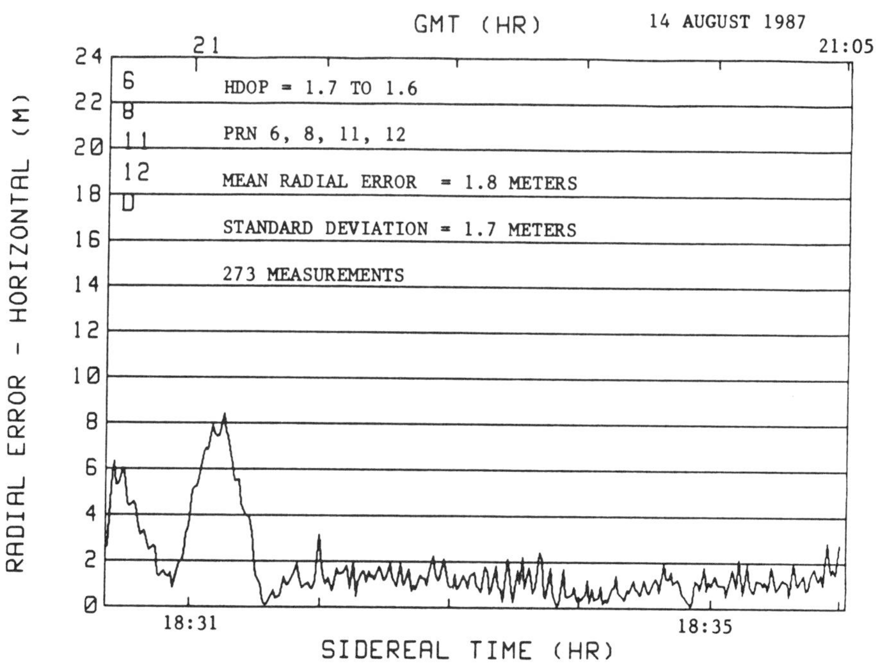

Fig. 10—Horizontal Radial Error of Dynamic Differential T-set Positions Relative to FALCON Positions During 20 kn Figure Eight

The objective of the U.S. Coast Guard program is to attain reliable 8 to 20 m accuracy (2 drms). The accuracy being observed throughout our tests suggests that the practical local area accuracy is better than 8 m. The geographic size of the local area will be the subject of future tests.

DGPS C/A-CODE VERSUS P-CODE PERFORMANCE IN PARALLEL TRACK SEARCH

The maneuver made during this test consisted of four parallel track lines separated by about 200 m. This test is designed to compare the performance of differential GPS C/A-code and normal P-code. The analysis provides a variety of interesting comparisons and exposes areas deserving of more development work.

Corrections from both the TAU and Magnavox monitor stations were broadcast during this test. Only the TAU monitor station was used to broadcast corrections during the figure eight maneuver data presented above. The dynamic P-code position measurements were made using the "84-PLUS" version of the Texas Instruments 4100. The parallel track pattern selected for this test maneuver is common in search and rescue and mine counter measure related work. This maneuver was performed on August 14 and took 36 min to complete. During the maneuver the boat speed was 7 kn.

Figures 11 and 12 show the error in the static receiver position calculation. The 11 to 3 m offset at the beginning of the test (Figure 12) occurred while the TAU monitor station was being used. This offset was observed each day after the TAU monitor acquired PRN 3 and added it to the constellation of PRNs 8, 11, 12, and 13. The static T-set's position solution followed approximately the

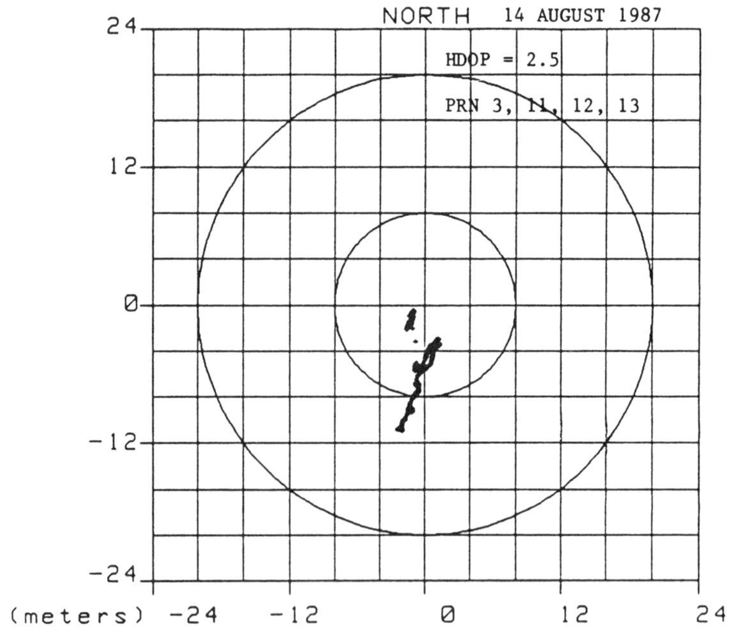

Fig. 11—Stationary Differential T-set Position Error During Execution of 7 kn Parallel Trackline Maneuver

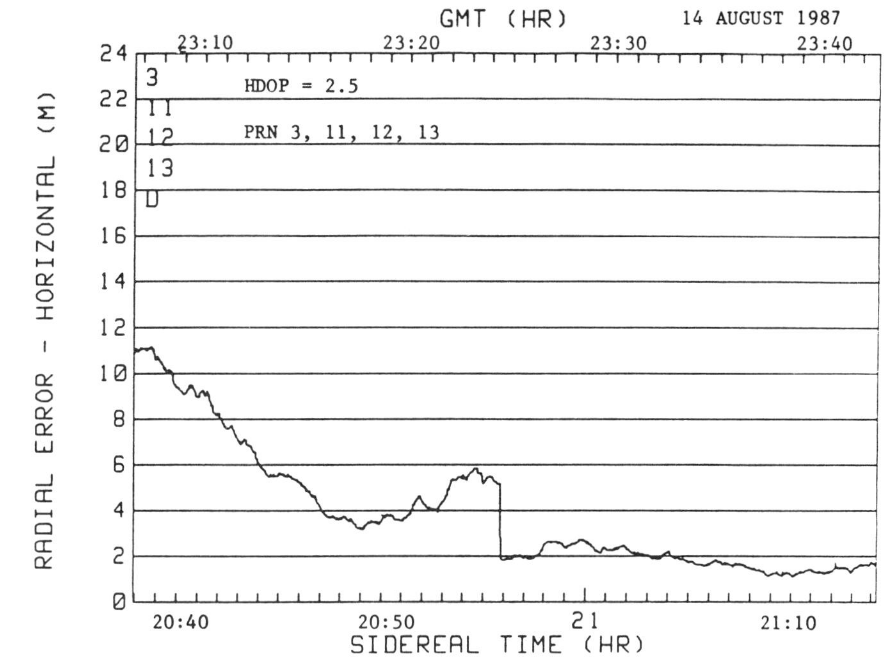

Fig. 12—Horizontal Radial Error of Stationary Differential T-set Positions During Execution of 7 kn Parallel Trackline Maneuver

same path of convergence each day. Although within specifications, this phenomenon is considered a problem and is being investigated. Unfortunately a comparison in performance with the Magnavox monitor during the first half of the test was not possible. The Magnavox monitor station didn't acquire PRN 3 until it was about 16 degrees above the horizon each day. Normally PRN 3 is acquired below 5 degrees above the horizon. This could be a problem with the acquisition software, and is being discussed with Magnavox. The sudden position error jump in the middle of Figure 12 occurred when the source of differential corrections was switched from the TAU monitor station to the Magnavox monitor station. The last half of the test was completed using the Magnavox monitor.

Figures 13 and 14 show the error in the dynamic T-set's position calculation. Again the increase in dynamic position error noise is easily seen by comparing Figures 11 and 13. The similarity of the scatter patterns suggests that the bias errors are in the differential corrections themselves. This can also be seen by comparing Figures 12 and 14. Both sets of errors follow the same general pattern over the test period.

The abrupt position jump, caused by the TAU to the Magnavox monitor switch, is clearly seen in Figure 12, but it seems to be absent from Figure 14. A review of the latitude and longitude plots for both data sets (not shown here) shows that both the static and dynamic differential solutions moved north almost 4 m at the time the source of the corrections changed. In the case of the dynamic T-set, the solution was about 2 m south of the true location prior to the monitor switch. The net change in radial error was less than 1 m as the

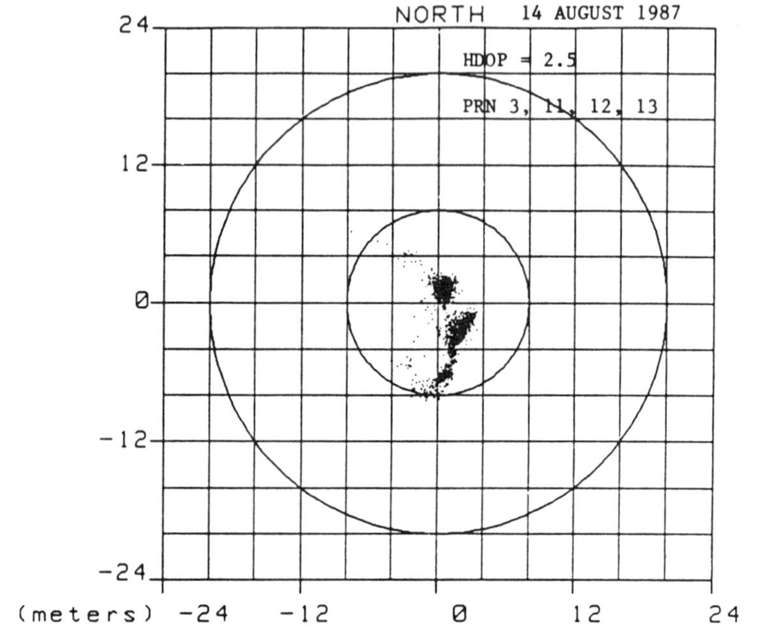

Fig. 13—Dynamic Differential T-set Positions Relative to FALCON Positions During 7 kn Parallel Trackline Maneuver

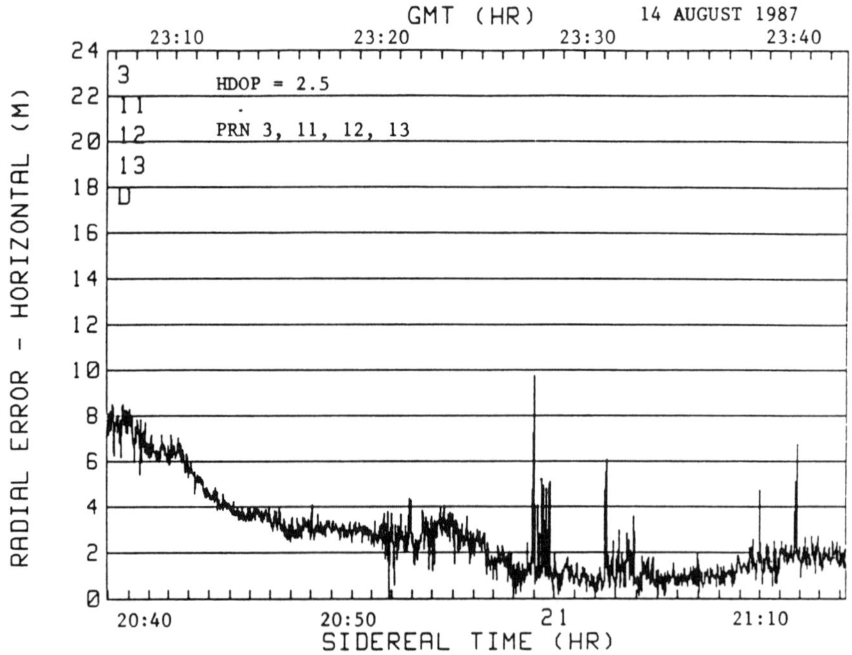

Fig. 14—Horizontal Radial Error of Dynamic Differential T-set Positions Relative to FALCON Positions During 7 kn Parallel Trackline Maneuver

solution jumped to about 2 m north of the true location. This abrupt change is visible in the scatter plot (Figure 13). The northern most cluster represents all the position solutions using corrections from the Magnavox monitor.

The P-code position errors are shown in Figures 15 and 16. Figure 15 shows the solution to be several meters south of the true antenna location for almost the entire test. The statistical analysis calculates the mean latitude at −12.3 m and mean longitude at 0.2 m. Figure 16 shows the amplitude of the short term noise level of the errors.

It is interesting to compare Figures 14 and 16 with respect to both error bias and noise. The truth system used to calculate the actual dynamic TI-4100 antenna position is based on WGS-84 geodetic position. This position was established by two separate DMA surveys. The truth system measurements are believed to be absolutely accurate to within 2 m. The approximate 12 m error in the P-code position solution probably is the true P-code error during this test. The error is a combination of the P-code broadcast and the processing done by the "84-Plus" TI-4100. The TI-4100 does not make use of carrier aiding. This could explain why the noise level is greater in the P-code position analysis than the carrier aided C/A-code analysis. These results make it necessary to do additional P-code performance testing in the future.

During this particular maneuver, the P-code position solutions would have met the 20 m accuracy goal of the project. It most certainly would not have met the 8 m goal. The differential GPS met both the 20 and 8 m goals despite obvious monitor problems. The high correlation between the static and dynamic differential GPS position stability suggests that improvement of the differential monitor performance will have the greatest direct impact on the performance of the differential GPS demonstration system.

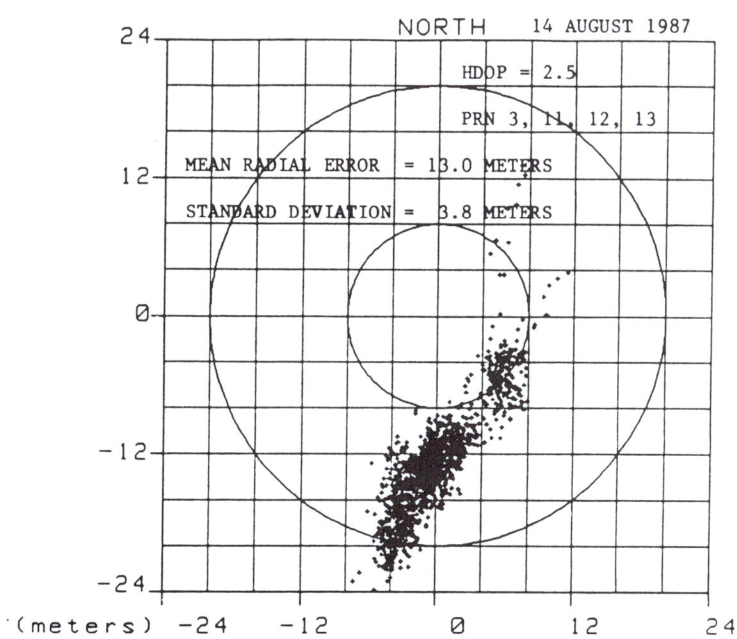

Fig. 15—Dynamic TI-4100 P-Code GPS Positions Relative to FALCON Positions During 7 kn Parallel Trackline Maneuver

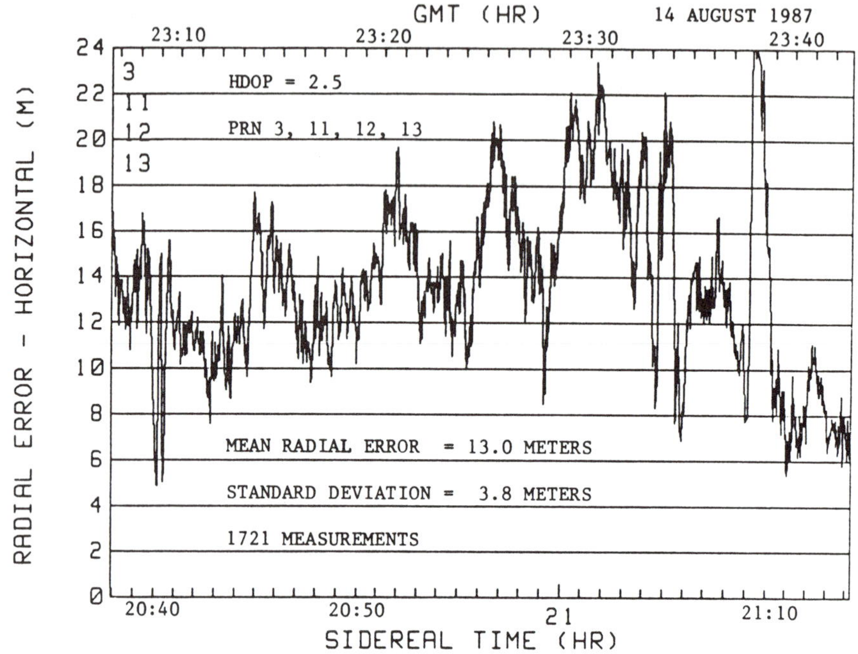

Fig. 16—Horizontal Radial Error of Dynamic TI-4100 P-Code GPS Positions Relative to FALCON Positions During 7 kn Parallel Trackline Maneuver

SENSITIVE AREAS EXPOSED BY TEST PROGRAM

1) No standard algorithm exists for "earth rotation correction." This was realized while studying bias offsets found during static testing. If differential GPS is expected to work using a combination of equipment designs, it is necessary that there be agreement on any critical user equipment and monitor algorithms. The people working on such an algorithm agreement should be the same ones responsible for algorithm selection at the different manufacturers. Perhaps the earth rotation correction is an exception, but it is difficult to know unless information about how calculations are being done is shared or at least tested for sufficient resolution compatibility. The accuracy of differential GPS is very good. Algorithms that may seem to be close enough for normal C/A-code, GPS may introduce a very detectable error into the position solution when differential corrections are added. More precision internal to the user equipment may be a requirement if adding the option of differential operation is anticipated. Certainly the switch between the TAU and Magnavox monitor was very detectable. Is this to be considered typical or can this observable problem be corrected?

2) Not all parameters broadcast by the satellite that affect the parametric synchronization of the differential user equipment and monitor can be identified with the IODE (issue of data, emphemeris). For example the clock parameters are labeled with the IODC (issue of data, clock). Obviously the user must be using the same clock information that the monitor is using for the pseudorange corrections to apply. Regardless of why the user and monitor parameters might differ, a mechanism should be built into the RTCM SC-104 format that can

correct or warn that such a condition may exist. The issue of parametric synchronization should be reviewed by the RTCM SC-104. This is an important side of the integrity built into differential GPS.

3) Carrier aiding in the user equipment is an important factor in maintaining the dynamic accuracy of the position solution. Carrier aiding and differential corrections enhance the performance of each other. The differential community needs to review the content of the proposed RTCM SC-104 messages. Have the necessary provisions correcting carrier-aided navigation solutions been included? Should more be done to enhance pertinent carrier information or correct for the effects that selective availability may have on the actual carrier? Remember that the pseudorange rate is not a carrier correction. It is simply the rate of change of the code correction to allow the user to adjust the pseudorange correction with time.

SUMMARY

A differential GPS demonstration system has been assembled using equipment from a variety of manufacturers. During August 1987, this system demonstrated that the predictable and relative navigation accuracy requirements for the Harbor Approach and Harbor Phases of marine navigation specified in the Federal Radionavigation Plan (FRP) can be satisfied by applying the differential GPS concept. Under the particular circumstances of our tests, the real time dynamic position solution of a Magnovox differential 5-channel T-set satisfied the 8 to 20 m, 2 drms accuracy goal of the FRP. Future test results are expected to be better or worse depending on which way key test conditions are altered. The significant test conditions that should be kept in mind when interpreting these results or attempting to extrapolate the performance to a specific application are the following:

1. navigation solution using four satellites
2. one satellite using a crystal clock
3. fixed altitude aiding not used
4. HDOP less than 3
5. full carrier aiding of navigation solution
6. marine maneuver dynamics up to 20 kn
7. variety of typical patrol boat maneuvers
8. no selective availability applied to satellite signals
9. differential monitor less than 3 mi away
10. use of RTCM SC-104 differential correction format
11. using TAU or Magnavox developed monitor stations.

Expanded testing around some of these constraints (such as distance to monitor) are part of the U.S. Coast Guard test program. Note that obvious bias errors in position solutions were not removed before determining the accuracy of the navigation solution as presented in this paper. The findings of this paper are based on position information actually calculated by the T-set during the dynamic maneuvers. No post-test adjustments of the T-set data were made.

The good performance of the truth positioning system used during the dynamic maneuvers in combination with the smooth performance of the T-set's carrier aiding helped greatly in isolating the remaining significant sources of error in the system. These errors are bias like position or range offsets in the user equipment and both monitor stations. Correlation of the bias problems between static and dynamic user equipment is very good. Research into elimination of the bias problems will continue without the need for extensive dynamic testing.

ACKNOWLEDGMENTS

The authors would like to thank the product development teams at TAU Corporation, Magnavox Advanced Products and Systems Company, Trimble Navigation, and Texas Instruments for their efforts to provide the latest version of their products for these tests. Thanks to Mr. Jerry Knight at Magnavox and Mr. Greg Kremer at TAU for being patient with our endless requests for software modifications. A special thanks to Kathleen, Irene, and Jeanine for being extra patient with their husbands during the unusual hours required by the test program.

REFERENCES

1. Quill, J., *U.S. Coast Guard Differential GPS System Development,* Radio Technical Commission for Maritime Services Assembly Meeting, Boston, Massachusetts, April/May 1986.
2. Denaro, R. P., Quill, J., and Kalafus, R. M., *Differential GPS Reference Station Design and Development,* paper presented at The Institute of Navigation National Technical Meeting, Ahaheim, California, January 1987.
3. Kalafus, R. M., Van Dierendonck, A. J., and Pealer, N. A., *Special Committee 104 Recommendations for Differential GPS Service,* NAVIGATION, Vol. 33, No. 1, Spring, 1986.

NAVIGATION: *Journal of The Institute of Navigation*
Vol. 37, No. 1, Spring 1990
Printed in U.S.A.

The Effect of Selective Availability on Differential GPS Corrections

GREGORY T. KREMER, RUDOLPH M. KALAFUS,
PETER V. W. LOOMIS, and JAMES C. REYNOLDS

Trimble Navigation Limited, Sunnyvale, California
Received September 1989
Revised December 1989

ABSTRACT

The newer Block II GPS satellites have the capability to control the accuracy of the Standard Positioning Service signals by a combination of dithering the satellite clock and manipulating the ephemeris data. This capability is called Selective Availability (SA). Differential GPS operation greatly reduces the effects of SA, in addition to the effects of ionospheric, tropospheric, and other slowly-varying errors, for users in a local area around a reference station.

A major technical issue in the design of reference stations has been the correction update rate. It has been assumed until recently that new differential GPS corrections are required about once every 10 s to maintain 5–10 m service accuracy. Until the recent launch of the new Block II satellites, there has been no firm evidence that this update rate would be adequate. This paper presents statistical analysis of recently obtained data sets from Block II satellites, specifically position and velocity variations and correlation times. Pseudorange corrections generated by differential reference station equipment using these signals are applied to the pseudoranges from a navigation receiver to obtain estimates of the effect of correction update rate on differential GPS performance.

BACKGROUND

The Rationale for Selective Availability

The fundamental design concept of GPS was to enable fast receiver startup and satellite signal acquisition with a coarse acquisition (C/A) signal before switching to a precise code (P-code) having a high positioning accuracy capability. While the P-code was to be encrypted and available only to authorized users, the C/A-code was to be unencrypted and thus generally available to civil and other unauthorized users. Originally, the C/A-code positioning accuracy was expected to be about 400 m; field tests indicated that the accuracy was more like 20–40 m, with velocity measurements good to a fraction of a meter per second.

This unexpected situation prompted the U.S. government to review the security consequences of making such positioning information available to everyone. The result was that a program, now called Selective Availability (SA), was implemented to incorporate in the Block II satellites accuracy control for unauthorized users. Authorized users would have a key to a classified algorithm that would undo the effects of SA and provide the full system accuracy of 10–20 m in three dimensions.

The service available to authorized users is called the Precise Positioning Service (PPS), while that available via the C/A-code to other users is called the Standard Positioning Service (SPS). When first announced, SPS was intended to have an accuracy of 500 m; later, the accuracy was improved to 100 m (two-dimensional radical error, 95th percentile). The implementation of SA is now known to be a combination of frequency dithering and manipulation of the ephemeris data. Dithering is intended primarily to degrade velocity information, and causes the satellite pseudoranges to vary quickly (period of minutes) with limited amplitude; ephemeris data manipulation causes a very large but slowly varying bias in the position solution. A stationary receiver under the effects of SA would see its position wander slowly within a circle of about 100 m over a matter of hours.

Differential GPS

Early in the development of GPS, the technique of differential GPS (DGPS) was recognized as a method of achieving high accuracies by removing most of the biases that make up the 10–20 m position error in the PPS. DPGS operations entail placing a receiver at a known location, determining position (or pseudorange) errors, and broadcasting corrections to nearby users. The resulting accuracy is generally 5–10 m or better. It was further recognized that C/A-code users could achieve similar accuracies using only the L1 frequency, since the effect of inosophere on position was relatively constant over separations of a few hundred kilometers or less.

Even though differential GPS also counters the effects of SA, the government determined that differential operation was acceptable. First, the major issues regarding security were found to be uncompromised by local-area differential corrections. Second, control of differential operations operated by foreign interests is virtually impossible. As a consequence, differential GPS was determined not be a security issue.

To counter the relatively faster error dynamics of frequency dithering, new DGPS corrections must be transmitted every few seconds. Without dithering, corrections issued every few minutes would be adequate. Consequently, one of the three major issues to be resolved in implementing DGPS operationally is the update rate employed by a differential service. The other two major issues, the area of effective coverage of the differential corrections and the message content in a differential GPS broadcast, are not considered here.

Previous Analyses of Selective Availability

The first issue, that of update rate, was addressed by the U.S. Department of Transportation (DOT), which asked the GPS Joint Program Office (JPO) for statistical information on the expected dithering. The JPO provided a set of scrambled 90 s segments, which were analyzed by the DOT Transportation Systems Center [1]. The results showed one-sigma pseudorange errors of about 30 m, range-rate errors of about 0.13 m/s, and acceleration errors of about 0.0037 m/s^2.

The JPO also provided DOT with a plot of the standard deviation of the growth of pseudorange errors (which included SA, atmospheric, and receiver effects) following an update, assuming that only pseudorange corrections, and not range-rate corrections, were available. Based on this plot, the dithering was modeled as a second-order Gauss-Markov process (exponential correlation function), and coefficients were selected to provide pseudorange error dynamics closely matching the released data [2]. In addition to providing a handy way to generate SA-like signals in a simulation, the dither model allowed analysis of the impact of range-rate corrections.

The Radio Technical Commission for Maritime Services (RTCM) established Special Committee (SC)-104 to examine differential GPS operation and to recommend standards for such a service. The standards were in large part driven by one particular form of differential system, a pseudolite, which is a ground transmitter with a C/A-code signal, but a different code from that of the GPS satellites. Pseudolites provide an "internal" data link to GPS receivers and obviate the need for a separate external differential data link, with its extra hardware and antenna. Since pseudolite data links are limited to the same data rate as the GPS satellite vehicle broadcast, SC-104 defined a data message format that would support high-accuracy service within a constraint of 50 bits/s.

The accuracy of the correction degrades as it gets older and the clock dither wanders. The update rate should be designed to provide correction accuracy consistent with typical one-sigma differential pseudoranging errors of better than 1 m. With pseudorange corrections only, and no range-rate corrections, the standard deviation of the pseudorange correction error would increase roughly linearly with time at the SA range rate error standard deviation of 0.13 m/s. Consequently, the differential pseudorange error standard deviation would be 2 m after 14 s, adding a meter every 8 s afterward. This led RTCM SC-104 to add a range-rate correction to the differential correction message. The rate correction reduced the growth in the pseudorange correction error over 10 s under expected SA conditions to a negligible quantity as compared with differential pseudorange error; growth over 30 s was limited to about 1.5 m. The increased message length was more than compensated by the lower update rate allowed by introducing the range-rate corrections.

To translate the above GPS range accuracy discussion into navigation accuracy, the differential pseudorange error can be converted to two-dimensional (horizontal) radial error statistics using HDOP. The typical differential pseudorange error of 1 m results in horizontal position errors at approximately the 95 percent confidence level of about 2 m $\times$ the HDOP. On average, the currently proposed 21 satellite GPS constellation will provide HDOPs of 1.5, and rarely will the HDOP be worse than 2.0 (assuming mask angles of 10 deg or less) [1]. This means that position errors will be about 3 m, and will rarely exceed 4 m.

In summary, the RTCM SC-104 message structure was designed to support differential service employing a 50 bits/s data link to provide 5 m typical position accuracy even after 30 s. If higher accuracies are required for a particular application employing an external data link, the data rate can be increased beyond 50 bits/s. The update rate is driven almost entirely by the SA dithering, since atmospheric effects and ephemeris manipulation tend to be constant over a long period of time. As a rough rule of thumb, correction accuracy improves with the square of the update rate. If no dithering were used, the data rate could be reduced to as low as 1–2 bits/s; it would only have to account for time variations in the time delay of the signal path through the ionosphere.

The U.S. Coast Guard is vitally interested in this subject, since the agency plans to implement a differential GPS capability and to use their existing radiobeacon system as a data link [3]. The data rate requirements strongly impact data link design.

The RTCM standards and differential GPS equipment have been designed around the above parameters, which are based on limited and perhaps outdated pseudorange error data and its one-sigma characteristics. As a consequence, the GPS community has been eager to observe SA in operation.

Recent Measurements with Block II Satellites

With the recent launch of PRNs 14, 2, 16, 19, and 17, it is expected that SA will soon be observed operationally. As of October 1989, PRNs 14 and 2 had both exhibited some SA-like behavior. However, no official announcement has been made on whether and when SA has been turned on, so the civil GPS community has had to content itself with guessing.

At different times, PRN 14, which was launched the earliest of the three in February 1989, has exhibited signal behavior similar to that expected with SA. In April, several episodes were observed in which large variations were noted in both code and carrier, during which the User Range Accuracy (URA) data variable was set to 64 m, with health indications of both "healthy" and "unhealthy." During this time, roughly sinusoidal variations were observed with a period of about 5 min. This is a somewhat higher level of dithering than expected with the peacetime SA. Pseudorange error levels of as much as 150 m have been recorded, with range-rate errors of up to 0.5 m/s variations.

There were several days in early May, however, when the URA was set to 32 m, the level consistent with the advertised 100 m (95 percent) SPS, and the satellite was declared healthy. The variations had the general shape and rate expected from SA. Consequently, we believe we were seeing SA during this time on PRN 14.

During the second week of September, dynamic behavior was observed on both PRN 14 and PRN 2. The URAs for both satellites were set to 64 m, and the satellites declared healthy. This gave us a unique opportunity to see the effects of variations on more than one satellite.

In the absence of any official announcement on the subject, we are making the assumption that SA was being exercised during the periods when the satellite URA exceeded 16 m. The results of this paper are based on this assumption. If we guessed wrong, our conclusions will have to be modified. In any case, it is encouraging to find that the observed statistical behavior of the signals is similar to that anticipated by the differential GPS community. It is especially encouraging to find that differential GPS equipment defined along the lines of RTCM SC-104 guidelines will operate very well in the presence of SA dynamics.

DESCRIPTION OF THE GPS DATA SELECTED FOR ANALYSIS

Data Acquisition and Processing

We have been collecting GPS measurements from a static GPS receiver at a known location on a regular basis. These measurements consist of the C/A-code pseudorange and the L1 carrier frequency offset (Doppler effect) for each GPS satellite in view. Pseudorange rate is derived from the frequency offset measurement by multiplying by the L1 wavelength of approximately 19 cm.

The pseudorange and range-rate measurements from the GPS receiver are converted to pseudorange and pseudorange-rate errors by first correcting the raw measurements for satellite clock errors, relativistic effects, and atmospheric effects, and then subtracting the calculated range and range-rate for each satellite. The resulting pseudorange and pseudorange-rate errors consist of transmitted satellite range and range-rate errors, plus the unknown receiver clock error and other, less significant, errors. A rubidium frequency standard was used in the data collection, so the receiver clock phase error is modeled with a best linear fit over all satellites. This estimate is subtracted from the pseudorange errors to produce the best estimate of transmitted range errors for a given constellation. Similarly, the receiver clock frequency error is modeled as the (constant) time derivative of the linear fit. Like the pseudorange errors, the pseudorange-rate errors are frequency-corrected to produce transmitted range-rate errors.

There are several small sources of error in the above process. These include receiver miscalibration, atmospheric model errors, and code multipath errors. Accurate calibration of the GPS receiver can reduce calibration error to less than 1 m, typically to about 0.2 m. Atmospheric models tend to degrade at low satellite elevations, when the model can be off by a factor of two at times. The ionospheric delay is the least well-behaved at low satellite elevations. However, the profile of the ionospheric delay over time can be observed by plotting the difference between integrated range-rate from the carrier and pseudorange from the code. This difference can be compared to the modeled ionospheric delay to determine whether the modeled profile matches the measured profile. This procedure can at least in part validate the ionospheric model, allowing data sets to be rejected if the anomalies are evident. Algorithms using the ionospheric rate delay estimate can limit errors due to ionospheric delay mismodeling to less than 1 m for a setting satellite.

The major source of unmodelable error is code (pseudorange) multipath, which can have amplitude and periods that are close to what might be expected under SA conditions. However, a good antenna environment will reduce multipath errors to less than 1 m. In addition, carrier multipath is negligibly small in the context of DGPS navigation applications, so carrier filtering of the code will attenuate the effect of multipath (as well as code noise) considerably.

Note that code-carrier filtering would not be possible if the timing for both code and carrier transmitted from the GPS SVs were not derived from a single master frequency generator. Indeed, this hypothesis has been borne out by examination of the code-carrier difference, which shows little or none of the short-term variation typical of SA.

Description of Observed Data

PRN 14 exhibited periods of high range and range-rate errors on Julian Days 88, 90, 124, 131, 132, and 133 of 1989. On Julian Days 124 (March 4) and 131 (May 11), PRN 14 was set "healthy" by the GPS control segment. Data for Day 124 was examined in detail, since the URA issued for PRN 14 was most consistent with expected SA levels. Figure 1 shows range errors derived from code measurements for PRN 14 and PRN 6 (a Block I satellite) over a period of about 3 h. An interesting feature to note is the large range error bias of about 80 m initially present on PRN 14. This is significantly larger than expected, based on both the URA of 32 m and the 100 m (95 percent) position accuracy criteria. At 21:00, the range error bias decreases dramatically to an average value of near zero. At first glance, this might be attributed to the update of the ephemeris parameters used to calculate satellite position; under SA conditions, these parameters can be systematically tampered with to produce long-term range error biases. It turns out that in this case, however, the same behavior is observed when ephemeris updates are suppressed. The amplitude of the higher-frequency variations possibly due to satellite clock dithering remains more or less constant.

Figure 2 shows the range error rate and range error acceleration derived from carrier measurements for PRNs 6 and 14. The variation exhibited by Block I PRN 6 is due mostly to the carrier measurement noise of 3–5 mm/s. PRN 14 has range-rate errors with an RMS value of about 12 cm/s (compared with an expected 14 cm/s), and a period of 5–7 min. The average range error acceleration of both satellites is essentially zero, with about a 2 mm/s^2 variation on PRN 14. The rate and acceleration variations are fairly close to the values

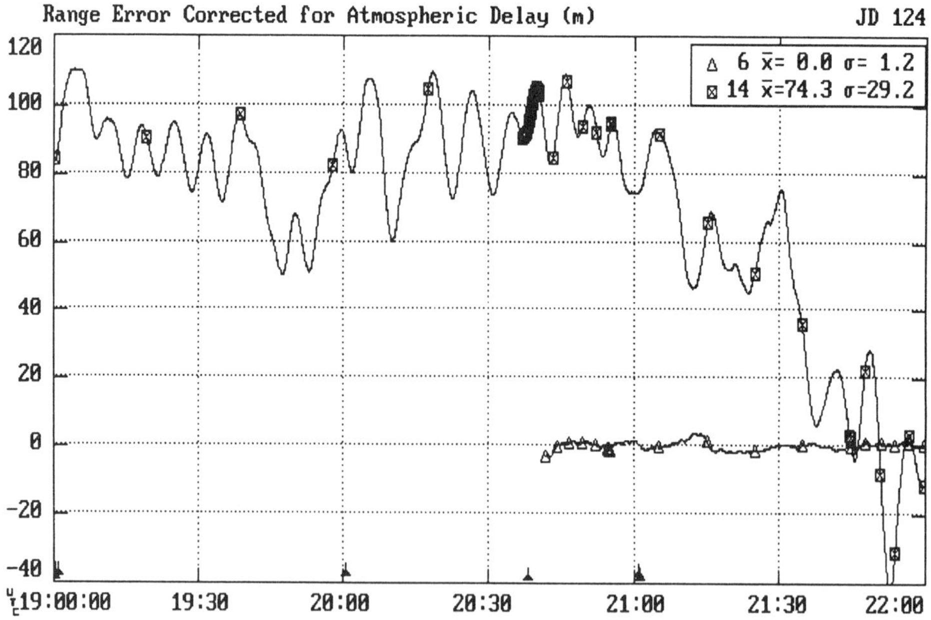

Fig. 1—Range Error (Julian Day 124)

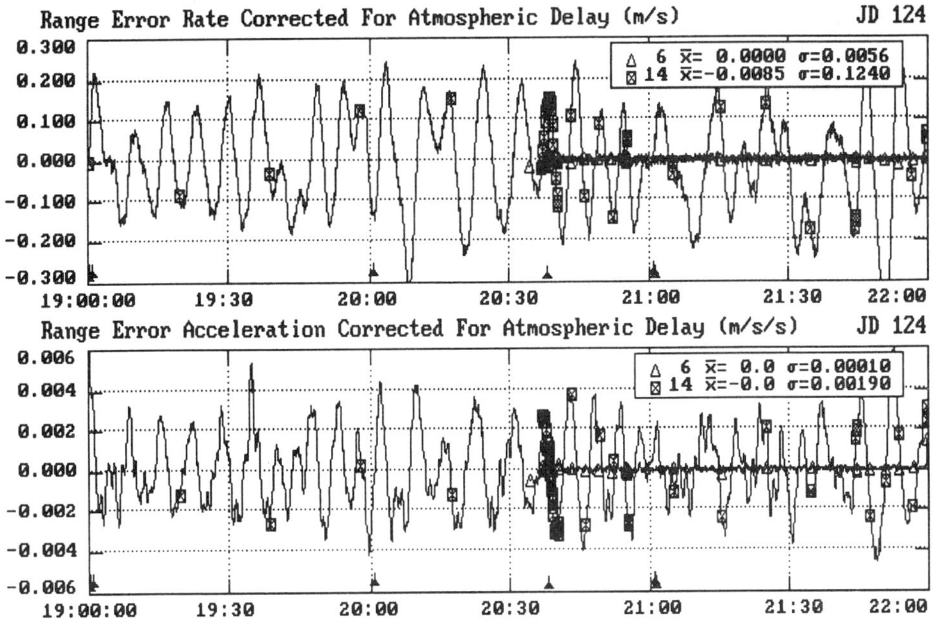

Fig. 2—Range Error Rate and Acceleration (Julian Day 124)

predicted for SA, with acceleration being somewhat less than expected (2 mm/s^{2}vs. 3.7 mm/s^2 predicted).

Beginning on Julian Day 257 (September 14), 1989, both PRN 2 and 14 showed dynamic behavior that continued for several days. Data for Day 259 is the most complete set currently available, so it is examined in detail. Figure 3

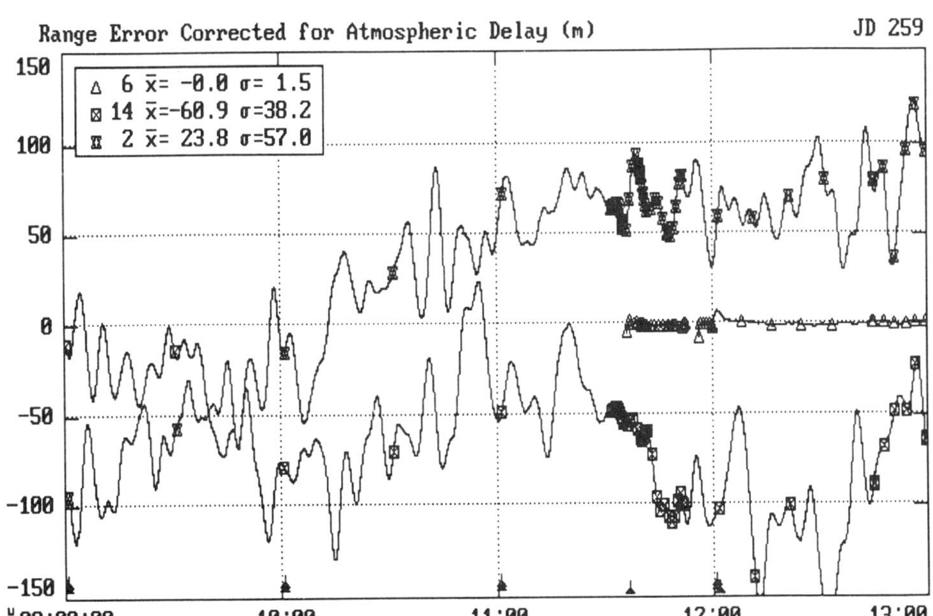

Fig. 3—Range Error (Julian Day 259)

shows the range errors for PRNs 2, 6, and 14. Again, PRN 6 has the very benign dynamics typical of the GPS Block I satellites. PRNs 2 and 14 both exhibit exceptionally large range error biases. The short-term variations are also larger than what would be expected under SA, but consistent with the URA of 64 m that was issued for PRNs 2 and 14.

Figure 4 shows the range error rate and acceleration for one of the Block II satellites (PRN 14) and PRN 6. PRN 2 is omitted from the plot for clarity; it exhibited behavior similar to that of PRN 14. The period of the range-rate errors transmitted by the Block II satellites is approximately the same as that observed on Day 124: about 5–7 min. The one-sigma variation in velocity is about 50 percent higher than what would be expected under SA, and significantly larger than that observed on Day 124.

DATA ANALYSIS TECHNIQUES AND RESULTS

There were three steps in analyzing the data:

1) Characterization of the pseudorange error, the pseudorange-rate error, and the pseudorange acceleration rate, as discussed in the previous section
2) Characterization of pseudorange correction error statistics as a function of update rate, using the RMS Range Error Difference Function
3) Examination of the effect of correction latency and update rates on differential GPS navigation

RMS Difference Plots

The RMS Range Error Difference Function

$$Y(\tau) = (E\{[x(t+\tau) - x(t)]^2\})^{1/2} \tag{1}$$

was used to parametrize the SA error contribution to pseudorange error [2].

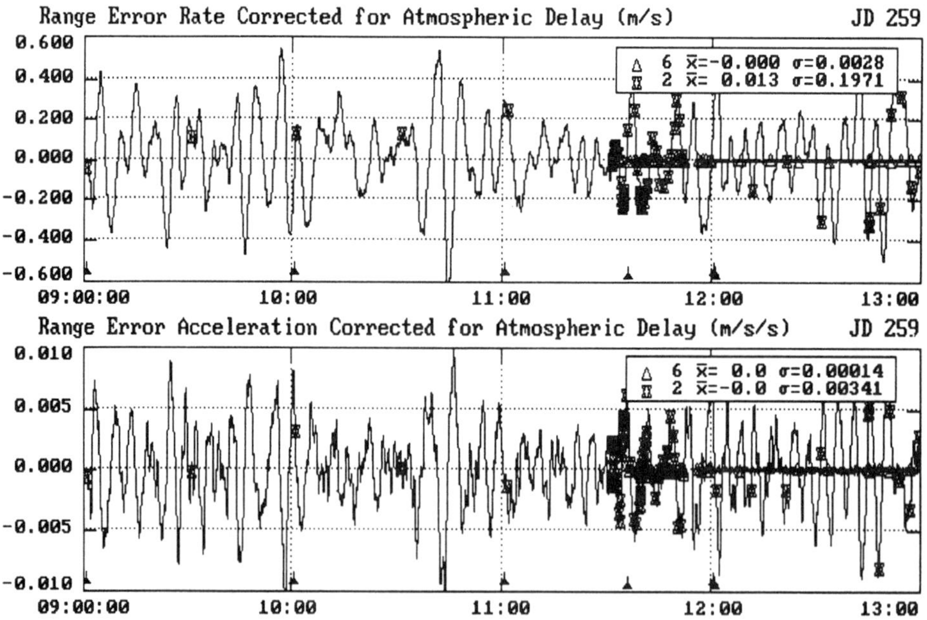

Fig. 4—Range Error Rate and Acceleration (Julian Day 259)

Y(τ) is a measure of the decorrelation of the function x(t) (in this case, x(t)is pseudorange error as a function of time) over an interval τ. Figure 5 shows the plot of RMS Range Error Difference of range errors provided by the GPS JPO to DOT. The main features to note are (1) the linear behavior over small update intervals, (2) the knee of the curve, and (3) the asymptotic behavior as the interval gets longer. The irregular behavior of the right-hand part of the curve is due to the fact that the curve is based on a limited data set. The features of the RMS Range Error Difference plot provide the following range error statistics:

1) The initial portion of the curve provides an estimate of the one-sigma temporal decorrelation following the calculation of a pseudorange correction (no range-rate correction term). For the JPO data, this is roughly 5 m after 30 s of delay (τ = 30 s). With an HDOP of 1.5, this corresponds to a position error of 15 m.
2) The knee of the curve gives an idea of the correlation time, estimated at 180 s.
3) The asymptotic behavior demonstrates the completely decorrelated value. Note that the asymptotic value is the standard deviation of the short-term variation in uncorrected range errors, multiplied by the square root of 2. This short-term variation is due almost exclusively to SA frequency dither.

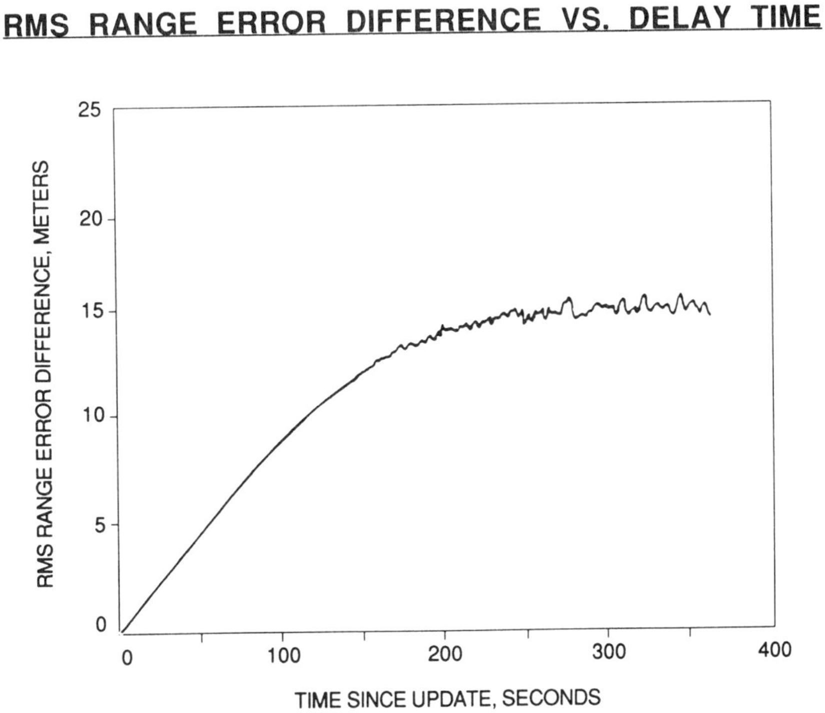

Fig. 5—RMS Range Error Difference (GPS Joint Program Office Data)

The range errors provided by the GPS JPO have a correlation time of approximately 180 s, and a maximum RMS Range Error Difference of approximately 15 m, due mostly to SA.

Figure 6 plots RMS Range Error Difference for Day 124. The curve is very similar to Figure 5. This is not surprising, since the one-sigma values of range error, error rate, and error acceleration observed on Day 124 correspond closely to those of the GPS JPO data. PRN 6 is included for comparison.

After 30 s, the RMS Range Error Difference is 3 to 4 m. This is significantly in excess of the 1.7 m range error uncertainty required for 5 m positioning as recommended by RTCM SC-104. Recall that as a rule of thumb, the 95th percentile position error is roughly 3 times the one-sigma range error (HDOP = 1.5).

SC-104 addressed this by including in the differential correction message range-rate corrections that are used to propagate range correction estimates over the latency interval associated with real-world data link delays. By modifying the RMS Range Error Difference in equation (1) as follows,

$$Y(\tau) = (E\{[x(t+\tau) - (x(t) + \tau\dot{x}(t))]^2\})^{1/2} \quad (2)$$

an estimate of RMS Extrapolated Range Error Difference, using range error rate, is generated as a function of delay time.

Figure 7 shows both the RMS Range Error Difference and RMS Extrapolated Range Error Difference for Day 259. Note that the use of range error rates to extrapolate the range error reduces the RMS Range Error Difference over the correlation interval. After that, errors build up rapidly, and the asymptotic error is greater than with the RMS Range Error Difference function.

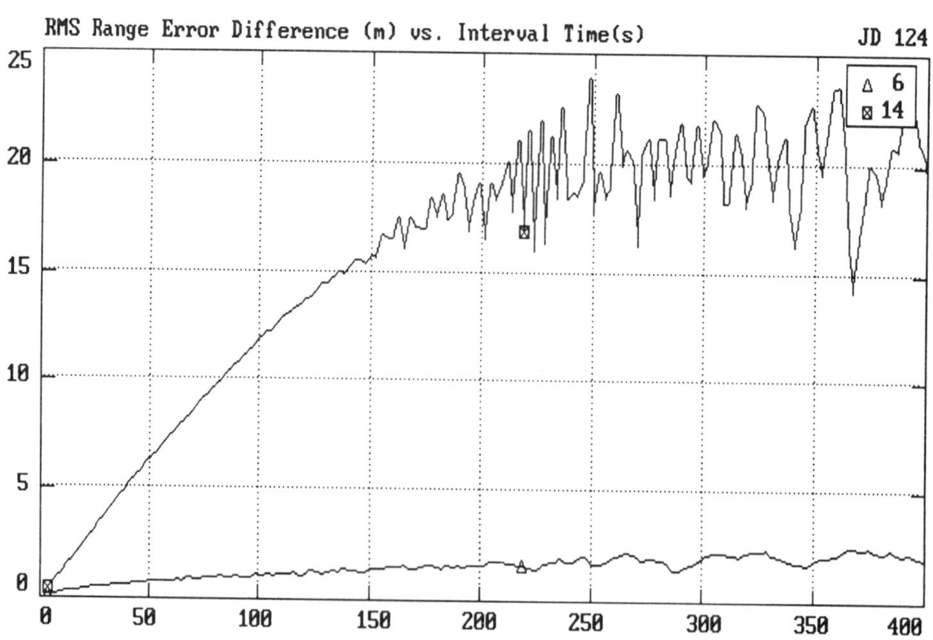

Fig. 6—RMS Range Error Difference (Julian Day 124)

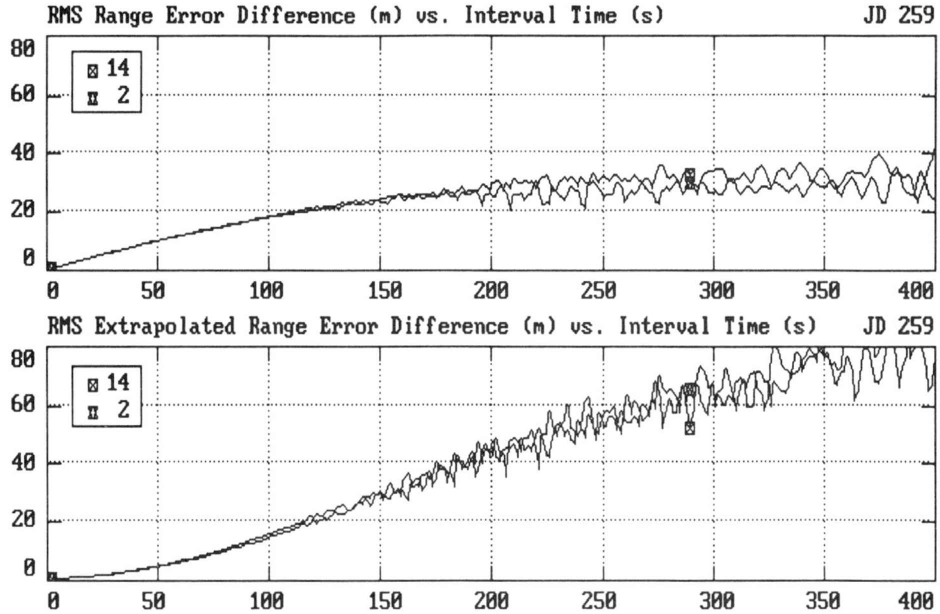

Fig. 7—Effect of Range Error Rate Propagation on RMS Range Error Difference

Figures 8 and 9 show the RMS Extrapolated Range Error Difference over a delay of twice the SC-104 baseline of 30 s. Note that the behavior of the data is approximately quadratic: doubling the delay increases the RMS Extrapolated Range Error Difference by a factor of 4. After a 30 s delay, the RMS Extrapolated Range Error Difference ranges from 1.2 m (Day 124) to about 2 m (Day 259).

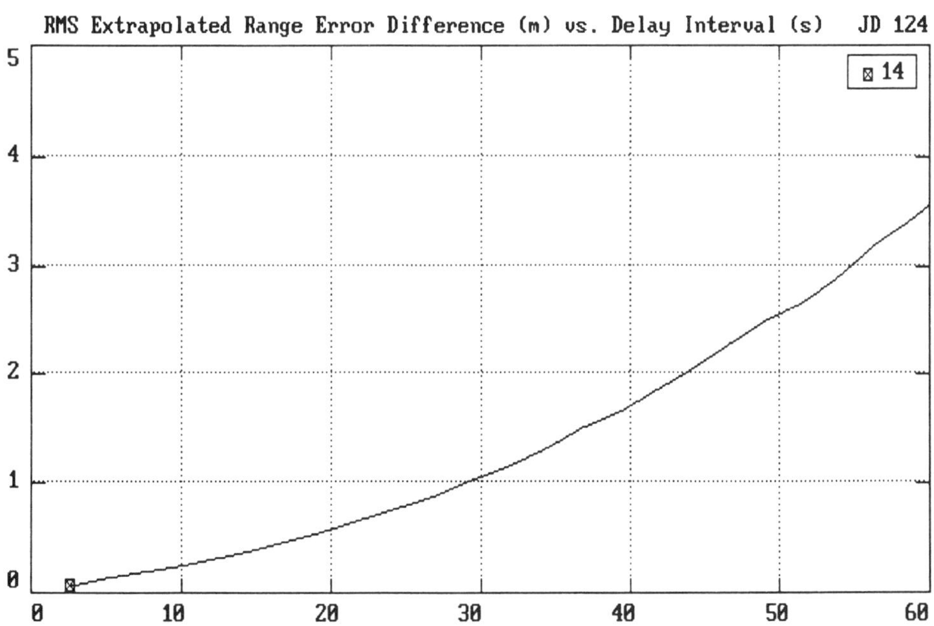

Fig. 8—RMS Extrapolated Range Error Difference (Julian Day 124)

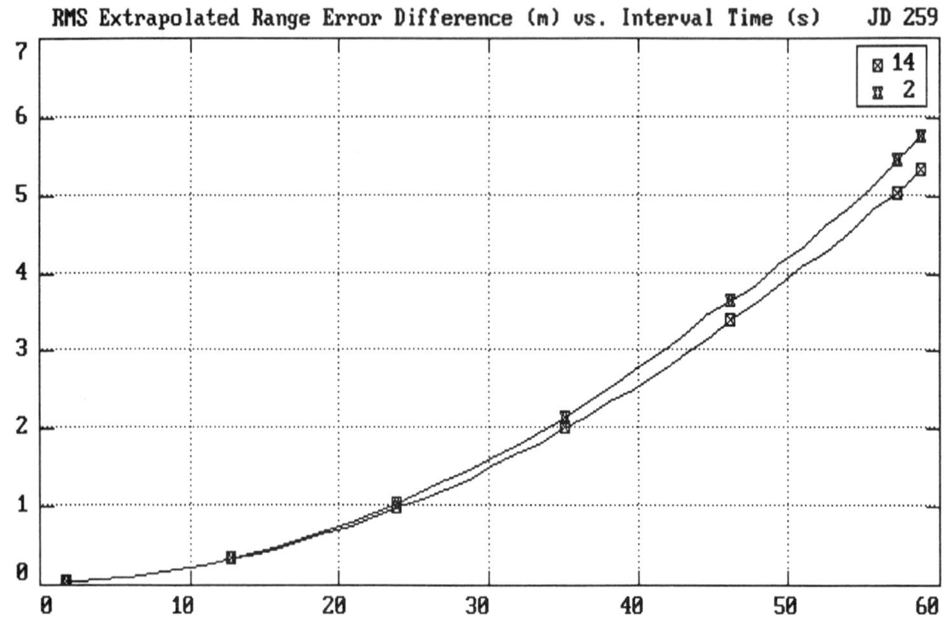

Fig. 9—RMS Extrapolated Range Error Difference (Julian Day 259)

Temporal Decorrelation and Data Latency Effects

The RMS Extrapolated Range Error Difference represents a worst-case estimate of range error difference. A differential user receiving corrections from a reference station would see the accuracy of the corrections degrade over the period of use. If a particular data link had a 10 s latency, there would be about 0.3–0.5 m uncertainty in the corrections when they were received. As the next set of corrections was being broadcast, the uncertainty would increase to something like a meter. Figure 10 shows how this error might appear over time to a differential user. For purposes of clarity, only one satellite, PRN 2, is plotted. The behavior of PRN 14 would be similar.

A piecewise fit of range error is generated by sampling the actual range error at 10 and 30 s intervals. The range error is propagated over either interval using the range-rate error as a differential user would. This corresponds to the somewhat unrealistic situation in which a user receives corrections every 10 (or 30) s over a data link with zero delay. The user's range error difference is therefore initially zero. During the propagation interval, the difference between the user's estimate of range error (propagated with range-rate error) and actual range error grows after the fashion predicted by the RMS Extrapolated Range Error Difference function, i.e., quadratically. This can be seen most clearly in the plot of the 30 s piecewise fit. In both cases, the variation of the range error difference over a given interval is approximately equal to the average value of the RMS Extrapolated Range Error Difference curve over that interval. For a 10 s interval, the one-sigma standard deviation of the user's predicted range error (using this particular satellite) is 0.11 m. Over the 30 s interval, the one-sigma variation grows to 0.92 m.

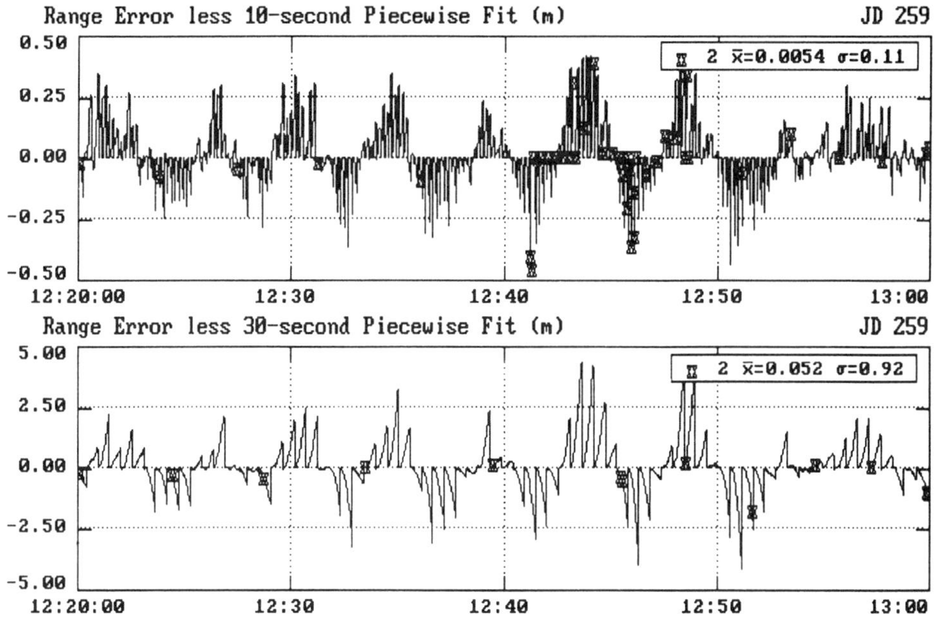

Fig. 10—Range Error Less Piecewise Fit

Note that after a 10 s interval, the errros rarely exceed 0.5 m; after a 30 s interval, they do not exceed 5 m. In practice, even this error would rarely be encountered, and would apply only to one differential measurement set. Expected errors for any particular data link latency and period of use can be determined by inspecting the appropriate segment of the RMS Extrapolated Range Error Difference plot.

SUMMARY

Table 1 summarizes the results of the Block II dynamic satellite signal behavior for specific days. Expected values are provided for comparison. It can be seen that both qualitatively and quantitatively, the observed behavior is similar to that predicted by analysis of statistical SA data previously made available to the civil community. It is also apparent that the assumptions on which RTCM SC-104 based its selection of message and data rates are adequate to deal with the observed behavior.

CONCLUSIONS

There has been no official confirmation at this time that the range errors observed on PRNs 14 and 2 are the result of SA. The range errors, range error rates, and accelerations observed are the same or somewhat larger than the levels anticipated. To the extent that the variations observed are consistent with SA levels anticipated during peacetime, the following conclusions can be drawn:

1) The levels of pseudorange errors, range-rate errors, and range acceleration errors are about those anticipated.

Table 1—Summary of Results

	PRN 6 (No SA)	EXPECTED (100 m, 95%)	PRN 14 (JD 124)	PRN 2, 14 (JD 259)
RANGE ERROR				
Mean	0	0	74 m	−61 to 38 m
1 σ	1 to 1.5 m	30 m	29 m	38 to 57 m
URA	2 to 4 m	32 to 64 m	32 m	64 m
RANGE ERROR RATE				
Mean	0	0	0.01 m/s	0.01 m/s
1 σ	0.0055 m/s	0.14 m/s	0.12 m/s	0.20 to 0.21 m/s
RANGE ERROR ACC				
Mean	0	0	0	0
1 σ	0.4 mm/s^2	3.7 mm/s^2	2.0 mm/s^2	3.4 to 3.6 mm/s^2
CORRELATION TIME				
90%	N/A	180 s	180 s	180 s
RMS RANGE ERR DIFF				
τ = 10 s	N/A	0.3 m	0.2 m	0.2 to 0.3 m
τ = 30 s	N/A	1.7 m	1.2 m	1.5 to 1.9 m

2) The RTCM SC-104 data rates appear adequate to provide 5 m positioning service (95 percent accuracy).
3) Our differential GPS system handles the variations very well, and provides highly accurate corrections.

ACKNOWLEDGMENTS

The authors wish to thank David Pietraszewski and Jay Spalding for their continuing interest in this subject, and for pointing out the satellite signal behavior on Julian Day 124.

Based on a paper presented at The Second International Technical Meeting of the Satellite Division of The Institute of Navigation, Colorado Springs, CO. This paper was the winner of the Satellite Division Best Paper Award.

REFERENCES

1. Kalafus, R., Knable, N., Kraemer, J., and Vilcans, J., *NAVSTAR GPS Simulation and Analysis Program,* DOT Report No. DOT-TSC-RSPA-83-11, December 1983.
2. Matchett, G., *Stochastic Simulation of GPS Selective Availability Errors,* Technical Memorandum, TASC Contract DTRS-57-83-C-00077, June 1985. This document can be obtained through the DOT/TSC Navigation Division, Cambridge, MA.
3. Pietraszewski, D., Spalding, J., Viehweg, C., and Luft, L., *U.S. Coast Guard Differential GPS Navigation Field Test Findings,* NAVIGATION, Journal of The Institute of Navigation, Vol. 35, No. 1, Spring, 1988, pp. 55–72.

NAVIGATION: *Journal of The Institute of Navigation*
Vol. 33, No. 4, Winter 1986–87
Printed in U.S.A.

Optimal Locations of Pseudolites for Differential GPS

BRADFORD W. PARKINSON, KEVIN T. FITZGIBBON
Stanford University, Palo Alto, CA
Received June 1986

ABSTRACT

Because of the announced DOD policy to degrade GPS accuracy, Differential GPS has assumed increasing importance. Furthermore, the delay in operational deployment of the NAVSTAR satellites has focused increased attention on the use of pseudo-satellites (pseudolites or PLs) to enhance accuracy and provide earlier operational capability. Pseudolites perform a double function: they both calibrate the local ranging errors and significantly improve the user's geometry as measured by GDOP (Geometric Dilution of Precision).

This paper examines the following problem: What is the best location for a PL when one considers the worst possible satellite outage in twenty-four hours? What is the resulting worst case GDOP? Both single and dual PL systems are considered, and the results are extrapolated to expected user receiver performance (horizontal and vertical errors).

The results indicate that a dual PL system will outperform the satellite-only GPS system, even with a worst case satellite outage. In addition, there should be negligible degradation during periods when the military intentionally reduces ephemeris or clock accuracy. Furthermore, the accuracy of such a system would provide aircraft positioning very close to Category I FAA landing requirements.

INTRODUCTION

Basic accuracy of GPS can be viewed as the product of the errors in ranging to the satellites and the diluting effects of the geometry. Thus the simplified relationship is:

$$\text{USER POSITION ERROR} = (\text{GEOMETRIC DILUTION}) \times (\text{RANGING ERROR})$$

where Geometric Dilution is a class of multipliers called DOP or Dilution of Precision. Examples: VDOP = Vertical Dilution of Precision, PDOP = Position Dilution of Precision. In the discussion which follows it is important to remember that the two effects are separate; reducing the ranging error does not affect the geometric dilution effects. While the ranging errors typically vary by a factor of two or three, the geometric dilutions may vary by a factor of ten or more.

The GPS system promises excellent overall global accuracy (typically 10 to 20 meters on C/A code), but there are some locations which have substantially greater errors. For example, at San Francisco International Airport (SFO) the vertical dilution of precision (VDOP)[7] can exceed 15 for a fully operational constellation. A breakdown of the ranging error sources and the impact on vertical accuracy is shown in Table 1. While the typical vertical error might be 14 meters, at times it is expected to reach 87 meters.

Table 1—Conventional C/A Code Accuracy.

Error Source	1 Sigma Error (Meters)		
	Bias	Random	Total
Ephemeris Data	3.5	0.0	3.5
Satellite Clock	1.5	0.7	1.7
Ionosphere	4.0	0.0	4.0
Troposphere	0.0	0.5	0.5
Multipath	0.0	1.0	1.0
Receiver	0.0	3.5	3.5
Calibration Site Residual	0.0	0.0	0.0
UERE (RMS)	5.5	3.7	6.7
Filtered UERE (RMS)	5.5	0.9	5.6
1 Sigma Vertical Axis Error	(VDOP = 2.5)		14.0
Maximum Vertical Error (SFO)	(VDOP = 15.6)		87.4

Impact of Selective Availability. If one considers the deliberate degradation of both ephemeris and clock-calibration (called Selective Availability-SA) which have been established by DOD for the operational system, the situation can become far worse than shown in Table 1. For example, consider a 20 meter deliberate ephemeris error (which is less than the maximum expected under these circumstances). As shown in Table 2, the vertical error can then be as large as 300 meters.

Table 2—The Impacts of Selective Availability on Conventional C/A Code Accuracy.

Error Source	1 Sigma Error (Meters)		
	Bias	Random	Total
Ephemeris Data	20.0	0.0	20.0
Satellite Clock	1.5	0.7	1.7
Ionosphere	4.0	0.0	4.0
Troposphere	0.0	0.5	0.5
Multipath	0.0	1.0	1.0
Receiver	0.0	3.5	3.5
Calibration Site Residual	0.0	0.0	0.0
UERE (RMS)	20.5	3.7	20.8
Filtered UERE (RMS)	20.5	0.9	20.5
1 Sigma Vertical Axis Error	(VDOP = 2.5)		51.2
Maximum Vertical Error (SFO)	(VDOP = 15.6)		319.4

Basic Differential Operations. A technique for improving the GPS system accuracy is to rely on some form of differential GPS operation.[2] The technique uses a calibration station at a known location which measures the ranging errors to all satellites in view. The corresponding corrections are then transmitted to nearby users with any convenient communications link. Note that *this does not improve the geometric dilution effects.*

Nonetheless, differential corrections to ranges (i.e. without PLs) can materially improve the situation for both the normal operation and when SA is being used (the differential results are virtually identical for each case). The estimated accuracy for this situation is shown in Table 3. A typical vertical error is 2.5 meters. This error estimate is very close to the experimental results obtained in one of the early tests of differential GPS[15]. These tests were performed at Yuma Proving Ground using five channel receivers.

Table 3—Differential C/A Code Accuracy with or without Selective Availability.

Error Source	1 Sigma Error (Meters)		
	Bias	Random	Total
Ephemeris Data	0.0	0.0	0.0
Satellite Clock	0.0	0.7	0.7
Ionosphere	0.0	0.0	0.0
Troposphere	0.0	0.5	0.5
Multipath	0.0	1.0	1.0
Receiver	0.0	3.5	3.5
Calibration Site Residual	0.0	1.9	1.9
UERE (RMS)	0.0	4.2	4.2
Filtered UERE (RMS)	0.0	1.1	1.1
1 Sigma Vertical Axis Error	(VDOP= 2.5)		2.6
Maximum Vertical Error (SFO)	(VDOP=15.6)		16.4

Requirements for Greater Accuracy. There are a number of users who would benefit from even greater accuracy. These applications include;

—Non-precision aircraft approach
—Maritime navaid location
—Aerial crop spraying
—Aerial surveying
—Marine oil surveying

One of the most stringent positioning requirements for the dynamic user is aircraft landing. The FAA navigation system standards for landing are shown in Table 4. These requirements are specified by positional accuracy only. However, the GPS system also can provide very precise velocity. For the automatic piloting of aircraft, velocity can provide greatly enhanced controllability over positioning alone; it might logically lead to some relaxation in the specification on positioning-only accuracy.

Table 4—The FAA Navigation Standards for Aircraft Landing System.

Operational Phase		Minimum Altitude (Ft)	Accuracy (2 DRMS)	
			Lateral	Elevation
Enroute/Terminal		500	4 NM	500 M
Approach and Landing	Non-Precision	250	2 NM	100 M
	Precision Category I (CAT)	100	±9.1 M	±3 M
	Precision Category II	50	±4.6 M	±1.4 M
	Precision Category III	0	±4.1 M	±0.5 M

Pseudolites. To further improve the user's accuracy, it is desirable to improve the geometry, i.e. to reduce the Geometric Dilution of Precision. This leads to an interesting variation of the differential method. This idea is to transmit the corrections on a link which is on the same frequency and uses the same format as the satellites. This saves the cost of an additional user communications receiver, and also can greatly improve the geometry.

A typical system configuration is shown in Figure 1. Standard formats and communications protocols have been developed by the Radio Technical Commission for Maritime Services (RTCM)[10]. These stationary transmitters are called Pseudolites or PLs. The definition of a PL is shown in Table 5. In a previous paper,[1] the use of PLs for improving GPS performance was described. It was shown that PLs could solve certain geometric problems caused by the proposed 21 satellite constellation.

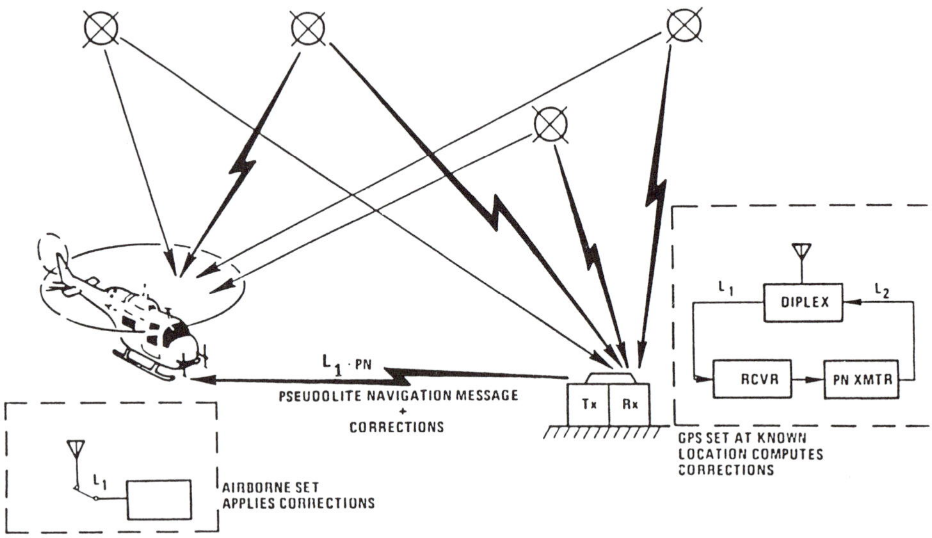

Fig. 1—Typical Configuration of a Differential GPS System Aided by Pseudolites.

Table 5—The Definition of Pseudolite (PL) Developed by the RTCM.

The RTCM Special Committee on differential NAVSTAR/GPS service has defined a Pseudolite in the following way:

1) *Receives GPS satellite signals and computes pseudo-range and range rate corrections,* etc., as needed to satisfy the requirements of any other differential GPS monitor station.
2) *Transmits the correction information at 50 bits per second on an L-band frequency,* preferably but not necessarily L_1.
3) The transmitted *signal is GPS-like,* in that a PN code is used to permit local user equipment to *obtain an additional pseudorange measurement* to the transmitting antenna.
4) The transmitted signal is designed to prevent interference with other equipment.

THE PROBLEM

The problem addressed by this research is: what is the best location for pseudolites if worst case satellite outage occurs? The criteria for "best" is that the *maximum* dilution of precision (due to geometry) is minimized. The answer to this question is obviously dependent on the location of the user, *but* is the *same for any 24 hour* period of the year, since the satellites are controlled to 12 hour orbits, i.e. repeated ground traces. This criterion is a *guaranteed accuracy* measure, since it yields the best answer for the worst assumed situation. For example, the user is more interested in the worst case accuracy then the RMS or the CEP.

Optimizing the Position for Single and Dual PLs. We are seeking the location for one and two pseudolites (PLs) that minimizes the worst DOP condition with one satellite outage.

This optimization is also dependent on a set of assumed conditions or parameters such as minimum elevation mask angle, PL altitude, etc. Specifically the best location gives a fixed upper bound on the DOP for the whole 24 hour satellite cycle. Average DOP or one sigma values are *not being used,* since they do not tend to guarantee accuracies.

In this paper the proposed operational constellation of 18 satellites plus 3 spares will always be assumed.

SOLUTION METHOD

The method used to solve this problem is a systematic search in the vicinity of the user by scanning in azimuth when the PLs and the user are at the same altitude, and in a rectangular grid when the PLs and the user are at different altitudes.

An alternative method using a random search was investigated but not used because it tended to find a local minimum rather than the absolutely best position. The random search method was also discarded because it did not give an indication of the solution degradation in the vicinity of the optimum position.

PARAMETERS AFFECTING THE SOLUTION

Several parameters affect the selection of the best site, namely:

—the user's position (latitude, longitude and altitude).
—the elevation mask angle.
—the altitude of the PLs with respect to the user.
—the number of PLs.
—the satellite selection (DOP) criterion.
—the discretization in time.

Because searching for the best PL location can become a lengthy task, a number of techniques were found to reduce computation time without the loss of precision and reliability.

Time Interval. The satellite system is a continuous time process which is sampled at fixed intervals (Ts). This discretization in time has a strong influence on the solution; the selection of too coarse an interval can lead to erroneous results. Figure 2 presents the effect of sample time intervals on the simulations. There is a marked loss of details in the DOP-time history for Ts = 8 minutes. For example, the peak at 9 hours with Ts = 2 minutes is significantly reduced if Ts = 8 minutes.

High DOP peaks which occur during short periods of time (1 minute) may be

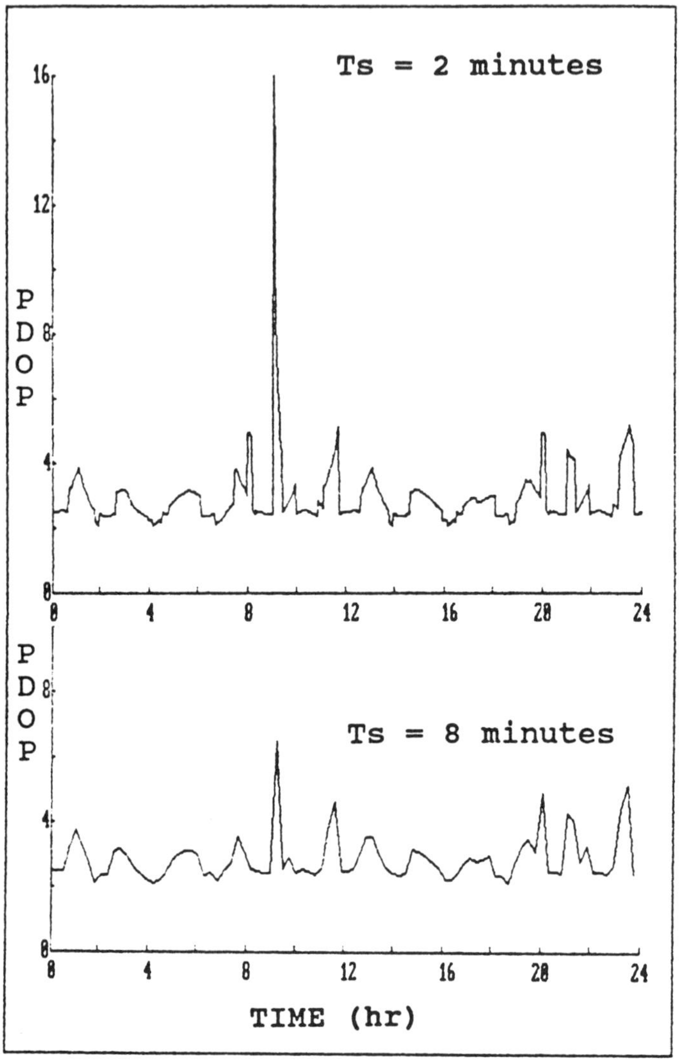

Fig. 2—Effects of Time Interval on PDOP Peaks. (The high peak at 9 hours with $T_s = 2$ minutes is significantly reduced with $T_s = 8$ minutes.)

well behaved in the adjacent time periods. The short periods of high DOPs are usually associated with a low number of visible satellites as shown in Figure 3. A small sample time (Ts) is necessary in the simulations to guarantee no loss of details. This leads to a penalty in terms of run speed. A selection of the best time interval was made by trial, and all final optimizations used Ts = 2 minutes.

Elevation Mask Angle. The elevation mask angle determines the region of the sky in which the satellites are visible and consequently the number of visible satellites. This angle applies to the user and can be limited by natural or artificial obstacles such as buildings, trees and hills above the local horizon or by the "visibility" of the antenna (say, if the antenna is located on the upper

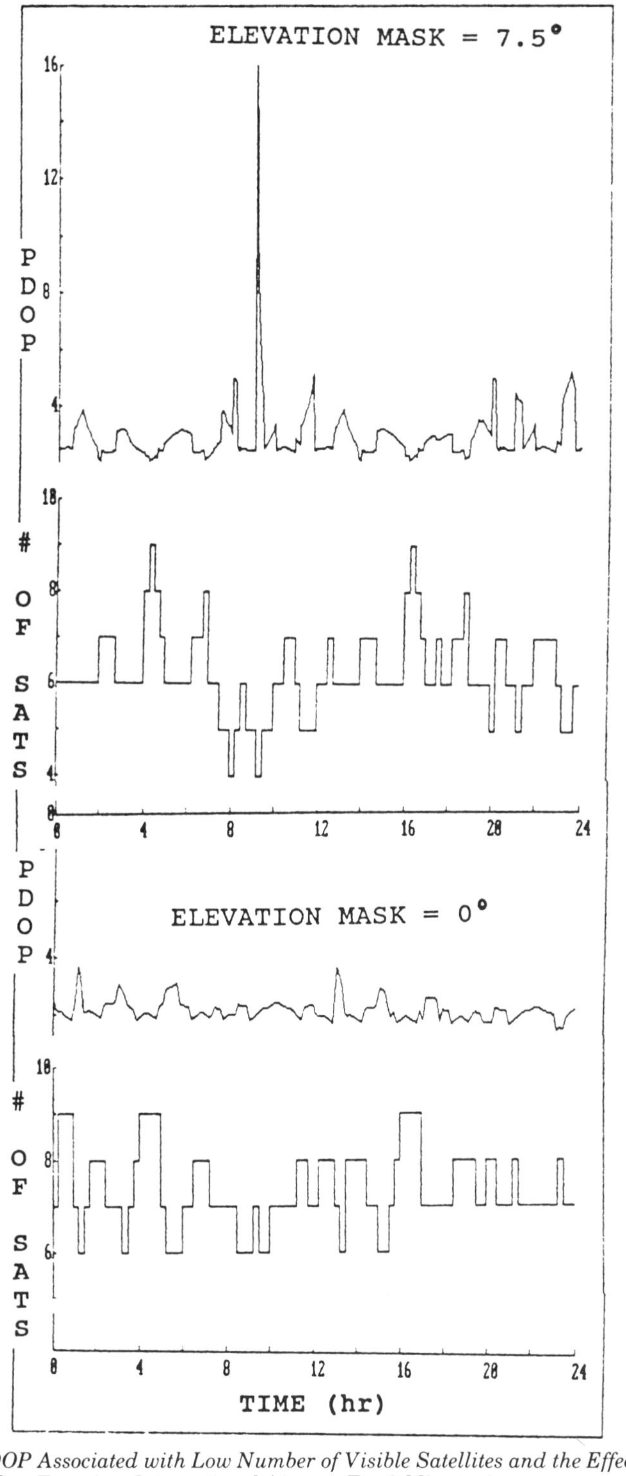

Fig. 3—High PDOP Associated with Low Number of Visible Satellites and the Effect of Mask Angle. (San Francisco International Airport, $T_s = 2$ Minutes.)

or lower surface of the fuselage of an airplane at a high altitude). In a practical implementation of the differential GPS system aided by PLs, the correction information is transmitted for *all* SV's in view of the ground station. The format would follow that specified by the RTCM-104 report.[10] It is assumed that all SV's which meet the elevation angle criterion for the user will be calibrated. Figure 3 presents the effect of the elevation mask angle on the number of visible satellites and the DOP history. Notice the elimination of the high peak on PDOP when using an elevation angle of 0 degrees.

Generally a 7.5 degree mask angle was used in this paper since it corresponds to previously published results[1] and also is the value generally used for studies by the FAA.

Dilution of Precision (DOPs). The DOP criterion was used to select four ranging signals from the available PL's and satellites (the visible ones excluding any designated failure). The selection is made by minimizing the DOP. The influence of the DOP criterion is presented in Figure 4. Two different criteria are used: VDOP and WDOP.

VDOP. The VDOP criterion increases the frequency of events of low VDOP values when compared to the PDOP criterion, as seen in the DOP frequency histograms, but creates many high peaks of PDOP. The PDOP criterion looks satisfactory for V, H and PDOP values.

WDOP. For the specific problem of an aircraft using the GPS and aided by PLs to perform a landing approach, a different criterion may be better. The FAA landing accuracy requirements are three to ten times more stringent on vertical than horizontal errors. Since the vertical position plays a more important role than the horizontal position, a Weighted DOP may be created by weighting the VDOP values as twice (this weighting could be changed if desired) as much as the HDOP values (in PDOP the VDOP and HDOP values have the same weight). The Weighted DOP (WDOP) criterion is

$$\mathrm{WDOP} = (2 * \mathrm{VDOP} + \mathrm{HDOP})/3$$

This criterion creates DOP values similar to the PDOP criterion. Unless otherwise specified, the WDOP criterion will be used in the results of this paper.

Distance Limitations in the Position of the PLs. Based on recommendations of the RTCM Special Committee 104 (1985) the PLs are expected to serve the community in a 50 kilometer (27 nautical miles) range and this will be the maximum distance between the user and one PL.

With two PLs the use of Gold codes to transmit the data signals from the PLs to the user impose constraints on the minimum distance between the PLs. If the same code is used, the minimum recommended distance between PLs is 130 kilometers (70 nautical miles). If different codes are used, which will be assumed here, the minimum distance between PLs is 54 kilometers (29 nautical miles)

This ensures that a well designed receiver can continue to operate in the central region of the pulse overlap. The overlap region is limited by hyperbolic curves with the PLs at the focus, a focal distance of 12 kilometers (6.5 nautical miles) and a minimum distance of 30 kilometers (16 nautical miles) between the hyperbolas. The hyperbolic curves are the locus of points at which the difference in distance to the two PLs is 30 kilometers. This subject is presented

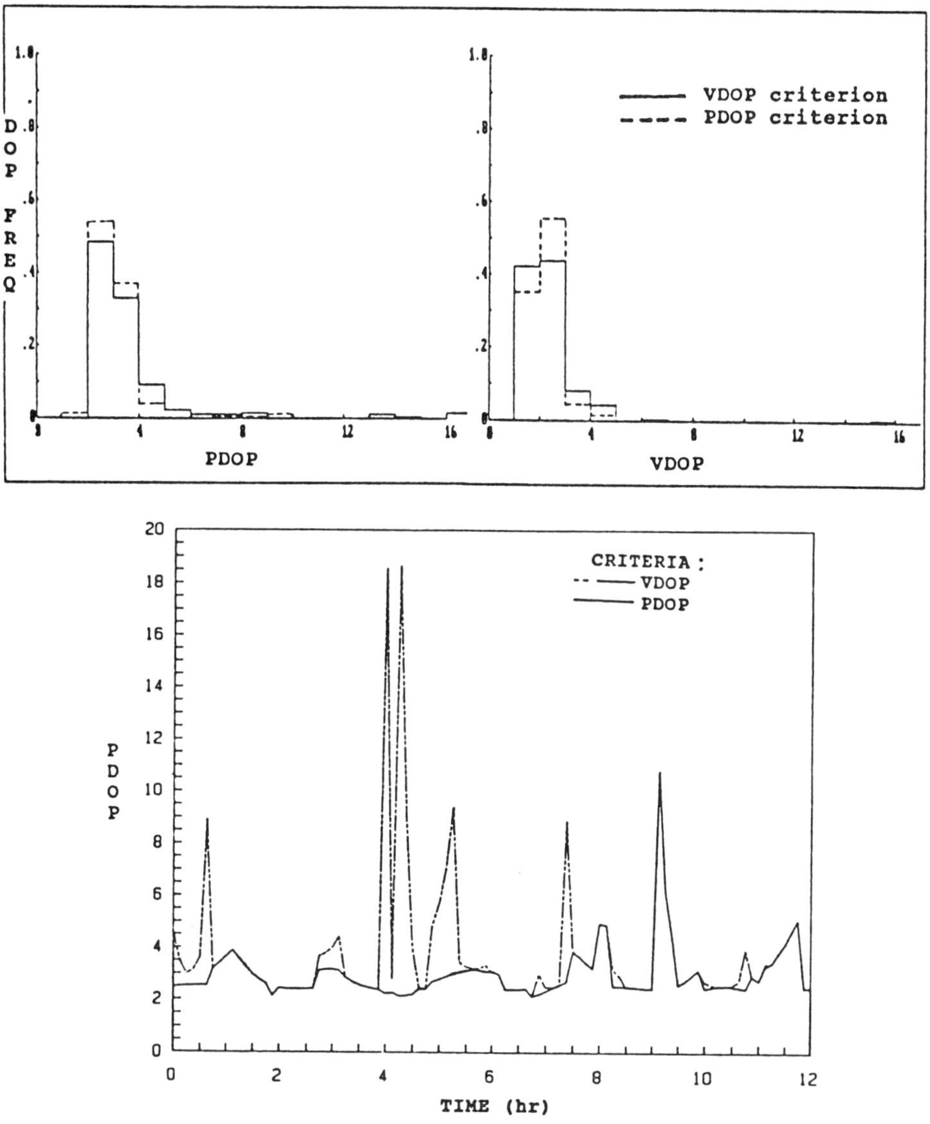

Fig. 4—Effects of DOP Criterion on PDOP and VDOP.

in detail in Ref. 10 and will not be discussed here. Figure 5 presents the characteristics of this configuration.

Description of the method. A systematic procedure has been used to find a solution for the absolute best PL positions with the given parameters. The elevation mask angle must be defined by the particular visibility constraints of the user. Then:

(a) Run a 24 hour simulation using a small time interval (start with one minute or less) and check critical situations when high DOP and low number of visible satellites occur. The objective is to define the time interval for the following simulations without losing detailed information of the DOP peaks.

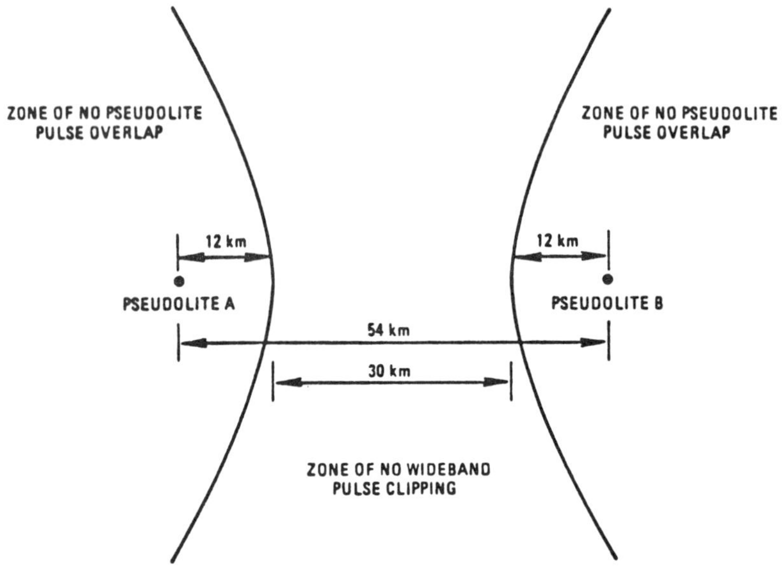

Fig. 5—Distance Limitations for the Dual Pseudolite System Recommended by RTCM Special Committee 104.

(b) Run a simulation with one outage and observe the high peak DOPs. These values are usually associated with a low number of visible satellites and are potential candidates to be critical times at which the position of the PL critically affects the maximization of the DOP criterion; for instance, if only 4 satellites are visible, as occurs some times in the SFO area, and one fails or is not reliable, then the PLs are essential to the solution of the position of the user (assuming that four fixes are needed). At this point the critical times are selected but the following simulations will still be run for the full 24 hour period. These critical times will be useful at the final stages of the problem.

(c) Use an azimuth scan (from 0 to 360 degrees) with a large interval (say 10 degrees) if the PL(s) is(are) at the same altitude of the user, or a rectangular grid scan in the vicinity of the user (say a square region of 60 by 60 nautical miles with steps of 10 nautical miles) if the user and the PL(s) is(are) at different altitudes, to have a rough initial guess about the direction of the best location for the PL(s). Observing the distance limitations select any radial distance between user and one PL and, if two PLs are being scanned, select the minimum distance between the PLs. At this point a first rough solution is available for the position of the PL(s).

(d) Run the one outage simulation with azimuth scan in the vicinity of the solution from (c) using an interval of 0.2 degrees. Now a more precise solution is achieved.

(e) Using the solution from (d) compute the DOP versus time history with one outage, during a 24 hour period and small time interval. Check the results. If any DOP unbalanced peaks are noticed, i.e., there is still one (not necessarily high) peak, or two or more peaks not with the same amplitude, then the best solution has probably not been achieved. Now using only the critical times when these peaks occur repeat from (c) until the peaks are balanced. This procedure will converge to the final absolute best positions for the PL(s).

Flowchart. The flowchart in Figure 6 and the schematics in Figure 7 show how the search for the best PL position is computed. Each PL position is associated with a set of values of VDOP, HDOP, PDOP and the DOP criterion value which maximizes the selected DOP criterion in a 24 hour period. First the PL(s) is(are) positioned in a node on the rectangular grid (or a point on the circle if azimuth scan is being used). Then for each time in the 24 hour period the DOP criterion is minimized by selecting four satellites and PL(s) after one satellite is failed among the set of those visible at that time. All visible satellites are failed, one by one, resulting in a set of minimized DOP criterion values at the specific time. The best set of remaining range signals is then selected (based

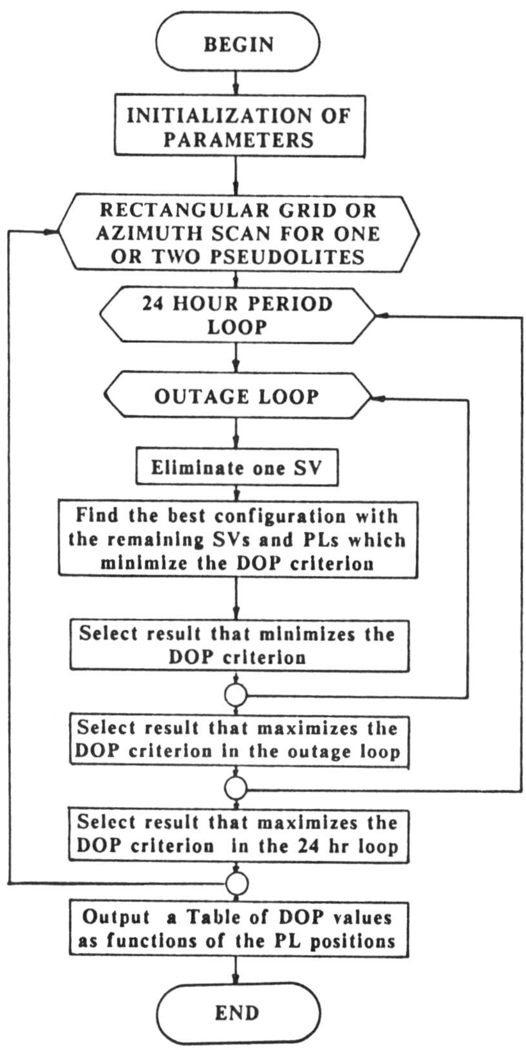

Fig. 6—Flowchart for Finding Best Pseudolite Position with Worst Case Outage.

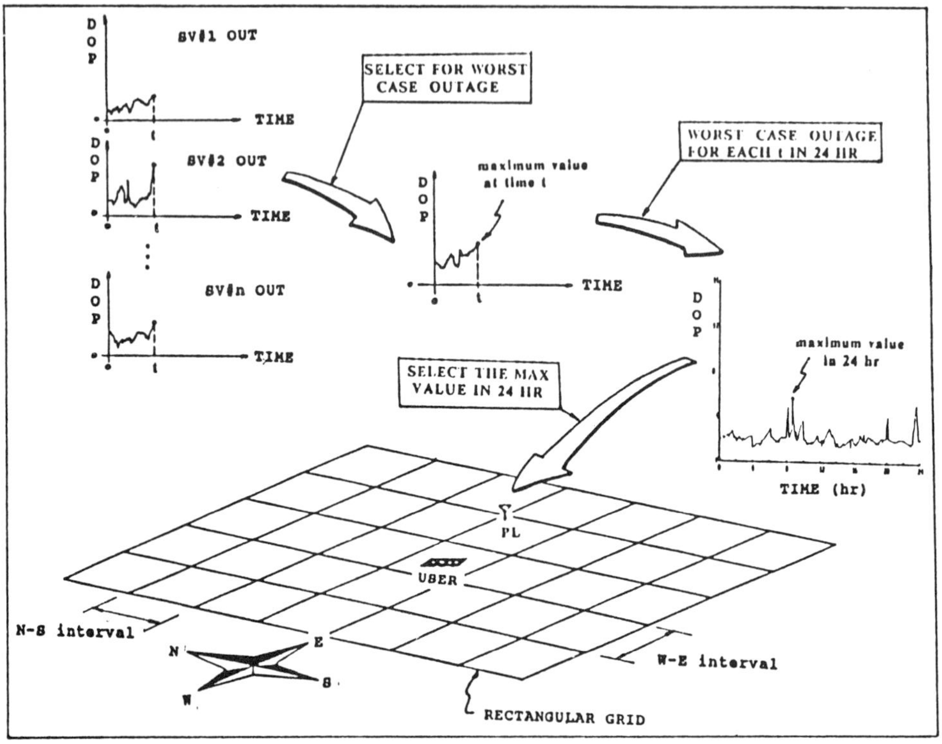

Fig. 7—Finding Best Pseudolite Position with Worst Case Outage.

on minimum DOP criterion). The highest DOP value among these minima for all possible sets of ranging signals is the "worst case outage" at that point in time.

The procedure is repeated for the 24 hour period and the maximum value is now selected to be the absolute worst or maximum DOP criterion value which will be used to select the best position for the PL(s).

The best PL position is the one with the lowest maximum DOP criterion value. This solution will guarantee that by locating the PL(s) at the selected position(s) the DOP criterion value will never exceed a maximum for the one outage SV with the selected set of parameters.

The scanning technique used for dual PL systems is implemented by scanning one PL and, while keeping it fixed, scan the other with the only restrictions being the distance limitations discussed before.

APPLICATION OF THE METHOD

Optimized locations were found for Denver, New York and San Francisco. The San Francisco International Airport will be discussed. For a 7.5 degree elevation mask angle, no PLs and no satellite outage, there is a high PDOP peak of more than 16 at approximately 9 hours as shown earlier in the paper. With the worst case single outage the PDOP values are above 16 during a total of more than 18% of the 24 hour period as shown in Figure 8.

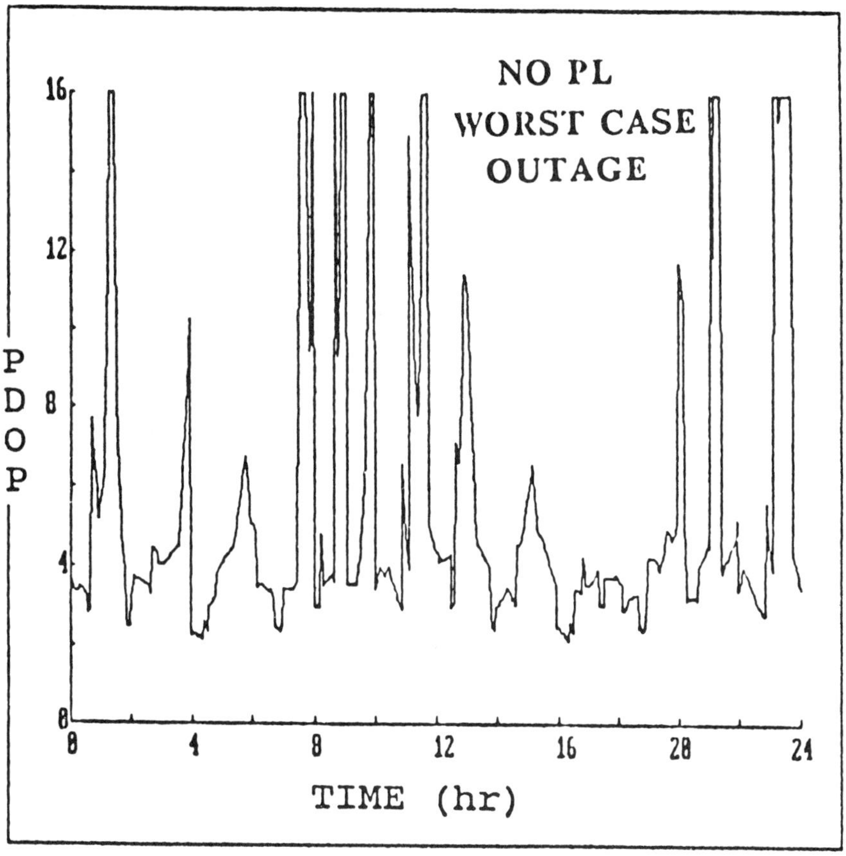

Fig. 8—Worst Case Outage at San Francisco International Airport with No Pseudolites, using WDOP Criterion and Mask Angle of 7.5 Degrees.

The results presented used the following set of parameters unless otherwise specified:

* user's position (San Francisco International Airport):
  Latitude: 37° 36.7′ North.
  Longitude: 122° 21.3′ West.
  Altitude: 5000 ft.
* elevation mask angle = 7.5 degrees above local horizon.
* DOP criterion: WDOP.
* one outage.
* 24 hour period with 2 minutes interval.
* constellation of 18 satellites plus 3 spares.

RESULTS

First the results for a single PL system will be presented. An azimuth scan resulted in 86.6 degrees azimuth as the best direction to locate the PL with the distance limited by the recommendations from the RTCM-104 Committee.

Effects of the Relative Altitude. A rectangular scan was used to analyze the effects of the relative altitude between the user and the PL. For increasing altitude differences the best PL location diverges from the 0 ft direction as shown in Figure 9, but the best position remains within a 0.75 nautical mile by 2.0 nautical mile region for altitude differences ranging from 500 to 4000 feet

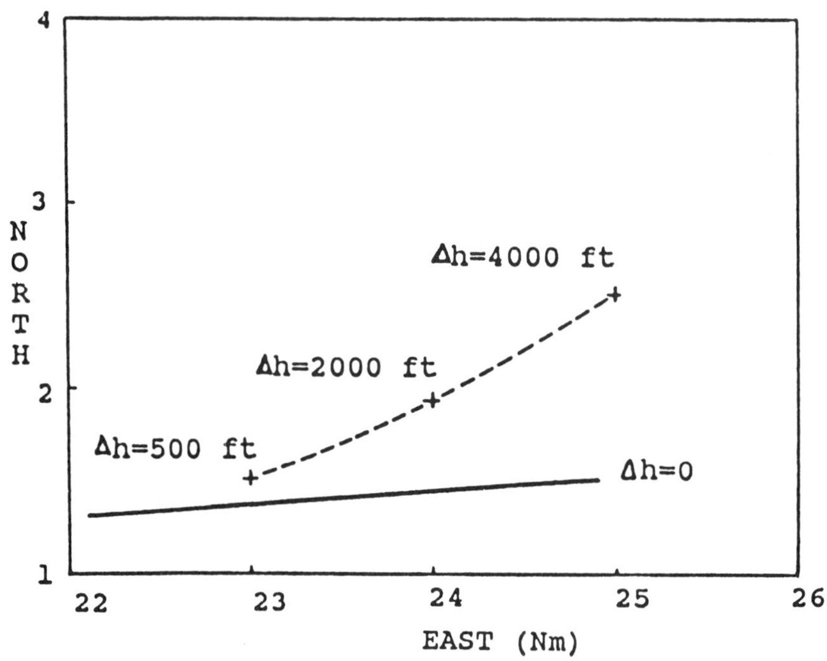

Fig. 9—Best Pseudolite Position as a Function of Altitude Difference between User and Pseudolite.

Effects of Covarying the Absolute Altitude. The absolute altitudes of both the user and the Pl were varied from 0 to 5000 ft and did not affect the best PL position within the grid interval accuracy (0.25 nautical miles).

A final decision on the best position must consider the geographical limitations on installing the PL. The best position of the PL assumes a fixed position of the user. For PLs located more than 25 miles from the user, variations of 1 or 2 miles in the user's location will not significantly change the results. For the SFO case, positioning the PL along the best azimuth direction and at the maximum distance of 27 nautical miles from the user, places the PL at 2000 ft above sea level. Under the point of view of direct LOS this is an excellent result, but corrections must be applied to compensate for the altitude difference.

The Best PL Positions. By assuming an altitude difference of 1000 ft, the best PL positions (with respect to the user) are

—for 1 PL: 25 nmi east, 2.5 nmi north;
—for 2 PLs: PL#1 at 0 nmi east, 10 nmi south; and PL#2 at 19.5 nmi east, 5.2 nmi north.

The locations for the best PL positions are shown in Figure 10.

Notice that the best position for a single PL system does not coincide with any of the best positions for the dual PL system although one of the positions

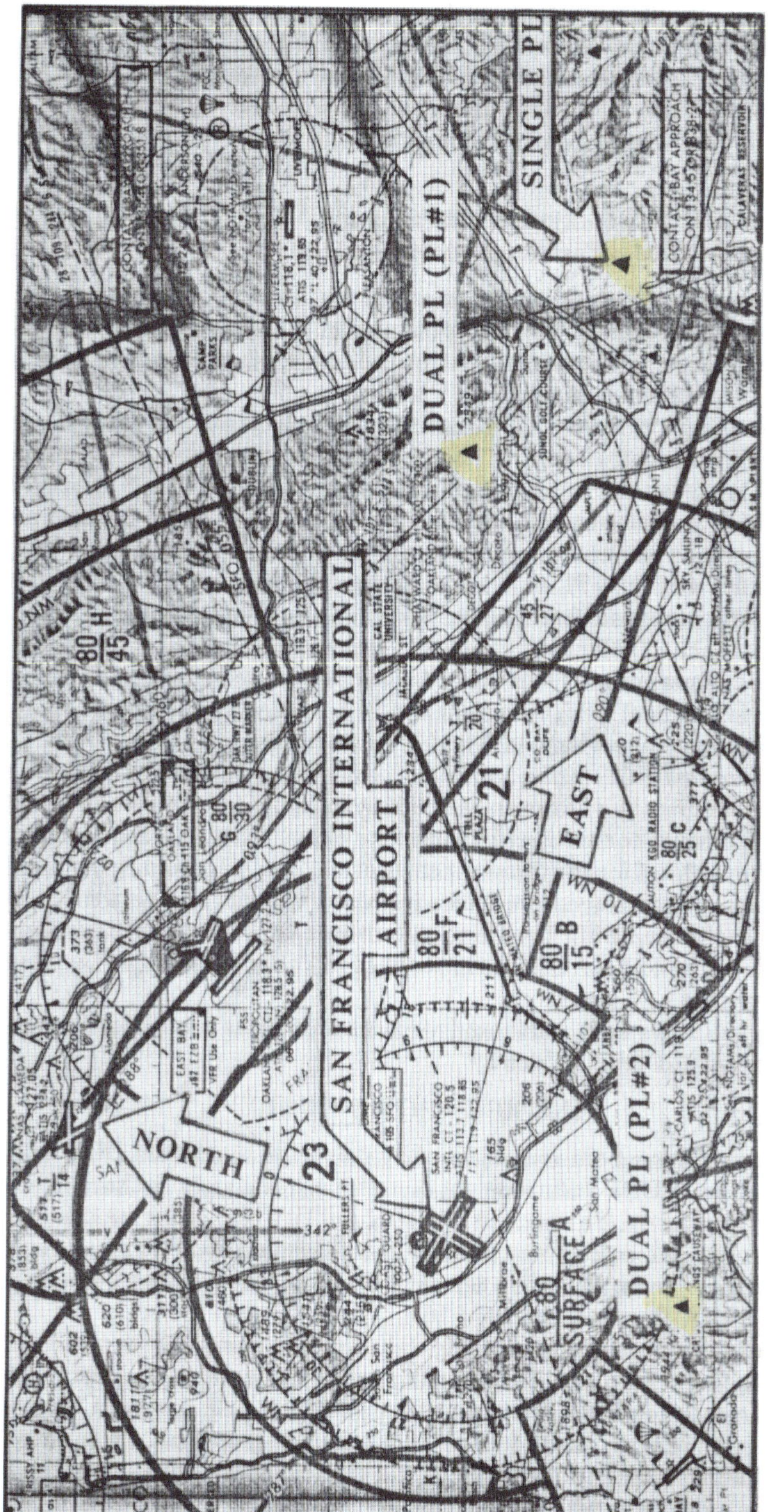

Fig. 10—Best Positions for the Single and Dual Pseudolite Systems for San Francisco International Airport.

is quite close. By placing PL#1 at the best position for one PL and selecting the best position for the second PL, maximum DOP criterion is decreased over the use of one PL, but does not optimize the use of two PLs.

DOP Versus Time. DOP time histories are presented in Figures 11 and 12 for the following cases (always comparing with the NO PL—NO OUTAGE situation)

(a) single and dual PL systems and no outage.
(b) single and dual PL systems and one outage.

By visual inspection qualitative improvements are noticed. The use of one or two PLs eliminates the high DOP peaks even for one satellite outage. Some DOP peaks above 5 are still present in the one PL-one outage case, but they are completely eliminated with two PLs.

DOP Frequency of Events. A quantitative analysis can be carried out with the results from Tables 6 and 7. The percentage of frequency of events for the various DOPs over the 24 hour period is presented in Table 6. Notice in the case for one PL and one outage that the placement of the PL at the best position does not increase the number of events for VDOP less than 2 when compared to the results with the PL at a non-best position (arbitrarily selected only 5 nautical miles northeast from the best position as a measurement of sensitivity) but eliminates the PDOP peaks above 8. So the statement that the best position minimizes the maximum DOP values is confirmed and it does not necessarily decrease the average. With the use of two PLs, even with one outage, the VDOP and PDOP values never exceed 2 and 3, respectively.

The Maximum DOP Values and the Associated Position Errors. A major concern is whether the Differential GPS system aided by PLs can satisfy the FAA requirements for an aircraft landing system. The FAA requirements are presented in Table 5 for the different categories. Table 7 presents the maximum DOP values and the respective errors in the east, north and vertical directions based on the filtered RMS User Equivalent Range Error (UERE) presented in Table 1. The filtered RMS UERE values used in Table 7 are for the Conventional and Differential C/A codes.

The errors in the east, north and vertical directions are related to the east, north and vertical DOP values by

$$[\text{ERROR}] = [\text{DOP}] * [\text{UERE}]$$

and the maximum errors are obtained by using the maximum DOP values.

The maximum DOP values do not occur simultaneously so there is no direct relationship between the maximum VDOP and HDOP and the maximum PDOP or WDOP. For one outage and no PLs the maximum DOP values are infinity and the errors are expressed in terms of the total time, in the 24 hour period, in which the DOP values exceeded 16.

A decrease in the maximum DOP values when using one or two PLs, compared to the case with no PLs, is conclusive by inspection. An improvement of 14 to 35 times is achieved with the Differential C/A code over the Conventional C/A code with one PL, and of 40 to 70 times with two PLs. An improvement greater than 250 times is achieved in the vertical error when using the Differential C/A code with two PLs and no outage over the C/A code with Selective Availability (see Table 3).

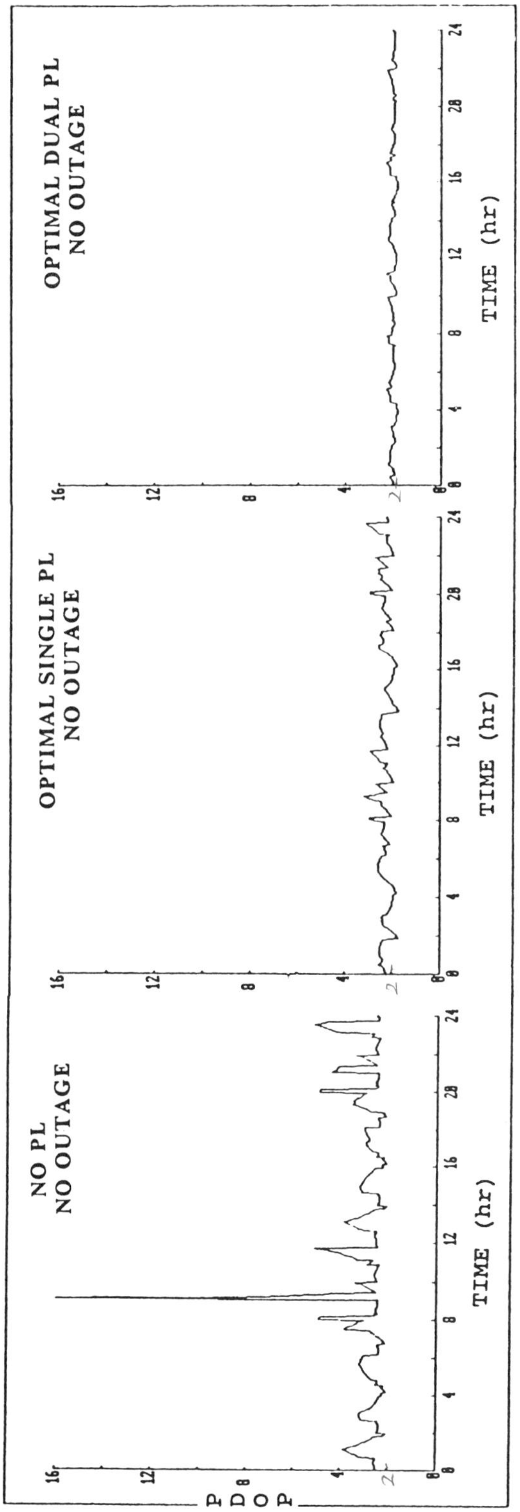

Fig. 11—PDOP Time History for One and Two Pseudolites Located at Their Best Positions for the No-Outage Case.

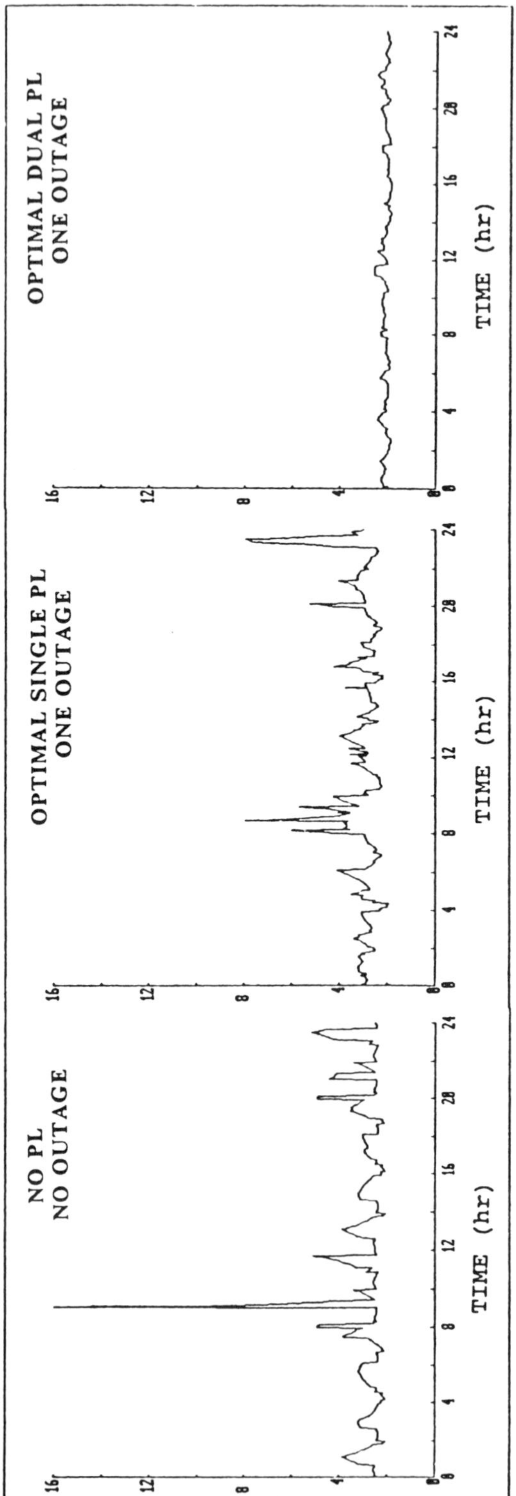

Fig. 12—DOP Time History for One and Two Pseudolites Located at Their Best Positions for the Worst Outage Case.

Table 6—DOP Frequency of Events.

# of OUT	# of PLs	Percentage of Events VDOP <2	HDOP <2	PDOP <3	VDOP >5	HDOP >5	PDOP >8
0	0	37	91	67	0.8	0.2	0.5
1	0	5	43	11	2.3	11	18
1	1	33	51	56	0.4	1.4	0
1	1*	36	52	55	1.0	1.7	1.0
0	1	95	96	99	0	0	0
1	2	100	96	100	0	0	0
0	2	100	100	100	0	0	0

*PL positioned at 5 Nm northeast from optimal single PL location.

Table 7—The Maximum DOP Values and Position Errors.

# of OUT	# of PL	Maximum Values VDOP	HDOP	PDOP	WDOP	σR	ERROR (m) VERT	EAST	NORTH
0	0	15.6	7.71	17.4	13.0	5.6	87	13	41
1	0		——— (*) ———			5.6	33 min	2 h 40 min	(**)
1	1	5.51	6.18	7.94	5.58	1.1	6.0	3.0	6.3
0	1	2.26	2.36	3.16	2.19	1.1	2.5	1.3	2.5
1	2	1.92	2.21	2.56	1.77	1.1	2.1	1.4	1.5
0	2	1.66	1.95	2.28	1.51	1.1	1.2	1.3	1.7

σR = Filtered UERE (RMS).
(*)—values are infinity.
(**)—total time intervals in which VDOP and HDOP are > 5.

Table 8—Differential C/A Code Accuracy for the Dual PL System with No Outage.

Error Source	1 Sigma Error (Meters) Bias	Random	Total
Ephemeris Data	0.0	0.0	0.0
Satellite Clock	0.0	0.7	0.7
Ionosphere	0.0	0.0	0.0
Troposphere	0.0	0.5	0.5
Multipath	0.0	1.0	1.0
Receiver	0.0	3.5	3.5
Calibration Site Residual	0.0	1.9	1.9
UERE (RMS)	0.0	4.2	4.2
Filtered UERE (RMS)	0.0	1.1	1.1
1 Sigma Vertical Axis Error	(VDOP = 1.42)		1.5
Maximum Vertical Error (SFO)	(VDOP = 1.66)		1.8

Tables 8 and 9, similar to Table 1, present the 1σ errors and the maximum vertical errors for the Differential C/A Code using the dual PL system with no outage and with the worst case outage. They show results that are better than the FAA requirements for Category I landings.

The Error Ellipsoids. Figure 13 presents the error ellipsoids in the east, north and vertical principal axes and compares the results of the no PL—no outage ellipsoid with the FAA requirements and the maximum errors for one and two

Table 9—Differential C/A Code Accuracy for the Dual PL System for the Worst Case Single Outage.

Error Source	1 Sigma Error (Meters)		
	Bias	Random	Total
Ephemeris Data	0.0	0.0	0.0
Satellite Clock	0.0	0.7	0.7
Ionosphere	0.0	0.0	0.0
Troposphere	0.0	0.5	0.5
Multipath	0.0	1.0	1.0
Receiver	0.0	3.5	3.5
Calibration Site Residual	0.0	1.9	1.9
UERE (RMS)	0.0	4.2	4.2
Filtered UERE (RMS)	0.0	1.1	1.1
1 Sigma Vertical Axis Error	(VDOP = 1.48)		1.6
Maximum Vertical Error (SFO)	(VDOP = 1.92)		2.0

PLs and no outage. With the Selective Availability the ellipsoids are 300 meters long in the vertical principal axis and are not presented for sake of clarity.

The FAA requirements for the Category I aircraft landing systems are satisfied for one and two PLs if no outage occurs. If two PLs are used, the criterion is satisfied for the single outage worst case. With one PL and one outage the vertical error exceeds the FAA admissible value while the horizontal error is satisfactory. A more careful choice of the DOP criterion and improvements on the receivers, or the use of a radar altimeter with the GPS system, may bring this vertical error below the FAA requirements.

Non-Optimal Locations. Table 10 presents the WDOP worst case outage values in the vicinity of the user (a grid of 60 by 60 nautical miles with 10 nautical mile intervals) and close to the optimal single PL position (a grid of 1.25 by 5 nautical miles with 0.25 and 1 nautical mile intervals). Locating a PL in a non-optimal position can lead to very high peaks of WDOP (and PDOP). For the SFO case, the optimal single PL position is not very sensitive in the vicinity of the optimal location.

The rough optimal position found in the first scan was approximately 5 nautical miles northeast of the final position and unbalanced peaks in the WDOP were noticed as shown in Figure 14. A step in the WDOP values from about 5 to 7 is due to a change in the visible constellation of satellites.

A PL located at 30 nautical miles south of the SFO International Airport was used by Parkinson[1] to eliminate the high PDOP peak at 9 hours with no outage. The situation was reproduced in agreement with the reference as presented in Figure 15. But for the worst case SV outage, the PL is not located at the optimal position: a PDOP peak higher than 300 is present.

THE VALUE OF PSEUDOLITES

The utility of an optimally located PL system is clear: both ranging accuracy and geometry are substantially improved. This combined effect *reduces a user's maximum positioning error by factors of 14 to 17.* If Selective Availability is being used, the improvement factor is likely to be even greater. In addition to

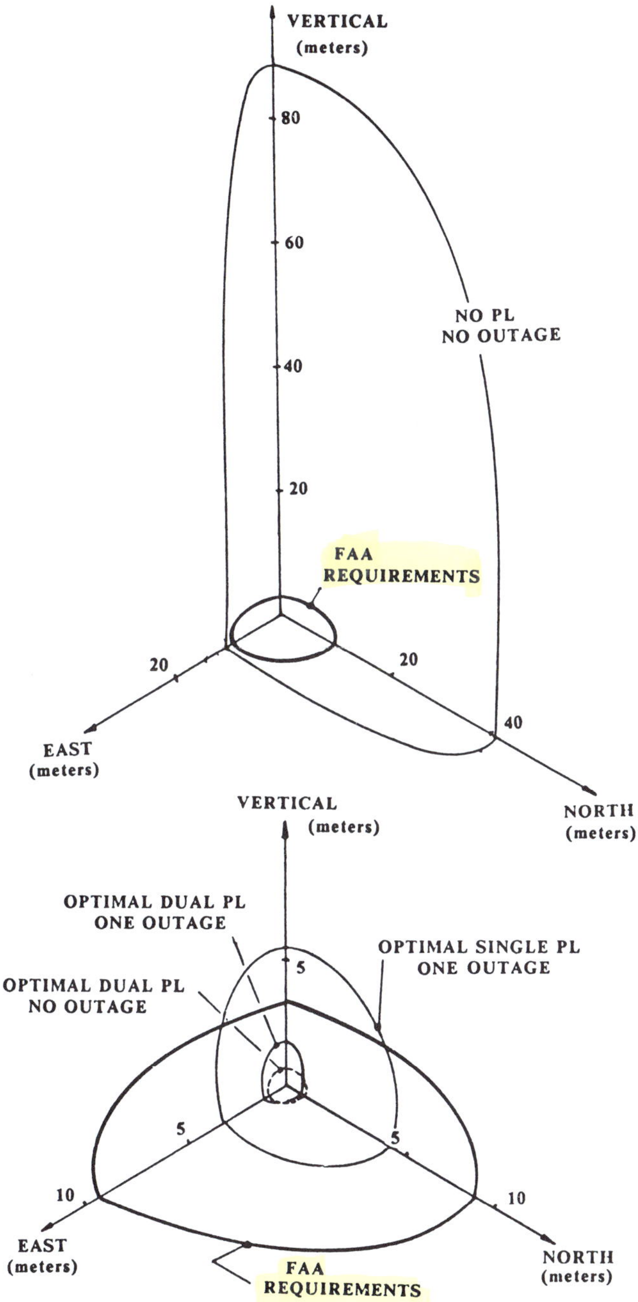

Fig. 13—The Ellipsoid Errors.

accuracy considerations, there are a number of powerful benefits to a PL implementation:

Availability. With known satellite outages, the system can continue to function with performance levels equal to or better than the unaugmented system.

Table 10—The WDOP Variation in the Vicinity of the User

		WEST		DISTANCE (Nm)			EAST	
		30	20	10	0	10	20	30
N	30			287	122	136		
O R	20		114	169	194	82	458	
T H	10	135	2462	348	250	284	27	12.4
	0	159	114	62	USER	8.3	7.3	7.0
S	10	61	33	*	750	52	7.6	13.5
O U	20		156	192	747	249	94	
T H	30			308	407	210		

		DISTANCE FROM THE USER (Nm)				
				EAST		
		23	24	25	26	27
	2.75	5.88	5.83	5.77	5.73	5.68
N	2.50	5.71	5.66	5.62	5.63	5.63
O R	2.25	5.64	5.64	5.64	5.65	5.65
T H	2.00	5.66	5.67	5.71	5.73	5.75
	1.75	5.79	5.81	5.84	5.79	5.88
	1.50	5.95	5.96	5.98	6.00	6.02

In addition, the system will effectively attain earlier operational status with the additional ranging signals.

Failure Notification. One of the FAA avionics requirements is that failures should be known to the users within 10 seconds. The redundancy provided by the pseudolites gives the user a very simple cross-check to determine failures. With the potential cost of additional receiver channels now under $200, the six or eight channel receiver can not only produce statistically better positioning, it can also determine which ranging signal has the inconsistency and quickly discard it from the overall solution.

Local Immunity to Selective Availability. The PL technique provides positioning accuracy that will be unaffected by the projected errors due to selective availability. However, because the pseudolites use C/A- code, they would not have wartime utility (due to spoofing and jamming). For use in a sophisticated electronic battlefield environment, a P-code version of the PL would have to be developed with appropriate security and encryption.

Reduced Sensitivity to Local Shading Problems. The statistical coverage offered by the full operational system can be very optimistic due to local shading problems. For example, airports located in the vicinity of tall mountains, can lead to extremely large GDOPs. Fortunately, these mountainous obstructions are excellent locations for supplemental PL transmitters. As has been shown, the optimal Dual PL configuration for SFO is about 2000 feet up the sides of adjacent mountains.

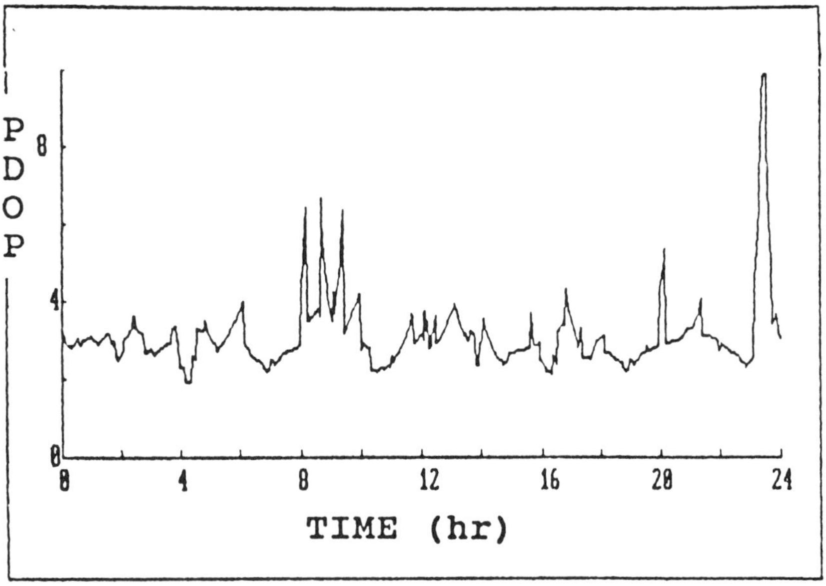

Fig. 14—PDOP for a Single Pseudolite Located 5 Nautical Miles Northeast of Optimal Single-Pseudolite Position.

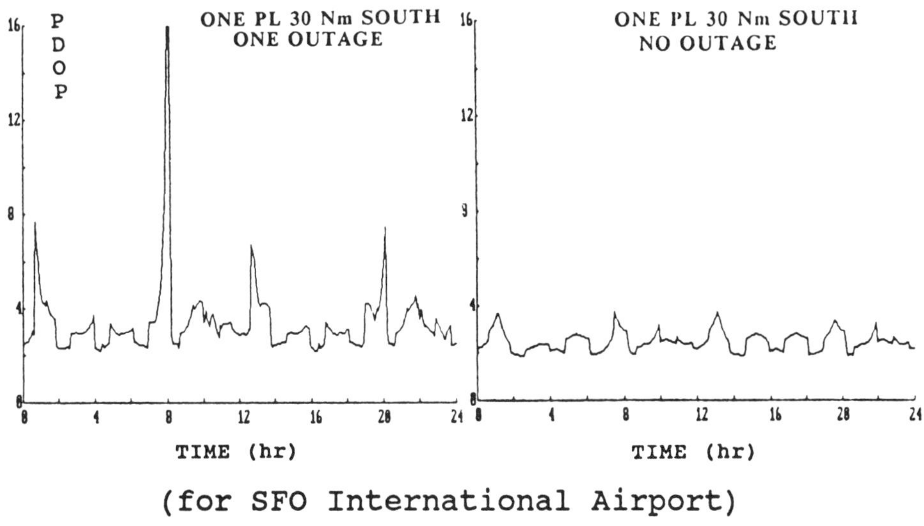

Fig. 15—PDOP for a Single Pseudolite Located 30 Nautical Miles South of Optimal Single-Pseudolite Position.

Low cost correction link. Because the differential corrections are being transmitted over the GPS link itself, no additional hardware is required. The user cost should not be appreciably greater than for a standard receiver. This differential technique should be particularly appealing for the aircraft community with their limitations on space, weight and, to some extent, power.

LIMITATIONS

A prototype PL of the type specified by RTCM 104 has not yet been built. While the specified general design seems very sound, careful testing is warrenanted. In particular, any problems of mutual interference must be resolved. Initial pricing by commercial electronics firms indicates that PLs can be built for $100K to $200K, depending on redundancy and the quality of the atomic clock. A prototyping process would help verify the economic viability of the system.

CONCLUSIONS

Placement of PLs has some strong impacts on worst case performance of the GPS system. For the critical vertical channel maximum errors of 87 meters (no outages, no SA) or over 300 meters (no outages, with SA), can potentially be reduced to 1.2 meters (two PLs, no outages) or 2.1 meters(two PLs, worst case outage) with or without SA. Thus, the optimal single or dual PL system, can produce virtual immunity to single satellite failures.

Since the delayed deployment of the operational satellite system can be viewed in the same light (i.e. a lesser number of available SVs), the use of PLs can potentially offer local three-dimensional coverage while the worldwide coverage is still two dimensional.

REFERENCES

1. Klein, D. and Parkinson, B.W. *The Use of Pseudo-Satellites for Improving GPS Performance,* Proceedings of the Fortieth Annual Meeting of the ION, Cambridge, MA, June 1984.
2. Beser, J. and Parkinson, B.W. *The Application of NAVSTAR Differential GPS in the Civilian Community,* Navigation, Journal of The Institute of Navigation, Vol. 29, No 2, pp. 107–136, 1982.
3. ANON., *Federal Radionavigation Plan,* Department of Defense and Department of Transportation, July 1980.
4. Bulban, E.J. *Civil Helicopter Growth Seen Through 1986,* Aviation Week and Space Technology, March 9, 1981.
5. Denaro, R.P. *Simulation and Analysis of Differntial GPS,* Proceedings of ION National Technical Meeting, San Diego, Janurary 1984.
6. Fried, W.R. *A Comparative Performance Analysis of Modern Ground-based, Air-based and Satellite-based Radio Navigation Systems,* Navigation, Journal of The Institute of Navigation, Vol. 24, No 1, Spring 1977.
7. *Global Positioning System,* The Institute of Navigation, Washington D.C., (1980) reprint of the Summer, 1978, issue of Navigation, Journal of The Institute of Navigation, Vol. 25, No 2 (together with additional articles concerning the system).
8. Jorgensen, P.S. *NAVSTAR/Global Positioning System 18-Satellite Constellations,* The Aerospace Corporation,El Segundo, CA.
9. Johnson, C. and Ward, P. *GPS, Applications to Seismic Oil Exploration,* Navigation, Journal of the Institute of Navigation *Vol. 26, No 2, Summer 1979.*
10. *Recommendations of Special Committee 104 Differential NAVSTAR/GPS Service,* Radio Technical Commission for Maritime Services, Washington D.C., November 20, 1985.

11. Kalfus, R. et al. *NAVSTAR/GPS Simulation and Analysis Program,* DOT-TSC-RSPA-83-2, May 1983.
12. Milliken, R.J. and Zoller, C.J. *Principle of Operation of NAVSTAR and System Characteristics,* The Institute of Navigation, Global Positioning System Special Issue, 1980.
13. Morgan, J.G. *The role of Navigation Satellites in Oil Exporation,* Navigation: Journal of The Institute of Navigation, Vol. 26, No 1, Spring 1979.
14. *Principles and Operational Aspects of Precision Position Determination System,* North Atlantic Treaty Organization, AGARDOGRAPH No 245, July 1979.
15. Teasley, S.P., Hoover, W.M. and Johnson, C.R. *Differential GPS Navigation,* Texas Instruments Incorporated, Dallas TX.

GDOP = Geometric - - - - -

DOP = Dilution of Precision (Factor)

VDOP

HDOP

PDOP

WDOP ≡ (2 * VDOP + HDOP)/3

weighted.

NAVIGATION: *Journal of The Institute of Navigation*
Vol. 36, No. 2, Summer 1989
Printed in U.S.A.

Correction Algorithms for Differential GPS Reference Stations

PETER LOOMIS, GREG KREMER, and JAMES REYNOLDS
Trimble Navigation, Sunnyvale, California
Received September 1988
Revised April 1989

ABSTRACT

The generation of differential GPS corrections for general navigation purposes must consider the effects of selective availability, ionosphere, data link latency, and processing time. The U.S. Coast Guard Differential GPS Harbor Entrance program has developed a differential GPS reference station system design that successfully addresses these concerns. This paper presents a high-level description of this design and some of the test data that led to the critical design decisions.

INTRODUCTION

The premise of differential GPS is that errors in GPS navigation are highly correlated over broad areas. With a reference receiver installed at a well-surveyed location, the errors in the GPS signal can be measured in real time and broadcast to local users, who can then generate very accurate differentially corrected GPS navigation solutions.

The concept is beguilingly simple, and a simple implementation can be quickly constructed as follows. The first step is to survey the position of the reference GPS receiver position relative to a recognized coordinate system, such as WGS 84 or that of a commonly used navigation chart. In operation, this receiver will measure pseudorange (distance in space-time) to each GPS satellite; the reference station will subtract from this the range to the satellite as computed by the orbital (ephemeris) parameters provided in the navigation message superimposed on the GPS ranging signal. The resulting difference is commonly termed the pseudorange error (PRE); this can be broadcast to local users, who subtract it from their measured pseudoranges to obtain differentially corrected pseudoranges. (Typically, the PRE is transformed into a pseudorange correction [PRC] before broadcast by changing its sign; to get the corrected solution, the PRC is added to the user's measured pseudorange.) The corrected pseudoranges are used as inputs to the user receiver's navigation algorithm to produce an accurate navigation solution relative to the reference station, with locally common biases removed.

This simple construction works reliably, eliminating the locally common biases (such as GPS navigation message ephemeris error and ionospheric modeling error) in the users' pseudorange measurements. However, it is not optimal because noncommon error sources unique to the reference receiver and its antenna placement (such as receiver thermal noise, multipath, and receiver clock phase errors) are not removed from the broadcast correction. The resulting error in the differentially corrected solution can easily be larger than the original locally common bias error, which may be on the order of a few meters. This would not meet the 5–8 m accuracy goal of the U.S. Coast Guard Differential GPS Harbor Entrance program.

The purpose of the pseudorange correction generator (PRCG) of the reference station is to filter out those errors unique to the reference receiver without filtering out the locally common bias errors, and to create a pseudorange correction for L1 C/A-code suitable for broadcast in the Radio Technical Commission for Maritime Services Special Committee-104 (RTCM SC-104) format [1]. (Carrier measurements are not corrected under this format.) This paper is a high-level description of the method that is currently used in the Coast Guard differential GPS reference station [2], with brief descriptions of some of the problems and the directions of more recent developments. Equation-level documentation of the PRCG is available for the preliminary design [3] and software implementation [4].

ALGORITHM DESCRIPTION

The main body of the PRCG, shown in Figure 1, consists of a sequence of three stages. The first collects the time-tagged pseudorange and delta-range data, and prepares them for processing by converting them to a PRE and a delta-range error (DRE), respectively. The PRE is the difference between the measured C/A-code pseudorange and the computed range to the satellite as calculated from the broadcast ephemeris. Similarly, the DRE is the difference between the measured change in the L1 carrier pseudorange over the dwell interval and the computed change in range. The second stage filters the noisy PRE measurements against the DRE measurements. The third stage takes the

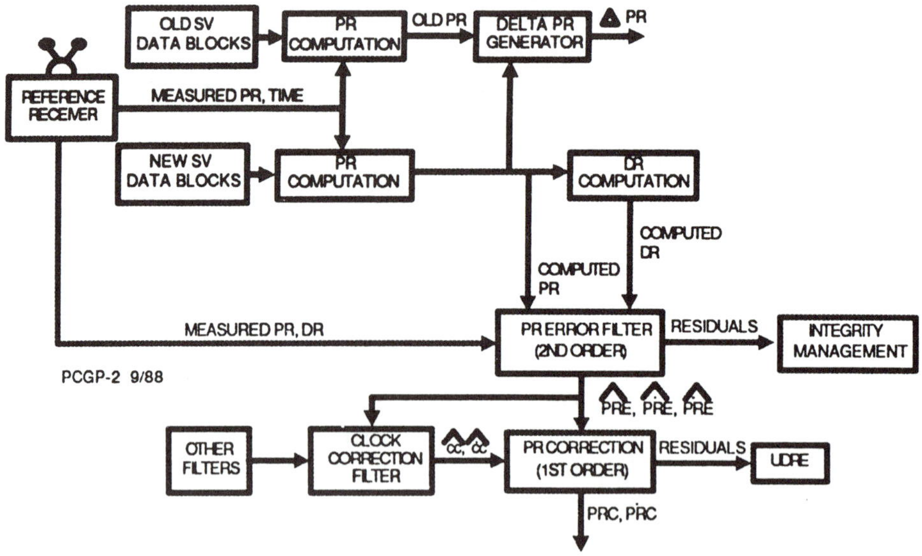

Fig. 1—Pseudorange Correction Generator

resulting best estimates of pseudorange error and pseudorange rate error and converts them into a pseudorange correction appropriate for broadcast in the RTCM SC-104 message format.

In addition to the above three stages, the PRCG provides ancillary functions. These include the integrity management function, limited currently to assessing filter health and reasonableness of data; computation of the User Differential Range Error (UDRE); and generation of the RTCM SC-104 Type 2 message, which will be defined and discussed later in this paper.

Measurement Preparation

The first stage, preparing the measurements, is identical to the ephemeris calculations that must be done in all GPS navigation receivers. In practice, different user receivers may use slightly different algorithms because of memory or throughput constraints of the navigation processor. In the reference station, the satellite orbit and clock computations in the reference station adhere exactly to the equations set down in the GPS Navstar Interface Control Document 200 [5], calculated to centimeter-level precision.

In accordance with RTCM SC-104 standards, ionospheric and tropospheric delay effects are not removed from the pseudorange, even though current "standard" models exist. The corrections as broadcast include the effects of the atmospheric delays. Users have the option of using their own models to remove ionospheric and tropospheric effects from both the broadcast PRCs and their own measured pseudoranges.

The measurement characteristics of the hardware must be considered in constructing the details of the algorithm. The Trimble 4000A used in this development, for instance, reports pseudorange and delta-range as averages over a known dwell period, with the measurement time tag reported as the middle of the dwell. Using the instantaneous computed range to the satellite at the time tag can cause an error of up to 5 cm for a 2 s measurement dwell, when compared with the correct average computed range over the interval. This small difference must be taken into account if accuracy requirements dictate; the difference between the two ranges is of course deterministic and can easily be corrected in the range computations.

Filter

The second stage of the PRCG is the filtering of the raw PRE and DRE data. This stage consists of a bank of independent Kalman filters, one for each satellite, operating on the measurements coming from the reference receiver. It is at this stage that errors unique to the reference receiver are separated from the locally common bias errors to the extent possible, by modeling the lower-frequency receiver-unique errors in the filter state and filtering out the higher-frequency ones.

The design of the PRCG filter was heavily influenced by the hardware available at the start of the project in early 1985. The Trimble 4000A reference receiver available at that time experienced occasional data dropouts when the 8 bit navigation CPU became overwhelmed. A Kalman filter was chosen for the PRCG estimator because of its ability to maintain an optimal estimate in the presence of numerous data dropouts. Two other factors weighed heavily in favor of the Kalman filter: its behavior during the first few seconds of data of a rising satellite was well understood; and the process model could be easily expanded to model the ionospheric effects on code and carrier pseudoranges, as described below.

To define the Kalman filter, one must specify the state model and the associated models for the propagation and measurement processes. Recalling that each filter in the PRCG deals only with a single satellite, the four-state filter model is defined as follows:

x_1 is the pseudorange error (PRE).
x_2 is the PRE rate.
x_3 is the PRE acceleration.
x_4 is the difference between the rates of change of the code and carrier pseudoranges.

The fourth state, the DRE bias state, was proposed early in the project as a cure for the group delay/phase advance phenomenon associated with the ionosphere [6]. This somewhat counter-intuitive phenomenon dictates that as total electron count (TEC, the integral of ionospheric electron density over the signal path) increases, the code is delayed, but the carrier actually advances an equal and opposite amount. Thus, the pseudorange rate as measured from the carrier is slightly biased relative to the pseudorange rate as measured from the code when the satellite is rising or setting, or the vertical TEC is changing.

The broadcast pseudorange correction is intended to be referenced to the C/A-code pseudorange. To prevent the extreme accuracy of the DRE measurements from pulling the filter estimate toward the carrier-derived pseudorange, the fourth state was implemented to account for the bias in the DRE measurement. Although originally intended to compensate for ionosphere, the DRE bias state actually encompasses all phenomena that affect code and carrier in different ways, and maintains the PRE estimate close to the center of the PRE data. In the section below on the ionosphere, some empirical results highlight the importance of considering these effects.

As mentioned above, the Kalman filter is defined by the propagation and measurement models. The propagation model is treated discretely as follows:

$$\mathbf{x}(t_{i+1}) = \mathbf{\Phi x}(t_i) + \mathbf{Gw}_i, \text{ where} \tag{1}$$

$\Delta t = (t_{i+1} - t_i)$,

$$\mathbf{\Phi} = \begin{bmatrix} 1 & \Delta t & \Delta t^2/2 & 0 \\ 0 & 1 & \Delta t & 0 \\ 0 & 0 & 1 & 0 \\ 0 & 0 & 0 & 1 \end{bmatrix}, \tag{2}$$

and $\mathbf{Gw}_i$ represents the random (mean zero) propagation error, or process uncertainty, in the ith time interval.

The process uncertainty accounts for the fact that the true dynamics of the pseudorange vary slightly from the mathematical models of state dynamics in the filter. The exact value of the propagation modeling error at each step is of course unknown, but it is assumed to be zero on the average. The covariance of the process uncertainty, denoted $\boldsymbol{\Gamma Q \Gamma^T}$, is used by the filter in assessing how quickly the memory of old data fades. In the PRCG pseudorange error models, there are two process uncertainties: the first is that the PRE acceleration is not constant, but rather changes in some unknown and random manner; the second is that the DRE bias state is not constant. Both are defined as slow random walks in the current PRCG, leading to a process noise covariance matrix:

$$\boldsymbol{\Gamma Q \Gamma^T} = \begin{bmatrix} q_r\Delta t^5/20 & q_r\Delta t^4/8 & q_r\Delta t^3/6 & 0 \\ q_r\Delta t^4/8 & q_r\Delta t^3/3 & q_r\Delta t^2/2 & 0 \\ q_r\Delta t^3/6 & q_r\Delta t^2/2 & q_r\Delta t & 0 \\ 0 & 0 & 0 & q_i\Delta t \end{bmatrix} \quad (3)$$

In the above equation, the two parameters q_r and q_i are spectral densities for the random walk processes. The first parameter, q_r, includes effects of orbit errors, propagation errors, and clock errors on both the satellite and the reference receiver. (The reference receiver clock is a rubidium frequency standard, chosen in part to keep the value of q_r low.) The second parameter, q_i, is tuned to low-frequency fluctuations in the TEC; it is set low enough to exclude multipath, which is not included in the DRE bias state because it is a receiver-unique error.

Two variations of the above model were examined during reference station testing to determine whether the PRCG model could be improved. In the first variation, the PRE acceleration and DRE bias states were modeled as Markov processes rather than random walks. The second variation centered around how the filter states must be defined to match the available measurements, specifically, whether the DRE measurement is instantaneous, backwards differenced over contiguous dwell intervals, or sampled over short discontinuous dwell intervals. Although the Trimble 4000A carrier measurement fits in the latter category, in the PRCG it is modeled as the simpler instantaneous Doppler measurement. No empirical evidence was found that either variation led to an improved PRCG filter solution.

In contrast to the filter process model, the filter measurement model is straightforward. At each measurement time, two measurements enter the filter together. The PRE as passed from the measurement preparation stage is a direct (and noisy) measurement of the PRE state. The DRE measurement is equal to the PRE rate state minus twice the DRE bias state. The two measurements are assumed to have uncorrelated noise for static receivers in low-noise conditions; the measurement noise variance for the code was on the order of $(1\text{-}2\ \mathrm{m})^2$, and for the carrier (pseudorange rate) was on the order of $(10^{-2}\ \mathrm{m/s})^2$.

The outputs of the filter are the pseudorange error estimate, and the pseudorange error rate and acceleration estimates. The fourth state is used only for modeling the DRE measurement and is not considered when generating the pseudorange corrections from the pseudorange error estimates.

Pseudorange Correction Computation

After the best estimates of all satellites being tracked have been computed by the Kalman filter, the pseudorange correction must be computed. Because of the constraints of the RTCM SC-104 broadcast format, the PRCs must be adjusted to roughly zero mean, and must account for the latency between the time at which they are computed and the time at which they will be received and implemented by the mobile GPS users. This latency can be on the order of 10 s when the signal is broadcast at 50 bits/s.

The first step in the PRC computation is to remove the receiver clock error. This is a major reference receiver-unique error common to the PRE estimates for all satellites. An error in the reference receiver clock estimate, provided that it affects all pseudorange corrections equally, does not appear as a differential positioning error; rather, it appears directly as an error in user differential time. It is necessary to remove the clock error only to reduce the magnitude of the PRCs, so that they can fit in the limited dynamic range (± 3000 m) allowed by the RTCM SC-104 format. In practice, we have found that although the accuracy of the clock correction is not critical, the smoothness of the clock correction is important. Jitter and jumps in the clock correction appear directly in the broadcast PRCs and can apparently impair navigation accuracy in some user receivers that have carrier aiding of the code. To ensure smoothness, the clock correction algorithm is a very heavily damped running average of the PREs.

Latency is a more difficult problem in the reference station design. By the time a new PRC has been broadcast and implemented in the user receiver, the previous PRC may be older than 12 s. The RTCM SC-104 format includes PRC rates as well as PRCs, so the user can project the current PRC forward in time until the next PRC arrives. Even so, accelerations due to satellite clock dithering (selective availability) may cause higher than desired errors in the linearly projected PRC over a seemingly short period of time. To improve the accuracy and remain within the guidelines of broadcasting PRC and PRC rate, the PRCG uses the quadratic formulation of the PRE estimate (error, rate, acceleration) to project the PREs ahead in time to the interval of application at the user. A linear best fit of the quadratic function of time is generated to yield the PRC and PRC rate. Figure 2 illustrates this concept. The PRC is generated at time t_{CALC1} and will be implemeted by the user receiver at t_{appl1}, with the next PRC expected to be implemented by the user receiver at t_{appl2}. The PRE is projected forward as a quadratic function of time, and a best fit (minimum maximum error) linear function is generated to yield the PRC and PRC rate for broadcast.

Filter Health

The PRCG allows examination of the measurement residual stream to detect filter divergence and general filter health. The measurement residual is defined as the difference between the value actually measured and the value predicted by the filter. In a properly tuned Kalman filter, the residual sequence will be white (uncorrelated). A significant bias in the recent measurement residuals can be an indicator of filter divergence, and in certain circumstances the PRCG filter will be reinitialized when this indicator reaches a predetermined threshold. In practice, this test is weaker than might be expected, because it is quite normal for the filter to show a statistically significant pseudorange error residual bias. This bias is often due to unmodeled multipath which appears as a sinusoidal error of periods on the order of minutes and amplitudes on the order

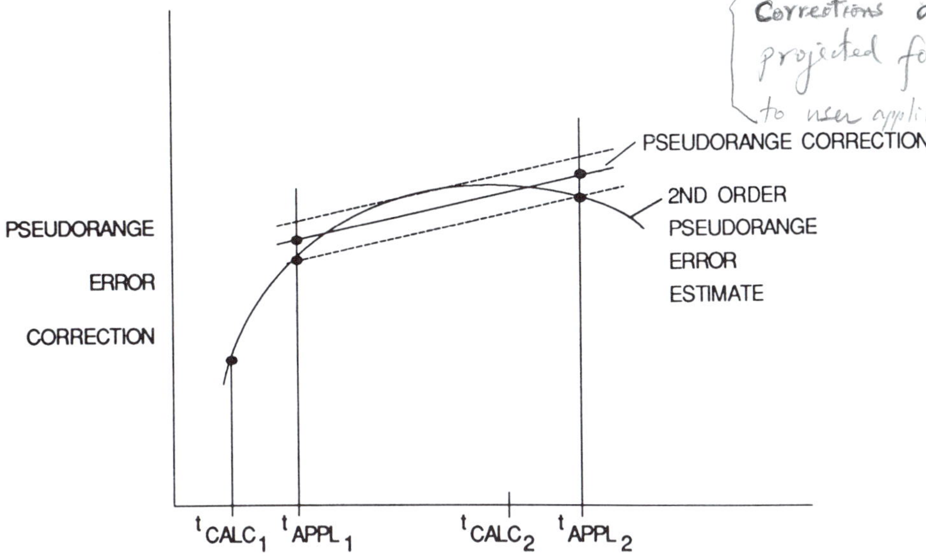

Fig. 2—Pseudorange Correction Computation from Pseudorange Error Estimate

of meters. The residual error statistics are made available to the operator of the reference station so that the filters can be monitored and tuned manually. These statistics include sample means, variances, and percentage of residuals rejected by a three-sigma data-editing threshold.

User Differential Range Error (UDRE)

The UDRE is intended as a measure of the standard deviation of the error attributable to the differential correction. It can be divided into four parts: reference receiver-unique errors (e.g., multipath, interchannel bias); algorithmic errors; temporally decorrelated errors (e.g., satellite clock dithering); and spatially decorrelated errors (e.g., atmospheric effects and navigation message orbit error). The last cannot be addressed properly without spatially separated multiple reference receivers to measure the spatially decorrelated error. The receiver-unique portion cannot be addressed without multiple (redundant) receivers at the reference site to detect and isolate the receiver-unique errors. Such multiple receiver setups are not currently proposed, so the only factors contributing to the UDRE that can be measured by a single reference station are the algorithmic and the temporally decorrelated errors. Consequently, the broadcast UDRE is computed as the square root of the sum of three terms: the current filter variance for the PRE; the square of the mean of the PRE measurement residuals over the past few minutes; and the square of the difference between the current PRC and the PRC as projected from the previously broadcast PRC and PRC rate.

Type 2 Message

This message contains the difference in correction when comparing the old and the new ephemeris parameters. A user receiver that has not picked up the new ephemeris in a timely manner can offset the broadcast PRCs with these broadcast deltas to match the old ephemeris. The Type 2 message allows the reference receiver to send only corrections relative to the new ephemeris, with occasional broadcast of the difference between those and the corrections based on the previous ephemeris. Although in general a change in ephemeris produces a negligible change in PRC, at uploads the change can be a quite substantial step change in range. As a rule, the new ephemeris information is applied to the PRE filters as a deterministic offset in the PRE state vector, so the resulting step in range is not passed through the filter, and the PRE estimate responds immediately to the change.

NOTES ON PERFORMANCE

The U.S. Coast Guard Research & Development Center in Groton, Connecticut has conducted sea trials using the reference station and a Magnavox user receiver aboard a small vessel. The results of the testing in general exceeded the original design requirements of the Coast Guard and are presented in [7].

The TAUNAV Systems Division of Trimble Navigation, Ltd. has participated in the development and test of the USCG reference station since 1985. During this period, we collected a substantial amount of data in measuring the performance of the reference station, and have continually examined the original design assumptions in a critical light. Two subjects in particular were of interest: the clock phase error treatment, specifically the removal of the clock phase error from the PREs to prepare them for the limited dynamic range allowed for the PRCs; and the treatment of ionosphere and general divergence of carrier and code in the filter. The next two sections discuss these in more detail.

Reference Station Clock Phase Error Correction

At the beginning of the development effort for the Coast Guard reference station, we came to the conclusion that a rubidium clock was a cost-effective solution to a number of problems facing the reference station. The rationale for this decision was twofold. First and foremost, the GPS receiver available at the time had irregular measurement-reporting behavior, even though it was a multichannel receiver. The GPS pseudorange measurements in each measurement packet were often not at the same time, with one lagging as much as 3 s behind the others. With a crystal clock, the error in the lagging measurement due to clock dynamics had the potential of degrading the pseudorange corrections. With a stable atomic frequency reference, on the other hand, the pseudorange data could be realigned with reasonable confidence. Also, since the atomic frequency standard acts as an extra satellite because of its frequency stability, it can be used as a failure detection/isolation device for the GPS signal. This is particularly important when there are not many visible satellites; when selective availability clock dithering is invoked; or when a truth system is required, as in the current feasibility program.

The difficulties with development of software for the crystal (quartz) clock center around the fact that the error dynamics of a crystal clock are at the same or higher frequency as some of the pseudorange error effects, selective availability in particular. Crystal clocks typically exhibit a general wandering behavior and infrequent, sudden jumps in frequency (popcorn), both of which have been observed in our laboratory.

As the crystal clock phase error passes through the pseudorange filter, a residual clock phase error remains in the filtered solution. Its magnitude depends heavily on the filter gains of the individual PRCG filter for the particular satellite, which is in turn a function of the time since that particular filter was initiated and the number of measurements that have been processed. If the filter gains are different in the current set of PRCG filters running in the reference station, a differential navigation error results. The size of the differential positioning error attributable to these effects is difficult to quantify; depending on filter tuning and the actual clock event, it may be on the order of a meter or more.

The error dynamics of atomic clocks, on the other hand, are reliably much lower than the pseudorange error dynamics. Atomic clock phase error dynamics are so low that they pass virtually unchanged through the PRCG pseudorange error filter. As a result, the residual clock phase error for an atomic clock is independent of the filter gain and is the same for each PRE filter. As previously noted, the residual clock phase errors in the filtered PREs have no effect on error in the differentially corrected position solution provided that they are identical, in which case they are absorbed into the user time offset.

Since the original Coast Guard reference station design was completed, alternative filter design concepts have been developed that remove the clock error from the pseudorange measurements before entering the filters. This improvement in the estimation software is partly due to the advances in available GPS hardware since the beginning of the project. The original Coast Guard reference receiver hardware had two idiosyncrasies that prevented this type of estimation from being done: the above-mentioned measurement packets that were not time-aligned, and measurement dropouts for a variety of reasons. Both these characteristics were caused basically by the limitations of the reference receiver navigation processing unit. With the current new reference receiver hardware supplying regular GPS carrier data on all channels, these new filter designs can operate on crystal clocks and achieve accuracy equivalent to that of a reference station driven by a rubidium frequency standard.

One aspect of the original reference station design that caused difficulty during testing was the jitter in the PRCs due to estimation and removal of the reference receiver clock error. Recall that the accuracy of the clock error estimate is not critical to navigation, since it is included only to reduce the dynamic range of the broadcast corrections. An inaccurate reference receiver clock error estimate results in an equally inaccurate user receiver clock estimate, but no positioning error. Depending on the design of the user receiver, however, excessive jitter in the PRCs caused by jitter in the reference receiver clock estimate can cause anomalous behavior in the navigation filter. In the current Coast Guard reference station, the ensemble of current PRE estimates is averaged to obtain a clock phase error estimate, which is then low-pass filtered and subtracted from the filter solutions. The original low-pass filter was not heavily damped and was sensitive to impulse errors from new (rising) or retiring (setting) satellites. By using a rubidium frequency and retuning the clock estimate damping, the jitter was eliminated, and the system was desensitized to adding or dropping satellites.

A new concept was developed in which clock steering tracks the clock phase error by integrating the measured frequency offset (the median or average among the available pseudorange rate error measurements), and a Butterworth filter steers the integrated Doppler toward the average of the filter solutions. Limiters are put on the clock correction frequency drift to prevent an overly quick response, which might adversely affect some user navigation filters. The resulting clock phase error and the associated frequency error are subtracted from the raw pseudorange and pseudorange rate measurements before filtering, rather than after. Note this technique would not have been suitably matched to previous versions of the GPS receiver; it stumbles badly if the measurements are asynchronous, because the median or average frequency error on which it is based is not always available.

The Divergence of Carrier and Code

Two measurements are available from the GPS receiver as implemented in the Coast Guard reference station: a measurement of the code pseudorange, and a measurement of the change in carrier pseodorange. The carrier pseudorange is not directly measurable; although the carrier phase is measured very accurately, the integer number of carrier wavelengths to the satellite cannot be known. In general, the change in carrier phase is exactly equal to the change in code phase over time, and is measured much more precisely. This fact is used in the PRCG filters and many GPS navigation receivers to smooth the code phase to a minimal level of jitter. There are, however, two phenomena which are widely recognized to have different effects on code and carrier: multipath and ionosphere.

The first of the two, multipath, results from the superposition of a second signal, typically reflected off the side of a building or a water surface near the receiver antenna, onto the original ranging signal from the GPS satellite. Such interference effects depend on wavelength; consequently, code and carrier respond differently because of their widely disparate frequencies. Code multipath tends to have signal delays on the order of meters and a characteristic sinusoidal signature with a period on the order of minutes when the reflective geometry is stationary, as it would be with a reference station. Carrier multipath has much smaller delay magnitudes and shorter periods, and can be considered negligible for the purposes of differential navigation. The error due to multipath is of course not a locally common bias error, being unique to the reflection geometry of the GPS receiver antenna placement; the PRCG filter should be constructed to remove multipath effects from the PRE estimate.

Ionosphere also acts differently on code and carrier, but with a different signature. As signals propagate through the ionosphere, they experience a delay proportional to the inverse square of the frequency. A classic result of physics states that while the code signal indeed experiences a delay, there is an equal and opposite apparent phase advance in the carrier (for a good explanation of this somewhat counterintuitive effect, see [8]). In contrast to the sinusoidal error signature of multipath, ionospheric delay tends to decrease or increase at certain times of the day and at rise or set of the satellites. Ionospheric delay, especially the more slowly-varying components of ionospheric delay, is generally common to local users and should be reflected in the broadcast pseudoranges.

Most other phenomena (troposphere, selective availability, receiver clock errors) affect code and carrier in a like manner. By subtracting integrated carrier from the code, one can isolate the multipath and ionospheric effects from other major GPS error sources. Appropriate filtering further isolates multipath from ionospheric effects.

The question arises whether code/carrier divergence under changing ionospheric conditions is important from the standpoint of the reference station. Until recently, significant data has been difficult to collect because ionospheric conditions have been in the quiet part of the sunspot cycle. However, a simple set of tests performed recently has demonstrated that there is an error associated with mismodeling the ionosphere; the data is presented in Figures 3 through 6. The conditions that exacerbate the ionospheric effect are expected to be extremely common: satellites rising at midday over the equator, in periods of heavy sunspot activity.

To highlight the role played by divergence between code and carrier, the pseudorange has been smoothed with the well-known Hatch method [9] and displayed with the original unfiltered data. The Hatch method is a simple one-state filter that uses the DRE measurement to propagate the pseudorange. Each measured PRE is propagated forward to the present time using the sum total of DREs from its measurement time (i) to the current time (k),

$$(\mathrm{PRE}')_{ik} = \mathrm{PRE}_i + \sum_{j=1}^{k} (\mathrm{DRE}_j)\ . \tag{4}$$

Then the average over the (k) propagated measurements is computed to yield the filtered value:

$$\text{average } \mathrm{PRE}_k = \sum_{i=1}^{k} (\mathrm{PRE}')_{ik} / k\ . \tag{5}$$

This method is very accurate in filtering the code noise and providing a smooth estimate of the pseudorange when the carrier (DRE) and the code (PRE) track in tandem. Indeed, this method was suggested by Hatch only for (dual-frequency) ionosphere-corrected code and carrier, but is commonly used with single-frequency code and carrier as well. As shown below, carrier smoothing works well in the short term for single frequency, but breaks down over long periods of time under the influence of changing TEC.

An example of the Hatch method carried *ad absurdum* on single-frequency data is shown in Figure 3. This is a data set showing SV PRN 6 over its complete pass, more than 6 h during the middle of a June day. At the end of the data set, the curve of filtered pseudorange data bears no resemblance even to the shape of the unfiltered data curve. The unique behavior of the Hatch filter at satellite set is typical when the ionosphere electron count is high. This data set and those in Figures 5 and 6 have been post-processed to remove most of the phase and frequency offset of the receiver clock, much as is done in generation of the PRCs. This has been done consistently with all data, so as not to color the results. Figures 5 and 6 have been corrected for tropospheric delay using a standard model; Figure 3 has not.

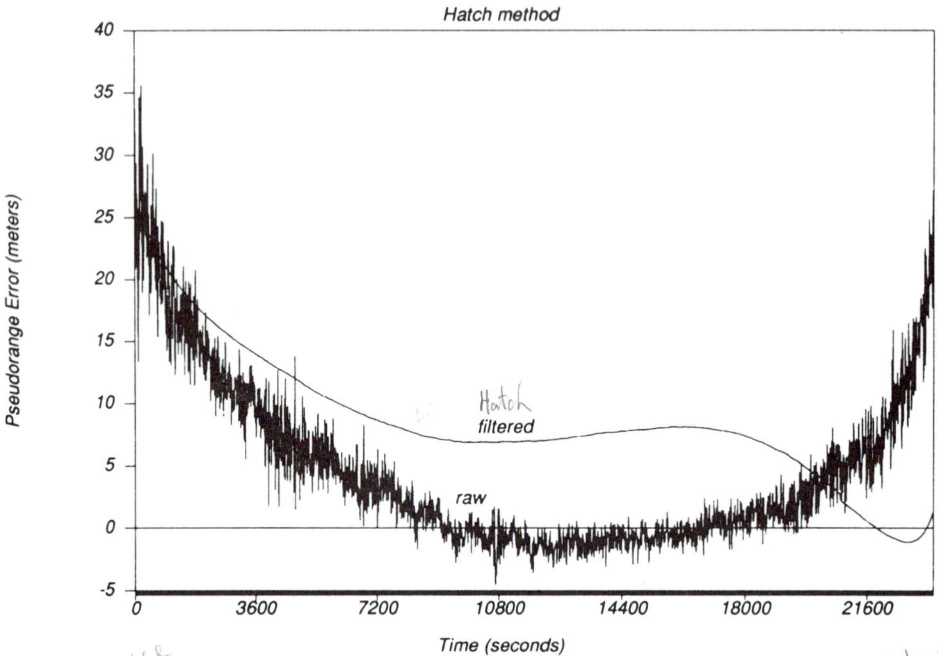

Fig. 3—Hatch Filter over a 6 h Data Set

Even seemingly reasonable lengths of time can end in a severe bias if the ionosphere is not taken into account. Figures 4 through 6 show data taken from SV PRN 13 a few days later. The satellite was rising in the south during the early afternoon, so the ionospheric electron density was at its highest, and the signal path was shortening quickly—ideal conditions for demonstrating code/carrier divergence, but quite realistic conditions nonetheless. The data shown here was typical of SV 13 for many days in mid-June.

Figure 4 shows two curves. The upper is the ionospheric delay, calculated from the SV Data Base I parameters downlinked by the satellite according to the ICD 200 TEC model. The lower curve is the half-difference between carrier and code, computed as the total change in carrier phase differenced from the change in code phase, then divided by two. In terms of the states of the PRCG filter, the latter plot is roughly equal to the integral of the DRE bias state. A smoothed version of the bottom curve is also shown, passed through a 20-point (post-process) triangular smoother. Note that the modeled ionospheric delay and the empirical integrated DRE bias curves agree in general shape but not in magnitude. Whereas the ionospheric model from the broadcast parameters shows the delay changing by only 4 m over the hour, the integrated DRE bias indicates it may have changed three times as fast. This disagreement was typical during our data collections, and may have been due to ionospheric TEC levels much higher than those predicted in June. A characteristic sinusoidal signature is also evident in the smoothed curves in Figures 4 through 6, which strongly points toward a multipath error.

Figures 5 and 6 use the same data set to compare the performance of the simpler Hatch model to that of the PRCG filter, which is essentially the Hatch

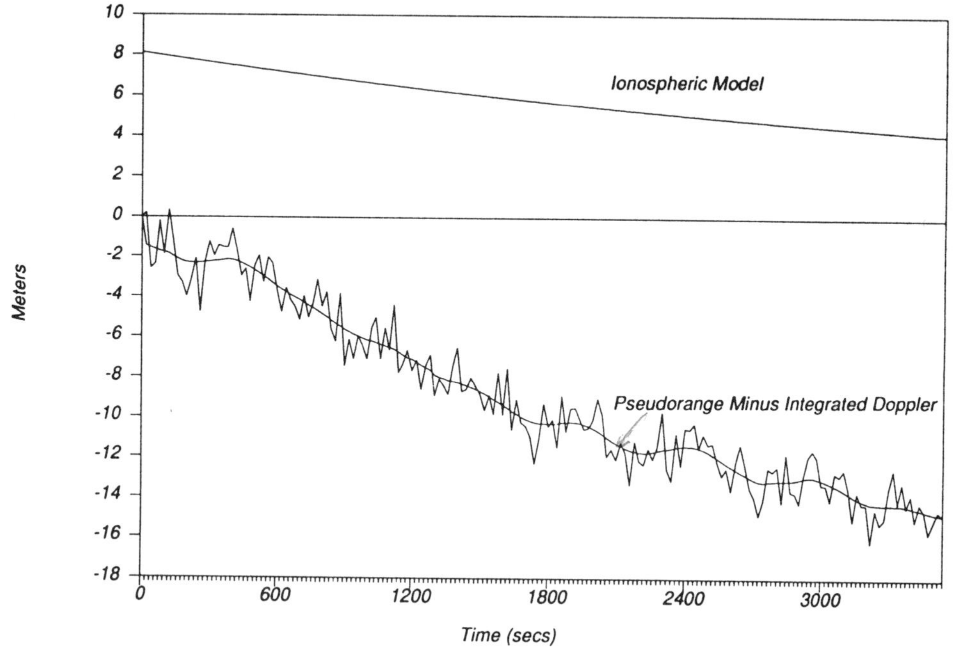

Fig. 4—Code Phase Minus Carrier Phase, Compared to the Ionosphere Model; 1 h Data Set

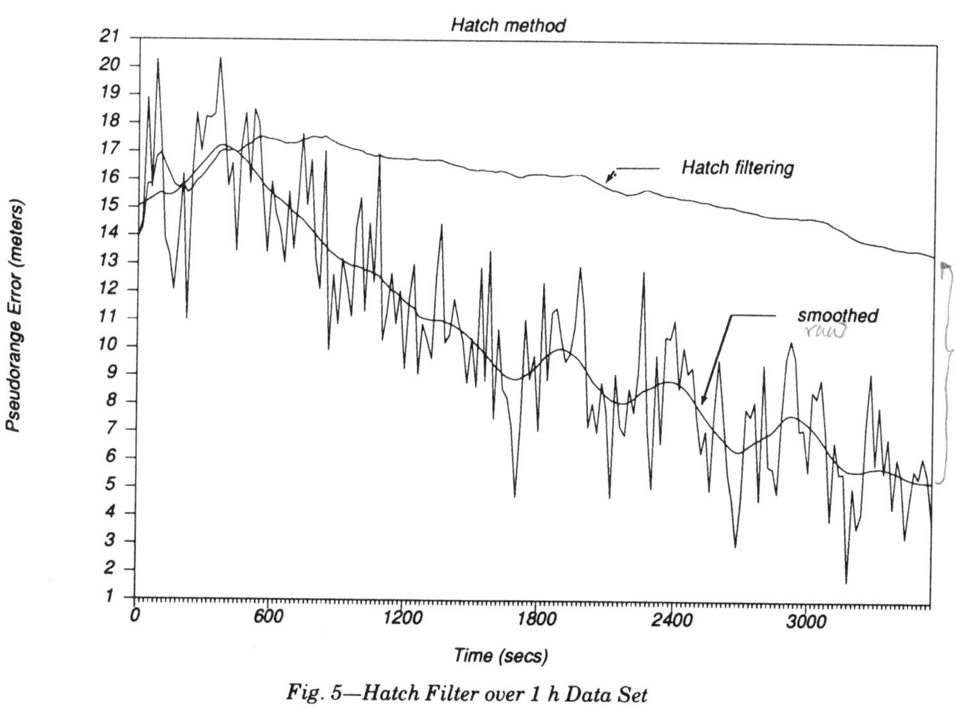

Fig. 5—Hatch Filter over 1 h Data Set

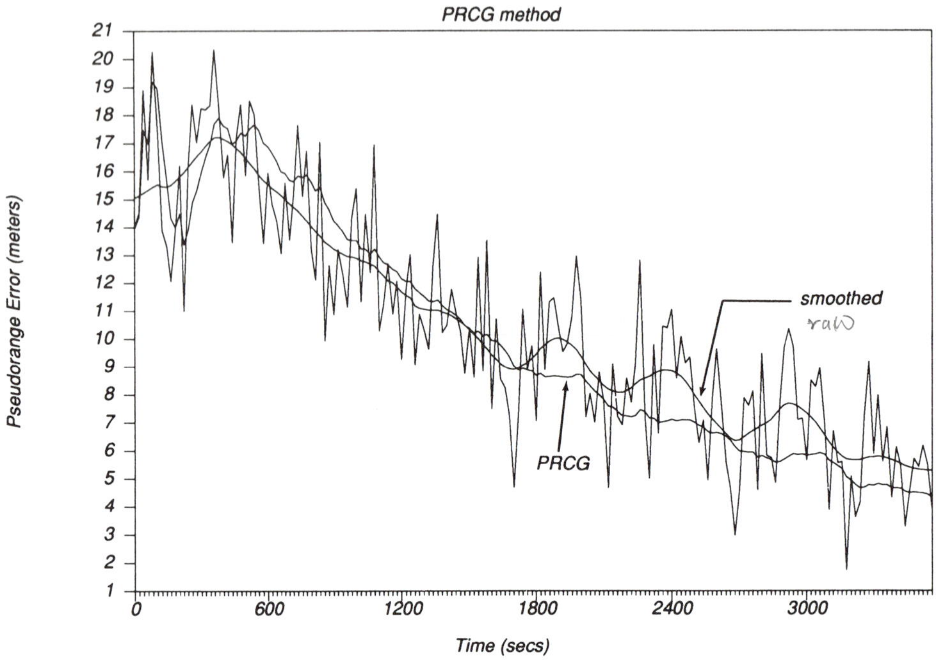

Fig. 6—PRCG over 1 h Data Set

model with an added bias between carrier and code rate (DRE bias, modeled as a slow random walk). Also shown are the raw PRE data and a (post-processed) 20-point triangular smoother of the raw PRE data for reference. Over the first hour of satellite visibility, as the satellite rises from 8 deg elevation to 35 deg, the simpler Hatch filter slowly builds an error in the PRE estimate that totals 9 m at the hour's end. The PRCG filter estimate is within the limits of the data noise, with a bias of perhaps a meter. The bias becomes an error in the pseudorange correction and translates directly to a range error in the differentially corrected user navigation filter.

The above data sets show that the ionospheric effect, which causes a dispersion between carrier and code phase measurements, can be responsible for gross errors in the pseudorange correction if treated naively. Most current implementations of Hatch filters, in fact, do not use the unweighted average shown in equation (5), but rather a weighted average with a time constant on the order of a few minutes. With these implementations, the bias does not have a chance to build up to the levels shown in these data sets. The PRCG filters, on the other hand, have been designed to model the ionospheric effect and are known to perform quite well from the standpoint of removing multipath and jitter, and avoiding over-damped behavior and large biases with respect to ionosphere. The error dynamics of the GPS signal due to ionosphere and multipath, respectively, are generally recognized to occupy different areas of the frequency spectrum and can be separated by filtering [10]. With proper tuning, the multipath (higher-frequency) can be filtered out, while the locally common ionospheric delay (lower-frequency) will pass through. Higher-frequency components of ionosphere, such as scintillation, are not a locally common bias except over a very small area, and should also be filtered from the pseudorange corrections.

CONCLUSION

In retrospect, the design of the pseudorange correction generator (PRCG) was well matched to the hardware capabilities at the inception of the project. Our original concern about ionospheric effects was well-founded, and the choice of a rubidium frequency standard proved to be invaluable in collecting and interpreting data sets. The development of the PRCG demonstrated that differential GPS could easily satisfy the rather modest 5–8 m positioning accuracy goal of the U.S. Coast Guard Harbor Entrance program with few exceptions, and that further evolution of differential GPS systems will continue to produce better accuracy.

Based on a paper presented at The Institute of Navigation Satellite Division Meeting, Colorado Springs, CO, September 1988.

REFERENCES

1. Kalafus, R., Van Dierendonck, A. J., and Pealer, N., *Special Committee 104 Recommendations for Differential GPS Service,* NAVIGATION, Journal of The Institute of Navigation, Vol. 33, No. 1, Spring 1986, pp. 26–41.
2. Quill, J., *U.S. Coast Guard Differential GPS System Development,* Radio Technical Commission for Maritime Services Assembly Meeting, Boston, MA, April/May 1986.
3. Loomis, P., *Design Document for Differential GPS Ground Reference Station Pseudorange Correction Generation Algorithm,* DOT-CG-N-03-86, December 1986.
4. Kremer, G., *Pseudorange Correction Generation and Satellite State Generation,* TAU Technical Report TR-2054-1/USCG, July 1988.
5. *Navstar GPS Space Segment/Navigation User Interfaces,* ICD-GPS-200, Rev. B., Rockwell International Corp., Nov. 30, 1987.
6. Klobuchar, J., *Ionospheric Effects on Earth-Space Propagation,* Scientific Report No. ERP 866, Air Force Geophysics Laboratory, Hanscom AFB, MA, December 1983.
7. Pietraszewski, D., Spalding, J., Viehweg, C., and Luft, L., *U.S. Coast Guard Differential GPS Navigation Field Test Findings,* NAVIGATION, Journal of The Institute of Navigation, Vol. 35, No. 1, Spring 1988, pp. 55–72.
8. Jorgensen, P., *Ionospheric Measurements from Navstar Satellites,* Technical Report No. TR-0079(4475-01)-2, The Aerospace Corporation, December 1978.
9. Hatch, R., *The Synergism of GPS Code and Carrier Measurements,* Proceedings of the Third International Geodetic Symposium on Satellite Doppler Positioning, DMA/NOS, pp. 1213–32.
10. Bishop, G., and Klobuchar, J., *Multipath Effects on the Determination of Absolute Ionospheric Time Delay from GPS Signals,* Radio Science, Vol. 20, No. 3, May 1985, pp. 388–96.

NAVIGATION: *Journal of The Institute of Navigation*
Vol. 36, No. 3, Fall 1989
Printed in U.S.A.

Extended Differential GPS

ALISON BROWN

The NAVSYS Corporation, Monument, Colorado
Received January 1989
Revised April 1989

ABSTRACT

The Radio Technical Commission for Maritime Services (RTCM) Special Committee (SC)-104 on Differential GPS has established a standard data format for providing differential GPS services. The satellite pseudorange corrections that are broadcast in the RTCM SC-104 differential GPS navigation message lump together the errors from a number of different sources. This results in a degradation of the differential navigation accuracy if the user is operating at a distance from the differential reference station. Generally, the user must be within 100 nmi of the reference station to benefit fully from the differential GPS corrections.

This paper presents an Extended Differential GPS concept that extends the range of operation of the differential GPS message. By using a network of differential GPS stations, one can compute a differential GPS message that provides corrections for the different components of the pseudorange error. This allows the differential GPS navigation accuracy to be extended to a range of 1000 nmi from the master differential station.

INTRODUCTION

The Extended Differential GPS concept is illustrated in Figure 1. With conventional differential GPS methods, a single differential reference station broadcasts differential GPS corrections to users in a local region. In the Extended Differential GPS concept, the differential reference stations are networked together to one or more master stations. The master differential station receives the differential GPS satellite error corrections from the individual monitor stations, then combines these to format a differential GPS message that will be valid over the extended range illustrated in Figure 1. The Extended Differential GPS message is then broadcast to users over the extended range through a satellite communications link. This approach provides a highly accurate differential GPS service over a large geographic area. In the illustration shown, the network is assumed to extend over the continental United States (CONUS) using two master stations, each with an individual satellite communications link. By linking differential networks together, continent-to-continent, the range could be extended even farther to encompass portions of the Pacific and Atlantic oceans.

The following sections include an analysis of the GPS errors that affect the accuracy of the differential GPS solution. Also included is a discussion of the additional models that need to be incorporated into the differential GPS message to extend the range of operation.

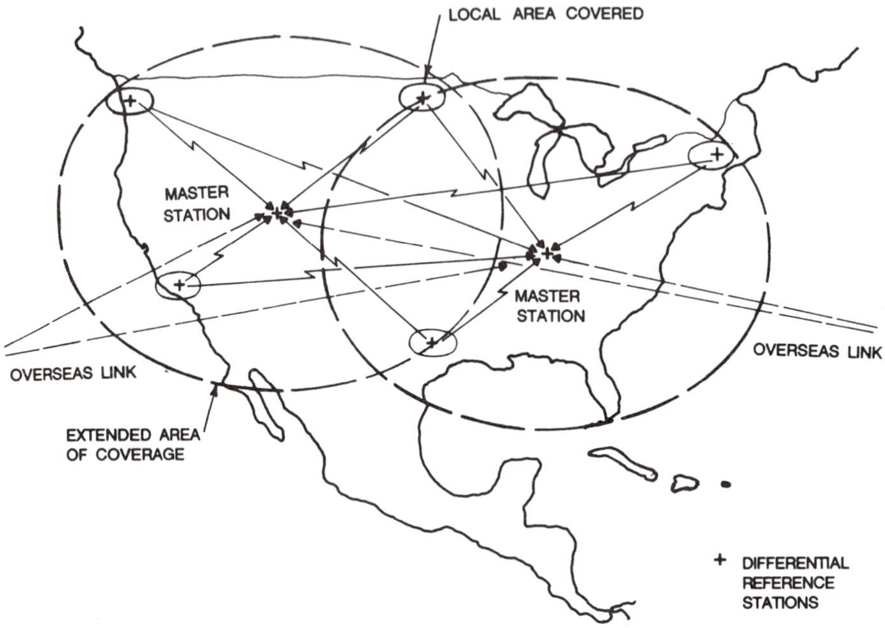

Fig. 1.—Extended Differential GPS Network

CONVENTIONAL DIFFERENTIAL GPS

With conventional differential GPS, a master station is used to track the GPS satellites, determine their range errors through comparison with a known reference solution, and broadcast differential GPS corrections to users in a local area. The accuracy of the final navigation solution is a function of the accuracy of the differential GPS satellite pseudorange error corrections.

The absolute navigation accuracy provided by GPS is a function of the accuracy of the pseudorange and delta-range measurements. The measurements are affected by the GPS system errors, such as ephemeris, clock, and selective availability (SA); atmospheric effects (ionosphere and troposphere); and receiver errors (noise, vehicle dynamics, and multipath). In some cases, differential GPS corrections can be used to reduce or eliminate these errors, resulting in improved performance. In the following paragraphs, the GPS absolute navigation error budget shown in Table 1 is discussed, and the error budget shown in Table 2 is derived for the differential GPS errors as a function of distance from the master station.

Clock and Ephemeris Errors

The errors allocated to the Space and Control Segments in Table 1 are 10 ft (1 σ) for the satellite clock timing error, and 9 ft (1 σ) for the error in the satellite ephemeris data. These errors are consistent with those specified for the GPS system in SS-GPS-300C [1].

The satellite clock timing error appears as a bias on the pseudorange measurement and so is totally removed by differential GPS corrections, as indicated

Table 1—GPS Absolute Navigation Error Budget

Source	Error Budget (ft)
Space Segment	
Clock Error	10
Control Segment	
Ephemeris Error	9
Selective Availability	90
Atmospheric Effects	
Ionospheric Delay	27
Tropospheric Delay	6
User Segment	
Receiver Noise	30
Multipath	10
User Equivalent Range Error (UERE) (RMS)	100
Navigation Accuracy 2 dRMS (HDOP = 1.5)	300

Table 2—Differential GPS Error Budget

Source/Range	0 nmi (ft)	100 nmi (ft)	500 nmi (ft)	1000 nmi (ft)	2000 nmi (ft)
Space Segment					
Clock Error	0	0	0	0	0
Control Segment					
Ephemeris Error	0	0.3	1.5	3	6
Selective Availability	0	0	0	0	0
Atmospheric Effects					
Ionospheric[1]	0	7.2	16	21	27
Tropospheric	0	6	6	6	6
Total (RMS)	0	9.4	17	22	28
User Segment					
Receiver Noise[2]	3	3	3	3	3
Multipath[3]	0	0	0	0	0
UERE (RMS)	3	9.8	17.4	22.2	28.5
Navigation Accuracy 2 dRMS (HDOP = 1.5)	9	30	52	66	86

Notes
[1]Ionospheric delay of 30 assumed.
[2]Receiver noise is filtered.
[3]Multipath rejection antenna used.

in the differential range error budget assigned in Table 2. However, the satellite position error does not cancel completely with differential navigation when the user and reference station are separated.

The satellite position error can be modeled as having along-track (ATK), cross-track (XTK), and radial (RAD) error vectors, as shown in Figure 2. Depending on the user's location, different components of these errors will subtend into the range from the user to the satellite. The satellite position errors will not, therefore, cancel completely unless the user and reference station are close to each other.

The contribution of the satellite position error components to the user range error (Pos_{err}) is given by equation (1). The angle α (RSO or USO) is between the line-of-sight vector from the reference station (**RS**) or user (**US**) to the satellite and the vertical from the earth's center to the satellite (**OS**). The angle β is between the along-track direction and the plane defined by the line-of-sight vector and the vertical to the satellite (ΔOSR or ΔOSU). The satellite position error vectors RAD, ATK, and XTK are in the radial, along-track, and cross-track directions, respectively. The relationship among the angle α, the radius of the satellite's orbit R_s, the radius of the user R_u, and the earth angle E (SOR or SOU) between the satellite and the user and reference station is given in equation (2).

$$Pos_{err} = \cos\alpha \, RAD + \sin\alpha \cos\beta \, ATK + \sin\alpha \sin\beta \, XTK \qquad (1)$$

$$\sin\alpha = R_u \sin E / \sqrt{R_u^2 + R_s^2 - 2R_uR_s \cos E} \qquad (2)$$

Experimental results have demonstrated typical satellite orbital errors of 3 ft in the radial direction (RAD), 21 ft along-track (ATK), and 10 ft cross-track (XTK) [2]. The pseudorange error, Pos_{err}, due to these satellite position errors can be computed using equation (1). Since the along-track error, ATK, is generally greatest, the maximum pseudorange error will occur when the line of sight to the user is in the along-track plane and the satellite is on the user's local horizon (zero elevation). It can be shown geometrically that this occurs

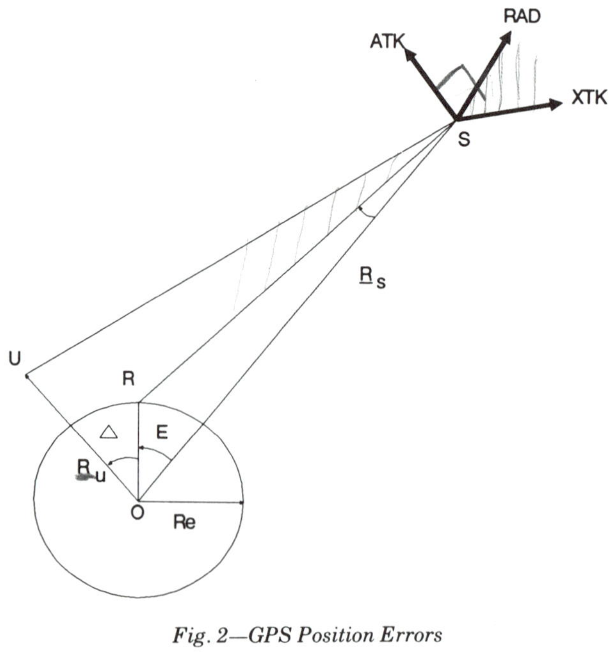

Fig. 2—GPS Position Errors

when the angle α satisfies the equation $\sin \alpha_m = R_e/R_s$, i.e., $\alpha_m = 13.9°$. Substituting this value for α into equation (1) with the errors RAD = 3 ft, ATK = 21 ft, and XTK = 10 ft results in a maximum pseudorange error, $Pos_{err,}$ of 8.5 ft. With these values for the ATK, XTK, and RAD satellite position errors, the pseudorange error, $Pos_{err,}$ will vary between 8.5 ft (at low elevation) and 3 ft (satellite directly overhead). These range errors are consistent with the 9 ft range error budget assigned in Table 1 to the satellite ephemeris errors.

If the GPS user and the differential reference station are separated by angle Δ, as shown in Figure 2, then the relative range error can be computed by substituting $E = E_0$ and $E = E_0 + \Delta$ into equation (2) and differencing the results of equation (1). The differential error is most sensitive to along-track (ATK) and cross-track (XTK) errors. In Figure 3, the differential error is shown as a fraction of the combined ATK and XTK errors in the plane of the line-of-sight vector and satellite radial vector. The error is shown for different angles, α (USO), between the line-of-sight vector and the satellite radial vector.

The differential error due to satellite position errors, $\Delta Pos_{err,}$ can be approximated using equation (3). This approximation is plotted in Figure 3, along with the exact differential errors. In equation (3), H is the vehicles' altitude, R_e is the earth's radius, R_s is the satellite's orbital radius, and Δ is the angular separation between the vehicle and the reference station in radians.

$$\Delta - Pos_{err}/\ (XTK \text{ or } ATK) = (R_e + H)\ \Delta\ /\ (R_s - R_e) \tag{3}$$

Substituting the values $(R_e + H) = 3{,}444$ nmi and $R_s - R_e = 10{,}898$ nmi into equation (3) results in a differential error growth rate of approximately 10 percent of the ATK or XTK error per 1000 nmi separation. Figure 3 shows

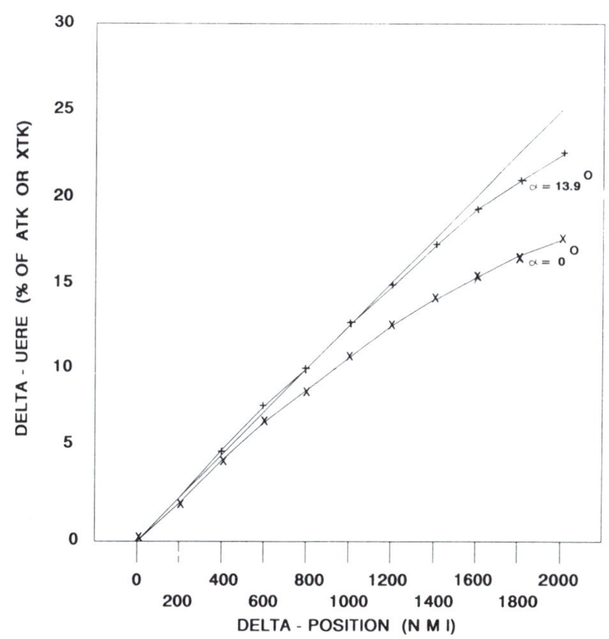

Fig. 3—Differential Errors Due to SV Position Errors

that this approximation is accurate to about 70 percent of the actual effect. If the along-track and cross-track errors result in a 25 ft error, the differential error will grow at approximately 3 ft/1000 nmi separation. In Table 2, the effect of this error is included in the differential error budget. It should be recognized that once the Standard Positioning Service (SPS) accuracy is degraded to 100 m (2 dRMS) (see the next section), the along-track and cross-track ephemeris errors may be significantly larger than 25 ft.

Selective Availability

The SA errors are introduced into the GPS signals on board the satellite to degrade the navigation accuracy provided by the SPS. Military user equipment includes a key that allows the SA errors to be removed from the pseudorange measurements. The SA errors are specified to degrade the horizontal navigation accuracy provided by the SPS to 100 m (2 dRMS). They can be modeled as a correlated noise process with standard deviation of 90 ft and a correlation time of a few minutes that acts as a slowly changing clock offset [3].

Since the SA errors act as a slowly changing bias on the pseudorange measurements, they can be completely removed using differential corrections. The only residual error remaining will be due to any time delay in applying the differential corrections. So long as the differential corrections are applied in near real time (within about 10 s), the residual error due to SA can be ignored.

Ionospheric Effects

The effect of the ionosphere on radio signals is proportional to the number of free electrons along the signal path. This value is known as the columnar electron density, N_c, or total electron content (TEC). N_c is computed from the integral of the electron density (electrons/m^3) along the signal path, and is generally given in total electron content units (TECU). One TECU is equivalent to 10^{16} electrons/m^2. If the signal path is in the vertical direction through the ionosphere, the columnar electron density is called the vertical electron content, N_{cv}.

The group delay introduced on the GPS satellite signals is proportional to the vertical TEC at the vehicle's location, scaled by an obliquity factor to account for slanted paths through the ionosphere for low-elevation satellites. The combined group delay ($c\Delta t$) is shown in equation (4), where f is the L1 frequency (1575.42 MHz) and El is the satellite's elevation angle.

$$\begin{aligned} c\Delta t &= \frac{40 \cdot 3}{f^2} N_{cv} \times Z(El)\ \text{m} = 0.53\ Z(El)\ \text{ft/TECU} \\ \text{where } Z(El) &= \sec\left[\sin^{-1}(0.94792 \cos El)\right] \end{aligned} \quad (4)$$

The vertical electron content in the mid-latitude regions (e.g., CONUS) is typically around 50 TECU [4]. It can, on occasion, reach values as large as 100 TECU during periods of high solar activity, such as can be anticipated during the solar maximum years 1989–90. The ionospheric obliquity factor Z(El) given in equation (4) varies between a factor of 1 when the satellite is directly overhead (vertical) and a factor of 3 when the satellite is near the horizon. If the vertical TEC is 50 TECU, from equation (4), the ionospheric group delay will vary between 27 ft and 80 ft, depending on the satellite elevation angle.

An ionospheric model is broadcast by the GPS satellites to allow a single-frequency user to compensate for the ionospheric group delay. This model, described in [5] and [6], allows a user to compensate for approximately 50 percent RMS of the total delay. This will typically result in a residual ionospheric delay over CONUS of about 27 ft, as shown in Table 1.

The accuracy of the differential navigation solution is strongly dependent on the spatial decorrelation of the ionosphere. When the master station and vehicle are separated geographically, the satellite signals are passing through different areas of the ionosphere and so will experience different group delays. The variability of TEC as a function of distance has been studied in [7]. TEC data from two sets of stations were used: one aligned approximately along an east-west direction, the other approximately along a north-south direction. No significant difference in correlation distance was found with season. The fractional reduction of the ionospheric group delay, F(r), due to common errors canceling is related to the correlation coefficient, r, through the equation $F(r) = \sqrt{1-r^2}$.

The studies documented in [7] showed that ionospheric delays have a correlation coefficient of 0.7 over 2000 km in latitude and over 3000 km in longitude at northern mid-latitudes. It should be noted that this data was taken at common local times. The large diurnal change in the ionosphere will introduce a much larger variation over longitude for the data sets used in the differential GPS process as the data is taken at the same UTC time and so at different local times.

In Table 3, the percentage of the ionospheric delay remaining after differential GPS corrections have been applied is shown as a function of separation distance between stations. At 1000 nmi separation, the ionospheric spatial correlation coefficient is 0.7, which means that the ionospheric delay will be reduced to 71 percent of its original value by differential corrections. The GPS ionospheric model will reduce the ionospheric delay, on average, to 50 percent of the total value. Table 3 shows that differential GPS will improve the navigation performance over the GPS ionospheric model only when the separation distance between stations is less than 500 nmi.

The differential GPS group delay is shown in Figure 4 for a 30 ft and 90 ft group delay, respectively. The ionospheric group delay will typically lie between these values over CONUS, depending on the magnitude of the vertical TEC and the obliquity factor Z(El). Differential GPS will provide improved accuracy over the GPS ionospheric model (50 percent reduction) when the separation distance between the vehicles is less than 500 nmi. The residual ionospheric delay will be between 15 and 45 ft at this point. As will be discussed later, an updated ionospheric model can further reduce the ionospheric delay.

Table 3—Differential Ionospheric Delay Reduction

Distance (nmi)	0	1	10	50	100	500	1000	2000
F(r) (%)	0	2	8	17	24	52	71	91

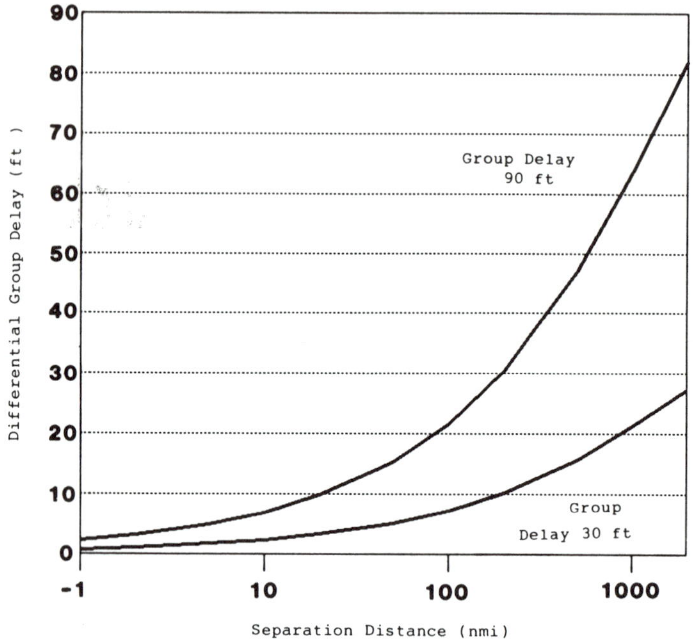

Fig. 4—Differential GPS Ionospheric Group Delay

Tropospheric Delay

The troposphere is the lower part of the atmosphere and is generally considered to extend up to an altitude of about 10 km. The tropospheric delay can typically be modeled using equation (5), where N_0 is the mean surface refractivity (3.2×10^{-4}), H is the vehicle's altitude, H_s is the tropospheric height (22,600 ft), and El is the satellite's elevation angle. The uncompensated delay can typically vary from 7 ft at high elevations to 80 ft at low elevations. The simple model given in equation (5) is adequate to reduce the residual tropospheric error to around 6 ft.

$$c\,\Delta t = N_0\, H_s\, e^{-H/H_s} \operatorname{cosec}(El) \qquad (5)$$

Since the tropospheric delay is a function of local humidity, temperature, and altitude, differential GPS corrections are ineffective for compensating for tropospheric errors. The error budget allocated for the differential tropospheric errors in Table 2 is, therefore, the same as that allocated for absolute navigation in Table 1.

User Segment Errors

The user segment errors are introduced at the user's receiver and are, therefore, unaffected by differential GPS corrections. Error budgets are assigned in Tables 1 and 2 for the contribution of receiver noise and multipath errors to the user equivalent range error (UERE).

With a C/A-code receiver, the noise introduced by the code-tracking loops is generally on the order of 30 ft. This can be significantly reduced by the use of filtering techniques, such as a Kalman filter. For differential GPS, it is essential to reduce the receiver noise by filtering to improve the navigation accuracy. By using a Kalman filter, it is possible to reduce the effective receiver noise to within 3 ft (1 σ).

Multipath errors are introduced at the receiver from reflected GPS signals received by the antenna. These vary depending on the types of reflective surfaces around the vehicle (e.g., water, concrete, buildings) and the type of antenna used. Typically, a C/A-code receiver will experience multipath errors on the order of 10 ft. The effect of multipath can be reduced significantly through the use of an antenna that will reject reflected signals received from directions below the local horizon. With a specially designed antenna, it is possible to reduce the multipath errors to insignificant levels [8]. This error source is, therefore, ignored in the differential GPS error budget shown in Table 2.

EXTENDED DIFFERENTIAL GPS

Table 2 shows that a number of different errors limit the accuracy of the differential GPS solution as the distance from the reference station grows. The most significant error source is ionospheric delay, which dominates the error budget at distances of 100 nmi or greater from the reference station. The next largest error of concern is tropospheric delay, which introduces an error of 6 ft when using the simple correction model shown in equation (5). Once the range extends more than 1000 nmi from the reference station, satellite ephemeris errors also affect the accuracy of the differential GPS solution.

To extend the range of the differential GPS operation, the differential GPS message will need to include an ionospheric model, a tropospheric model, and an estimate of the satellite ephemeris errors. The following sections address the accuracy of available models, and ways of applying them to this problem.

Ionospheric Modeling

There exist a variety of ionospheric models that predict, with varying degrees of accuracy, the TEC in different parts of the world. The features of some of the most appropriate of these models are described in this section. The accuracy of predictions by all ionospheric models, however, is limited by the high temporal variability of the ionospheric delay.

Temporal Variability

The TEC exhibits both diurnal and periodic variations due to perturbations of the ionospheric F region from geomagnetic substorms, meteorological sources such as weather fronts, shock waves from supersonic aircraft, volcanic explosions, rocket launches, and other miscellaneous sources. While these short-term variations in TEC cover a large range of periods and amplitudes, common periods range from 20 to over 100 min, with amplitudes of a few percent of the background TEC.

Studies documented in [9] concluded that the useful prediction time for mid-latitude stations measuring the ionosphere was a function of local time, season, and long-term sunspot activity. However, in most cases there was no significant improvement over the use of monthly mean TEC predictions when actual data more than 3 h old was used. These studies also predicted the TEC data for the same geographic location and direction in the sky as their measurements. If the prediction is for a different location, the temporal correlation will be lower.

Because of the low temporal correlation of the ionosphere, a model must be updated in near real time (within at least 20 min) with current measurements of the ionospheric profile over the geographic region of interest to be effective in removing the group delay in the highly variable low-latitude region. This can be achieved with the extended differential GPS concept by making dual-frequency ionospheric measurements at the differential reference stations throughout the network shown in Figure 1. Dual-frequency P-code receivers may be used for this purpose, but are not required. Codeless receivers are now available which make measurements of the ionospheric delay by tracking the relative P-code phase between the L1 and L2 frequencies, without knowledge of the P-code. This provides a measurement of the ionospheric delay accurate to within 3 ft, with an uncertainty of the integer number of P-code cycles involved, ± 48 ft. Since the ionospheric delay can be estimated to within 48 ft using a rough model, this measurement is sufficient for accurately measuring the delay to each satellite. Codeless dual-frequency receivers are preferred for civil applications since they do not require periodic updates to be able to track the encrypted Y-code.

The dual-frequency ionospheric measurements from the reference stations will be transmitted to the differential master station along with the observed satellite range errors. At the master station, the measurements from the stations will be combined and used to update a real-time ionospheric model. A number of models that can be adapted for this purpose are described in the following sections.

GPS Ionospheric Model

To allow single-frequency C/A-code users to compensate partially for ionospheric group delay errors, an ionospheric model is broadcast by the GPS satellites [5,6]. This model uses eight coefficients to provide a correction for approximately 50 percent RMS of the total ionospheric range error. This algorithm is a severely truncated version of a much larger empirical model of TEC developed by Bent [10] over 13 years ago. The algorithm consists of a cosine representation of the diurnal curve, allowed to vary in amplitude and in period with user geomagnetic latitude. The algorithm coefficients were computed from an empirical model of worldwide ionospheric behavior, derived by Bent, for each 10 day period of the year and for several values of average solar flux conditions. Coefficients broadcast by the GPS satellites are updated once each 10 days, or sometimes more frequently if the 5 day running mean solar flux changes by a large amount during that period.

Because of the high degree of day-to-day variability in the TEC from the monthly mean value, the GPS model generally removes only 50 percent of the total delay. The eight coefficients are used to model the amplitude of the cosine term and its period as a function of geomagnetic latitude using a third-degree polynomial. The nighttime constant level term and the phase of the cosine curve are assumed constant.

Bent Model

The main purpose of the Bent model is to determine the TEC of the ionosphere to determine the group delay experienced by a radio signal. Documentation and a description of a computer program that implements this model are included in [10].

The parameters input to the Bent model are the date, universal time, locations of the transmitter and receiver, rate of change in elevation and altitude of the satellite or vehicle, operating frequency, solar flux (10.7 cm flux), and sunspot number. The Bent computer model [10] outputs the vertical electron content above the transmitter, the profile of vertical electron density with altitude, and the TEC along the path between the satellite and the ground. The model operates using an empirical database of 50,000 topside ionospheric soundings, 6,000 satellite measurements of electron density, and 400,000 bottomside soundings of the ionosphere. The data extends from 1962 to 1969 to cover the maximum and minimum of the solar cycle. The model can predict the ionospheric group delay at the mid-latitudes with an accuracy of 75–80 percent using the empirical database of ionospheric data.

The model also has the option to update the ionospheric profile using current data from local observation stations. Up to eight observations of the TEC separated by different amounts in time and space from the evaluation time and master station can be accepted. To obtain the best possible update, the observation times and stations should be the closest to the evaluation condition available and must be within 1000 nmi of the master site. Using current data, the accuracy of the model at mid-latitudes is improved to 90 percent of the total effect.

Differential GPS Ionospheric Model

To minimize the impact on the GPS receiver navigation software, a combination of the GPS ionospheric model and the updated Bent model is proposed for the extended differential message. The concept is illustrated in Figure 5. Ionospheric measurements from the differential reference stations to the different GPS satellites are used to update a computer implementation of the Bent model at the master station. Presently, the model allows for only eight measurement updates to be applied. This could be extended if it were demonstrated that this would improve the model accuracy. The updated Bent model provides a real-time model of the ionospheric group delay accurate to 10 percent of the total effect over a region of 1000 nmi from the central master station. A set of eight coefficients is then generated to provide a best fit to the model averaged over CONUS. The approximations made in the eight-coefficient GPS model reduce the accuracy of the final model by an estimated 10 percent. The updated eight-coefficient model is broadcast over the extended differential data link to users of the service.

By using the extended differential GPS ionospheric model instead of the model broadcast by the GPS satellites, single-frequency C/A-code users can remove approximately 80 percent of the ionospheric delay when they are within 1000 nmi of the master station. The extended differential GPS service should, therefore, be able to reduce the ionospheric errors to below 6 ft over a ±1000 nmi region by broadcasting an updated eight-coefficient GPS model. If a more sophis-

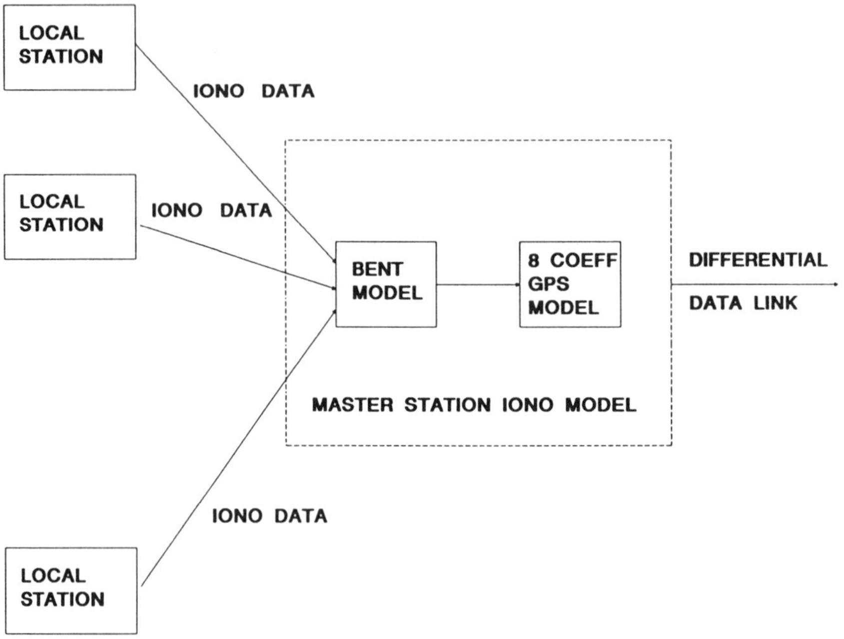

Fig. 5—Updated Ionospheric Model

ticated model were broadcast for the ionospheric delay and the differential GPS reference stations were placed every 600 nmi (10 deg) in latitude and longitude, the accuracy of the model and the range of operation could be further improved.

Tropospheric Modeling

An accurate, real-time tropospheric model can also be broadcast by the master station to reduce the differential GPS error from the residual tropospheric delay. A special-purpose model suitable for this purpose has already been designed by the Air Force Cambridge Research Laboratory (AFCRL) [11]. This model computes the tropospheric group delay as a function of surface refractivity, altitude, and elevation angle. The model can be updated in real time through a three-coefficient fit to the surface refractivity, N_s, averaged over latitude (L) and altitude (H). With measurements of the surface refractivity, the AFCRL tropospheric model is accurate to 4 percent of the total delay.

The extended differential tropospheric model can be updated in real time using surface refractivity measurements made over the region of coverage by the differential reference stations. The accuracy of the residual tropospheric delay using the extended differential GPS tropospheric model can, therefore, be expected to lie between 0.3 and 3 ft, depending on the elevation angle of the satellite.

Satellite Ephemeris Computation

The satellite ephemeris errors can also be observed by the extended differential GPS network using measurements from the reference stations. The reference stations over the region of coverage are observing the satellite range measurements. This data can be processed in a number of different ways to solve for the satellite ephemeris errors.

Networks of GPS monitor stations already exist which are used to compute precise ephemeris data for the GPS satellites through postprocessing. One such network is the Cooperative International GPS Network (CIGNET), which has tracking sites located at Mojave, Westford, Richmond, Yellowknife, Tromso, Onsala, Wettzel, Tsukaba, and Kokee Park, as illustrated in Figure 6. The accuracy of the precise ephemeris provided by CIGNET is about 1 part in 10^7, resulting in an estimated ephemeris error of 8 ft. It would be possible to implement a similar process in real time using the observed satellite range errors from the extended differential network to compute a set of precise orbital parameters. By broadcasting ATK, XTK, and RAD corrections for the ephemeris errors observed for each satellite, the orbital accuracy could be improved to within 8 ft. This would reduce the differential error introduced by the ephemeris data over ± 1000 nmi to below 2 ft.

It becomes significantly more complicated to compute the precise ephemeris data once SA is implemented. To be effective, the ephemeris determination program must be capable of separating out clock dither effects from errors in the broadcast ephemeris data. This can be achieved by using the extended differential GPS network to compute the satellite ephemeris data and SA clock errors using the method illustrated in Figure 7.

The range errors from four or more reference stations can be used to perform a least squares solution for position and time errors for each satellite, just as range measurements from four or more satellites to a user can solve for the user's position and time errors. The station range measurements observe the

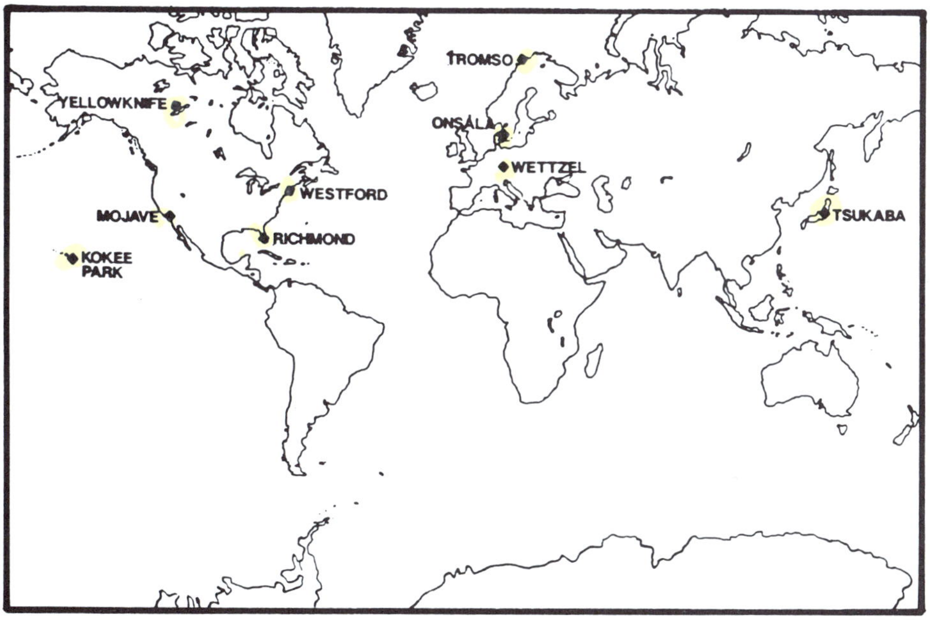

Fig. 6—CIGNET Network (9 stations)

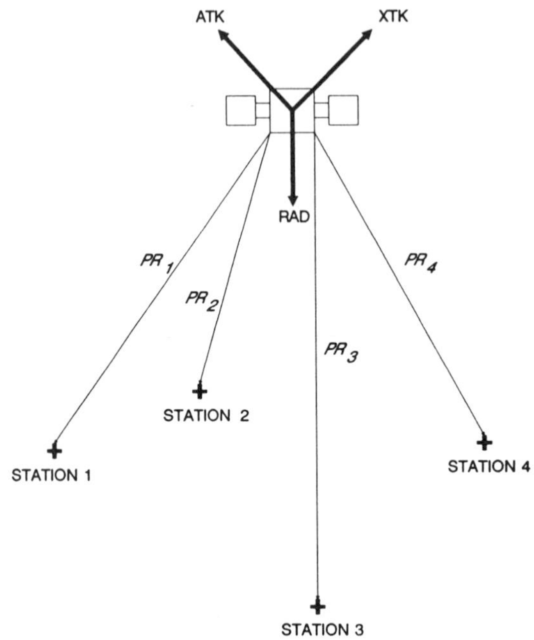

Fig. 7—Ephemeris Error Observation

satellite ATK, XTK, RAD, and clock errors (B_s) through the line-of-sight vectors from the reference stations to the satellite (x_i, y_i, z_i). The clock error term, B_s, will include both the clock offset in the satellite ephemeris data and the error introduced by SA. The least squares solution for the satellite errors using these measurements is shown in equation (7). The matrix H consists of the line-of-sight vectors from the differential stations to the satellite.

$$\begin{bmatrix} ATK \\ XTK \\ RAD \\ B_S \end{bmatrix} = \mathbf{G}\mathbf{H}^T \begin{bmatrix} PR_1 \\ PR_2 \\ \cdot \\ \cdot \\ PR_n \end{bmatrix} \tag{7}$$

$$\text{where } \mathbf{G} = (\mathbf{H}^T\mathbf{H})^{-1} \text{ and } \mathbf{H} = \begin{bmatrix} x_1 & y_1 & z_1 & 1 \\ x_2 & y_2 & z_2 & 1 \\ \cdot & \cdot & \cdot & \cdot \\ \cdot & \cdot & \cdot & \cdot \\ x_n & y_n & z_n & 1 \end{bmatrix}$$

The accuracy of the final satellite position solution will be a function of the geometry provided by the ranges from the reference stations and the accuracy of the range error measurements. If the line-of-sight vectors from the satellite are defined in along-track (x), cross-track (y), and radial (z) directions, dilutions of precision can be computed for the accuracy of the ATK, XTK, RAD, and time/SA estimates, respectively. These dilutions of precision are defined as follows: ADOP = G_{11} is the along-track dilution of precision, XDOP = G_{22} is the cross-track dilution of precision, RDOP = G_{33} is the radial dilution of precision, and TDOP = G_{44} is the time/SA dilution of precision.

Since the differential GPS errors are most sensitive to the along-track and cross-track errors combined, a space vehicle horizontal dilution of precision term (SVHDOP) is defined which is the combination of the ADOP and XDOP terms.

$$\text{SVHDOP}^2 = \text{ADOP}^2 + \text{XDOP}^2 \tag{8}$$

If the range measurements to the satellites are accurate to 3 ft (1 σ), the combined along-track and cross-track residual errors will be given by SVHDOP $\times$ 3 ft after the satellite position corrections have been applied from the extended differential GPS message. The residual position error will result in a differential GPS range error of approximately 10 percent of the along-track or cross-track errors over 1000 nmi. The residual differential range error can, therefore, be approximated by equation (9).

$$\Delta\,\text{Pos}_{err} = \text{SVHDOP} \times 0.3\text{ ft} / 1000\text{ nmi} \tag{9}$$

The RDOP and TDOP terms represent how well the extended differential network can differentiate between SA (clock dither errors) and satellite position errors in the radial direction. This is not critical since the radial error differs from a clock error only as a $(1 - \cos\alpha)$ effect (see Figure 2). Over the region in which the satellite is visible, this causes a variation of 3 percent = $(1 - \cos 13.9°)$ between the SA errors and a radial satellite position error. Even if the satellite radial position error is as high as 30 ft, the variation in the UERE over the region of satellite visibility will be only 1 ft. The satellite radial position errors and clock errors, including SA, can therefore be lumped together without affecting the extended differential GPS navigation accuracy.

Performance Simulation

The solution geometry provided by the CIGNET stations in observing the satellite ephemeris was simulated for the 6-plane, 24-satellite GPS constellation. The ground tracks of the first 8 GPS satellites in this constellation are shown in Figure 8. These ground tracks repeat at different times of day for the remaining 16 satellites in the 24-satellite constellation. The SVHDOP provided by the CIGNET geometry for observing the satellite along-track and cross-track errors using equation (7) is illustrated in Figure 9 for satellites 0 through 7. Because of the symmetry in the constellation, the pattern repeats for satellites 8 through 23. The periods of visibility for the GPS satellites over the United States are also illustrated in Figure 9.

An extended CIGNET network of stations is illustrated in Figure 10. It includes additional stations at Portland, Denver, Houston, and Anchorage in the United States, and at Gander (Newfoundland), Churchill (Canada), Bermuda, and Acores. This provides the improved satellite observation geometry illustrated in Figure 11.

Equation (9) shows that an SVHDOP of less than 10 is required before the differential network can improve on the accuracy of the GPS satellite ephemeris

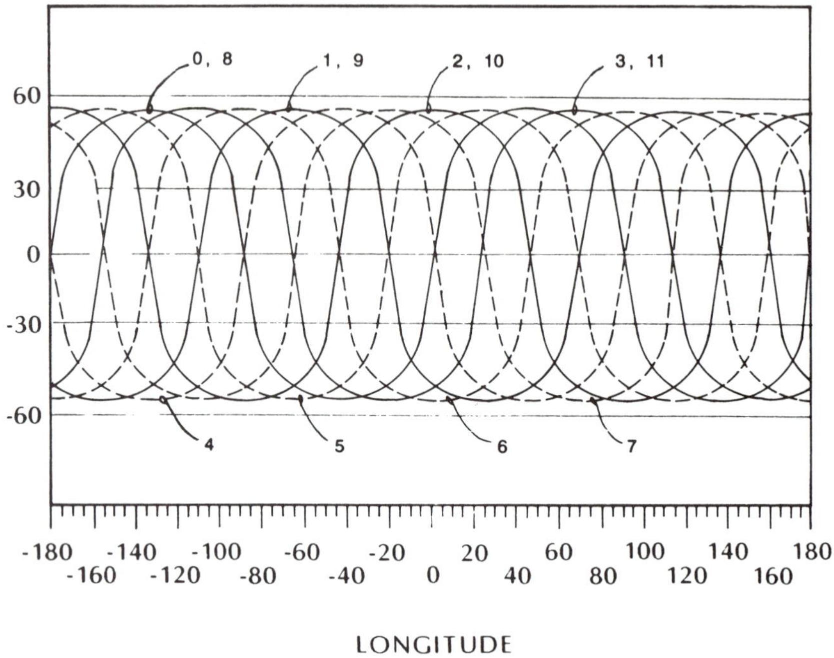

Fig. 8—24-Satellite Constellation Ground Tracks

data. Figures 9 and 11 show that both the CIGNET and extended CIGNET networks provide SVHDOPs as small as 4 for periods of time. The extended CIGNET network provides this geometry for a larger percentage of time when the satellites are visible over CONUS. With an SVHDOP of 4, the extended differential network can broadcast ATK, XTK, RAD, and time/SA corrections that will reduce the satellite ephemeris errors to 1.2 ft over a 1000 nmi range. If additional stations are included in the network, particularly on the Asian continent, it should be possible to estimate the satellite ephemeris errors accurately before the satellites are visible over the United States. The simple least squares solution shown in equation (7) makes it very easy to process data from large numbers of stations all over the world.

Extended Differential Navigation Accuracy

The navigation accuracy that can be achieved using the extended differential GPS method is summarized in Table 4.

It was shown earlier that the satellite ephemeris, clock, and SA errors can be estimated using an inverse ranging technique from the different reference stations in the differential network. A separate correction can then be broadcast for each satellite for the combined clock/SA/radial satellite errors and for the along-track and cross-track satellite position errors. This reduces the range error contribution from the satellite along-track and cross-track errors to below 3 ft over a ±1000 nmi range. The residual clock, SA, and radial position error introduces a 1 ft range error due to the difficulty in resolving between these different error sources.

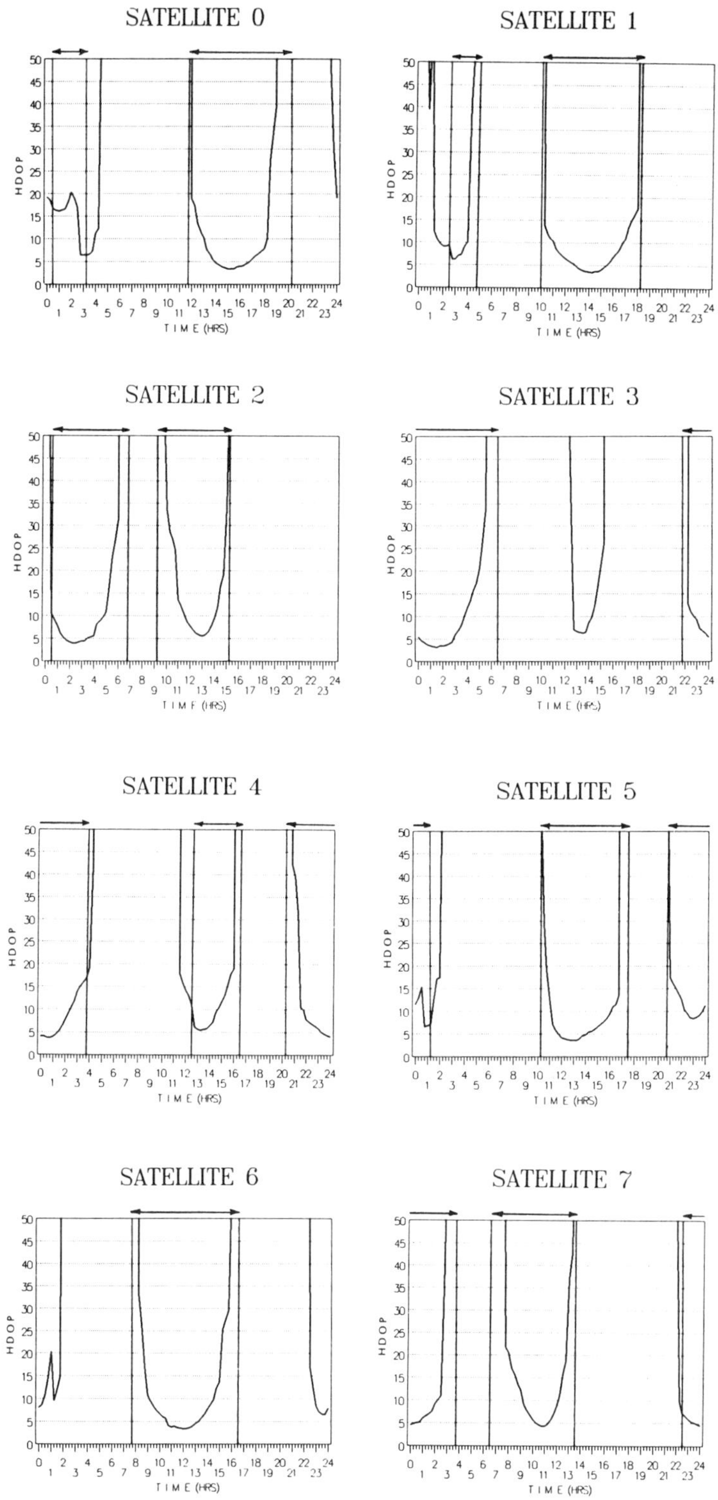

Fig. 9—CIGNET Network Geometry

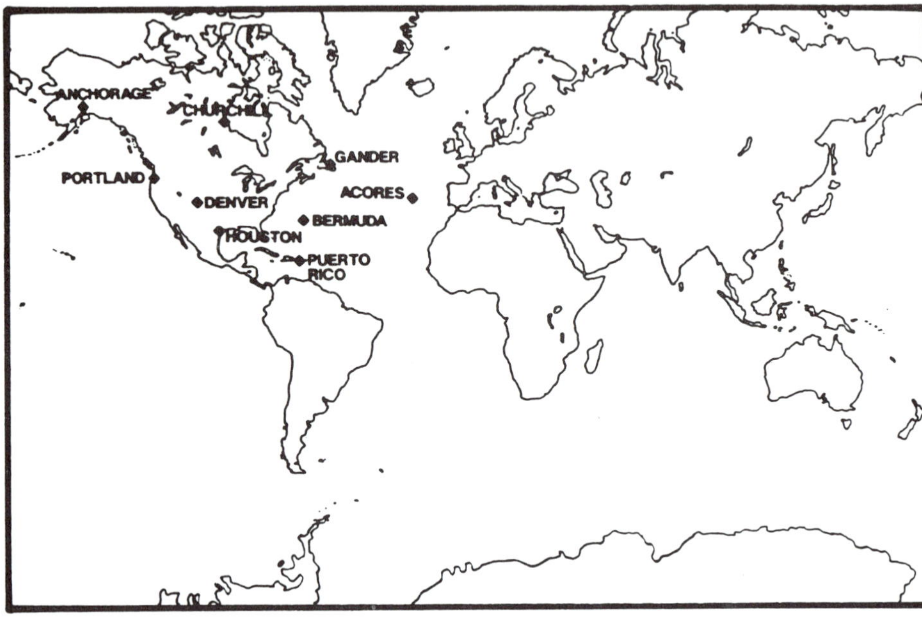

Fig. 10—Additional Differential Stations

The extended differential GPS message must also correct for atmospheric effects. As described earlier, an updated eight-coefficient GPS ionospheric model can be computed from dual-frequency GPS measurements throughout the extended differential GPS network. Using this model instead of the model broadcast by the GPS satellites will reduce the residual ionospheric delay to below 6 ft over a ±1000 nmi range. The residual tropospheric delay can also be reduced by broadcasting an improved model. A three-coefficient updated AFCRL tropospheric model should reduce the residual error to below 1 ft over the extended range.

The additional message types that are required for the extended differential GPS broadcast are listed in Table 5. RTCM SC-104 has tentatively allocated message types 15 and 17 for the broadcast of ionospheric, tropospheric, and satellite ephemeris data. The suggested content for these message types is shown in Table 5.

Message type 15 is reserved in the SC-104 format for ionospheric and tropospheric models. Using the models described earlier, this message would contain the updated eight-coefficient GPS ionospheric model and an updated three-coefficient tropospheric model. Message type 17 is tentatively reserved for satellite ephemeris data. Instead of broadcasting the complete satellite ephemeris data, it is suggested that only an along-track and cross-track error be included for each satellite with the time of applicability (t_{oe}). This error should then be applied as a correction at the receiver to the broadcast satellite ephemeris data. This approach significantly reduces the size of the message while still providing accurate corrections.

The pseudorange corrections broadcast in message type 1 will include the combined satellite clock/SA/radial errors computed by the differential GPS

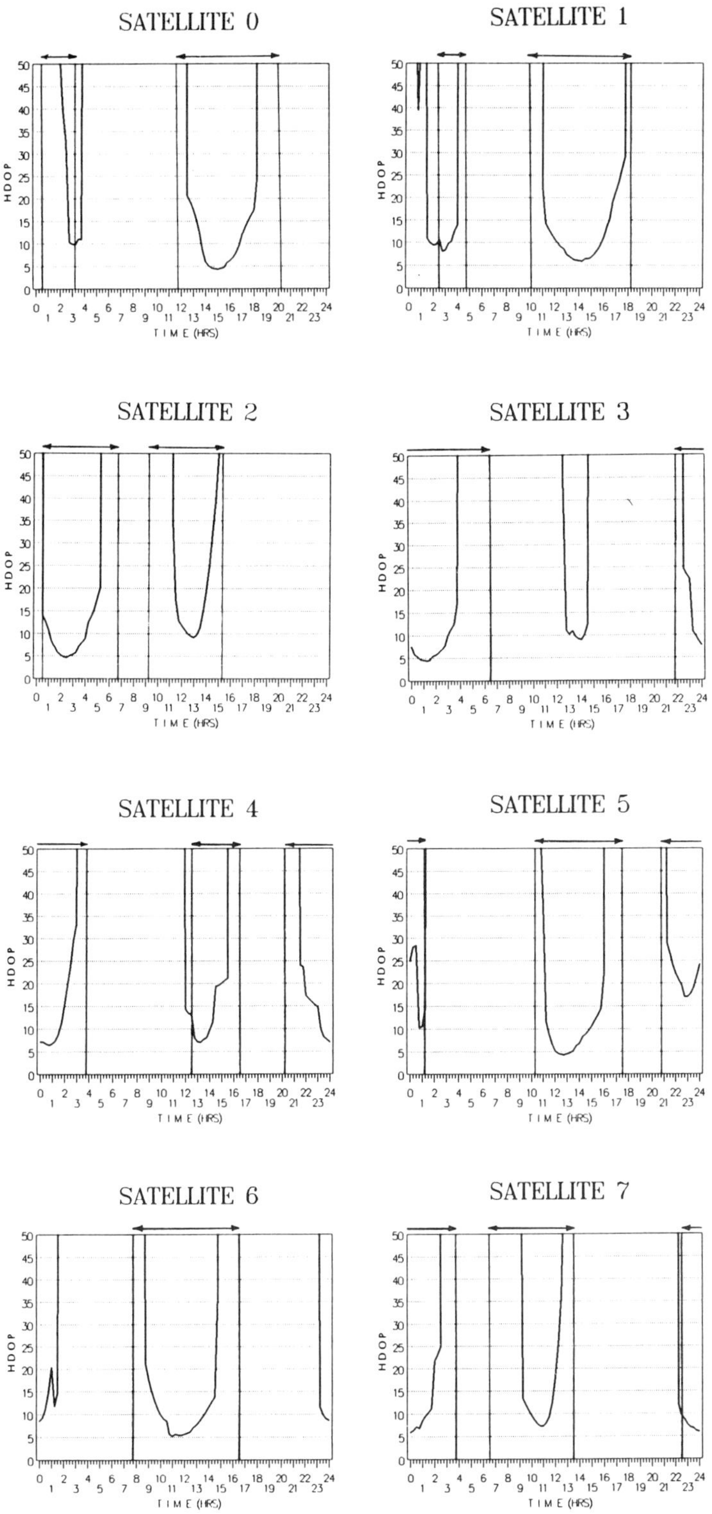

Fig. 11—Extended CIGNET Network Geometry

Table 4—Extended Differential Navigation Accuracy

Source	Error Budget 2000 nmi range (ft)
Space Segment	
Clock Error	0
Control Segment	
Ephemeris Error[1]	3
Selective Availability[2]	1
Atmospheric Effects	
Ionospheric[3]	6
Tropospheric[4]	1
Total (RMS)	7.4
User Segment	
Receiver noise[5]	3
Multipath[6]	0
UERE (RMS)	8
Navigation Accuracy	
2 dRMS (HDOP = 1.5)	24

Notes
[1]Residual ATK and XTK errors from ephemeris estimates.
[2]Residual radial/clock/SA ephemeris errors estimation.
[3]Residual error in updated eight-coefficient model.
[4]Residual error in updated tropospheric model.
[5]Receiver noise is filtered.
[6]Multipath rejection antenna used.

Table 5—Extended Differential Messages

Message Type	Data Content
1—Range errors	SV clock, SA, and radial position errors
15—Iono/Tropo	8—coefficient ionospheric model 3—coefficient tropospheric model
17—Ephemeris	ATK and XTK errors at t_{oe} for each satellite

master station for each satellite. This is the only message that needs to be broadcast by the extended differential service at a high rate since the SA errors change with time.

SUMMARY

The conventional differential navigation error budget is shown in Table 2. Conventional differential GPS using Radio Technical Commission for Maritime Services Special Committee (SC)-104 message type 1 provides a navigation accuracy of 30 ft (2 dRMS) over a range of 100 nmi from the reference station. With the Extended Differential GPS concept, separate corrections are broadcast in the message for the clock and selective availability (SA), ephemeris, ionospheric, and tropospheric errors on each satellite's signal. This allows the differential GPS service to be extended over a range of ±1000 nmi, providing a navigation accuracy of 24 ft (2 dRMS) or better.

Figure 1 illustrates an extended differential GPS network operating over CONUS. Each individual differential station in the network provides the satellite range error corrections to a master differential station. The master station uses these and other measurements from the reference stations to compute an extended differential GPS message that separates out the different error sources. This message is then broadcast to users over a 1000 nmi region using a satellite communications link. By using two differential master stations and satellite links, an extended differential GPS service could be provided that would cover the entire continental United States.

Differential GPS has already proven to be a valuable service for many applications. Once SA has been implemented and the accuracy of the Standard Positioning Service has been degraded to 100 m (2 dRMS), more applications will require differential GPS corrections. The Extended Differential GPS method will allow an economical, nationwide, differential GPS service to be implemented which will provide high-accuracy navigation to a large number of users.

Based on a paper presented at The Institute of Navigation National Technical Meeting, San Mateo, CA, January 1989.

REFERENCES

1. *GPS System Specification,* SS-GPS-300C, USAF.
2. Russel, S.S. and Schaibly, J.H., *Control Segment and User Performance,* Global Positioning System, papers published in NAVIGATION, reprinted by The Institute of Navigation, Vol. I, 1980. pp. 74–80.
3. Kalafus, R.M., *Special Committee 104 Recommendations for Differential GPS Services,* Global Positioning System, papers published in NAVIGATION, reprinted by The Institute of Navigation, Vol. III, 1986, pp. 101–16.
4. Jursa, A.S., *Handbook of Geophysics and the Space Environment,* Air Force Geophysics Laboratory, AFSC, USAF 1985.
5. Klobuchar, J.A., *A First Order, World-wide, Ionospheric, Time-Delay Algorithm,* AFCRL-TR-77-0502, AD A018862, 1975.
6. Klobuchar, J.A., *Design and Characteristics of GPS Ionospheric Time Delay Algorithm for Single Frequency Users,* IEEE Position Location and Navigation Symposium (PLANS) 86, NV, 1986, pp. 169–76.
7. Klobuchar, J.A. and Johanson, J.M., *Correlation Distance of Mean Daytime Electron Content,* AFGL-TR-77-0185, AD A048117, 1977.
8. Rocken, C. and Meertens, C.M., *GPS Antenna and Receiver Tests: Multipath Reduction and Mixed Receiver Baselines,* Fifth International Geodetic Symposium on Satellite Positioning, Las Cruces, NM, March 1989. pp. 375-85.
9. Donatelli, D.C. and Allen, R.S., *Time Cells for Adaptive Predictions of Total Electron Content,* Radio Science, V16, 1981, pp. 261–69.
10. Llewellyn, S.K., et al., *Documentation and Description of the Bent Ionospheric Model,* AFCRL-TR-73-0657, AD 772733, July 1973.
11. Altschuler, E. and Kalaghan P., *Tropospheric Range Error Corrections for the NAVSTAR System,* AFCRL-TR-0198, AD 786928, April 1974.

NAVIGATION: *Journal of The Institute of Navigation*
Vol. 38, No. 2, Summer 1991
Printed in U.S.A.

Wide Area Differential GPS

CHANGDON KEE, BRADFORD W. PARKINSON
Stanford University, Stanford, California

PENINA AXELRAD

Stanford Telecommunications, Santa Clara, California
Received September 1990
Revised April 1991

ABSTRACT

GPS has proven to be an extremely accurate positioning sensor for a wide variety of applications. However, in some situations, such as aircraft approach and landing, higher accuracy is required. Wide Area Differential GPS (WADGPS) is a system that could be used to meet such requirements. The WADGPS system comprises a master station, and local monitor stations distributed across the United States. The system calculates and transmits a vector of error corrections to users. This correction vector consists of parameters describing the three-dimensional ephemeris errors, satellite clock offsets, and ionospheric time delay parameters. The master station gathers GPS measurements made at each of the local stations, and estimates the errors using a combination of nonlinear static estimation and batch least squares. The performance of a 15-station WADGPS network was investigated by simulation for users at sites across the United States. The monitor stations were located at existing Loran or VOR stations. Simulation results indicate that normal GPS positioning errors can potentially be reduced by more than 95 percent using WADGPS.

INTRODUCTION

GPS is expected to greatly improve the accuracy and reduce the cost and complexity of navigation for land, marine, and aircraft users. Under normal operating conditions, it is able to provide positioning accuracies in the range of 15–25 m. However, with Selective Availability (SA), the errors incurred by typical civilian users have been found to be 100 m or more. In some situations, for example, harbor navigation or precision approach of an aircraft, these accuracies are not sufficient.

Differential GPS (DGPS) is a means for improving navigation accuracy in a local area. A single DGPS monitor station at a known location can compute a range error correction for each GPS satellite in view. These error corrections are then broadcast to users in the vicinity as depicted in Figure 1. By applying the corrections to the signals received, a user can typically improve the accuracy down to the 5 m level [1]. However, as the distance between the user and the monitor station increases, range decorrelation occurs, and accuracy degrades. The error increases because projection of the ephemeris error onto the user-satellite line of sight is no longer the same as the monitor station-satellite line of sight. This is illustrated in Figure 2.

The maximum range error difference, δR_{error}, between the monitor and user is given by

$$\max(\delta R_{error}) \approx \frac{d}{D}\,\delta R \tag{1}$$

where δR is the magnitude of the satellite ephemeris error, d is the separation between the user and the monitor station, and D is the distance from the user to the GPS satellite.

In addition, if the two receivers are widely separated, the lines of sight through the ionosphere are also different, resulting in differences in the ionospheric delay observed. A similar, but smaller effect occurs for the tropospheric delay.

Beyond a separation distance of 100 km, a range error correction is not sufficiently accurate to realize the full potential of DGPS. In fact, over 500 monitor stations would be required to provide standard single-station DGPS aiding across the entire United States. Wide Area DGPS (WADGPS) provides a powerful means for bridging the gap between unaided performance and high-accuracy navigation in the vicinity of a correction station. Various WADGPS techniques have been suggested by Brown [2] and Loomis et al. [3]

Wide Area Differential GPS (WADGPS)

Instead of calculating a scalar range error correction for each satellite, as is done in DGPS, WADGPS provides a vector of error corrections comprising a

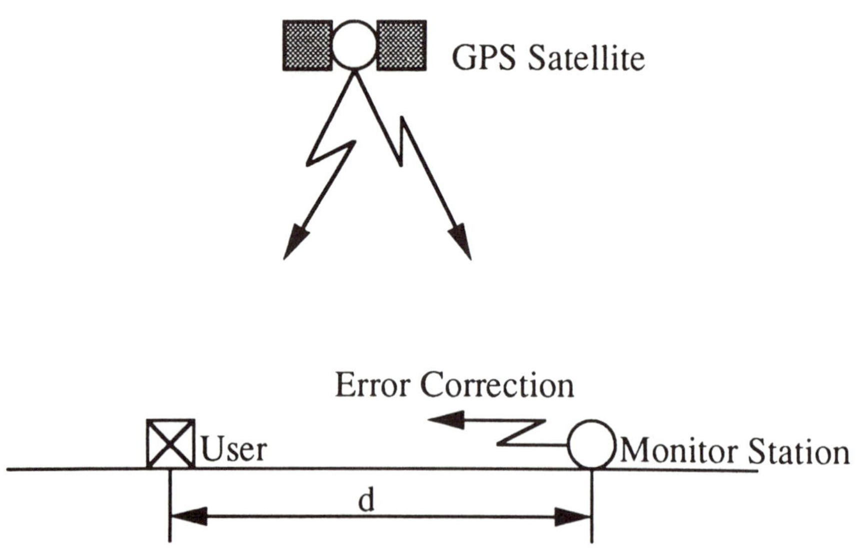

Fig. 1—Overview of Differential GPS

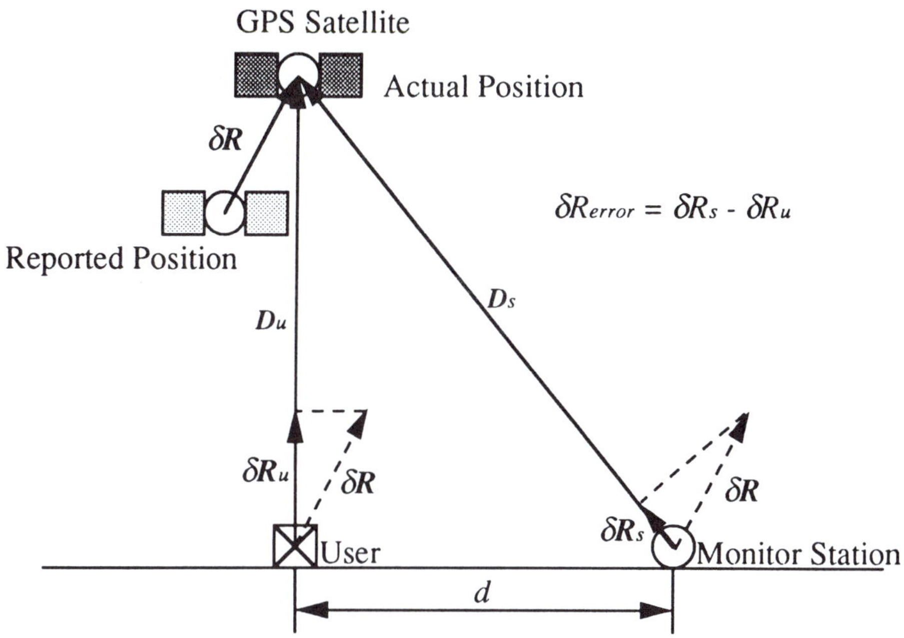

Fig. 2—Degradation of DGPS Accuracy with Distance (The accuracy of the range correction broadcast by a DGPS monitor station degrades with distance. For example, the figure shows an ephemeris error δR which produces a small range error, δR_s, at the monitor station, but a larger range error, δR_u, at the user location. If the user were to employ the range correction broadcast by the monitor station, a residual range error of $\delta R_s - \delta R_u$ would remain.)

three-dimensional ephemeris error and clock bias for each GPS satellite, plus ionospheric time delay parameters. The accuracy of the WADGPS correction is consistent within the monitored region, and degrades gracefully on the perimeter.

The WADGPS network comprises one master station, a number of local monitor stations, and communication links. In this concept, 15 local monitor stations are located at existing Loran or VOR stations across the United States (including Alaska and Hawaii). Each local monitor station is equipped with a high-quality rubidium clock and a GPS receiver capable of tracking all satellites within the field of view. GPS measurements taken at each local station are sent to the master station. The master station estimates ionospheric time delay parameters, plus the satellite ephemeris and clock errors, based on the known monitor station locations and the information collected. The computed error corrections are transmitted to users via any convenient communication links, such as satellite, phone, or radio.

Figure 3 provides an overview of the WADGPS, and Figure 4 shows the flow of information among the system components. The process can be summarized as follows:

1) Local monitor stations at known locations collect GPS pseudoranges from all satellites in view.

2) Pseudoranges and dual-frequency ionospheric delay measurements are sent to the master station.

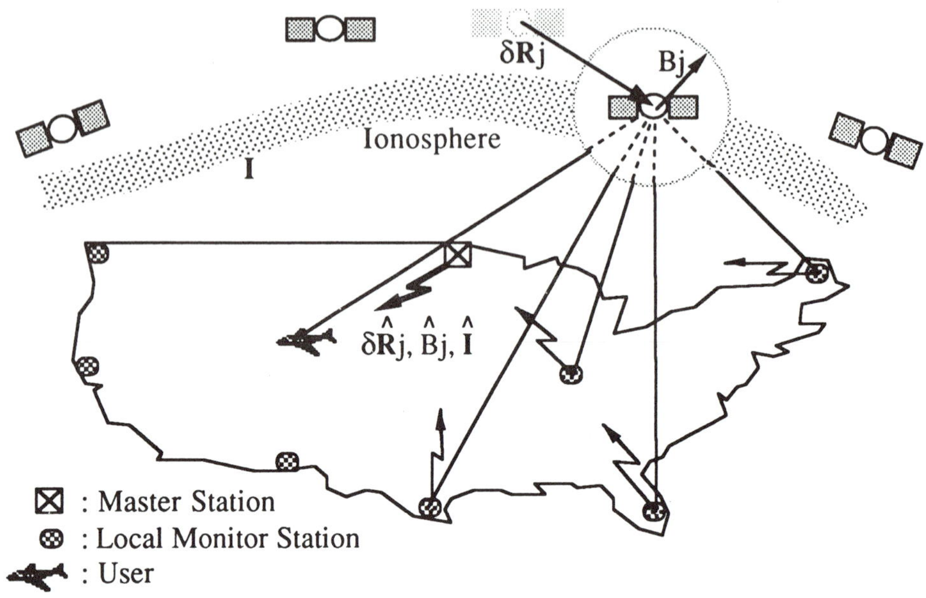

Fig. 3—Wide Area Differential GPS Concept

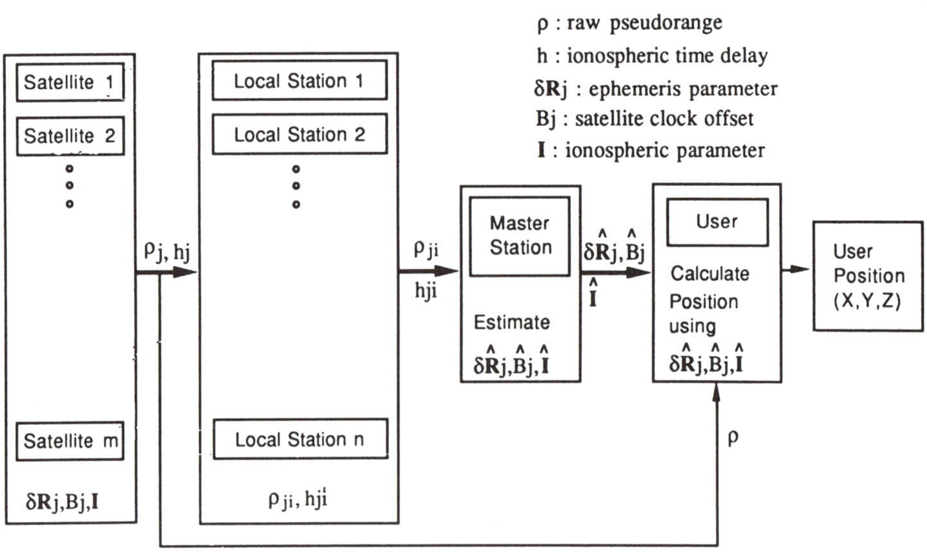

Fig. 4—Block Diagram of WADGPS Components

3) The master station computes an error correction vector.

4) The error correction vector is transmitted to users.

5) Users apply the error corrections to their measured pseudoranges and collected ephemeris data to improve navigation accuracy.

The following subsections briefly describe the local monitor stations and the typical WADGPS users. The functioning of the master station is then described in detail.

Local Monitor Stations

For this study, we have assumed 15 local monitor stations, located at Loran or VOR stations across the United States, including Alaska and Hawaii. Figure 5 illustrates the location and Table 1 the latitude and longitude of each station. The position of each station is known extremely well, and is situated such that satellites can be reliably tracked down to an elevation mask angle of 5 deg.

Each monitor station is equipped with a rubidium clock, and a receiver capable of providing dual-frequency ionospheric error measurements. Raw pseudorange measurements and these local ionospheric errors are transmitted to the master station in or near real time. This could be done via a communication satellite data link as suggested in [3].

Users

A typical user that would benefit from the WADGPS has a single-frequency, C/A-code GPS receiver and a quartz oscillator. We have modeled 81 stationary users at locations distributed uniformly across the United States (Figure 6). All users are assumed to have an elevation mask angle of 6.5 deg, which is typical of aircraft antenna visibility.

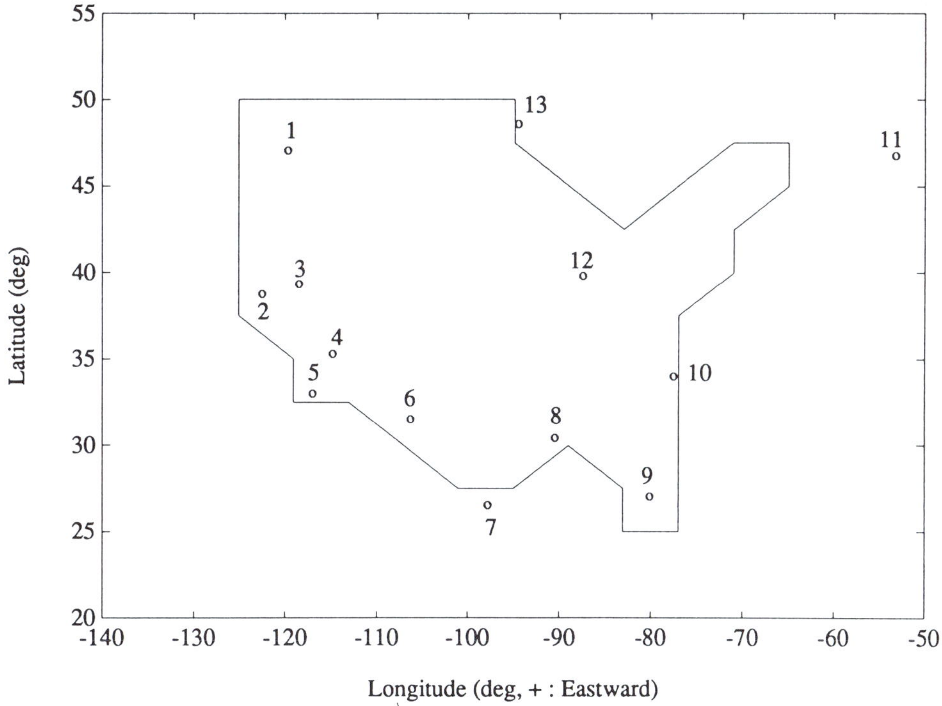

Fig. 5—Locations of Local Monitor Stations (Narrow Cape and Upolo Pt. are not shown on the map.)

Table 1—Locations of Local Monitor Stations

Location	Latitude	Longitude	Loran Site	VOR Site
1. George (WA)	47:04 N	119:45 W	√	
2. Middletown (CA)	38:47 N	122:30 W	√	
3. Fallon (NV)	39:33 N	115:50 W	√	
4. Searchlight (NV)	35:19 N	114:48 W	√	
5. San Diego (CA)	33:00 N	117:00 W		√
6. El Paso (TX)	31:30 N	106:20 W		√
7. Raymondville (TX)	26:32 N	97:50 W	√	
8. Grangeville (LA)	30:43 N	90:50 W	√	
9. Jupiter (FL)	27:02 N	80:07 W	√	
10. Carolina Beach (NC)	34:04 N	77:55 W	√	
11. Cape Race (Newfoundland, Canada)	46:47 N	53:10 W	√	
12. Dana (IN)	39:51 N	87:29 W	√	
13. Baudette (MN)	48:37 N	94:33 W	√	
14. Narrow Cape (Kodiak Is., AK) (not shown on the map)	57:26 N	152:22 W	√	
15. Upolo Pt. (HI) (not shown on the map)	20:15 N	155:53 W	√	

Each user collects pseudoranges and ephemeris parameters for all GPS satellites in view. In addition, it receives the error correction vector sent by the master station via phone link or satellite transmission. These corrections are applied directly to the raw measurements. The adjusted measurements are then used to form a least-squares solution for the user position, and clock bias.

Master Station Error Modeling

The key to WADGPS is the formulation and computation of the error correction vector by the master station. This correction consists of a three-dimensional ephemeris error and clock bias for each GPS satellite in view of one or more of the local monitor stations, plus eight ionospheric time delay parameters. These parameters are estimated based on the information gathered by the local moni-

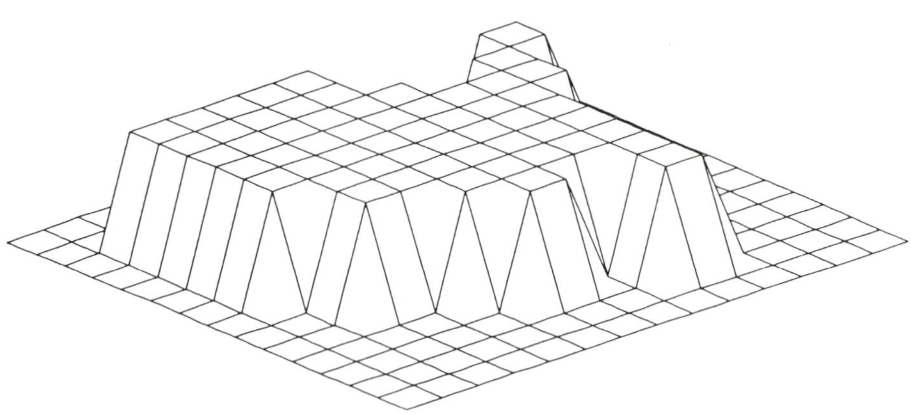

Fig. 6—Mesh Plot of the Continental U.S. (Each point on the grid represents one of the 81 simulated user's positions.)

tor stations. In addition to the error correction vector, the master station must also estimate the offset of each local monitor station clock from a set reference.

The following subsections describe the sources of error, the models used by the master station, and the techniques for estimating the model parameters. The master station computes the correction vector in a two-step process. In the first step, the parameters in the ionospheric model are identified by a nonlinear static estimation (NSE) algorithm. (We also evaluated the performance of an extended Kalman filter, but did not observe a substantial difference in the results.) The estimated ionospheric delays are then used to adjust the raw measurements from each of the local stations. The second stage solves for the ephemeris and clock errors for each of the GPS satellites observed by the network, using a batch least squares (BLS) solution.

Ionospheric Delay

As GPS satellite signals traverse the ionosphere, they are delayed by an amount proportional to the number of free ions encountered (total electron content). The ion density is a function of local time, magnetic latitude, sunspot cycle, and other factors. Its highest peak occurs at 2:00 pm local time.

Klobuchar developed a simple analytical model for ionospheric time delay, which we have used as the basis for the ionospheric correction model [4]. His model yields an RMS correction of at least 60 percent for the entire Northern Hemisphere [5]. By performing a parameter fit optimized for the United States, we can improve this accuracy.

In Klobuchar's model, the vertical ionospheric time delay is expressed by the positive portion of a cosine wave plus a constant nighttime bias, as follows [4]:

$$T_{ji} = A_1 + A_2 \cos\left[\frac{2\Pi\,(\tau - A_3)}{A_4}\right] \tag{2}$$

where T_{ji} is ionospheric time delay in vertical direction at subionospheric intersection from j-th station to i-th satellite; A_1 is 5×10^{-9} s (nighttime value); A_3 is 14:00 local time (phase); A_2 is $\alpha_1 + \alpha_2\,\phi_M + \alpha_3\,\phi_M{}^2 + \alpha_4\,\phi_M{}^3$ (amplitude); A_4 is $\beta_1 + \beta_2\,\phi_M + \beta_3\,\phi_M{}^2 + \beta_4\,\phi_M{}^3$ (period); ϕ_M is geomagnetic latitude of ionosphere subpoint; α_ι, β_ι are ionospheric parameters (**I**); and τ is local time.

In addition to the delay predicted by this model, we have included two terms in our truth model to account for higher-frequency ionospheric variations that have been observed in experimental data. The first is a sinusoidal error with amplitude of 5 percent of the cosine peak, and period of ⅕ of the cosine period of equation (2). The second is a random error with zero mean and standard deviation equal to 5 percent of the sum of the cosine terms.

Ionospheric parameters (**I**) are varying from the nominal values with time constant of 6 h, and result in a maximum vertical ionospheric time delay of 30 m and a minimum delay of 1.5 m.

A typical vertical time delay profile generated by this model is shown in Figure 7. The delay shown corresponds to an L1 signal coming from a satellite directly above the observer. To represent the actual ionospheric time delay (h_{ji}) for a given satellite elevation angle, one must scale T_{ji} by the appropriate

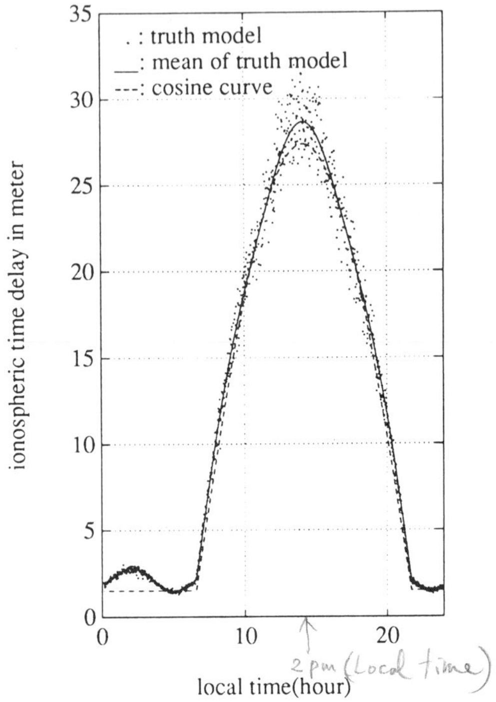

Fig. 7—Truth Model of Ionospheric Time Delay (at Stanford, elevation angle = 90 deg)

obliquity factor Q_{ji}, which is defined as the secant of the zenith angle at the mean ionospheric height, as follows [4]:

$$h_{ji} = T_{ji}(\mathbf{I}) \cdot Q_{ji}(\theta) \tag{3}$$

where τ and ϕ_M are constant at each time step, h_{ji} is ionospheric time delay from j-th station to i-th satellite, Q_{ji} is obliquity factor from j-th station to i-th satellite, θ is elevation angle, and $\mathbf{I}$ is $[\alpha_1, \ldots, \alpha_4, \beta_1, \ldots, \beta_4]^T$: ionospheric parameter. An average ionospheric height of 190 nmi was assumed in this paper.

The task of the master station is to generate the eight parameters. $[\alpha_1, \ldots, \alpha_4, \beta_1, \ldots, \beta_4]$, which, when substituted in the Klobuchar model, will yield the best ionospheric delay estimate for single-frequency users across the United States. We have investigated an NSE technique for forming these estimates.

Ionospheric Time Delay Measurement Equation

By collecting h_{ji} for satellite i = 1, . . . ,m, and station j = 1, . . . ,n, an ionospheric time delay measurement equation can be obtained:

$$\mathbf{d} = \mathbf{h}(\mathbf{I}) + \mathbf{v} \tag{4}$$

where $\mathbf{d}$ is $[d_{11}, \ldots, d_{1m} \vdots, \ldots, \vdots\, d_{n1}, \ldots, d_{nm}]^T$; d_{ji} is measured ionospheric time delay from j-th station to i-th satellite using dual-frequency technique; $\mathbf{h}(\mathbf{I})$ is $[h_{11}(\mathbf{I}), \ldots, h_{1m}(\mathbf{I}) \vdots, \ldots, \vdots\, h_{n1}(\mathbf{I}), \ldots, h_{nm}(\mathbf{I})]^T$; and $\mathbf{v}$ is measurement noise.

The linearized form of equation (4) is

$$\delta\mathbf{d} = \mathbf{H}\cdot\delta\mathbf{I} + \boldsymbol{\nu} \tag{5}$$

where

$$\mathbf{I} = \mathbf{I}_0 + \delta\mathbf{I}$$

$$\mathbf{I}_0 = [\alpha_{1_0}, \ldots, \alpha_{4_0}, \beta_{1_0}, \ldots, \beta_{4_0}]^T : \text{nominal ionospheric parameters}$$

$$\delta\mathbf{I} = [\delta\alpha_1, \ldots, \delta\alpha_4, \delta\beta_1, \ldots, \delta\beta_4]^T : \text{increment of ionospheric parameters}$$

$$\delta\mathbf{d} = \mathbf{d} - \mathbf{h}(\mathbf{I}_0)$$

$$\mathbf{H} = \left.\frac{\partial\mathbf{h}}{\partial\mathbf{I}}\right|_{\mathbf{I}=\mathbf{I}_0}$$

$$= [\mathbf{T}_{11}^T\cdot\mathbf{Q}_{11}, \ldots, \mathbf{T}_{1m}^T\cdot\mathbf{Q}_{1m} \vdots, \ldots, \vdots\ \mathbf{T}_{n1}^T\cdot\mathbf{Q}_{n1}, \ldots, \mathbf{T}_{nm}^T\cdot\mathbf{Q}_{nm}]^T$$

$$\mathbf{T}_{ji} = [T_{ji_{\alpha 1}}, \ldots, T_{ji_{\alpha 4}}, T_{ji_{\beta 1}}, \ldots, T_{ji_{\beta 4}}]^T$$

$$T_{ji_{\alpha k}} = \left.\frac{\partial \mathbf{T}_{ji}}{\partial \alpha_k}\right|_{\mathbf{I}=\mathbf{I}_0}$$

$$T_{ji_{\beta k}} = \left.\frac{\partial \mathbf{T}_{ji}}{\partial \beta_k}\right|_{\mathbf{I}=\mathbf{I}_0}$$

$$\boldsymbol{\nu} = \text{measurement noise}$$

Nonlinear Static Estimation of Ionospheric Parameters

A nonlinear static estimation technique [6] developed by Bryson was applied to the problem of fitting the ionospheric parameters to the data collected by the local monitor stations. Define the state **x** and measurement **z** as

$$\mathbf{x} = \mathbf{I} = [\alpha_1, \ldots, \alpha_4, \beta_1, \ldots, \beta_4]^T \tag{6}$$

$$\mathbf{z} = \mathbf{d} = [d_{11}, \ldots, d_{1m} \vdots, \ldots, \vdots\ d_{n1}, \ldots, d_{nm}]^T \tag{7}$$

The algorithm to find the solution may be formulated as follows:

1) Guess **x**.
2) Evaluate $\mathbf{h}(\mathbf{x})$ and $\mathbf{H} = \mathbf{h}'(\mathbf{x})$.
3) $\mathbf{P} = (\mathbf{M}^{-1} + \mathbf{H}^T\mathbf{V}^{-1}\mathbf{H})^{-1}$.
4) $\frac{\partial \mathbf{J}}{\partial \mathbf{x}} = \mathbf{M}^{-1}(\mathbf{x} - \bar{\mathbf{x}}) - \mathbf{H}^T\mathbf{V}^{-1}[\mathbf{z} - \mathbf{h}(\mathbf{x})] \equiv \mathbf{GR}$.
5) If $|\mathbf{GR}| < \epsilon$, then set $\hat{\mathbf{x}} = \mathbf{x}$ and stop.
6) Replace **x** by $(\mathbf{x} - \mathbf{P}\cdot\mathbf{GR})$.
7) Go to (2).

This algorithm is explained in greater detail in Appendix A.

Ephemeris and Satellite Clock Errors

The GPS navigation message broadcast by the satellites provides a means for computing the satellite positions in the WGS-84 coordinate system [7].

These reported positions are in error because of the limitations of the GPS control segment in predicting the satellite motions, and potentially also because of intentional degradation of the reported parameters under SA. These GPS satellite ephemeris errors can be estimated using a network of local monitor stations, essentially by using GPS upside down. In the same way a user can determine his position and clock bias based on the ranges to the known locations of four or more GPS satellites, four or more monitor stations viewing the same satellite from known locations can be used to estimate the satellite position and clock offset.

The measured pseudorange (ρ_{ji}), from monitor station (j) to GPS satellite (i) (already adjusted for ionospheric errors) is modeled by

$$\begin{aligned}\rho_{ji} &= \mathbf{D}_{ji}\cdot\mathbf{e}_{ji} + B_j' + B_i + n_{ji} \\ &= [(\mathbf{R}_i + \delta\mathbf{R}_i) - \mathbf{R}_j']\cdot\mathbf{e}_{ji} + B_j' + B_i + n_{ji}\end{aligned} \tag{8}$$

where ρ_{ji} is measured pseudorange, $\mathbf{D}_{ji}$ is range vector from station to satellite, $\mathbf{e}_{ji}$ is range unit vector, $\mathbf{R}_i$ is satellite location calculated from GPS message, $\delta\mathbf{R}_i$ is ephemeris error vector of i-th satellite, $\mathbf{R}_j'$ is known local monitor station location, B_j' is monitor station clock offset, B_i is satellite clock offset, and n_{ji} is measurement noise. This is illustrated in Figure 8.

The master station uses a batch least-squares technique to estimate the three-dimensional ephemeris error vector and clock bias for each GPS satellite within view of the network. If more than four monitor stations can see the same satellite, the observation equations for that satellite are overdetermined, and the solution is picked to minimize the measurement residual sum of squares. If fewer than four stations view the satellite (Figure 9), the solution is underdetermined, and the optimal estimate minimizes the 2-norm of the ephemeris error solution. The details of the minimum norm solution are given in Appendix B.

It is assumed that both the satellite ephemeris and clock errors are slowly varying, so that a constant estimate of each is sufficiently accurate for the

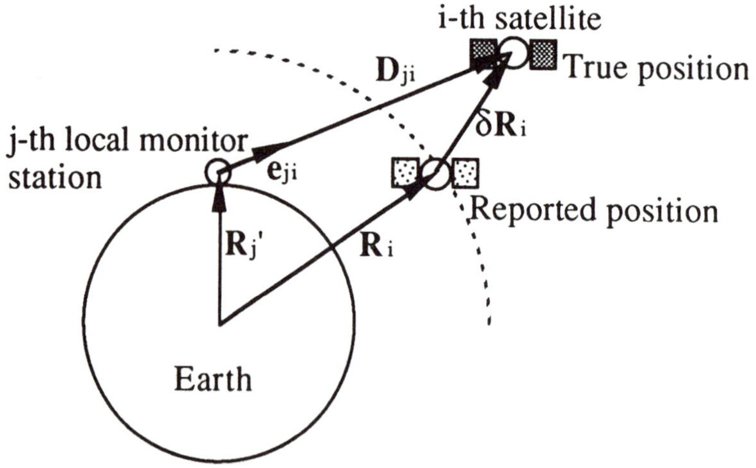

Fig. 8—GPS Ephemeris Errors

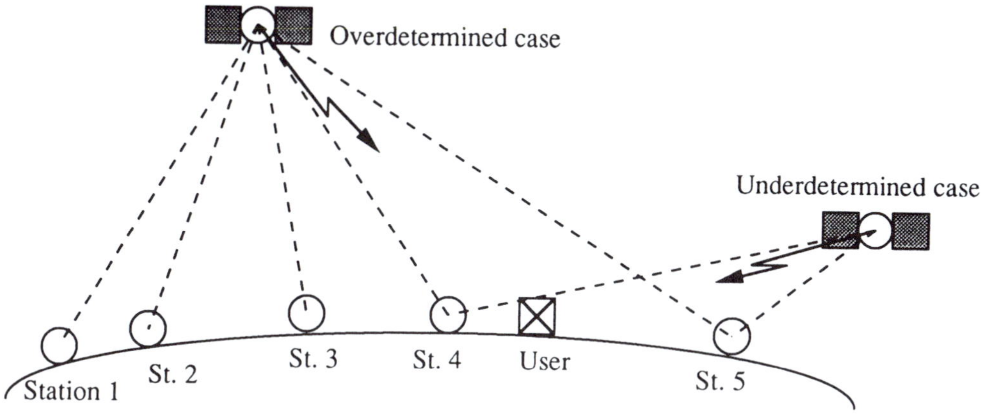

Fig. 9—Example of Overdetermined and Underdetermined Cases of Estimating Ephemeris Errors

WADGPS error correction vector. Ephemeris and clock errors have been observed to have variations with time constants on the order of 30 and 3 min, respectively; thus with a proposed update rate of 0.2 Hz, we believe this assumption to be valid.

SIMULATION

The performance of the WADGPS network described was evaluated using a computer simulation. This simulation comprises four modules describing the GPS satellites, the local monitor stations, the master station, and the users. A block diagram is shown in Figure 10. The truth model error specifications are listed in Table 2. The simulation was run for 12 h starting at 6:00 am Pacific Standard Time (PST).

GPS Satellite Module

The GPS 21 Primary Satellite Constellation is modeled in the simulation [8]. The ephemeris reported by the GPS module to the local monitor stations and the users is equal to the true ephemeris corrupted by an error vector. Each ephemeris error vector component is produced by passing white noise through a first-order shaping filter with time constant of 1800 s and standard deviation of 20 m [9].

Each satellite clock offset is also modeled by white noise input to a first-order shaping filter, this time with $\tau = 200$ s [10], and standard deviation of 30 m. These values are somewhat conservative to account for possible effects due to SA.

The ionospheric delay is modeled according to the Klobuchar model, plus the random elements described earlier.

Local Monitor Station Module

The local monitor station module generates the pseudorange measurements and ionospheric delays observed by the monitor receivers. The local monitor

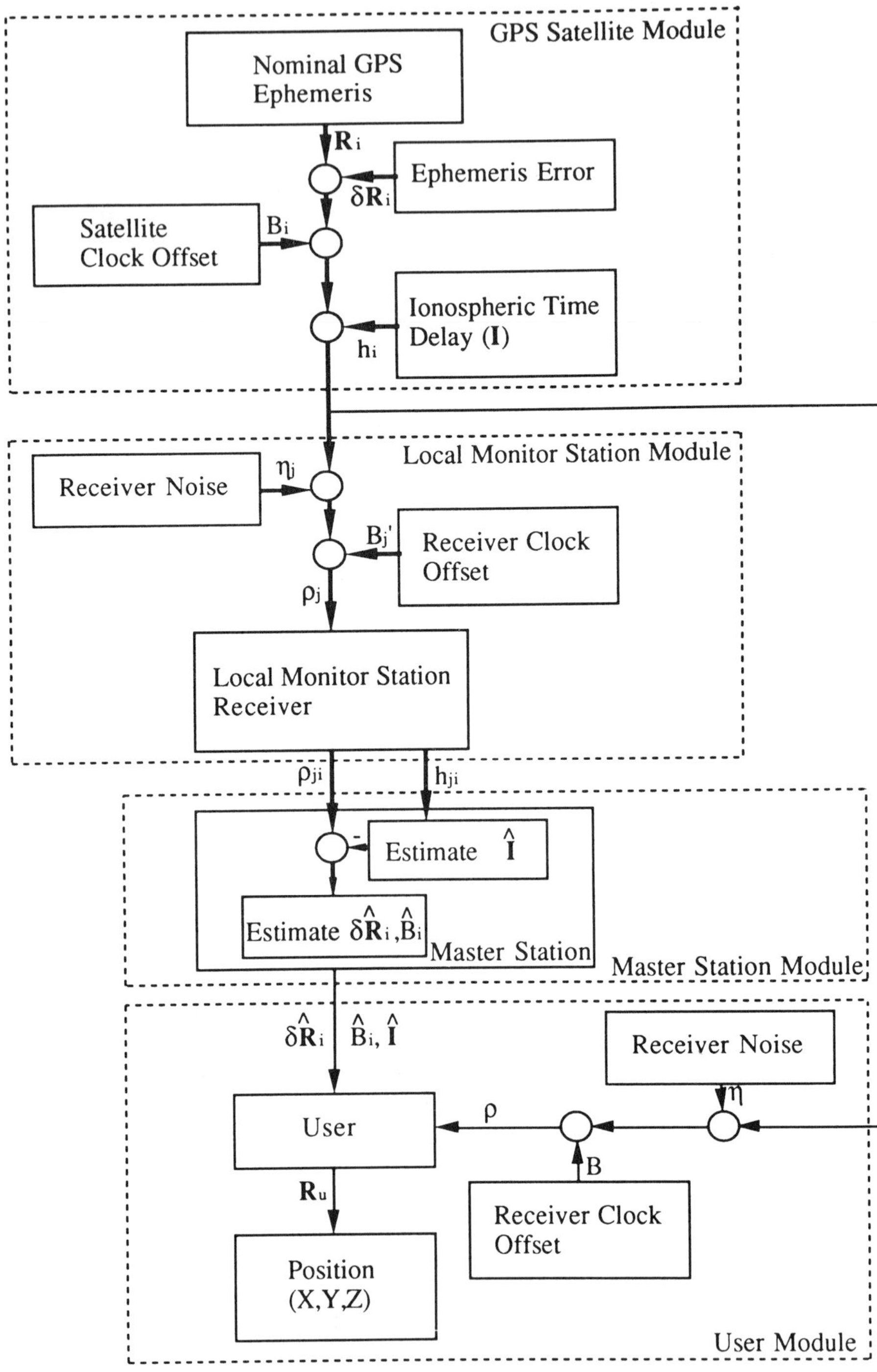

Fig. 10—Block Diagram of WADGPS Computer Simulation

Table 2—Truth Model Error Specifications

Error Source	Error Model	Time Constant	Min. (m)	Max. (m)	RMS (m)
3-D Satellite Ephemeris Errors	1st-order Markov process	1800 s			20
Satellite Clock Offset	1st-order Markov process	200 s			30
Ionospheric Time Delay*	Klobuchar's model, plus spatial sinusoidal bias and white noise	6 h	1.5 at zenith	30 at zenith	
Tropospheric Error	modeled as receiver noise**				
Local Monitor Station Receiver Clock Offset	2nd-order Markov process	$h_0 = 1.8 \times 10^{-22}$ $h_{-1} = 6.5 \times 10^{-26}$ $h_{-2} = 1.37 \times 10^{-33}$			
User Receiver Clock Offset	2nd-order Markov process	$h_0 = 9.4 \times 10^{-20}$ $h_{-1} = 1.8 \times 10^{-19}$ $h_{-2} = 3.8 \times 10^{-21}$			
Receiver Noise	white noise				0.2
Multipath	modeled as receiver noise				

*Ionosphere is varying with time constant of 6 h, and 5 percent of ⅕ period spatial sinusoidal bias and 5 percent of white noise were added.
**A new sophisticated model is under development.

station receiver clock error is modeled by white noise input to a second-order Markov process based on [11], as follows:

$$\begin{bmatrix} x_1 \\ x_2 \end{bmatrix}_{k+1} = \begin{bmatrix} 1 & \tau \\ 0 & 1 \end{bmatrix} \begin{bmatrix} x_1 \\ x_2 \end{bmatrix}_k + \begin{bmatrix} w_1 \\ w_2 \end{bmatrix}_k \tag{9}$$

$$\mathbf{Q} = E\{\mathbf{w}\ \mathbf{w}^T\} = \begin{bmatrix} Q_{11} & Q_{12} \\ Q_{12} & Q_{22} \end{bmatrix}$$

$$Q_{11} = \frac{h_0}{2}\tau + 2\,h_{-1}\,\tau^2 + \frac{2}{3}\pi^2\,h_{-2}\,\tau^3$$

$$Q_{12} = 2\,h_{-1}\,\tau + \pi^2\,h_{-2}\,\tau^2$$

$$Q_{22} = \frac{h_0}{2\tau} + 2\,h_{-1} + \frac{8}{3}\pi^2\,h_{-2}\,\tau$$

$$h_0 = 1.8\times10^{-22}$$

$$h_{-1} = 6.5\times10^{-26}$$

$$h_{-2} = 1.37\times10^{-33},$$

where x_i is clock offset (second), x_2 is average frequency, τ is sampling time, and the receiver clock offset values are based on a typical rubidium standard [12].

The receiver noise is assumed to be white with zero mean and standard deviation of 0.2 m. This is based on averaging over 10 measurements at 1 s intervals. The local monitor station locations were shown in Figure 5.

Master Station Module

The master station module collects inputs from the local monitor station module, and implements the ionospheric and ephemeris error estimation algorithms described in the previous section. Ionospheric delay parameters and estimated ephemeris and clock errors are provided to the user module.

User Module

The user module simulates the operation of the user receiver. The user clock error is assumed to be white noise input to a second-order Markov process based on equation (9), with $h_0 = 9.4 \times 10^{-20}$, $h_{-1} = 1.8 \times 10^{-19}$, $h_{-2} = 3.8 \times 10^{-21}$, and the receiver clock offset values are based on a typical quartz standard.

The receiver noise is assumed to be white with zero mean and standard deviation of 0.2 m. This is based on averaging over 10 measurements at 1 s intervals.

The user applies the eight parameters in the Klobuchar model to the raw pseudorange and adjusts the ephemeris parameters received from the GPS satellite module by the correction vector sent by the master station. He then forms a least-squares position solution using measurements to all the satellites within his field of view. The performance of the WADGPS is evaluated by comparing the error in this solution with the error that would have been obtained by using the raw measurements directly.

SIMULATION RESULTS AND DISCUSSION

A 12 h simulation was run starting at 6:00 am PST and ending at 6:00 pm PST. The NSE technique was used to determine the ionospheric delay parameters. In this section, we present and discuss the results in terms of ionospheric model estimation accuracy, and navigation performance.

Ionospheric Model Accuracy

The first step in evaluating the performance of the WADGPS is to see how well it determines the ionospheric errors. Figures 11 and 12 show contour plots of the actual and estimated vertical ionospheric delays, superimposed on a map of the United States. These represent snapshots from the 12 h simulation at 5:00 pm PST and 2:00 pm PST, respectively. The contour lines are labeled by the ionospheric delay in 3 m increments. Notice that the actual values of the ionospheric delay increase from east to west as we get closer to local noon.

In the 5:00 pm plot we can see that the NSE algorithm does well at estimating the delay since the estimated contours are within 1.5 m of the true error contours. The performance results are summarized in Table 3.

Figure 11 also shows the improvement in ionospheric delay estimates toward the center of the country. This can be attributed to the larger number of local monitor stations that can observe satellite signals passing through the central

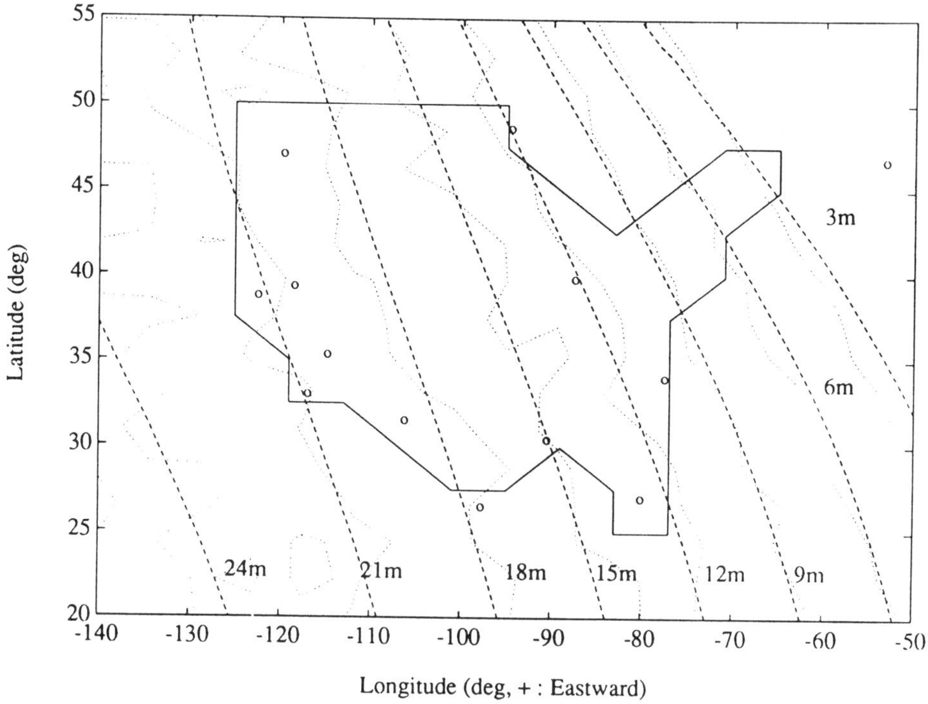

Fig. 11—Ionospheric Time Delay Estimates, 5:00 pm PST (This figure shows the vertical ionospheric delay in meters as generated by the truth model and the NSE. A map of the United States and the monitor station locations is also shown. The dotted line is the true delay contour; the dashed line is the NSE estimate.)

Table 3—RMS and Maximum Errors in Ionospheric Estimates

PST Local Time		7:00–8:00 am	10:00–11:00 am	1:00–2:00 pm	4:00–5:00 pm
Nonlinear Static	max* (m)	1.0	1.4	1.3	1.3
Estimation	RMS** (m)	0.3	0.5	0.5	0.4

*max (abs($z_i - \hat{z}_i$)), where z_i is true vertical ionosphere measurement, and $\hat{z}_i$ is estimated value.
**RMS ($z_i - \hat{z}_i$).

region as compared with satellites in the far eastern or western parts of the sky.

Figure 12 shows the same type of results for 2:00 pm PST. In this case, however, we notice local contours of varying heights that are not estimated. These small areas of variation in the ionospheric delay are generated by the random noise introduced in the truth model. Because the standard deviation of this variation is set at 5 percent of the nominal value for the local time of day, the maximum random error is as much as 4.5 m at 2:00 pm. There is no way for the WADGPS network model to estimate these random, high-frequency, localized components of the ionospheric error.

Navigation Performance

The objective of the WADGPS system is to improve the navigation accuracy for users across CONUS. The simulation results indicate that this goal can be achieved using the proposed system.

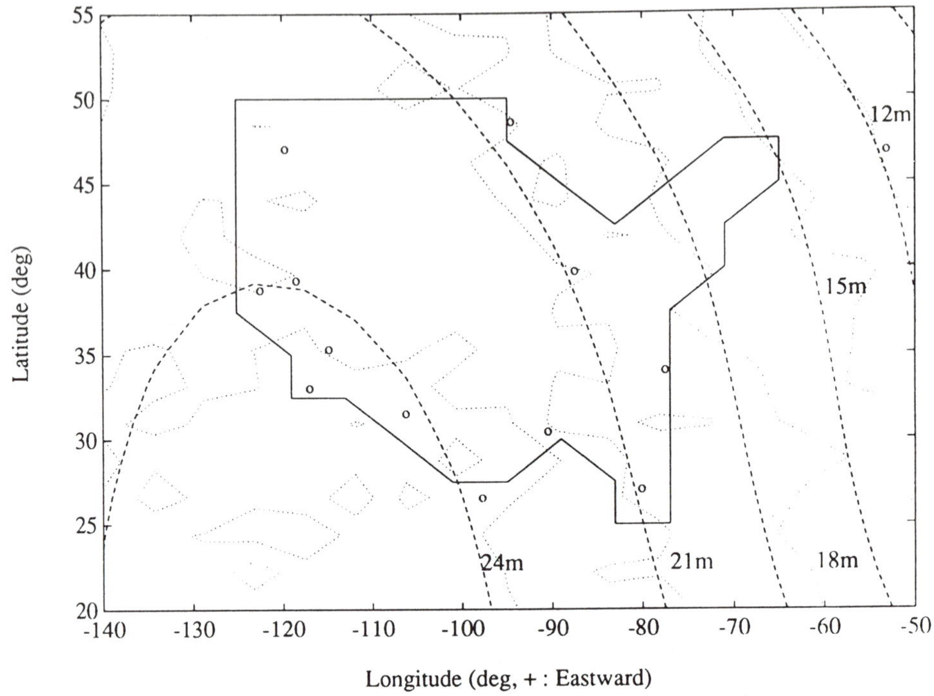

Fig. 12—Ionospheric Time Delay Estimates, 2:00 pm PST (This figure shows the vertical ionospheric delay in meters as generated by the truth model and the NSE. A map of the United States and the monitor station locations is also shown. The dotted line is the true delay contour; the dashed line is the NSE estimate.)

Figures 13 through 18 provide a very compact summary of the simulation results. The (a) and (b) sets of figures show mesh plots and contour plots of the error distributions across the country. In the mesh plots, (a), each grid point within the outline of the United States represents one of the 81 user locations we considered. The height of the grid point above the surface corresponds to the magnitude of the error at the grid location. These heights reflect the RMS or maximum error for that user over the entire 12 h simulation period. The (b)

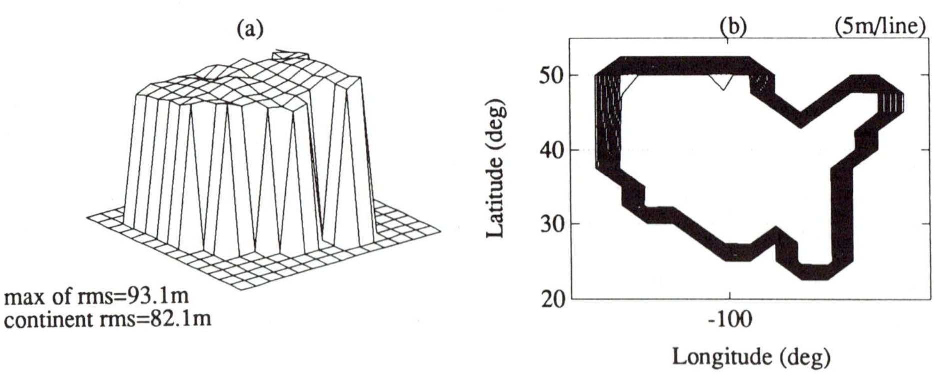

Fig. 13—RMS Value of Vertical Positioning Error Without Correction

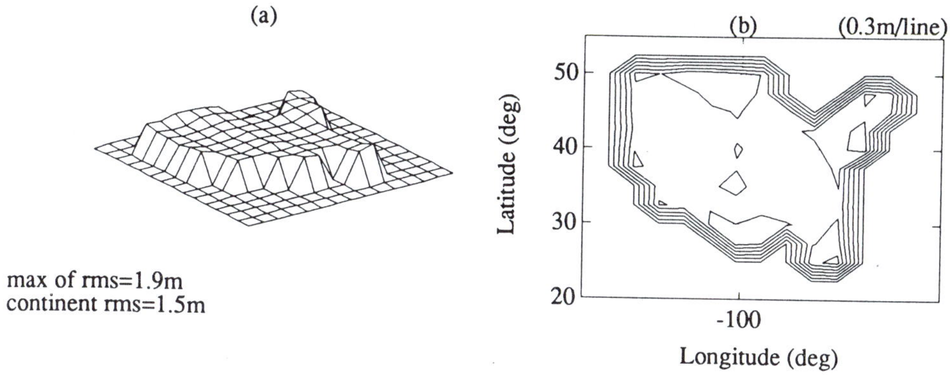

Fig. 14—RMS Value of Vertical Positioning Error with Correction

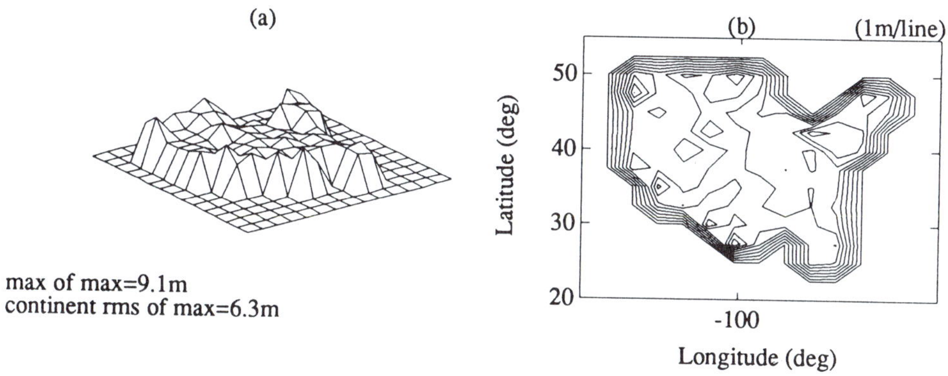

Fig. 15—Maximum Vertical Positioning Error with Correction

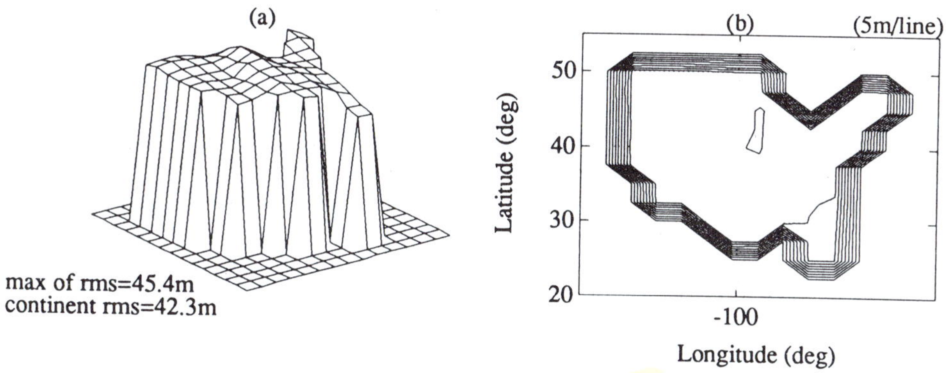

Fig. 16—RMS Value of Horizontal Positioning Error Without Correction

figures are contour plots that show lines of constant error magnitude corresponding to the mesh plots on the left.

Uncorrected Navigation Performance

Figures 13 and 16 show the uncorrected navigation performance of typical users across the United States. The RMS of the positioning errors for all user

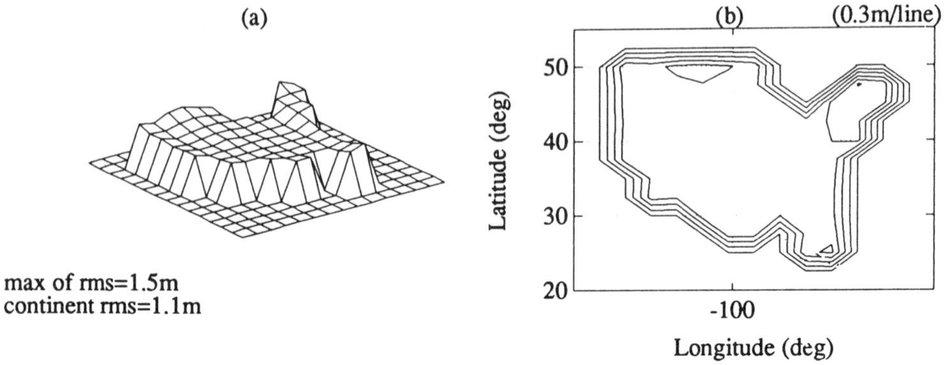

Fig. 17—RMS Value of Horizontal Positioning Error with Correction

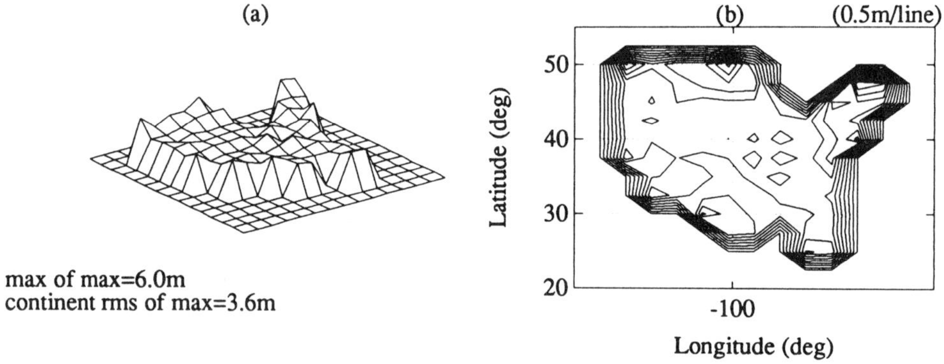

Fig. 18—Maximum Horizontal Positioning Error with Correction

locations over the entire 12 h period is 82 m in the vertical direction and 42 m in the horizontal direction. As one might expect, the error magnitudes are fairly uniform over the entire area. As is common in GPS navigation, the vertical error is approximately two times larger than the horizontal error because of the larger geometric dilution of precision in the vertical direction (VDOP).

Navigation Performance with WADGPS

Figures 14–15 and 17–18 show the significantly improved navigation accuracy achieved using WADGPS with the NSE ionospheric estimation algorithm, and ephemeris and clock bias BLS algorithm. Continental RMS averages of the vertical and horizontal position errors have been reduced to 1.5 m and 1.1 m, respectively. Figures 15 and 18 show the maximum values of the errors. The largest vertical and horizontal errors anywhere in the United States over the entire 12 h period are 9.1 m and 6.0 m, respectively. Table 4 lists the simulation results for the NSE.

One of the most striking features of the plots of WADGPS corrected errors is the concave shape of the error mesh. In general, the navigation performance in the center of the United States is better than along the coasts. This is because the satellites viewed by users in this region are also visible from a large number of local monitor stations, and with better geometry (lower GDOP) than their coastal counterparts.

One exception to this observation occurs in the southwestern United States. This region exhibits better positioning accuracies than other edges of the country because of the high density of monitor stations (Figure 14–15 and 17–18). Likewise, the north central part of the country is noticeably worse than average because of the relative sparsity of monitor stations.

Ionospheric Effect on Navigation Performance

The ionospheric delay, and our ability to model it in the master control station algorithm, has a profound effect on the navigation performance using WADGPS. Figure 19 illustrates the RMS vertical and horizontal positioning errors over all users, as a function of time. The peak errors occur between 12:00 and 1:00 pm PST—the time of day when the largest ionospheric delays, accompanied by the greatest variations in the delay, occur over the center of the WADGPS network.

Another indication of the influence of ionospheric errors and their estimation is given by the greater improvement in RMS vertical positioning accuracy due to WADGPS, as compared with the horizontal direction. This is because measurements from lower on the horizon have a larger obliquity factor, and consequently an amplified random error component. Thus, since the unmodeled random part is larger, the horizontal positioning error is not improved as much as the vertical positioning error by the ionospheric correction.

If ephemeris errors were the only source of error included in the truth model, the horizontal positioning errors would be reduced by a much greater percentage than the vertical errors. This is because the viewing geometry between the monitor stations and the satellites is stronger in the horizontal plane. When ionospheric delays are introduced, the corrected vertical error is reduced to 1.9 percent of its uncorrected value, whereas the horizontal error is reduced to 2.5 percent.

Table 4—Summary of Positioning Errors

			RMS Value of Positioning Error Without Correction (m)	RMS Value of Positioning Error With Correction (m)	(With Error Correction)/(Without Error Correction) (%)	Max. Value of Positioning Error With Correction (m)
Nonlinear	vertical	max*	93.1	1.9	2.4	9.1
Static		RMS**	82.1	1.5	1.9	6.3
Estimation	horiz.	max*	45.4	1.5	3.5	6.0
		RMS**	42.3	1.1	2.5	3.6

*continent maximum.
**continent RMS value.

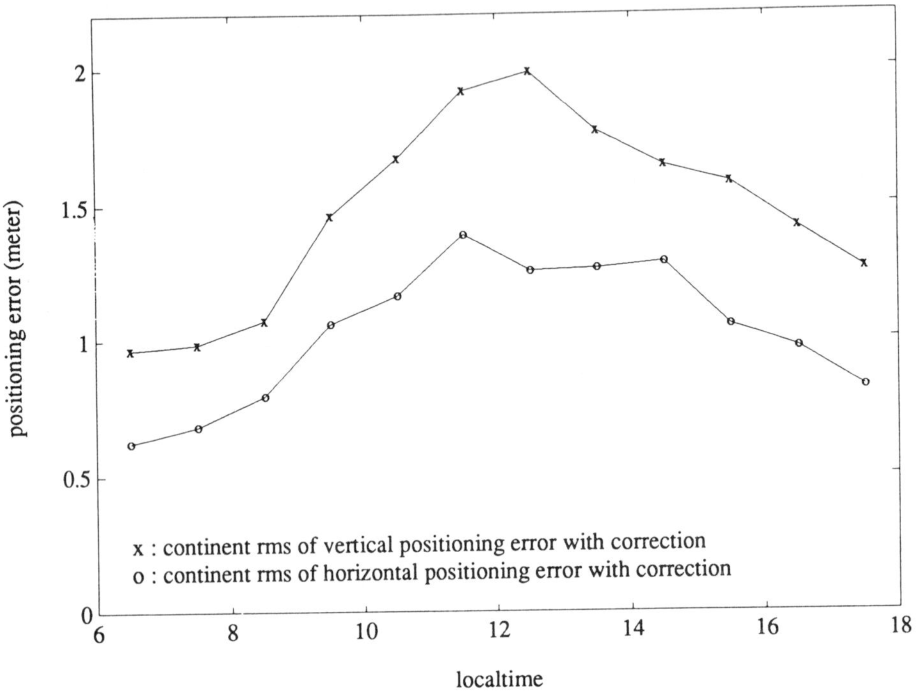

Fig. 19—Continental RMS Values of Vertical and Horizontal Positioning Errors for Each Hour Period (with correction, 6:00 am–6:00 pm)

Summary of Results

These results indicate that WADGPS can provide accurate ionospheric delay estimates, as well as positioning errors typically on the level of standard DGPS operations. The largest errors occur during the peak of ionospheric activity in the center of the WADGPS network.

SUMMARY AND FUTURE WORK

A concept for Wide Area Differential GPS (WADGPS) has been proposed. This technique makes use of a network of 15 local monitor stations distributed across the United States to improve navigation accuracy of users within the United States. A central WADGPS master station collects pseudoranges and ionospheric delay measurements from the local stations and generates an error correction vector to be broadcast to users. This vector comprises a three-dimensional ephemeris error and clock bias for each satellite observed by the network, and estimates of the eight parameters of the Klobuchar ionospheric model optimized for use in the United States.

Simulation results over a 12 h period indicate that stand-alone GPS positioning errors can be reduced by over 95 percent using WADGPS without degradation because of separation between the monitor station and users. The largest errors occur during the peak of ionospheric activity in the center of the WADGPS network.

In the future, we plan to evaluate the performance of the system by using actual GPS data collected at a variety of locations around the country. This will permit us to investigate realistically the effects of data latency and modeling errors, especially in the presence of Selective Availability (SA). Improvements to the estimation algorithms will also be explored, for example, separation of the daytime and nighttime ionospheric delay models.

Based on a paper presented at ION GPS-90, Third International Technical Meeting of the Satellite Division of The Institute of Navigation, Colorado Springs, Colorado, September 1990.

REFERENCES

1. Kremer, G. T., Kalafus, R. M., Loomis, P. V. W., and Reynolds, J. C., *The Effect of Selective Availability on Differential GPS Corrections,* NAVIGATION, Journal of The Institute of Navigation, Vol. 37, No. 1, Spring 1990, pp. 39–52.
2. Brown, A., *Extended Differential GPS,* NAVIGATION, Journal of The Institute of Navigation, Vol. 36, No. 3, Fall 1989, pp. 265–85.
3. Loomis, P. V. W., Denaro, R. P., and Saunders, P., *Worldwide Differential GPS for Space Shuttle Landing Operations,* IEEE Position Location and Navigation Symposium, Las Vegas, NV, March 1990.
4. Klobuchar, J., *Design and Characteristics of the GPS Ionospheric Time Delay Algorithm for Single Frequency Users,* IEEE Position Location and Navigation Symposium, Las Vegas, NV, November 1986, pp. 169–76.
5. Feess, W. A. and Stephens, S. G., *Evaluation of GPS Ionospheric Time-Delay Model,* IEEE Transactions on Aerospace and Electronic Systems, Vol. AES-23, No. 3, May 1987, pp. 332–38.
6. Bryson, A. E., Lecture Notes, *Optimal Estimation and Control Logic in the Presence of Noise* (Stanford Course AA278B), Spring 1989.
7. *NAVSTAR GPS Space Segment/Navigation User Interfaces,* Rockwell International Corporation, ICD-GPS-200, November 30, 1987.
8. Green, G. B., Massatt, P. D., and Rhodus, N. W., *The GPS 21 Primary Satellite Constellation,* NAVIGATION, Journal of The Institute of Navigation, Vol. 36, No. 1, Spring 1989, pp. 9–24.
9. Well, D., *Guide to GPS Positioning,* Canada GPS Associates, 1987.
10. Chou, H., *An Anti-SA Filter for Non-differential GPS Users,* Proceedings of ION GPS-90, Colorado Springs, CO, September 1990, pp. 535–42.
11. Van Dierendonck, A. J., McGraw, J. B., and Brown, R. G., *Relationship Between Allan Variances and Kalman Filter Parameters,* Proceedings of the 16th Annual PTTI Applications and Planning Meeting, MD, November 1984, pp. 273–93.
12. Efratom Division, *Precision Time and Frequency Handbook,* 6th edition, 1987.

APPENDIX A

NONLINEAR STATIC ESTIMATION

Given $\mathbf{x} = N(\bar{\mathbf{x}}, \mathbf{M})$, $\mathbf{z} = \mathbf{h}(\mathbf{x}) + \mathbf{v}$, $\mathbf{v} = N(\mathbf{0}, \mathbf{V})$, and $\mathbf{M}, \mathbf{V}$ are symmetric covariance matrices.

Find $\mathbf{x}$ which minimizes,

$$\mathbf{J} = \frac{1}{2}(\mathbf{x} - \bar{\mathbf{x}})^T \mathbf{M}^{-1}(\mathbf{x} - \bar{\mathbf{x}}) + \frac{1}{2}[\mathbf{z} - \mathbf{h}(\mathbf{x})]^T \mathbf{V}^{-1}[\mathbf{z} - \mathbf{h}(\mathbf{x})].$$

Solution:

Define $\mathbf{H} = \mathbf{h}'(\mathbf{x})$.

$$\begin{aligned} \mathbf{J} &= \mathbf{J}_0 + d\mathbf{J} \\ d\mathbf{J} &= (\mathbf{x} - \bar{\mathbf{x}})^T \mathbf{M}^{-1} d\mathbf{x} + [\mathbf{z} - \mathbf{h}(\mathbf{x})]^T \mathbf{V}^{-1}[-\mathbf{h}'(\mathbf{x})]\, d\mathbf{x} \\ &\quad + \frac{1}{2} d\mathbf{x}^T \left[\mathbf{M}^{-1} + \mathbf{H}^T \mathbf{V}^{-1} \mathbf{H} - [\mathbf{z} - \mathbf{h}(\mathbf{x})]^T \mathbf{V}^{-1}[\mathbf{h}''(\mathbf{x})]\right] d\mathbf{x} + \text{h.o.t.} \end{aligned} \tag{A1}$$

If $[\mathbf{z} - \mathbf{h}(\mathbf{x})]^T \mathbf{V}^{-1}[\mathbf{h}''(\mathbf{x})]$ and h.o.t. are small,

$$\begin{aligned} d\mathbf{J} &= (\mathbf{x} - \bar{\mathbf{x}})^T \mathbf{M}^{-1} d\mathbf{x} + [\mathbf{z} - \mathbf{h}(\mathbf{x})]^T \mathbf{V}^{-1}[-\mathbf{h}'(\mathbf{x})]\, d\mathbf{x} \\ &\quad + \frac{1}{2} d\mathbf{x}^T [\mathbf{M}^{-1} + \mathbf{H}^T \mathbf{V}^{-1} \mathbf{H}]\, d\mathbf{x} \end{aligned} \tag{A2}$$

Since the solution $\mathbf{x}$ can not be expressed in analytic form, $\mathbf{x}$ should be found numerically. $\mathbf{J}$ can be minimized by making equation (A2) a perfect square in $d\mathbf{x}$. In each numerical step, $d\mathbf{x}$ should be found to make $\frac{\partial \mathbf{J}}{\partial \mathbf{x}} = 0$, i.e., to minimize $\mathbf{J}$.

The algorithm for nonlinear static estimation is

1) Guess $\mathbf{x}$.
2) Evaluate $\mathbf{h}(\mathbf{x})$ and $\mathbf{H} = \mathbf{h}'(\mathbf{x})$.
3) $\mathbf{P} = (\mathbf{M}^{-1} + \mathbf{H}^T \mathbf{V}^{-1} \mathbf{H})^{-1}$.
4) $\frac{\partial \mathbf{J}}{\partial \mathbf{x}} = \mathbf{M}^{-1}(\mathbf{x} - \bar{\mathbf{x}}) - \mathbf{H}^T \mathbf{V}^{-1}[\mathbf{z} - \mathbf{h}(\mathbf{x})] \equiv \mathbf{GR}$.
5) If $|\mathbf{GR}| \langle \epsilon$, then set $\hat{\mathbf{x}} = \mathbf{x}$ and stop.
6) Replace $\mathbf{x}$ by $(\mathbf{x} - \mathbf{P} \cdot \mathbf{GR})$.
7) Go to (2).

APPENDIX B

MINIMUM NORM PROBLEM

Given $\mathbf{Ax} = \mathbf{b}$, where $\mathbf{A}$ is m by n matrix ($m \leq n$) and has full rank, find the solution $\mathbf{x}$ which satisfies the above linear system and minimizes $\mathbf{J} = \frac{1}{2}\mathbf{x}^T\mathbf{x}$.

Solution: $\mathbf{x} = \mathbf{A}^T(\mathbf{A}\,\mathbf{A}^T)^{-1}\mathbf{b}$

Proof:

Define Hamiltonian $\mathbf{H} = \frac{1}{2}\mathbf{x}^T\mathbf{x} + \boldsymbol{\lambda}^T(\mathbf{b} - \mathbf{Ax})$ where $\boldsymbol{\lambda}^T$ is a Lagrangian multiplier. To get a mininum norm solution, $\mathbf{x}$ should satisfy two equations,

$$\frac{\partial \mathbf{H}}{\partial \mathbf{x}} = 0,$$

$$\frac{\partial \mathbf{H}}{\partial \boldsymbol{\lambda}} = 0.$$

Those equations give

$$\mathbf{x} = \mathbf{A}^T\boldsymbol{\lambda}, \tag{B1}$$

$$\mathbf{Ax} = \mathbf{b}. \tag{B2}$$

By substituting equation (B1) into equation (B2), we obtain

$$\mathbf{A}\,\mathbf{A}^T\boldsymbol{\lambda} = \mathbf{b}.$$

Because $(\mathbf{A}\,\mathbf{A}^T)$ has full rank, the inverse of $(\mathbf{A}\,\mathbf{A}^T)$ exists. Premultiplying by the inverse of $(\mathbf{A}\,\mathbf{A}^T)$ gives

$$\boldsymbol{\lambda} = (\mathbf{A}\,\mathbf{A}^T)^{-1}\mathbf{b}. \tag{B3}$$

Substituting equation (B3) into equation (B1) gives the desired minimum norm solution,

$$\mathbf{x} = \mathbf{A}^T(\mathbf{A}\,\mathbf{A}^T)^{-1}\mathbf{b}.$$

NAVIGATION: *Journal of The Institute of Navigation*
Vol. 36, No. 1, Spring 1989
Printed in U.S.A.

High-Precision Kinematic GPS Differential Positioning and Integration of GPS with a Ring Laser Strapdown Inertial System

GÜNTER W. HEIN, GERALD BAUSTERT, BERND EISSFELLER, and HERBERT LANDAU

University FAF Munich, Neubiberg, F.R. Germany
Received September 1988
Revised January 1989

ABSTRACT

Many fields, such as geodetic surveying, positioning of airborne sensors (photogrammetric cameras), laser bathymetry, airborne gravimetry, and hydrographic surveys, require accuracies in the decimeter or even centimeter range in differential positioning. Using the full potential of the GPS in a dynamic environment, the Institute of Astronomical and Physical Geodesy (IAPG) has carried out several kinematic tests with respect to land-based, shipborne, and airborne applications.

After reviewing the main concepts in differential GPS kinematic positioning, this paper presents experiences with and practical results of kinematic GPS tests. Results of a Kalman filter covariance analysis are shown for the integration of GPS phase and pseudorange data with ring laser strapdown inertial systems in order to recover cycle slips and short-time loss of lock in GPS. In particular, the question is discussed of the extent to which inertial strapdown systems can be used to bridge these times and to maintain centimeter accuracy.

INTRODUCTION

Geodesy and surveying are presently in a transition state with respect to positioning. Although the GPS Block II satellites are not yet in orbit, experiences in the past years have led to a considerable replacement of theodolites and electronic distance meters by GPS (phase) tracking receivers. A second change is also becoming noticeable. In many cases, the static GPS observation scenario can be replaced by a kinematic differential positioning technique without losing the high precision in the centimeter range.

Moreover, new applications in geodesy and surveying are becoming possible, in which sensors have to be positioned during motion. Examples are the determination of 3-D coordinates of a photogrammetric airborne camera during exposure, laser digital profiling, or airborne gravimetry where the effect of the acceleration of the aircraft on the gravimeter has to be determined. Even satellite orbit determination can be done via GPS observations between the GPS satellites and the satellite being positioned. Moreover, geodetic kinematic

GPS techniques can be used for precise aircraft landing approaches of Category II (CAT II) or even CAT III. In marine applications, GPS can be used in bathymetric surveying (echo sounding) to achieve an accuracy of 1 m in horizontal position and 0.3 m in vertical position.

Actually, Remondi [1–3] was the first to present results of a terrestrial kinematic survey, in which he occupied stations of a small network for a few seconds, instead of 15 min to some hours as is done in the stationary mode. In fact, we must differentiate between kinematic surveying methods in which the receiver is in motion (no stop possible) and those (as in terrestrial land surveying) in which a stop of a few minutes (~ 5 min) is still required [4]. Methods of the last type are excluded from this paper because they are categorized according to short-time static observations and not kinematic surveying. In the last two years, many results have been reported in marine kinematic positioning and other fields, using a combination of pseudorange and phase measurements [5–14]. Airborne applications for aerotriangulation without ground control have also been investigated [15–20].

The discussion below first reviews the main methods in differential GPS positioning and the cycle slip problem in phase observations. It then describes the experiences of the Institute of Astronomical and Physical Geodesy (IAPG) with terrestrial, shipborne, and airborne applications. Finally, results from a feasibility study of an integration of GPS with a ring laser gyro strapdown inertial system for high-precision kinematic surveying are discussed.

GPS DIFFERENTIAL HIGH-PRECISION NAVIGATION AND KINEMATIC SURVEYING

There may be only one difference between high-precision navigation and kinematic surveying. In the latter, a real-time solution is in most cases not necessary. However, growing needs of the customer and new geodetic applications may also change this picture in the future.

Depending on the level of accuracy required, three main concepts can be used. These are discussed below.

Differential Position Corrections

The principle is simple. Differences between known and GPS-determined position or related parameters of a permanently tracking receiver at a base station are transmitted to the nearby moving receiver, where they are used to improve the (navigation) algorithm. The observables are pseudoranges. Accuracies of the moving receiver are in the range of 5 to 15 m for the Standard Positioning Service (SPS) and 3 to 5 m for the Precise Positioning Service (PPS).

The transmitted differences can be of the following form:

1) *Errors in ellipsoidal coordinates (φ,λ,h) between known and GPS-determined position*—The disadvantage of this approach is that both receivers, stationary and moving, have to track the same satellites.
2) *Delta-range corrections*—Errors in pseudoranges to all visible satellites are determined and transmitted. The advantage is that the moving receiver can track different satellites since it obtains delta-range corrections to all

satellites (depending only on the number of available channels of the static receiver).

3) *Pseudolite corrections*—A pseudolite is merely a simulated GPS satellite on the ground. The corresponding instrument receives GPS signals and computes pseudorange and range-rate corrections, which are transmitted at 50 bps on L-band frequency. This transmitted signal is GPS-like. The local user equipment can obtain additional pseudorange measurements. This concept has several advantages: the baseline to the pseudolite is very precise; the moving receivers do not need a particular hardware component to receive the corrections, as in approaches (1) and (2); and no additional channel for communication is necessary. On the contrary, the stationary working receiver has to fulfill functions similar to those of a satellite, and thus needs appropriate hardware. In addition, visibility between the moving and stationary receivers is required. To our knowledge, this concept has until now not been realized.

Phase-Smoothed Pseudoranges (Carrier-Smoothed Code)

With the differential position correction concept, described above, only pseudoranges are used as observations. In contrast, this concept requires receivers capable of observing carrier phases as well. A smoothed pseudorange at epoch t_i is a linear combination of a pseudorange at t_i with weight w_1 and the smoothed pseudorange at t_{i-1} calculated for epoch t_1, using the phase difference over $(t_i - t_{i-1})$ with weight w_2 ($w_1 + w_2 = 1$). The weight scheme looks like this:

Start of observation:	$w_1 = 1.00$	$w_2 = 0.00$
.	.	.
.	.	.
.	.	.
After approx. 100 measurements	$w_1 = 0.01$	$w_2 = 0.99$

If a cycle slip occurs in the phase observations, the weights are reset to

$$w_1 = 1.00 \qquad w_2 = 0.00$$

Cycle slips have to be detected, but not corrected. Several modifications in the algorithm above are possible, such as wide laning [21]. The accuracy of the method is on the order of 1 to 3 m for SPS and 0.5 to 1 m for PPS. Whereas differential position correction is seldom applied in geodesy and surveying because of its low accuracy, many experiences with phase-smoothed pseudoranges, especially for marine applications, are reported in the geodetic literature [5–8, 10–13, 21–24].

Phase Differencing

As in the stationary application of GPS in surveying and geodesy, the high accuracy potential of GPS in the centimeter (or even millimeter) range can be realized only when using carrier phases. Here also the term kinematic surveying may have its origin. As mentioned in the introduction, Remondi first showed that the kinematic case can in principle use the same algorithms as those used in stationary baseline determination.

Consider a double difference $\phi_{i\,j}^{1\,2}(t)$ between two satellites and two stations:

$$\phi_{i\,j}^{1\,2}(t) = -\frac{f}{c}(\rho_i^1 - \rho_i^2 - \rho_j^1 + \rho_j^2) + N_{i\,j}^{1\,2} \tag{1}$$

where

$$\phi_{i\,j}^{1\,2}(t) = \phi_i^1 - \phi_i^2 - \phi_j^1 + \phi_j^2 \tag{2}$$

and Φ_i^k is the phase observable from station i to satellite k, ρ_i^k is the distance from station i to satellite k, f is the carrier frequency, c is the speed of light, and $N_{i\,j}^{1\,2}$ is the integer double-difference bias.

Once $N_{i\,j}^{1\,2}$ is known, $\Phi_{i\,j}^{1\,2}$ is only a difference of ranges, allowing precise determination of position. Thus, the task is to solve for the initial double-difference integer unknown $N_{i\,j}^{1\,2}$ at lock on during the solution.

$N_{i\,j}^{1\,2}$ can be determined by one of the following means:

1) Starting from two stations with known coordinates.
2) Observing an initial baseline in stationary mode for 10 to 30 min.
3) Exchanging the antennas between two stations at the beginning of the survey ("antenna swap"; see [1,25]). The apparent movement of the antennas yields twice the initial baseline vector.

In addition to the double-difference approach, the use of single differences in corresponding algorithms is also possible. As long as the clock errors are modeled epochwise as white noise, this is identical to the double-difference approach. However, stochastic clock error modeling can be used to improve the solution. The initial integer unknown $N_{i\,j}^{1\,2}$ remains the same.

In principle, differences between the moving receiver and the stationary one are determined. From each difference, a position solution can be achieved independently of all the other observations. Therefore, this approach has been entitled the "no dynamics" model [26]. A model such as use of a Kalman filter that takes into account all measurements along the trajectory is doubtless superior to the "no dynamics" approach. In addition, there are many applications in air- or shipborne positioning where the trajectory is also of interest.

The use of phase measurements requires that there be no cycle slips in the data, or that they can be detected and removed.

CYCLE SLIP FIXING OF CARRIER PHASE OBSERVATIONS

It seems that geodetic GPS receivers of the last generation have cycle slips because of failures in hardware and internal software (such as register overflowing, internal cross-talk, aliasing, self-jamming, and truncation), and that new sophisticated carrier phase receivers will never (?) have cycle slips. Nevertheless, such phenomena will happen if the antenna's view of the sky is obstructed. This can be the case during a terrestrial kinematic survey if trees or buildings along the traverse are shadowing satellites. It can also occur during marine applications in the harbor area or in airborne positioning when parts of the ship or aircraft may obstruct the intervisibility between antenna and satellites.

From our own experiences, we can state that it is therefore absolutely necessary to deal with cycle slip correction methods or algorithms, even though pragmatic-thinking colleagues may argue that an appropriate observation plan should avoid cycle slips.

The following approaches for cycle slip repair are possible:

1) *Repeat the survey*—This approach would obviously be used for the worst case, which may happen with single-frequency receivers having only a small number of tracking channels (four to five).
2) *Use instruments with more than four channels* to get redundant observations. For this case also, single-frequency receivers are suited.
3) *Use the dual-frequency phase ratio method* [4]—For cases in which two-frequency phase data are observed, advantage can be taken of the so-called ionospheric residual $\delta\phi$ [4]:

$$\delta\phi(t) = \phi_1(t) - \frac{f_1}{f_2}\phi_2(t) \tag{3}$$

where Φ_1, Φ_2 are the carrier phases of the corresponding frequencies f_1, f_2 (L_1, L_2). This quantity has the property that it contains no terms depending on the station-to-satellite range and on the clock errors, and thus has a linear, nearly constant behavior. The ionospheric residual can be used for overbridging loss of lock and corresponding cycle slip correction.
4) *Integrate GPS with INS or other sensors*—When the kinematic GPS positioning technique is used in urban areas, in woodlands, etc., an interpolator is especially needed during times of loss of lock. For this purpose and for cycle slip detection and correction, the integration of GPS/INS is the appropriate surveying and navigation tool of the future. A later section of this paper presents the results of a corresponding feasibility study of the integration of a GPS receiver and a ring laser gyro strapdown system with regard to high-precision applications in the centimeter range. It may be noted also, however, that additional sensors, such as a laser or radar altimeter over sea, or an atomic clock or the inner orientation of stereo images of a photogrammetric GPS flight, may help to compensate cycle slips.

RESULTS OF TERRESTRIAL, SHIPBORNE, AND AIRBORNE DIFFERENTIAL GPS KINEMATIC POSITIONING APPLICATIONS

The following summarizes the experiences of IAPG with respect to kinematic GPS positioning tests in the last two years. Some of the terrestrial and airborne tests have already been described [27, 28, 29].

Hardware, Data Reduction Algorithm, and Software

For all tests, the dual-frequency Texas Instruments TI 4100 receiver was used. It is a P-code multiplex receiver that can measure both pseudorange and continuous phase. The operating software includes TI 4100 Navigator [30], as well as special version 32.1 of the GESAR (Geodetic Satellite Receiver) software [31]. The main difference between these two softwares is the capability of GESAR 32.1 to select the measurement rate and to change the user dynamics class at any time during the mission. This is an advantage in terrestrial surveying applications, in which it is usually possible to bring the antenna to a short rest, and to switch the dynamics class to the stationary case. Lowering the user dynamics (class) improves the system accuracy. On the other hand, the TI 4100 Navigator software provides the two-phase measurements over a 160 ms time interval before and after the code Fundamental Time Frame (FTF), which may be helpful for detecting cycle slips.

For the cycle slip correction, the dual-frequency phase ratio method was applied. Approximate values were found using the phase-range combination

$$\Delta\rho_i(t) = \lambda_i[\phi_i(t) - \phi_i(t_1)] - [\rho(t) - \rho(t_1)] \tag{4}$$

where i indicates L_1 or L_2, ρ is the observed range derived from code measurements, t is the actual epoch, and t_1 is the initial epoch. Final values are obtained by the ionospheric residual (equation (3)). In contrast to the linear regression algorithm used in [17] and [18], a Kalman filter was applied [29, 32], where the dynamic system matrix was approximated by use of a simple quadratic polynomial over time [4]. The system driving noise matrix Q is a 3×3 matrix defined empirically for the phase-range combination in equation (4) by

$$Q_1 = \text{diag}(10^{-8}, 10^{-10}, 10^{-12})\ [\text{m}^2, -, \text{m}^{-2}] \tag{5}$$

and for the ionospheric residual in equation (3) by

$$Q_2 = \text{diag}(10^{-6}, 10^{-8}, 10^{-10})[\text{cycles}^2, -, \text{cycles}^{-2}] \tag{6}$$

The measurement noise matrix, in our case defined as a simple scalar (only one measurement per epoch), is given for the phase-range combination by

$$r_1 = 2.25\ [\text{m}^2] \tag{7}$$

and for the ionospheric residual by

$$r_2 = 2.5 \times 10^{-3}\ [\text{cycles}^2]. \tag{8}$$

After computing the approximate cycle slip values by equation (4), the final corrections are found by equation (3) through varying the approximations within a small search window, e.g., [−4, 4], and looking for the minimum in $\delta\phi$. It has been stated that in order to guarantee a unique result, the approximate values have to be better than 3 cycles [32]. This is a very high requirement, which may limit the reliability of the method in the presence of a low signal-to-noise ratio.

Again, in the relative positioning mode, a Kalman filter was applied for the combination of single-difference carrier phases and pseudoranges [29]. For further modeling of clock errors and of ionospheric effects, the reader is referred to [33].

The briefly described data reduction algorithm is implemented in the software "DYNAMITE" (Dynamic Trajectory Estimation), written by H. Landau in FORTRAN 77 for the DEC VAX family under VMS.

Terrestrial Kinematic GPS Trials

For land kinematic GPS surveying applications, the procedure is as follows. We start from a station (with known coordinates) where the antenna is first put on a tripod centered over the station. After a short while, the antenna is carried to a car. Usually, it is mounted on the roof of the car during the move in order to keep continuous tracking. After arriving at the next station to be determined, the antenna is taken from the car and mounted on the tripod, which was put over the station beforehand. Again, for a while the antenna is at rest before being transported to the next station. Theoretically, the period of the rest should guarantee only that at least one GPS measurement is recorded. This means something like 5 s when assuming a time interval of 3 s between subsequent measurements. However, in practice the period will be a little bit extended, say, to 20 s or even 1 min, in order to check for gross errors or failures, or simply to allow time to make a notation in the field books.

Two land tests were carried out in an area called Werdenfelser Land, approximately 50 km south of Munich along an east-west traverse. Seven stations with separations between 400 m and 1000 m were observed over 3 km (Figure 1).

The test at the Sydney airport was conducted during the sabbatical of the first author with A. Stolz from the University of New South Wales [28]. The observed traverse around the runway (Figure 2) has a length of about 5 km, with 6 stations belonging to the precise airport geodetic 3-D network, with an estimated station accuracy of approximately ±5 mm.

The scenarios for all three land tests were identical. The control stations to be observed in kinematic mode were equipped with survey tripods carefully centered over the survey marks. Before and after the move (double run), a 10 to 15 min static positioning interval took place, in which the moving receiver

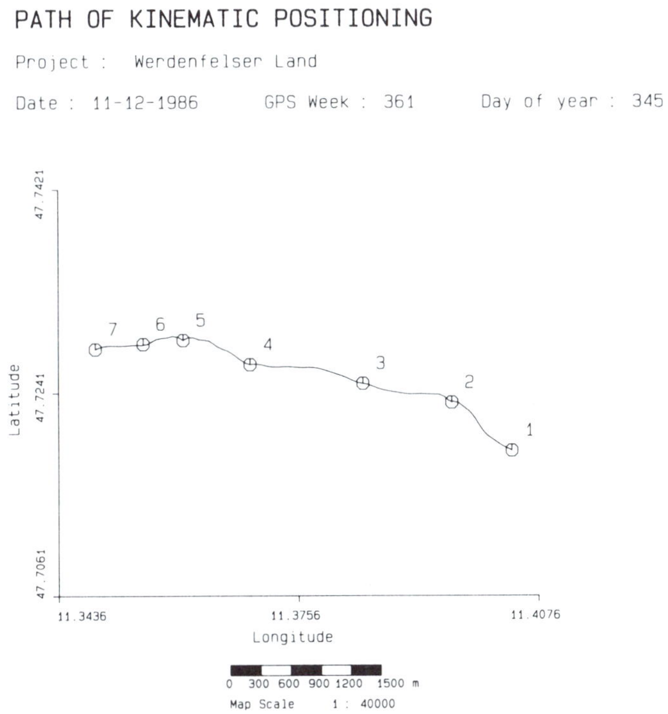

Fig. 1—Location of the Traverse Used in Werdenfelser Land 1 & 2 Tests

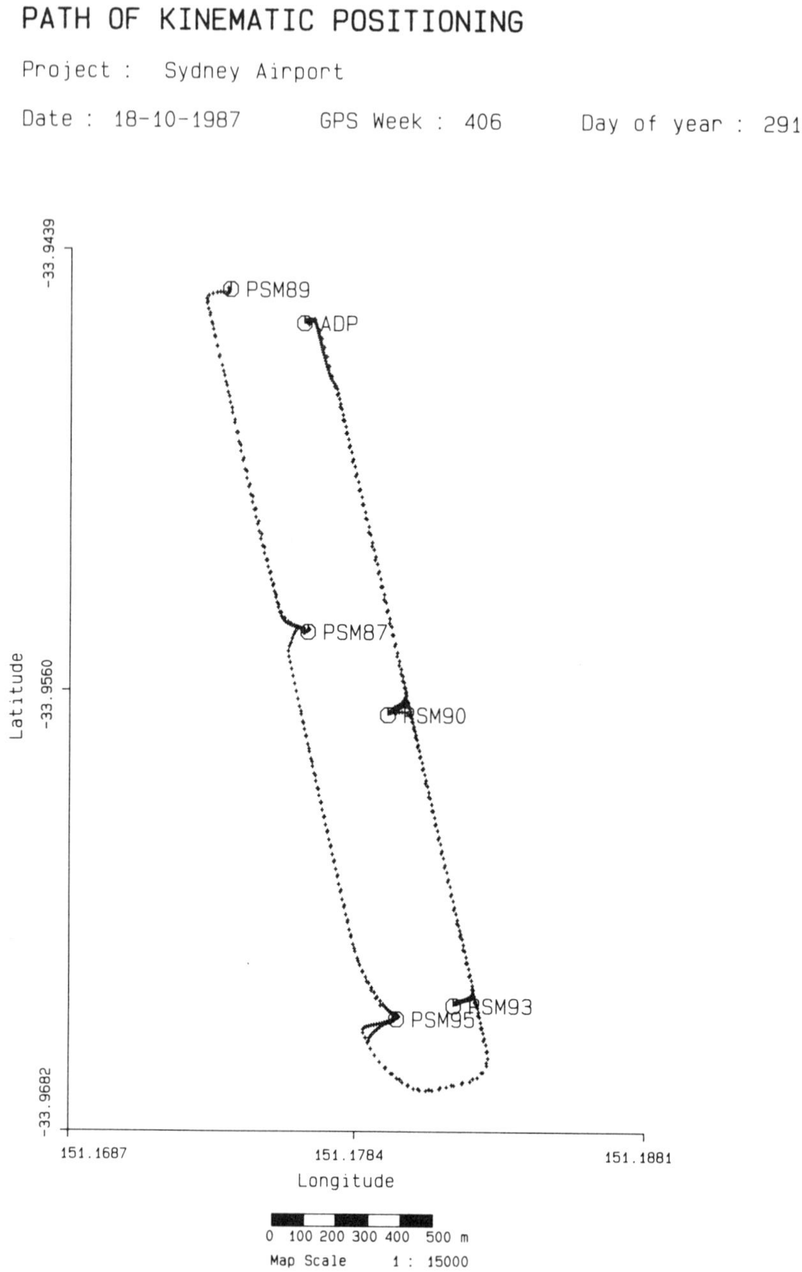

Fig. 2—Location of the Traverse Used in Sydney Airport Tests

was observed at the first station to get initial positioning values with respect to the static receiver located a few kilometers away. The traveling speed in the Werdenfelser Land 1 + 2 test was up to 75 km/h, whereas at the Sydney airport it was less than 25 km/h. In the Werdenfelser Land 2 test, the tracking bandwidth of the receiver was set to the stationary mode (0.7 Hz) during the stop at the stations, and changed again before traveling.

Table 1 summarizes the error statistics of the terrestrial kinematic trials when using the most precise phase differencing method discussed earlier. From the analysis, we found that 5 to 10 cycle slips/channel/h can be expected during kinematic operation with the TI 4100.

Table 1—Summary of Statistics for the Relative Kinematic GPS Positioning Tests on Land

	Werdenfelser Land 1		Werdenfelser Land 2		Sydney Airport	
GPS Frequency	Horizontal (cm)	Height (cm)	Horizontal (cm)	Height (cm)	Horizontal (cm)	Height (cm)
(1) L_1	1.3	1.9	0.9	1.5	2.3	2.4
(2) L_2	1.8	2.9	1.1	1.4	4.6	2.7
(3) True errors L_1/L_2 single run	—	—	—	—	2.3	1.7
(4) True errors L_1/L_2 double run	—	—	—	—	0.6	0.4

Note: RMS errors (1) and (2) computed from differences between forward and backward run. RMS errors (3) and (4) computed from residuals to precise terrestrial coordinate system.

For comparison, the Sydney airport test was analyzed with the three different differential GPS positioning algorithms outlined earlier. The RMS errors (true errors against the terrestrial reference or phase solution) are shown in Table 2. All derived GPS positions (every 3 s) along the traverse were considered. The RMS errors in parentheses in the row marked "Phase-smoothed pseudoranges" considered only the differences at the geodetic network stations.

Table 2—RMS Errors for the Three Different Differential GPS Kinematic Positioning Methods

Method	Northing (m)	Easting (m)	Height (m)
Differential corrections (pseudorange)	± 2.9	± 4.3	± 4.9
Phase-smoothed pseudoranges	± 1.38 (± 0.90)	± 1.98 (± 0.92)	± 1.99 (± 1.45)
Phase differencing	± 0.02	± 0.01	± 0.02

Shipborne Kinematic GPS Trials

A kinematic GPS positioning test of the survey hovercraft BEASAC (Belfotop Eurosense Acoustic Sounding Air Cushion Platform) (see Figure 3) was carried out in the harbor area of Zeebrugge. This test was conducted in cooperation with the Ecole Royale Militaire, Chair d'Astronomie-Geodesie (Prof. Dr. van Twembeke), Brussels, Belgium; the Survey Department of Rijkswaterstaat, Delft, Netherlands; and the Belgian firm Eurosense. The hydrographic survey platform permits bathymetric surveys in coastal and off-shore areas at survey speeds of up to 50 km/h. The system is employed for monitoring changes to access channels of the major Belgian seaports, such as Zeebrugge, and for supervising the extensive dredging activities [34]. The navigation and posi-

Fig. 3—Survey Hovercraft BEASAC

tioning system is usually a combination of a hyperbolic (Toran) and a circular (Trisponder) system. The so-derived horizontal position accuracy might be in the 1–3 m range.

In order to have a comparison of the kinematic differential GPS-derived position, an automated infrared instrument called MINILIR (precision in the 5 cm range) observed the hovercraft permanently from the dam at the harbor's entrance. This was achieved by placing appropriate lights directly centered under the TI 4100 antenna at the main mast, where the Toran and Trisponder antennas were also mounted.

Figure 4 shows the path of the hovercraft on June 2, 1988, when the differential GPS kinematic positioning test was carried out over 2.25 h. The hovercraft was flying along profiles as is the case in production work.

Although the detailed analysis is still ongoing, we can state already that the test was successful. The rough environment of the hovercraft (wind strength in the range of class 4 to 6) produced no more than the usual number of cycle slips in the TI 4100 data set. The inner accuracy (RMS value) of the determined positions from L_1 and L_2 frequencies is on the order of 3 cm.

Airborne Photogrammetric GPS Flights

On August 17, 1987, one of the flight tests (München-Freimann) out of a series of trials was carried out within a joint project on aerotriangulation

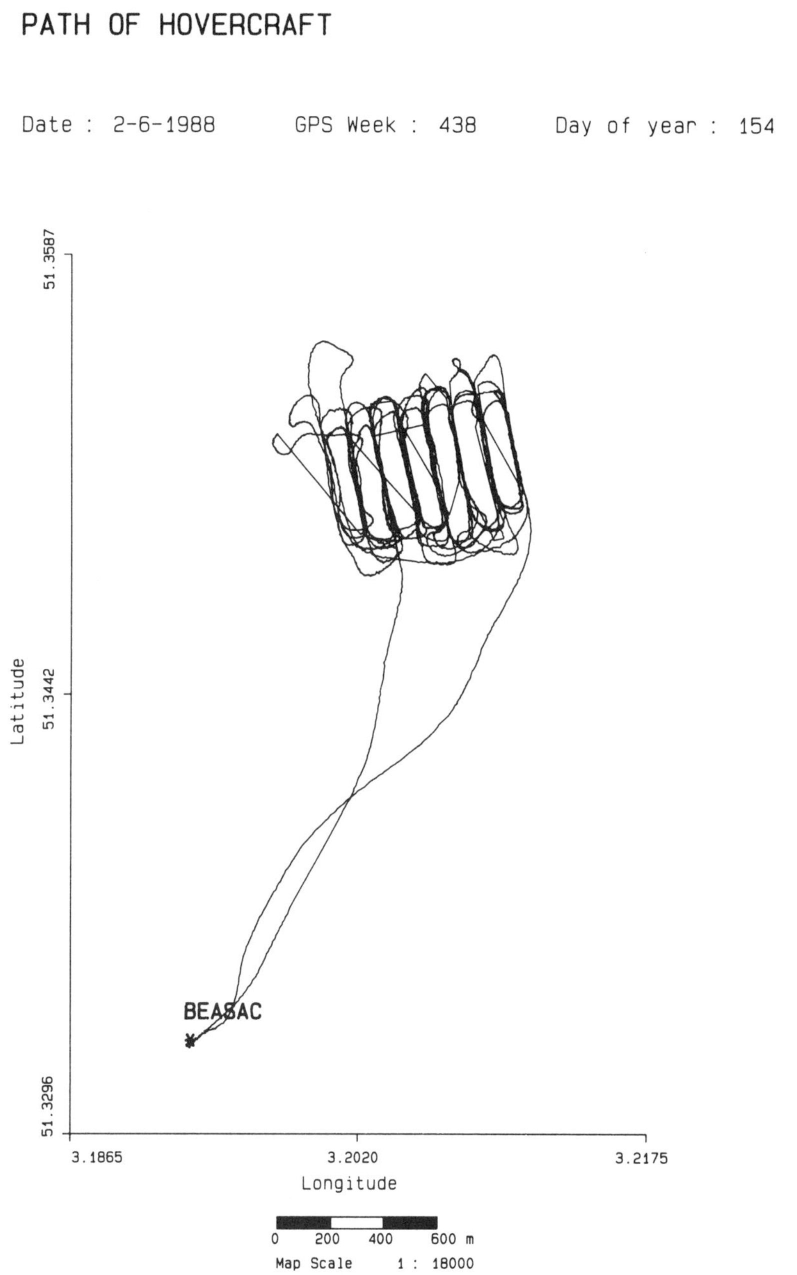

Fig. 4—Path of Hovercraft as Determined by GPS

without ground control with the Institute of Photogrammetry and Cartography of the University FAF Munich. A DORNIER DO 28 aircraft was used. A dual-frequency GPS flight antenna DM C146-2-1 from Dorne & Margolin Inc., N.Y., was mounted on top of the aircraft midway between the two wings (Figure 5). The photogrammetric camera RMK 15/23 from Zeiss, Oberkochen was time-

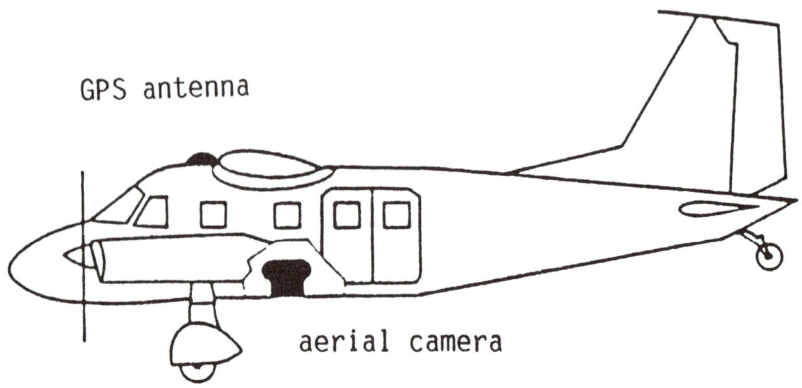

Fig. 5—Aircraft DO 28 Used for Photogrammetric Flight Tests

tagged with the one-pulse-per-s (1 pps) of the TI 4100 via a frequency generator (accuracy 0.25 msec). The time offset between the 1 pps impulse and the GPS time was recorded from the J2 port of the TI 4100 using an RS-232C interface; the offset between the electronic impulse of the camera shutter and the frequency counter was kept via an HP-IB interface. Both are incorporated in a recording HP-IL (interface loop) monitored by a simple HP 71B pocket computer [27].

Figure 6 shows the path of the München-Freimann 1 flight. The aircraft started at Neubiberg airport and was flying over a photogrammetric test area located in the north of Munich city. The DO 28 climbed up to approximately 920 m altitude above ground (corresponding to an orthometric height of 1420 m) and reached a velocity of 54 m/s (~ 190 km/h). Because of bad satellite geometry and the low signal-to-noise ratio in L_2 of the flight antenna used, only limited positioning results were achieved which could be compared with the camera positions derived from the photogrammetric block adjustment [35]. However, the RMS values of the differences between GPS-determined and photogrammetric-determined positions indicate an accuracy potential of GPS for such applications in the decimeter level or even better.

Recently, the flight experiments were continued with a new flight antenna from Adams-Russel. Excellent data sets for five photogrammetric flights were collected (Figure 7), which are presently being analyzed.

INTEGRATION OF GPS AND AN INERTIAL RING LASER GYRO STRAPDOWN SYSTEM: RESULTS OF A FEASIBILITY STUDY

As outlined earlier, there are several problematic factors that affect the reliability, versatility, and applicability of differential GPS kinematic positioning, in particular when using (only) the carrier phases. Again, these include reception of GPS signals in urban areas, interpolation due to the low GPS data rate of approximately 1 Hz in airborne applications, detection and fixing of cycle slips, and complete loss of lock in all GPS receiver channels.

There is no question that in avionics and military applications, the integration of GPS and INS is mandatory to guarantee the reliability of the positioning system. However, an integrated GPS/INS instrument also seems to be the tool

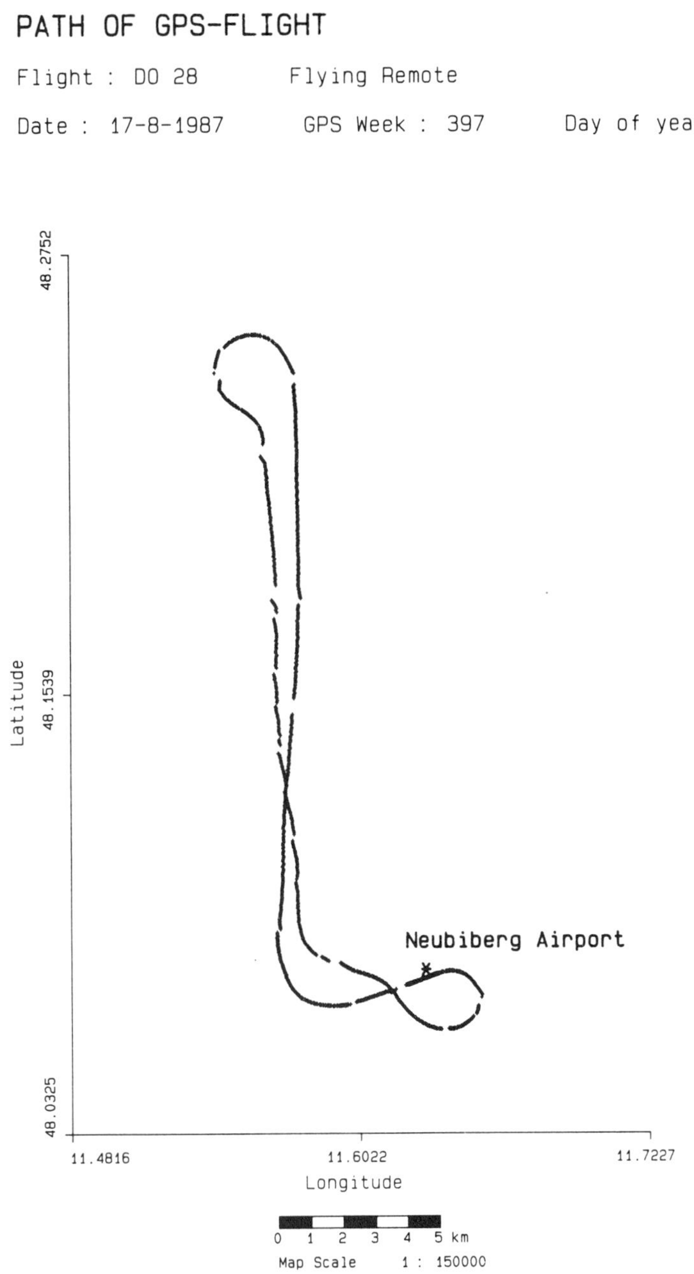

Fig. 6—Path of GPS Flight in München-Freimann 1 Test

for the future for geodesy and surveying in order to solve the application limitations of kinematic positioning mentioned above. Although mechanical inertial platform systems (Ferranti FILS II, Litton LASS II, Honeywell GEOSPIN II) have in the past met a wide range of surveying applications in the 10 to 50 cm range, their high price has limited their practical importance.

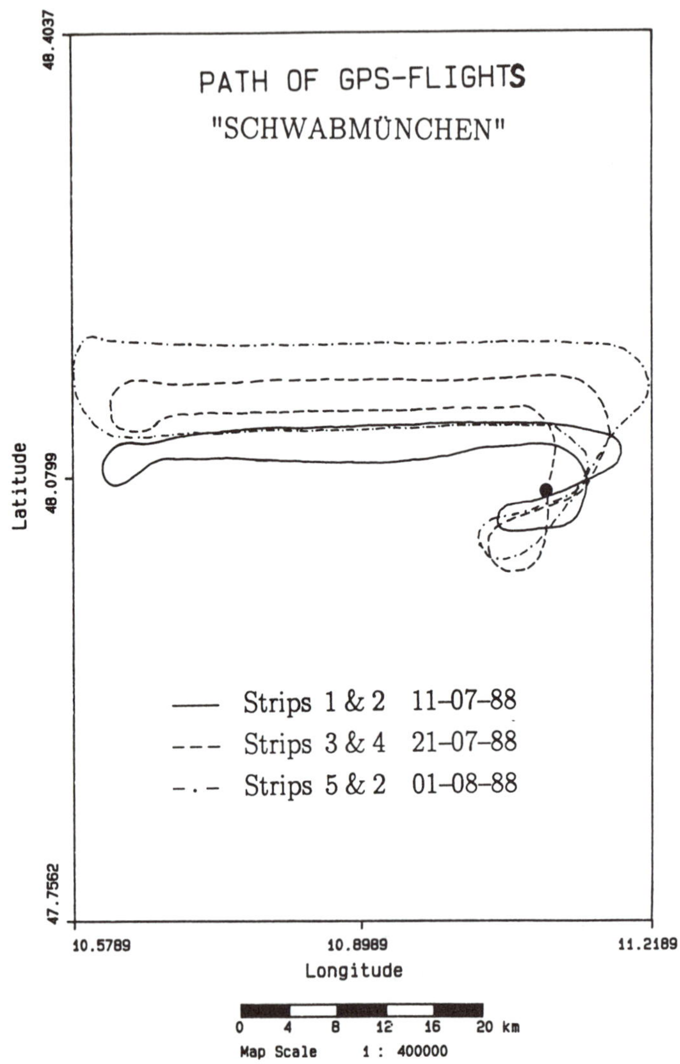

Fig. 7—Path of New GPS Flights

With the development of ring laser gyro strapdown systems (laser INS, or LINS), an ideal candidate for GPS/INS integration seems to be available, one which has a reasonable price and size. From the specifications of such a system, we know that it cannot be applied as a stand-alone system in geodesy and surveying because of its low accuracy level (which may be not better than in the meter range when using 0 velocity updates every 1 to 3 min in terrestrial applications).

Thus, the question we wanted to answer by a feasibility study was this: Can a LINS in combination with GPS solve the problems mentioned earlier? In particular, how long can be the time for coasting during loss of lock of GPS signals, without losing the high accuracy potential of kinematic GPS phase observations? For this purpose, a covariance analysis of a hybrid Kalman filter was carried out, the main results of which are presented here. The reader interested in more details is referred to [36].

The linear dynamic model for the LINS/GPS integration is of the following form:

$$\begin{bmatrix} \dot{\mathbf{x}}_1 \\ \dot{\mathbf{x}}_2 \\ \dot{\mathbf{x}}_3 \\ \dot{\mathbf{x}}_4 \end{bmatrix} = \begin{bmatrix} \mathbf{F}^{free} & \mathbf{F}^{rw} & \mathbf{F}^{bias} & \mathbf{0} \\ \mathbf{0} & \mathbf{0} & \mathbf{0} & \mathbf{0} \\ \mathbf{0} & \mathbf{0} & \mathbf{0} & \mathbf{0} \\ \mathbf{0} & \mathbf{0} & \mathbf{0} & \mathbf{F}^{GPS} \end{bmatrix} \begin{bmatrix} \mathbf{x}_1 \\ \mathbf{x}_2 \\ \mathbf{x}_3 \\ \mathbf{x}_4 \end{bmatrix} + \begin{bmatrix} \mathbf{G}_1 & \mathbf{0} & \mathbf{0} \\ \mathbf{0} & \mathbf{G}_2 & \mathbf{0} \\ \mathbf{0} & \mathbf{0} & \mathbf{0} \\ \mathbf{0} & \mathbf{0} & \mathbf{g}_4 \end{bmatrix} \begin{bmatrix} \boldsymbol{\epsilon}_1 \\ \boldsymbol{\epsilon}_2 \\ \boldsymbol{\epsilon}_4 \end{bmatrix} \tag{9}$$

where

$$\mathbf{x}_1 = [\delta\boldsymbol{\epsilon},\ \delta\mathbf{v},\ \delta\mathbf{r}]^T \qquad \text{(9 states)} \tag{10a}$$

$$\mathbf{x}_2 = [\boldsymbol{\epsilon}^{rw}]^T \qquad \text{(3 states)} \tag{10b}$$

$$\mathbf{x}_3 = [\mathbf{d}_0,\ \mathbf{k}_0,\ \mathbf{b}_0,\ \boldsymbol{\epsilon}_0]^T \qquad \text{(12 states)} \tag{10c}$$

$$\mathbf{x}_4 = [\eta_0,\ \eta_1,\ \mathbf{m}]^T \qquad \text{(6 states)} \tag{10d}$$

where $\delta\boldsymbol{\epsilon}$ is orientation errors, $\delta\mathbf{v}$ is velocity errors, $\delta\mathbf{r}$ is position errors, $\boldsymbol{\epsilon}^{rw}$ is random walk error angles, $\mathbf{d}_0$ is the fixed bias drift of ring laser gyros (RLGs), $\mathbf{k}_0$ is scale factor errors of RLGs, $\mathbf{b}_0$ is accelerometer bias, $\boldsymbol{\epsilon}_0$ is the accelerometer scale factor, η_0 is the GPS receiver clock bias, η_1 is the GPS receiver clock drift, $\mathbf{m}$ is ambiguities in four channels, $\mathbf{F}^{free}$ is the dynamic matrix of free INS, $\mathbf{F}^{rw}$ is RLG random walk dynamics, $\mathbf{F}^{bias}$ is the dynamics of sensor bias states, $\mathbf{F}^{GPS}$ is the dynamics matrix of GPS parameters, and $\boldsymbol{\epsilon}_1, \boldsymbol{\epsilon}_2, \boldsymbol{\epsilon}_3$, are disturbing process noise.

In the linear observation model

$$\mathbf{l}(t) = \mathbf{H}(t)\ \mathbf{x}(t) + \boldsymbol{\epsilon}(t) \tag{11}$$

and

$$\mathbf{l}(t) = \mathbf{L}^{Ext}(t) - \mathbf{L}^{INS}(t) \tag{12}$$

where $\mathbf{L}^{Ext}$ is the externally derived reference observation, $\mathbf{L}^{INS}$ is the observation derived by the INS, $\mathbf{H}$ is the measurement matrix, $\mathbf{x}$ is the state vector (comprised), and $\boldsymbol{\epsilon}$ is the measurement noise.

The following types of observations for updating the filter were considered:

1) GPS range differences
2) GPS single differences
3) GPS single-difference rates

The well-known discrete Kalman equations can be found in many textbooks (e.g., [37]). They are of the form

$$\tilde{\mathbf{P}}_k = \boldsymbol{\Phi}_{k-1}\ \hat{\mathbf{P}}_{k-1}\ \boldsymbol{\Phi}_k^T + \boldsymbol{\Gamma}_{k-1}\ \mathbf{Q}_{k-1}\ \boldsymbol{\Gamma}_{k-1}^T \tag{13}$$

$$\hat{\mathbf{P}}_K = (\mathbf{1} - \mathbf{K}_k\ \mathbf{H}_k)\ \tilde{\mathbf{P}}_k \tag{14}$$

$$\mathbf{K}_k = \tilde{\mathbf{P}}_k\ \mathbf{H}_k^T\ (\mathbf{R}_k + \mathbf{H}_k\ \tilde{\mathbf{P}}_k\ \mathbf{H}_k^T)^{-1} \tag{15}$$

where **P** is the variance-covariance matrix of the states, **Φ** is the transition matrix, **Γ** is the discrete disturbance input matrix, **Q** is the covariance matrix of discrete white noise, **K** is the Kalman gain matrix, **R** is the covariance matrix of observational noise, **H** is the measurement matrix, "˜" is the predicted value, and "^" is the updated value.

Here the 1 σ simulation results for the integration of an LINS/GPS in a land vehicle application are presented. An open-loop Kalman filter working as a simple output estimator was used. It was assumed that the filter, equation (15), was updated by GPS range differences, single differences, and single-difference rates. The filter cycle and the update frequency were selected as 1 Hz. The motion of the measuring car was simulated in the same way as that described in [37]. The traverse, which had a length of about 9.8 km, was moved from east to west in 1700 s with velocities up to 10 m/s. Every 180 s, the vehicle was stopped for 20 s in order to determine the coordinates of a geodetic station. All specifications for the LINS were similar to those of the Litton LTN-90.

The initial 1 σ values for the $\mathbf{P}_0$ matrix at $t = t_0$ are specified in Table 3. The precision of the measurements in the **R** matrix is shown in Table 4.

Table 3—$\mathbf{P}_0$ Values

State	1 σ
$\sigma\epsilon$	± 8″
σv	± 0.003 m/s
σr	± 0.03 m
ϵ_{rw}	± 0″
d_0	± 0.007 °/h
n_0	± 5 ppm
b_0	± 8 μg
ϵ_0	± 10 ppm
η_0	± 10 m
η_1	± 2.5 m/s
m	± 0 cycles

Table 4—**R** Values

Observation	1 σ
Range difference	± 1 m
Single difference	± 0.02 m
Single-difference rate	± 0.02 m/s

The 1 σ results for the LINS/GPS combination with respect to the position errors are plotted in Fig. 8. The position errors are bounded, are smaller than 30 mm, and show a smoother behavior due to the effect of the velocity error integrations. Of course, the high stability of the errors, here over the 1700 s interval, and their very low values are mainly due to the low GPS observational noise.

In order to get some insight into the problem of whether a LINS can help to solve the cycle slip problem, we have done two different simulations:

1) The 1 σ prediction error of the single-difference observable was investigated under the assumption that cycle slips occur in channels 1 to 4, one

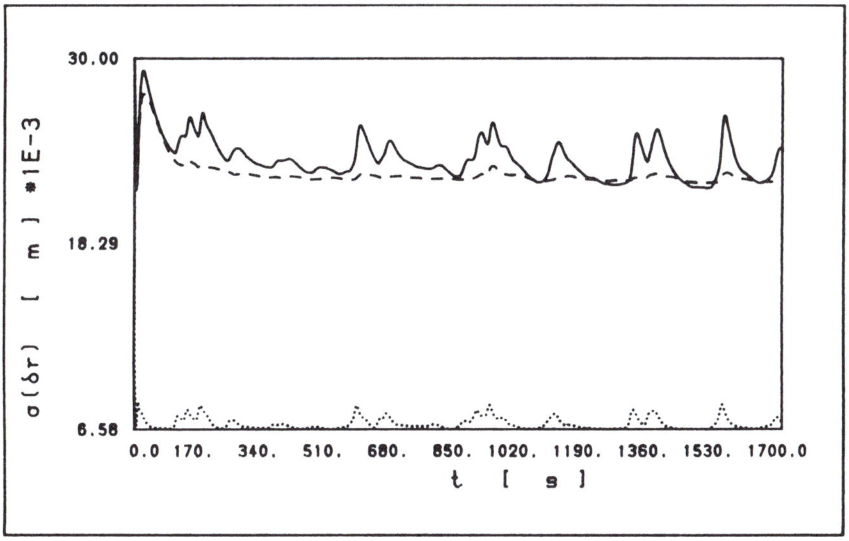

Fig. 8—Hybrid One-Sigma Position Error

after another, for the duration of 20 s. After 20 s, the specific channel has lock again.

2) The 1 σ position error was investigated under the assumption that a complete loss of lock in all channels occurred for about 60 s.

Figures 9 and 10 show that a loss of lock in a specific channel leads to a relatively large degeneration of precision in the same channel, and to a lower degeneration of precision in the other channels. However, the increase in the prediction error over 20 s is so small (< 5.3 cm) that we will succeed in estimating the change in ambiguity by equation (16):

$$\Delta m_j = m_j(t_k') - m_j(t_k) = \frac{\delta\{\Delta \psi(t_k')\} - \delta\{\psi(t_k)\}}{\lambda} \quad (16)$$

where $\delta(\Delta \psi)$ is the single difference, λ is the GPS carrier wave length, $t = t_k$ is the time when a cycle slip occurred, and t_k' is the time when the corresponding receiver channel has lock again.

In Figure 11, a complete loss of lock occurs at t = 30 s. At this time, the GPS signal is lost in all channels, and the LINS works as a free inertial system. In the period of about 60 s of signal loss, the LINS 1 σ position error increases to 0.50 m. This relatively large value indicates the error rate of a free RLG-INS. At t = 90 s, the receiver has lock again. However, during the 60 s, the ambiguity may change theoretically by an infinite number of cycles. Therefore, the standard deviations of the ambiguities **m** are reset to $\pm 10^6$ cycles. When the GPS observations are available again (t = 90 s), their accuracy potential is used at

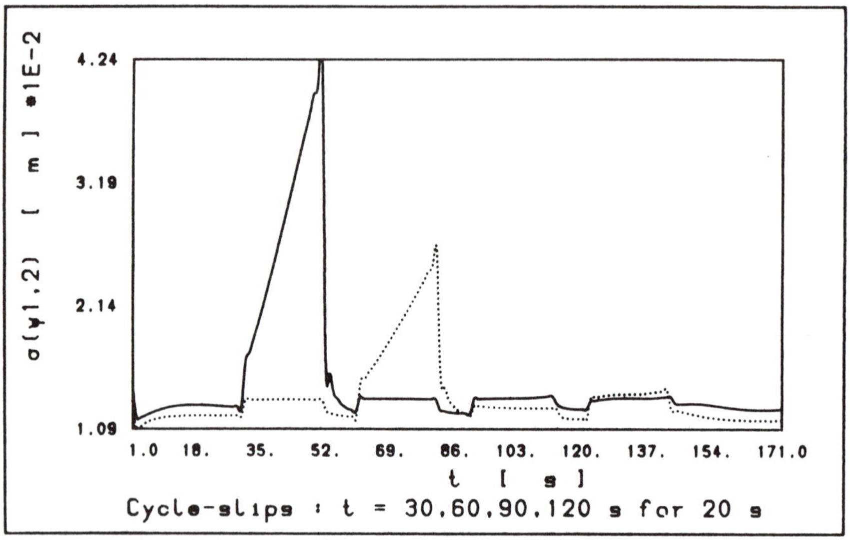

Fig. 9—One-Sigma Prediction Error of Single Difference in Channels No. 1 and 2

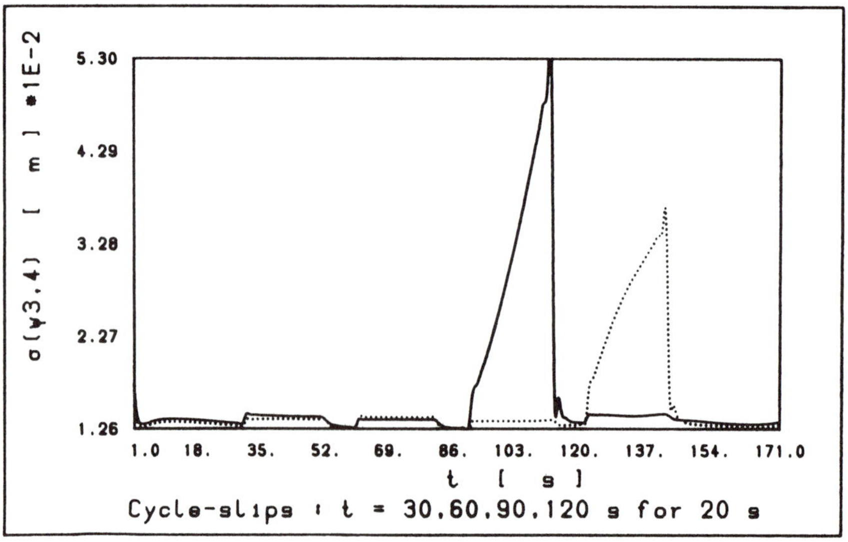

Fig. 10—One-Sigma Prediction Error of Single Difference in Channels No. 3 and 4

first to decrease the variances of the ambiguities. This is the main reason for the loss of precision in the positioning information in Figure 11. Although the degradation of accuracy is large in this solution, the stand-alone kinematic GPS method will give absolutely no information on position with errors of ± 0.5 m during these 60 s. A complete signal loss will result in accuracy degradation by a factor of ~ 10, which cannot be improved later on. However, in terrestrial applications the LINS can in such a situation make a 0 velocity update to bound the errors.

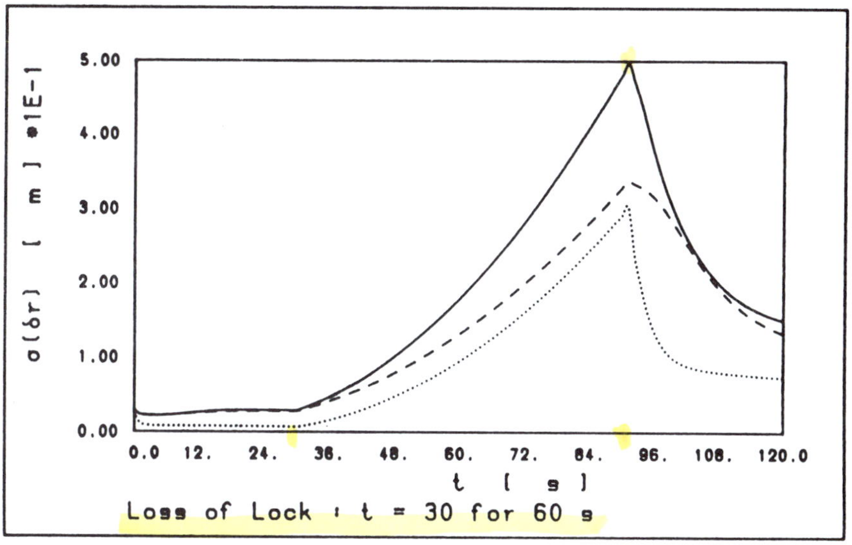

Fig. 11—One-Sigma Position Error, If Complete Loss of Lock Occurs

If we have to answer the question of how long we have a complete loss of lock in all GPS receiver channels without losing the high accuracy of the differential GPS kinematic phase observations (1–2 cm), we derive a time of approximately 20 to 25 s. This results from the consideration that the error of the free LINS should not be larger than half of the GPS wavelength (see Figure 11).

CONCLUSIONS

From our experience with practical high-precision differential GPS kinematic positioning and the results of a feasibility study for a GPS/LINS integration, we may conclude the following:

1) Accuracies of 1 to 2 cm are attainable if dual-frequency GPS receivers are used, and the phase ratio method is applicable for solving for cycle slips. With respect to single-frequency receivers with redundant channels, some software development still has to be done.
2) With the new generation of GPS phase tracking receivers, with a phase measurement noise < 1 mm, an improvement in accuracy still can be expected.
3) The integration of GPS/LINS provides a reliable system that can also be used in problem situations involving partial or short-time shadowing of the GPS signals.
4) Assuming a LINS of type Litton LTN-90, the position error rate is bounded to ± 50 cm/60 s if a complete loss of lock occurs.
5) If the high-accuracy potential of GPS phase observations should be met, a complete loss of lock in all GPS channels can be accepted only over approximately 20 s for such a GPS/LINS system.

ACKNOWLEDGMENT

The support of the German Research Foundation (Deutsche Forschungsgemeinschaft), Prof. Dr. Stolz (University of New South Wales, Australia), and Prof. van Twembeke (Ecole Royale Militaire, Brussels) is gratefully acknowledged.

Based on a paper presented at The Institute of Navigation Satellite Division Meeting, Colorado Springs, CO, September 1988.

REFERENCES

1. Remondi, B.W., *Performing Centimeter Accuracy Relative Surveys in Seconds Using GPS Carrier Phase,* Proceedings of the First International Symposium on Precise Positioning with the Global Positioning System, Rockville, MD, 1985, 789–98.
2. Remondi, B.W., *Performing Centimeter-Level Surveys in Seconds with GPS Carrier Phase: Initial Results,* NOAA Technical Memorandum NOS NGS-43, 1985.
3. Remondi, B.W., *Performing Centimeter-Level Surveys in Seconds with GPS Carrier Phase: Initial Results,* Proceedings of the Fourth International Geodetic Syposium on Satellite Positioning, April 28–May 2, 1986, Austin, TX, Vol. 2, 1229–50.
4. Goad, C.C., *Precise Positioning with the Global Positioning System,* Proceedings of the Third International Symposium on Inertial Technology for Surveying and Geodesy, September 16–20, 1985, Banff, Canada, 1986, 745–56.
5. Cannon, M.E., *Kinematic Positioning Using GPS Pseudo-range and Carrier Phase Observations,* UCSE Report No. 20019, The University of Calgary, 1987.
6. Cannon, M.E., Schwarz, K.P., and Wong, R.V.C., *Kinematic Positioning with GPS: An Analysis of Road Tests,* Proceedings of the Fourth International Geodetic Symposium on Satellite Positioning, April 28–May 2, 1986, Austin, TX, Vol. 2, 1251–67.
7. Kleusberg, A., Quek, S.H., Wells, D.E., Lachapelle, G., and Hagglund, J., *GPS Relative Positioning Techniques for Moving Platforms,* Proceedings of the Fourth International Geodetic Symposium on Satellite Positioning, April 28–May 2, 1986, Austin, TX, Vol. 2, 1299–1310.
8. Kleusberg, A., *Kinematic Relative Positioning Using GPS Code and Carrier Beat Phase Observations,* in Marine Geodesy, Vol. 10, No. 3/4, 1986, 257–74.
9. Lachapelle, G., Casey, M., Eaton, R.M., Kleusberg, A., Tranquilla, J., and Wells, D., *GPS Marine Kinematic Positioning Accuracy and Reliability,* The Canadian Surveyor, Vol. 41, No. 2, 1987, 143–72.
10. Lachapelle, G., Falkenberg, W., Hagglund, J., Kinlyside, D., Casey, M., Kielland, P., and Boudreau, H., *Shipborne GPS Kinematic Positioning: with Emphasis on Survey Launch Applications,* presented at the General Assembly of the International Association of Geodesy, Vancouver, August 1987.
11. Lachapelle, G., Falkenberg, W., and Casey, M., *Use of Phase Data for Accurate Differential GPS Kinematic Positioning,* Bulletin Géodésique, Vol. 61, 1987, 367–77.
12. Lachapelle, G., Hagglund, J. Jones, H., and Eaton, M., *Differential GPS Marine Navigations,* PLANS 84, San Diego, November 28–30, 1984.
13. Lachapelle, G., Lethaby, J., and Casey, M., *Airborne Single Point and Differential GPS Navigation for Hydrographic Bathymetry,* The Hydrographic Journal, Vol. 34, October 1984, 11–18.
14. Nard, G.P., *GPS Geodesy and Kinematic Topography Measurements and Data*

Processing, presented at the International GPS-Workshop Darmstadt, April 10–13, 1988.

15. Krabill, W.A. and Martin, C.F., *Aircraft Positioning Using Global Positioning System Carrier Phase Data,* NAVIGATION, Journal of The Institute of Navigation, Vol. 34, No. 1, 1–21.
16. Lucas, J.R. and Mader, G., *Successful Demonstration of Aerotriangulation without Ground Control,* Manuscript, NOAA/NGS, 1987.
17. Mader, G., *Decimeter Level Aircraft Positioning Using GPS Carrier Phase Measurements,* Proceedings of the Fourth International Geodetic Symposium on Satellite Positioning, April 28–May 2, 1986, Austin, TX, Vol. 2, 1311–25.
18. Mader, G., *Dynamic Positioning Using GPS Carrier Phase Measurements,* Manuscripta Geodaetica, Vol. 11/4, 1986, 272–77.
19. Mader, G., Carter, W., and Douglas, B., *Decimeter Precision Aircraft Positioning Using GPS Carrier Phase Measurements,* NAVIGATION, Journal of The Institute of Navigation, 1986.
20. van der Vegt, J.W., *Differential GPS: An Efficient Tool in Photogrammetry,* Manuscript Survey Department Rijkswaterstaat, Delft, 1988.
21. Hatch, R., *Dynamic Differential GPS at the Centimeter Level,* Proceedings of the Fourth International Geodetic Symposium on Satellite Positioning, April 28–May 2, 1986, Austin, TX, Vol. 2, 1287–98.
22. Hatch, R., *The Synergism of GPS Code and Carrier Measurements,* in Proceedings of the Third International Geodetic Symposium on Satellite Doppler Positioning, Las Cruces, NM, February 1982, 1213–32.
23. Lachapelle, G., Hagglund, J., Falkenberg, W., Bellameare, P., Casey, M., and Eaton, M., *GPS Land Kinematic Positioning Experiments,* Proceedings of the Fourth International Geodetic Symposium on Satellite Positioning, April 28–May 2, 1986, Austin, TX, Vol. 2, 1327–44.
24. Seeber, G., Schuchardt, A., and Wübbena, G., *Precise Positioning Results with TI 4100 GPS Receivers on Moving Platforms,* Proceedings of the Fourth International Geodetic Symposium on Satellite Positioning, April 28–May 2, 1986, Austin, TX, Vol. 2, 1269–86.
25. Hofmann-Wellenhof, B. and Remondi, B.W., *The Antenna Exchange: One Aspect of High-Precision GPS Kinematic Survey,* presented at the International GPS-Workshop Darmstadt, April 10–13, 1988.
26. Schwarz, K.P., Cannon, M., and Wong, R.V.C., *The Use of GPS in Exploration Geophysics—A Comparison of Kinematic Models,* presented at the IUGG Meeting, Vancouver 1987.
27. Hein, G.W., Landau, H., and Baustert, G., *Terrestrial and Aircraft Differential Kinematic GPS Positioning,* presented at the International GPS-Workshop Darmstadt, April 10–13, 1988.
28. Hein, G.W. and Stolz, A., *Kinematic GPS for Local Crustal Deformation Monitoring,* EOS Trans. AGU, 69, 324, 1988.
29. Landau, H., *Precise Kinematic GPS Positioning, Part 1: Experiences on an Land Vehicle Using TI 4100 Receivers and Software,* paper presented at the IUGG Meeting, Vancouver 1987.
30. *Texas Instruments TI 4100 Navstar Navigator, Owner's Manual,* Texas Instruments, GPS Navigation Systems, Lewisville, TX.
31. Darnell, A.R. and Hawkins, L.L., *User's Guide to GESAR Version 1.1* (with Amendments from G. Jones of June 19, 1986), NSWC, Dahlgren, VA, 1986.
32. Bastos, L. and Landau, H., *Fixing Cycle Slips in Kinematic GPS-Applications Using Kalman Filtering,* Manuscripta Geodaetica, Vol. 13/4, 1988, 249–56.
33. Landau, H., *Zur Nutzung des Global Positioning Systems in Geodäsie und Geodynamik: Modellbildung, Softwareentwicklung und Analyse,* Ph.D. Thesis,

Schriftenreihe Studiengang Vermessungswesen, Universität der Bundeswehr München, No. 36, 1988.

34. Kerckaert, P., Grobben, A., and de Candt, P., *Measurements from a Fast-Moving Air-Cushion Platform,* manuscript available from Eurosense Belfotop Wemmel, Belgium, 1986.
35. Dorrer, E. and Schwiertz, Ch., *Experiments with GPS-Supported Aerotriangulation,* presented at the International Symposium on Instrumentation, Theory and Analysis for Integrated Geodesy, Sopron, Hungary, May 16–20, 1988.
36. Eissfeller, B. and Spietz, P., *Basic Filter Concepts for the Integration of GPS and an Inertial Ring Laser Gyro Strapdown System,* submitted to Manuscripta Geodaetica, 1988.
37. Eissfeller, B., *Kalman-Filter und andere Echtzeitverfahren,* in Caspary, Hein, Schödlbauer (Eds.): Beiträge zur Inertialgeodäsie, Schriftenreihe des universitären Studiengangs Vermessungswesen, Universität der Bundeswehr München, No. 22, 1987, 139–96.
38. Gelb, A. (Ed.), *Applied Optimal Estimation,* The M.I.T. Press, Massachusetts Institute of Technology, Cambridge, MA, 1974.
39. Goad, C.C., *Investigating of an Alternative Method of Processing Global Positioning Survey Data Collected in Kinematic Mode,* presented at the International GPS-Workshop Darmstadt, April 10–13, 1988.

NAVIGATION: *Journal of The Institute of Navigation*
Vol. 37, No. 2, Summer 1990
Printed in U.S.A.

GPS Navigation: Combining Pseudorange with Continuous Carrier Phase Using a Kalman Filter

PATRICK Y. C. HWANG

Rockwell International Corporation,
Cedar Rapids, Iowa

R. GROVER BROWN

Iowa State University, Ames, Iowa
Received September 1989
Revised April 1990

ABSTRACT

The notion of aiding GPS code-ranging information with carrier phase measurements has been around for quite some time. The integration of these two complementary forms of data can provide highly precise position solutions corrupted by very little high-frequency error, which is advantageous particularly for high-dynamical situations. This paper presents two schemes, based on the Kalman filter, for integrating GPS code and carrier measurements efficiently. One of these schemes involves a general combination of the two types of measurements, where the vehicle motion is described as a random process. In the other, the continuous carrier phase measurements provide a dynamically exact reference trajectory in much the same manner as in an inertial system, and the pseudorange data is then used to update this reference trajectory via a Kalman filter. This method of integration looks especially attractive as a means of achieving good dynamical response while still retaining the benefits of filtering the pseudorange data.

INTRODUCTION

Over the past decade or more of GPS development, there has been a slowly emerging emphasis on processing methodologies that involve the GPS carrier signal. This trend, in uncovering better and better ways to take advantage of the intrinsic stability and accuracy of the carrier, has led to the opening up of new possibilities with improved GPS performance not originally envisioned. Hardware cost reductions, as well as precision surveying requirements, have driven the focus of high-performance receiver designs toward the multiple channel configuration, thereby allowing users access to continuous carrier tracking. At the same time, GPS technology has matured to the point where carrier phase measurements can be obtained with improved reliability, except for poor signal conditions due to external factors. It is for these reasons that carrier phase now plays an important role in present and future advances in system performance.

In conventional GPS models, the range equivalent of the propagation time delay between the satellite and receiver, known as pseudorange, gives the user the principal measurement of position. Additionally, an estimate of the Doppler frequency related to the GPS carrier, sometimes called delta range, provides a measurement of velocity. As a minimum, most standard GPS receivers are set up to process pseudorange and delta range measurements made from at least four satellites. It has long been recognized that, apart from velocity, a form of information that pertains to relative positioning can also be extracted from the GPS carrier [1]. The phase of the GPS carrier (conceptually equivalent to integrated Doppler) is useful as a measurement only if the carrier is continuously tracked over time and the number of accumulated cycles is accurately accounted for. Use of the continuous carrier phase measurement to enhance receiver performance and its processing in combination with conventional pseudorange are jointly explored in this paper.

Positioning accuracies provided by GPS receivers are limited by two types of noise processes: a low-frequency (quasi-bias) and a high-frequency (white) noise. The former is made up primarily of atmospheric delays, ephemeris errors, and also, perhaps, Selective Availability (SA) distortions. It is possible to reduce the impact of the above on position error by measuring the ionospheric component of atmospheric delays using an L_1/L_2 receiver, correcting ephemeris errors on a postprocessing basis, and decrypting SA with the proper equipment. However, it is generally recognized that these enhancements are not available to all users, and without such access, the accurate estimation of these low-frequency noise components is nearly impossible.

On the other hand, effects of the high-frequency noise component on the solution can be reduced by data processing. In the case of a stationary receiver, the high-frequency noise can easily be smoothed out by averaging the data over time. Quite simply, the longer the averaging, the less noisy the estimated solution. In a dynamic situation, though, the trajectory of the receiver itself resembles a random process, unless it is known in advance and modeled accordingly. The problem here is one of estimating a random signal immersed in unwanted noise. In a spectral sense, high-order dynamics equate to high frequencies and large bandwidths, the consequence of which is that it then becomes more difficult to separate the random signal from the noise. As a rule, estimation noise in the solution increases with an increase in the modeled dynamics, such as a fast rate of turn.

In comparison with pseudorange errors, measurement (high-frequency) noise in the GPS carrier phase is virtually nonexistent, while its low-frequency noise has nonstationary characteristics that lead to instability in the position solution. Consequently, by efficiently blending the information contributed by carrier phase with that of the pseudorange, a significant reduction in the high-frequency estimation noise can be achieved while ensuring a stable solution.

PERSPECTIVE ON MEASUREMENT INTEGRATION

Two different methodologies for combining GPS pseudorange and carrier phase data are presented in this paper. Both involve stochastic least-squares filter theory, but they are quite different in their basic philosophies.

Recall that a GPS receiver provides essentially independent measurements of pseudorange and carrier phase (integrated Doppler) via the code and carrier tracking loops. In a multiple channel set, simultaneous samples of these quantities for at least four satellites are made available for further processing at a relatively rapid rate. The resulting estimation (filter) problem is one involving multiple inputs and multiple outputs because of the measurements available and the need to estimate numerous state variables, including three components of position, receiver clock bias, and clock drift. The filter problem is relatively complex, but nevertheless quite manageable using Kalman filtering methods. We wish to explore here two different approaches to the optimization problem. Before doing so, however, it is informative for tutorial purposes to look at a simpler situation of estimating one signal for which two noisy scalar measurements are available.

Consider the two different methods of combining the noisy measurements as shown in Figures 1 and 2. The merits of each of these methodologies are discussed in detail by reference [2], so we will only summarize the results here. Looking first at Figure 1, we simply say that the estimate of s(t) (denoted $\hat{s}(t)$) will be formed from a general linear operation on the two noisy measurements. Here, we do not insist on any constraint between the transfer functions G_1 and G_2. The optimization problem is posed as follows: Given the mutual correlation structure of the signal and noises (or, equivalently, their spectral characteristics), what G_1 and G_2 will minimize the mean-square error? This is a standard Wiener (or Kalman) filter-type problem, and methods for solving it are well-

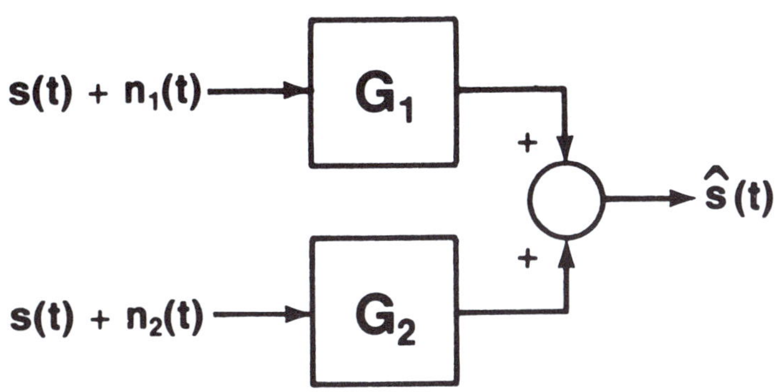

Fig. 1—General Linear Measurement Combination

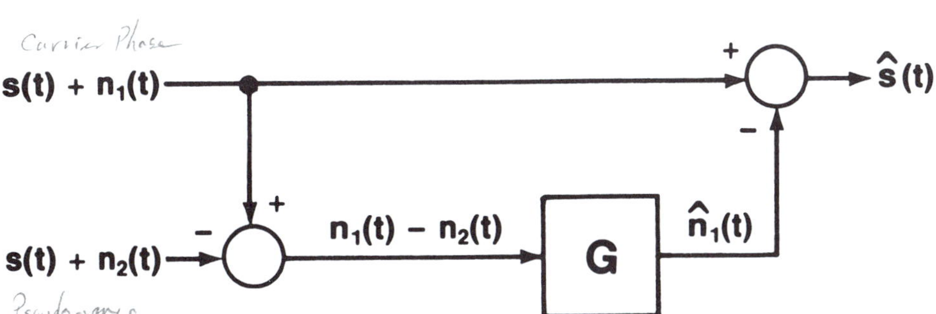

Fig. 2—Reference Trajectory Measurement Combination

known for both the discrete- and continuous-time cases [3,4]. It is important to note that, in this general setting, we are forced to assume that the signal s(t) is a stochastic process with a known correlation structure. That is, we must think of the signal as being noiselike in character. Clearly, this will not fit all physical situations. For example, in a navigation application, a vehicle may be making an intentional turn. In this case, the vehicle's position, velocity, and acceleration are better described as deterministic rather than random, and the noiselike assumption does not then fit the situation at hand. Thus, one might expect strange things to happen to the filter's output during deterministic maneuvers if the filter design is based on a stochastic model for the signal. It suffices to say that this strange behavior has been observed under such conditions in real life.

In order to avoid problems with deterministic inputs, the less-general method of combining two signals, shown in Figure 2, is often used. Here we think of one of the redundant inputs, say, $s(t) + n_1(t)$, as providing a reference trajectory, albeit a noisy one, and the other noisy measurement as assisting the first by providing corrections to the reference trajectory. Note that the filter's job in this formulation is simply to separate $n_1(t)$ and $n_2(t)$. The filter does not operate on s(t) in any way. For example, should s(t) undergo a step function, this would come through the system completely undistorted because the filter would be operating only on the noise terms. Therefore, in optimizing the transfer function G in Figure 2, one need make no rash assumptions about the stochastic character of the signal s(t). Philosophically, the methodology of Figure 2 is conservative in design; it guards against unusual events in s(t). As a result, it is suboptimal if s(t) is truly a benign, well-behaved random signal. However, a small degree of suboptimality is often a small price to pay for the additional robustness in being able to cope with unpredictable nonrandom inputs. (The resultant estimates of Figures 1 and 2 are not the same, even though the symbol $\hat{s}(t)$ is used for both.)

In the GPS problem, we will let the set of carrier phase measurements play the role of $s(t) + n_1(t)$, and the pseudorange measurements that of $s(t) + n_2(t)$. Of course, in the GPS case we have vector inputs and outputs, and the problem is further complicated by time-varying geometry and nonlinearity in the measurement relationships. Nevertheless, the two basic philosophies shown in Figures 1 and 2 for combining redundant measurements can be carried over to the GPS setting. This will be pursued in some detail in the remainder of this paper.

PROCESSING SCHEMES

The geometric relationship in the GPS measurement model which links the pseudorange (ψ) from a satellite to a user location $\mathbf{x}$ in some orthogonal coordinate frame of reference is nonlinear. To simplify the problem, this relationship can be linearized about some nominal point, $\mathbf{x}'$, such that $\mathbf{x} - \mathbf{x}'$ is small with respect to ψ. Note that all prime (′) quantities tacitly imply their association with the nominal point ($\mathbf{x}'$). This gives rise to the general GPS pseudorange measurement equation

$$\psi(t_k) - \psi'(t_k) = -\mathbf{H}(t_k)\mathbf{\Delta x}(t_k) + c\Delta t(t_k) + n_\psi(t_k), \tag{1}$$

where $\mathbf{\Delta x}(t_k)$ is $\mathbf{x}(t_k) - \mathbf{x}'(t_k)$, $\mathbf{H}(t_k)$ is unit direction vector (direction cosines), c is speed of light, $\Delta t(t_k)$ is receiver clock error, and $n_\psi(t_k)$ is pseudorange noise.

A measurement of carrier phase is, by itself, not entirely useful unless it is compared against another measurement that serves as a point of reference. By treating the problem in the context of fixed-point smoothing, we begin by looking at the following one-dimensional model:

$$\begin{aligned}\delta\phi(t_k) - \delta\phi'(t_k) &= [\phi(t_k) - \phi(t_0)] - [\phi'(t_k) - \phi'(t_0)] + n_\phi(t_k)\\ &= -H(t_k)\Delta x(t_k) + H(t_0)\Delta x(t_0) - c\Delta t(t_o) + n_\phi(t_k)\\ &\text{for } k = 1,2,\ldots \end{aligned} \tag{2}$$

where $\Delta x(t_k)$ is $x(t_k) - x'(t_k)$, $H(t_k)$ is direction cosine at t_k, c is speed of light, $\Delta t(t_k)$ is receiver clock error at t_k, and $n_\phi(t_k)$ is carrier phase noise.

Note that the problem here is to estimate $x(t_k)$ based on the sequence of code and/or carrier measurements. The initial position $x(t_0)$ is an additional state variable that is constant.

In the formulations that follow, measurement models are developed separately for the two approaches cited in the previous section. For convenience of reference, the first scheme (after Figure 1) is called the general integrated filter, and the second (after Figure 2) the complementary filter. The latter name arises from reference [2], where it is shown that the block diagram of Figure 2 is the equivalent of blending two inputs through filter transfer functions that are complements of each other. In comparison, the conventional filter model refers to the standard 8-state filter that processes pseudorange and delta range.

THE GENERAL INTEGRATED FILTER

To form a general integrated solution, equations (2) can be combined with a one-dimensional version of equation (1). Furthermore, we also choose to add range error state, ρ, to take care of the errors in the carrier phase, particularly the nonstationary ones. The resulting form of the noiseless measurement model is as follows:

$$\begin{bmatrix}\psi(t_k) - \psi'(t_k)\\ \delta\phi(t_k) - \delta\phi'(t_k)\end{bmatrix} = \begin{bmatrix}-H(t_k) & 1 & 0 & 0 & 0\\ -H(t_k) & 1 & H(t_0) & -1 & 1\end{bmatrix}\begin{bmatrix}\Delta x(t_k)\\ c\Delta t(t_k)\\ \Delta x(t_0)\\ c\Delta t(t_0)\\ \rho(t_k)\end{bmatrix} \tag{3}$$

From this, the extension to the four-dimensional GPS model is straightforward. Each of the position states, $\Delta x(t_k)$ and $\Delta x(t_0)$, from equation (3) becomes a 3-tuple state vector in the GPS model, and velocity states are added to accommodate spatial as well as clock dynamics. The range error part of the state vector, on the other hand, consists of random walk states, one assigned for each of the satellites contributing to the solution. To summarize, this optimal model is made up of 16 states and processes a pair of pseudorange and carrier phase measurements from each of the four satellites. These states are as follows: x_1, x_2, x_3, the orthogonal position (integrated random walk); x_4, the range bias (integrated random walk); x_5, x_6, x_7, the orthogonal velocity (random walk); x_8, the range drift (random walk); x_9, x_{10}, x_{11}, the initial position (random constant); x_{12}, the initial range bias (random constant); and x_{13}, x_{14}, x_{15}, x_{16}, the independent range error states associated with each one of four satellites (random walk). (Additional range error states may be added for more satellites.)

Although it may seem at first glance that combining the set of states x_9 through x_{12} with the set x_{13} through x_{16} is trivial, the resulting reduced-order 12-state filter is, in fact, not the same as the original 16-state one. The difference lies in the linear measurement connection H, which is time-varying for GPS, and hence is not the same at the initial point in time for $[x_9..x_{12}]$ as it is at the present, to which $[x_{13}..x_{16}]$ corresponds. The reduced-order filter is, therefore, suboptimal. By removing the constant constraint and allowing the initial states $[x_9..x_{12}]$ to random walk, we could, in a loose way, take care of the function served by the range error states $[x_{13}..x_{16}]$, which could then be dispensed with. The compromise on optimality here is a good one given the computational savings achieved by using a reduced-order filter.

THE COMPLEMENTARY FILTER

In contrast to the general integrated scheme, the complementary filter processes the information derived from the carrier phase and that derived from the pseudorange separately. To take advantage of the low white noise level of the carrier phase data, no filtering is performed on these data. A position solution is obtained with respect to a nominal point by directly solving the linearized simultaneous equations deterministically.

$$\begin{aligned}\delta\phi_i\,(t_k) - \delta\phi_i'(t_k) &= [\phi_i(t_k) - \phi_i(t_{k-1})] - [\phi_i'(t_k) - \phi_i'(t_{k-1})]\\ &= h_{ix}\,\Delta x(t_k) + h_{iy}\,\Delta y(t_k) + h_{iz}\,\Delta z(t_k) + \Delta t(t_k),\\ &\text{for } i = 1, 2, 3, 4\end{aligned} \tag{4}$$

where $\phi_i\,(t_k)$ is carrier phase measurement at t_k; h_{ix}, h_{iy}, h_{iz} is components of the ith unit direction vector; $\Delta x(t_k)$, $\Delta y(t_k)$, $\Delta z(t_k)$ is position error with respect to the reference location; and $\Delta t(t_k)$ is receiver clock error.

Since only the three components of position are of interest, a simplification intended to reduce the order of the system of equations was made by differencing out the Δt. Note that the differencing scheme assumes that the four measurements were made simultaneously. Hence, for time t_k, we difference across satellites and obtain

$$\begin{bmatrix}(\delta\phi_1-\delta\phi_2) - (\delta\phi_1'-\delta\phi_2')\\ (\delta\phi_2-\delta\phi_3) - (\delta\phi_2'-\delta\phi_3')\\ (\delta\phi_3-\delta\phi_4) - (\delta\phi_3'-\delta\phi_4')\end{bmatrix} = \begin{bmatrix}h_{1x}-h_{2x} & h_{1y}-h_{2y} & h_{1z}-h_{2z}\\ h_{2x}-h_{3x} & h_{2y}-h_{3y} & h_{2z}-h_{3z}\\ h_{3x}-h_{4x} & h_{3y}-h_{4y} & h_{3z}-h_{4z}\end{bmatrix}\begin{bmatrix}\Delta x\\ \Delta y\\ \Delta z\end{bmatrix} \tag{5}$$

The position vector $(\mathbf{x} - \mathbf{x}') = [\Delta x\ \Delta y\ \Delta z]^T$ can then be solved from the system of simultaneous equations in equation (5).

This carrier phase-derived position is then used as the position reference trajectory. It is dynamically exact except for noise effects, and the Kalman filter simply uses the pseudorange measurements to estimate the deviation of the true trajectory from the carrier-derived reference. This estimate is then used to correct the reference trajectory (see Figure 2). Without pseudorange updates, the uncertainty in the reference position will increase without bound, not unlike the case for a pure inertial system. This is the effect of nonstationary errors that result from integrating errors in the GPS Doppler in the process of obtaining position.

what then appears to the Kalman filter as just the static part. Thus, any unusual event that affects the signal relating to the dynamical motion is masked out. The processing model takes advantage of this fact by allowing only for the 3 position- and the 2 clock-related states. This results in a simple 5-state model. Reducing the order of the estimation, where it is proper to do so, likewise improves the accuracy of the estimation.

The navigation solution obtained from the complementary filter does not directly yield velocity, simply because the dynamical contribution to the signal was eliminated in the process. To obtain velocity, simple smoothing can be carried out on single-step finite differences of the carrier phase (equivalent to estimates of the Doppler frequency). The details are not pursued here.

SIMULATION RESULTS

The simulation experiments carried out to test the processing schemes were designed with the use of live GPS data collected with a stationary receiver. Also, to verify that the carrier phase data collected were free of cycle slips, additional data were also collected simultaneously by a second receiver. Then, using the double-differencing principle (a concept from precision GPS surveying), the two sets of carrier phase data were compared with the assumption that single slips occur, but rarely under good observation conditions. It did not take long to obtain several clean data records for use in the simulation experiments.

In Figure 3, the plots show the three-dimensional navigation solution processed by a general integrated filter for stationary data. In contrast, Figure 4 shows a noisier solution produced by a conventional 8-state Kalman filter processing pseudorange and delta range from the same data set. The comparison shows an unmistakable reduction of high-frequency noise content in the estimated solution by about a factor of four. The Kalman filter measurement noise parameters used here were $(20\text{ m})^2$ for pseudorange and $(0.02\text{ m})^2$ for carrier phase. The amount of "smoothing" that can be achieved invariably depends on the tuning of the associated filter. Not surprisingly, the performance results obtained for the complementary filter are almost identical to those shown in Figure 3 for the general integrated filter in these stationary circumstances.

In order to assess performance with dynamics, stationary data properly adjusted by correct amounts for a receiver traveling along a specified trajectory were used. In this case, the chosen trajectory laid out on a locally horizontal plane was perfectly circular, with a radius of 250 m. With a period of 30 s, the effect of a constant 1-g acceleration that changes in orientation on the moving receiver was simulated. In each simulation run, the receiver remained motionless for the first 30 s. The circle was then traversed five complete times before the receiver again came to a halt at the very point at which it had started.

The two-dimensional traces seen in Figures 5(a), (b), and (c) show the estimated trajectories processed respectively by an 8-state conventional filter, a 16-state general integrated filter, and a 5-state complementary filter. With the benefit of using the carrier phase, the overlapping trajectories estimated by the general integrated and complementary schemes are unquestionably cleaner than those of the conventional solution. Upon closer examination of the two-

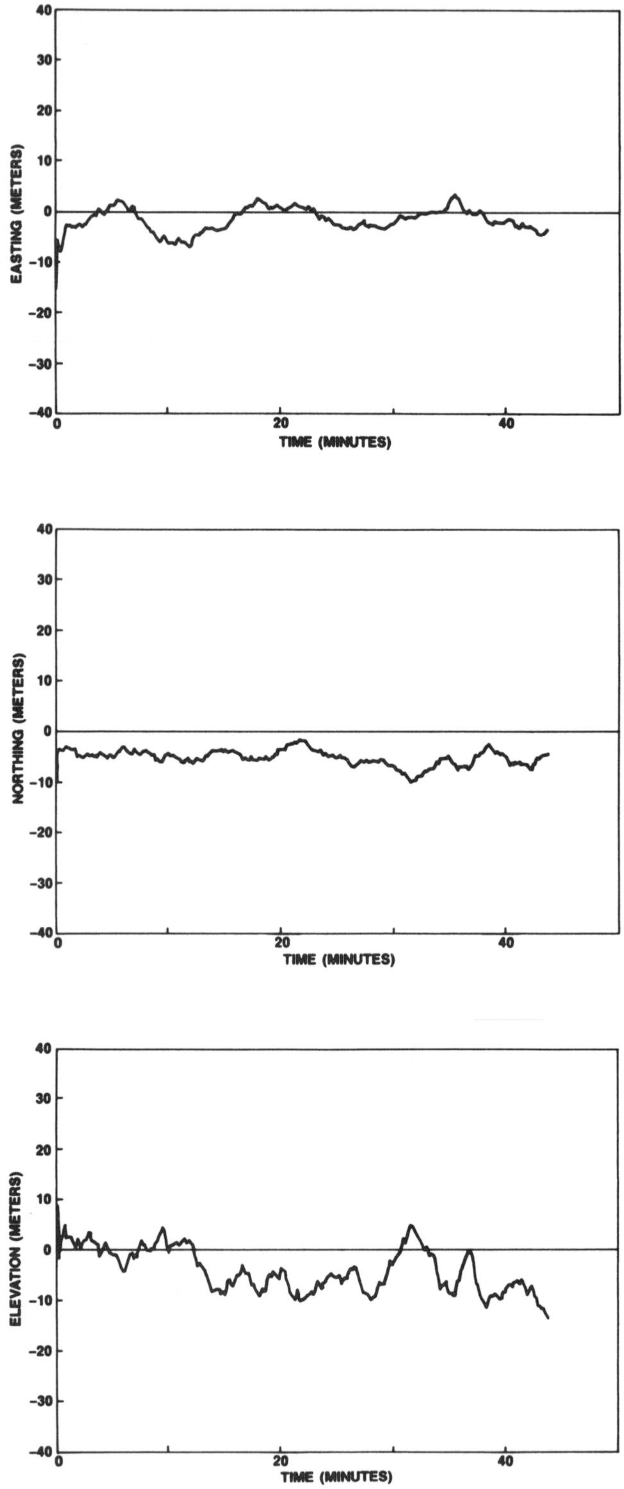

Fig. 3—Integrated Filter 3-D Stationary Position Solution

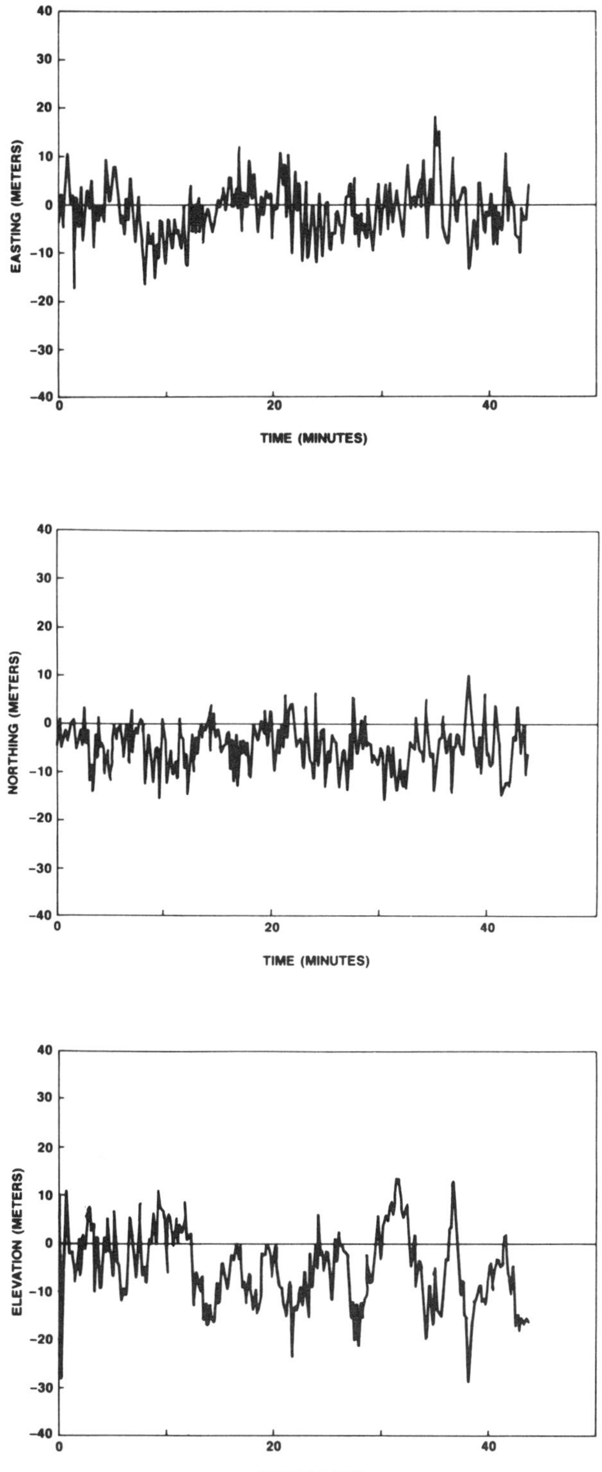

Fig. 4—Conventional 3-D Stationary Position Solution

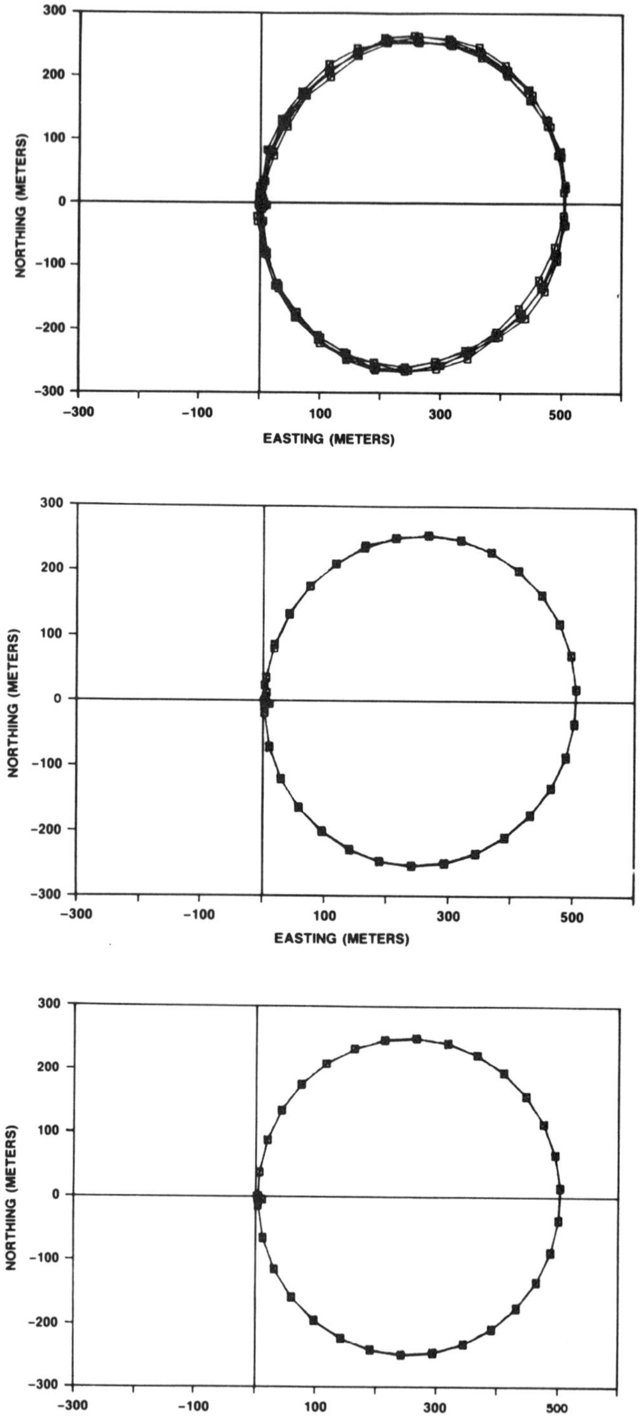

Fig. 5—2-D Position Estimation in 1 g Dynamics
a) Conventional Filter
b) Integrated Filter
c) Complementary Filter

dimensional position errors, they show distinct oscillations in the conventional as well as the general integrated filter solutions (Figures 6(a) and (b)). These errors are the consequence of processing filter lags arising from assuming models that do not accommodate acceleration (and higher-order dynamics) well. Because of the accuracy of the carrier phase over the pseudorange, the lag-related errors are about four times smaller for the general integrated filter than for the conventional filter. The spikes in the general integrated filter solution at the start and the end of the dynamical motion are due to the simplistic step function changes in the velocity profile.

On the other hand, the 5-state complementary filter exhibits none of the above acceleration-dependent effects and produces a very clean error solution, as shown in the results of Figure 6(c). These results seem to affirm the tradeoff philosophy behind the complementary design. It is much less sensitive to the unpredictable nature of the dynamics while being less formal about the stochastic structure of the combined measurements than the general integrated filter. In real-life navigation, this would be a prudent tradeoff.

CONSIDERATIONS OF CYCLE SLIPS

An important issue that is almost always related to the processing of carrier phase data concerns cycle slips, the problem associated with an undetected loss or introduction of an integral number of cycles to the cumulative count. Since the observability of a pure carrier tone is restricted to some fraction of one cycle, the undetected occurrence of a cycle slip is obviously not self-correcting. The effect of a cycle slip on the navigation solution processed by either of the schemes presented above is explored in this section.

Figure 7 shows the difference in the position solution between a simulation run with and without a contrived slip of two full cycles inserted at the 20 min time mark. The example chosen here involves data processed by the complementary filter. The deviation between the two solutions begins at the point of the cycle slip. The size of the initial jump at that instance depends on the size of the slip and the satellite geometry. At first, immediately following the cycle slip, the solution is pulled off sharply by the corrupted carrier phase data. Then, with the passage of time, the cycle slip that exists in the carrier phase information gets absorbed by the random walk processes that are built into the Kalman filter model.

This causes information originally dominated by the cycle slip to be deemphasized in deference to the more stable pseudorange data. The gradual and eventual readaptation of the erroneous solution to follow the correct one indicates the self-correcting nature of the blending of pseudorange and carrier phase information. Clearly, the navigation solution is quite robust to the effect of slipping even a few cycles, if the frequency of occurrence is low.

To be practical, however, there is need for a detection and possibly a correction scheme to guard against very large slips. When signal conditions are poor, a receiver may have trouble tracking the carrier of a satellite signal properly. This may even lead to a momentary loss of track. In such situations, slips of perhaps several thousands of cycles may occur. A conservative strategy may well be to terminate the processing of carrier phase data and resume when the conditions improve. If we consider, however, that such situations usually

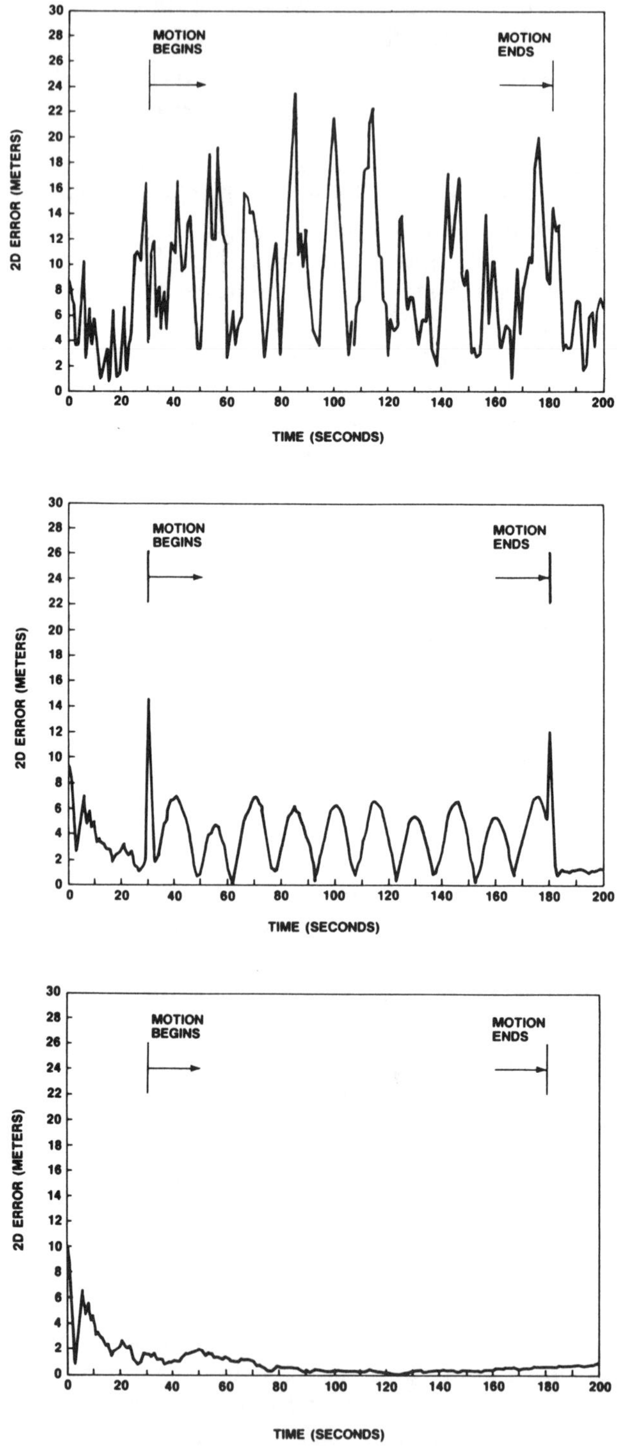

Fig. 6—2-D Position Error Magnitude in 1 g Dynamics
a) Conventional Filter
b) Integrated Filter
c) Complementary Filter

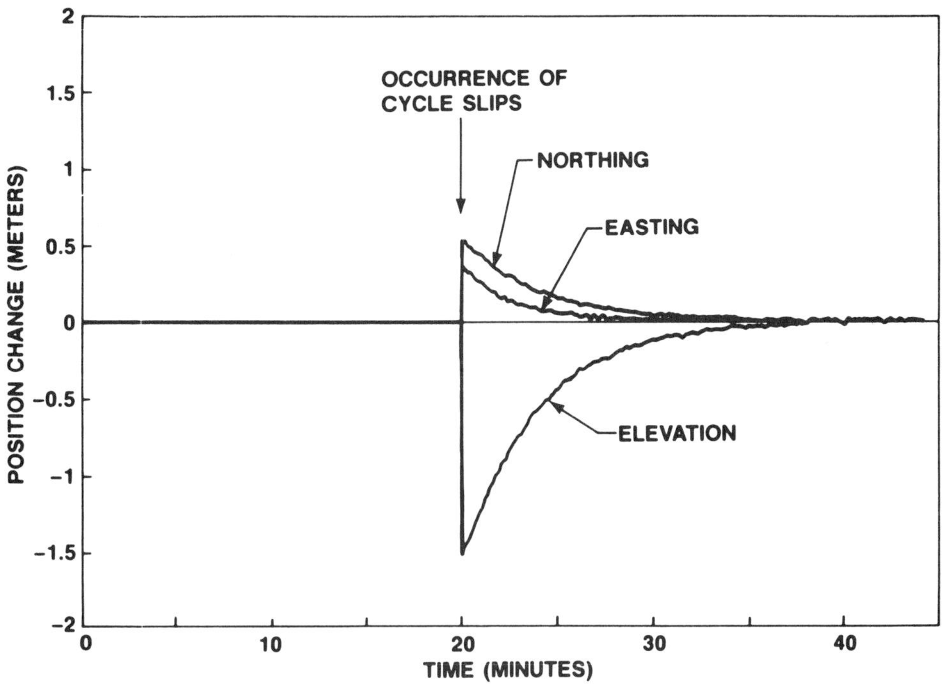

Fig. 7—Position Difference with and without Slip of 2 Cycles

involve just one satellite at a time, as in the case of signal obscuration, then there may be better strategies available.

One such strategy calls for using the redundancy of more than four satellites to facilitate the detection of cycle slips. Much research work on the use of redundancy for failure detection has been dedicated to the issue of GPS integrity for flight safety [5–7]. Fortuitously, the results of this collective research effort can be put to good use here for cycle slip detection. The results summarized by Figure 8 show a chi-square detection statistic used to flag the occurrence of a cycle slip with a jump. The size of the jump clearly varies depending on the satellite with which the slip is associated. Naturally, the higher the jump, the greater the detectability of the slip. The example used in Figure 8 shows that the detectability of all five satellites is ensured with respect to the 0.1 percent false alarm threshold chosen.

In addition to the size of the cycle slip, this detectability factor is also a function of satellite geometry. For the case of five satellites, suppose that the satellite associated with the slip is removed from consideration. Then the detectability of that slip is dependent on the geometry of the remaining four satellites. Poor geometry here will hamper the detection scheme in its task of distinguishing between a poor four-satellite solution due to the geometric conditions and a poor five-satellite solution due to bad data. Although this problem of unreliable geometry is something that continually unsettles the flight safety issue, it does not seem as critical here only because the solutions we are considering appear to be tolerant of a relatively larger error (one cycle

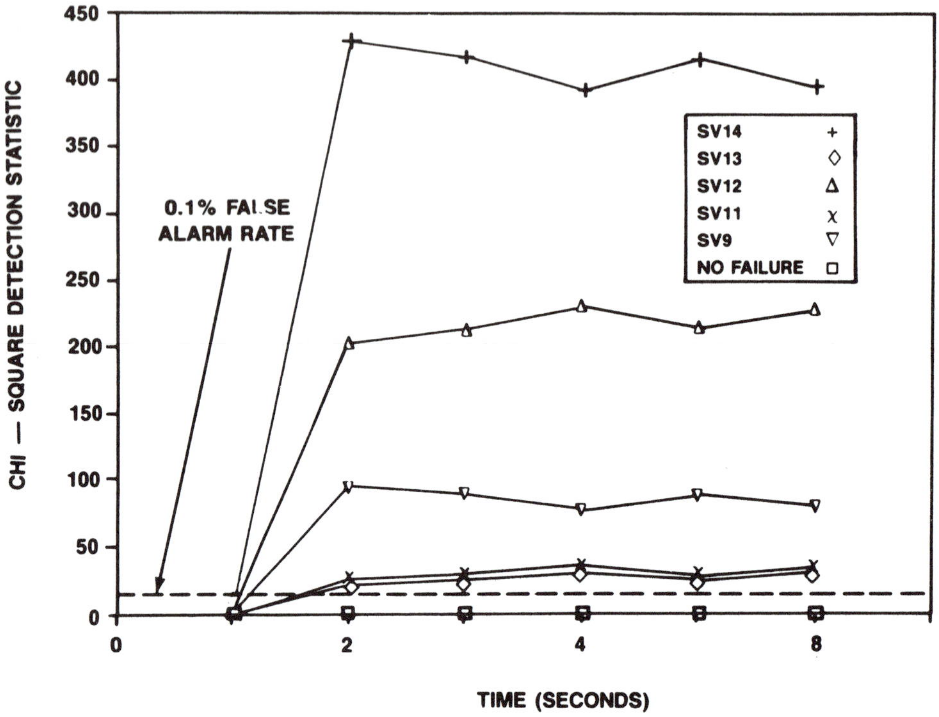

Fig. 8—Dependence of Chi-Square Detection Sensitivity on Satellite Geometry

is larger than 20 times the carrier phase noise). And the larger the slip, the better the detectability.

A brief discussion on the definition of poor geometry is in order here. While GDOP is a measure of estimation accuracy by code ranging, it is only indirectly related to the measure of accuracy for estimation by carrier phase. Previous work with carrier phase in the precision surveying field has established the rate of GDOP change, rather than GDOP itself, as the more relevant measure of estimation accuracy here [8]. In other words, the observability conditions improve the faster the GDOP changes with time.

In general, a sixth satellite is quite often needed for the isolation and correction of the failure. More specifically in our case, there is an easier alternative for addressing the catastrophic situation of loss of lock in carrier tracking. Indicators (hardware or software) built to monitor the carrier tracking loop conditions for each channel in the receiver should be just as effective in providing information to isolate the failure. This negates the requirement to track a sixth satellite for the purpose of cycle slip correction.

CONCLUSIONS

In this paper, two alternative schemes for processing carrier phase with pseudorange data are presented, together with results from simulation runs. The choice between the contenders is fairly clear. The complementary filter has an edge over the general integrated filter in performance for dynamical navigation applications and also calls for a lower-order Kalman filter better suited for implementation.

Although both schemes work in much the same way in cases of constant velocity, with about a four-to-one reduction in the high-frequency estimation noise, the general integrated scheme exhibits a significant amount of error due to the processing filter lag in the solution when exposed to a 1 g acceleration scenario. This error is also about one-fourth the size of that associated with a conventional processing scheme undergoing the same dynamics. On the other hand, the complementary filter exhibited little or no error due to dynamics. It is expected that, under conditions of extremely high dynamics, small errors due to tracking loop filter lags are unavoidable. Even so, with the complementary approach, there should be no such contribution coming from the processing filter itself.

It is apparent that, with the proper strategy for detecting and even correcting for cycle slips, uninterrupted operation of code and carrier navigation can be realized for long periods of time. One particularly promising strategy entails the use of a fifth satellite to detect cycle slips. This can be augmented further either by additional satellites or by some loss-of-lock tracking loop monitoring scheme to identify the failure source and permit correction of the anomalous data associated with it.

In summary, the authors suggest that the complementary filter scheme presented here can be used to achieve many of the same benefits that are derived from the addition of an inertial system to GPS, but without the additional expense. The carrier-derived position trajectory plays the same role in the complementary filter scheme as does the inertially-derived position in the integrated INS/GPS system. In both cases, the pseudorange data are simply used, via a Kalman filter, to update the reference trajectory at periodic intervals. The analogy is, of course, not perfect because the carrier phase data cannot provide vehicle attitude or low-noise instantaneous velocity, as does the INS. However, if position is the quantity of prime interest, the complementary filter can provide an effective means of mitigating the effect of high dynamics, and it can be used to obtain highly accurate instantaneous position information.

Based on a paper presented at The Institute of Navigation Satellite Division Meeting, Colorado Springs, CO, September 1989.

REFERENCES

1. Hatch, R. R., *The Synergism of GPS Code and Carrier Measurements,* Proceedings of the Third International Geodetic Symposium on Satellite Doppler Positioning, Las Cruces, NM, February 1982, pp. 1213–32.
2. Brown, R. G., *Integrated Navigation Systems and Kalman Filtering: A Perspective,* NAVIGATION, Journal of The Institute of Navigation, Vol. 19, No. 4, Winter 1972–73, pp. 355–62.
3. Brown, R. G., *Introduction to Random Signal Analysis and Kalman Filtering,* Wiley, New York, 1983.
4. Brown, R. G. and Nilsson, J. W., *Introduction to Linear Systems Analysis,* Wiley, New York, 1962.
5. Lee, Y. C., *Analysis of Range and Position Comparison Methods As a Means to Provide GPS Integrity in the User Receiver,* Proceedings of the Forty-Second

Annual Meeting of The Institute of Navigation, Seattle, WA, June 24–26, 1986, pp. 1-4.

6. Parkinson, B. W. and Axelrad, P., *Autonomous GPS Integrity Monitoring Using the Pseudorange Residual,* NAVIGATION, Journal of The Institute of Navigation, Vol. 35, No. 2, Summer 1988, pp. 255–74.
7. Brown, R. G. and McBurney, P. W., *Self-Contained GPS Integrity Check Using Maximum Solution Separation,* NAVIGATION, Journal of The Institute of Navigation, Vol. 35, No. 1, Spring 1988, pp. 41–53.
8. Ashjaee, J. M., *GPS Receiver Technologies,* Proceedings of the Forty-Second Annual Meeting of The Institute of Navigation, Seattle, WA, June 24–26, 1986, pp. 99–105.

NAVIGATION: *Journal of The Institute of Navigation*
Vol. 37, No. 3, Fall 1990
Printed in U.S.A.

Optimal Filtering of Pseudoranges and Phases from Single-Frequency GPS Receivers

CLYDE C. GOAD

The Ohio State University, Columbus, Ohio
Received February 1990
Revised June 1990

ABSTRACT

The filtering of observed pseudoranges with concurrent measures of phase or integrated Doppler can benefit significantly both absolute and differential positioning. The argument proceeds rather simply. If the position recovery uncertainty is predicted to be equal to the position dilution of precision (PDOP) times the range uncertainty, then reducing the range error will in turn reduce the recovered position uncertainty. With typical PDOP values of, say, 4 m/m, then for every meter improvement of the pseudorange, the recovered position uncertainty will also be reduced by 4 m. It has been argued that very significant improvements can be made when incorporating millimeter-level phases into pseudorange filtering algorithms. Here the filtering algorithm is examined, with emphasis on the ionospheric contribution. It will be argued that some degradation in the filter performance must be tolerated in order to accommodate this error source when using single-frequency receivers. During periods of peak solar activity or in geographic areas where unpredictable ionospheric activity is common, it will be seen that the optimal filter is one that averages the pseudorange and phase range measures, which then causes an improvement in the filtered pseudoranges of one-half the raw pseudorange noise. This improvement is significant. However, it is not as spectacular in its reduction as others have claimed who considered a cancellation of ionospheric contribution to the 10^{-6} level for relative or differential positioning.

INTRODUCTION

Since about 1985, geodetic quality receivers that deliver to the user carrier phase measures in addition to the code phase measures (pseudoranges) in less expensive coarse acquisition (C/A) versions have been used for a host of scientific experiments and engineering applications. Naturally, as the public consumption has increased, accompanied by associated competition in the marketplace, there have been improvements in product reliability, as well as major reductions in cost. In this environment, one continually discovers new applications of this technology where spatially referenced information is required. The first group to recognize the usefulness of this technology was the geodetic/survey community. Since 1983, even though satellites were not available except for a few hours each day, centimeter-level surveying still could be accomplished over distances that were essentially unaffordable before GPS. Today, as more

GPS satellites are launched, the number of uses begins to increase as we approach 24 h coverage. For example, the daily window of opportunity is now almost 8 h in the United States. This makes it possible for one project, at the Ohio State University Center for Mapping, to begin to develop a prototype system for mapping the center lines of roadways. While possible in the past, this is now also economically viable. The system software is being written to accept data from any manufacturer's receiver so long as that manufacturer provides measures of pseudorange, carrier phase, and broadcast orbit information in the Receiver Independent Exchange Format (Rinex)[1].

A major ingredient in the Ohio State University development is to provide reliable position recoveries. The optimal combination of phase and pseudorange is a very attractive opportunity. This would allow one to use a lower-cost C/A receiver and to combine phase and pseudorange, producing filtered C/A pseudoranges that would potentially have the error characteristics of the more expensive P-code receiver pseudoranges. It has been argued that the almost perfect carrier phase measures, or alternatively integrated Doppler measures, can be used to reduce single-frequency pseudorange noise to 10 cm [2] in differential tracking mode. This is possible only when major error sources are similar at each receiver in the differencing scheme. Here a generalized filter that accommodates both noise and differential error combinations is given. With this implementation, there is no need for restarting or "ramping" because of oppositely signed ionospheric effects on code and phase, as others have proposed. Passing through the data in both forward and backward directions can be done in a post-mission mode that allows for optimal smoothing of the data. Emphasis is placed on the terms filtered and smoothed: filtering refers to optimal estimation when no future data are available (real time); smoothed results are available only in post-mission mode, where the optimal estimate can be obtained from past, present, and future measurements. The implementation presented here allows for both types. Extension to dual-frequency receivers as originally suggested by [3] is easily accomplished.

SINGLE-FREQUENCY CODE AND CARRIER

Much development has taken place over the last five years, based mainly on precise survey requirements, to provide low-cost receivers that can take advantage of the broadcast information, such as the orbital prediction models, ionospheric prediction model, and pseudorange. This class of receivers generally has been developed to track the C/A signal, which is available only on the L_1 frequency. For precise surveying, the carrier phase is measured in addition to the pseudorange (code phase). To a great degree, the uncertainty of each of these measures depends on the dwell time and subsequent smoothing, which is very much manufacturer-dependent. However, for the sake of discussion, it will be assumed that the generic C/A-code receiver delivers to the user pseudoranges at the 1 m level and 1 mm phase ranges. P-code receivers exhibit better pseudorange resolution, usually at 70 cm or lower. The mathematical models of these measurements are as follows:

$$\text{(Pseudorange)} \qquad R = \rho + \frac{I}{f^2} + v_R \tag{1a}$$

(Phase Range) $$\phi = \rho - \frac{I}{f^2} + k\lambda + v_\phi \tag{1b}$$

where ρ is the geometrical distance between transmitter and receiver, plus nondispersive contributions such as tropospheric refraction and clock drift; I is the contribution of ionospheric refraction; λ is the L_1 carrier wavelength (19 cm); f is transmission frequency; k is an integer; and v denotes noise.

Here the carrier phase measurements have been scaled by the transmission wavelength to yield a measure with length units like those of the pseudorange. Notice that the phase range and pseudorange have oppositely signed ionospheric contributions (known as phase advance/group delay), and that the phase ranges are measured only to a fraction of a wavelength at satellite acquisition, which accounts for the (integer) bias $k\lambda$. The I values are positive, which makes the observed pseudoranges longer. The presence of these dispersive ionospheric contributions causes very real problems in processing of single-frequency (usually L1) data. The GPS control segment includes model parameters in the broadcast information to be used to eliminate the I terms in equations (1a) and (1b). These models are believed to be good to 50 percent of the correction [4]. So now, one can rewrite equations (1a) and (1b) as follows:

$$R = \rho + \frac{\delta I}{f^2} + v_R \tag{2a}$$

$$\phi = \rho - \frac{\delta I}{f^2} + k\lambda + v_\phi \tag{2b}$$

where the I terms have been replaced with δI to represent the residual contribution not contained in the ionospheric refraction value computed from the broadcast model.

The problem now is one of estimating the δI values from single-frequency data. First, it is easily shown that any offset in the δI cannot be distinguished from the bias $k\lambda$ based solely on the two single-frequency measurements of pseudorange and carrier phase. One possible additional piece of information is the expected value of zero on the model correction. However, since the model correction is in the 2–10 m range, an uncertainty of 50 percent offers very little improvement relative to the noise values. Also, the 50 percent statistic does not represent sudden (large) variations due to abrupt change in the ionospheric contribution, usually of solar origin.

However, what can be measured is the time-varying character of δI. Differencing equations (2a) and (2b) yields twice the ionospheric contribution, plus a bias. Since the bias remains the same when lock on the signal is maintained, the change from one epoch to another is available. Thus, equations (2a) and (2b) are now rewritten in terms of the quantities that can be distinguished. Letting n denote the epoch number, equations (2a) and (2b) are rewritten as follows:

$$R_n = \left(\rho_n + \frac{\delta I_1}{f^2}\right) + \left(\frac{\delta I_n - \delta I_1}{f^2}\right) + v_{R_n} \tag{3a}$$

$$\phi_n = \left(\rho_n + \frac{\delta I_1}{f^2}\right) - \left(\frac{\delta I_n - \delta I_1}{f^2}\right) + \left(k - 2\frac{\delta I_1}{\lambda f^2}\right)\lambda + v_{\phi_n} \tag{3b}$$

or equivalently

$$R_n = \rho_n^* + s_n + V_{R_n} \tag{4a}$$

$$\phi_n = \rho_n^* - s_n + k^*\lambda + V_{\phi_n} \tag{4b}$$

where ρ^* represents the combination of ρ plus the error in the initial ionospheric correction, s_n is the difference in ionospheric error at the nth epoch with the error at the first epoch, and k^* is now a modified bias which is no longer an integer.

These representations are not very much different from those presented in [3]. However, here the use of phase is emphasized, along with potentially harmful effects of the ionosphere when incorrect filter values are used. If the ionospheric corrections were without error, then equations (4a) and (4b) would represent the ideal, with $\rho_n^* \equiv \rho_n$, $s_n \equiv 0$, and $k^* \equiv k$.

Now that the stage has been set, the following question is posed: How and to what extent can the measures of phase be used to improve the meter-level measure of pseudorange? This is then an estimation problem. From equations (4a) and (4b), it is seen that ρ^* is the quantity to be estimated (improved) using ϕ and R. One could choose to use the sequential differences in time of phase, which is mathematically equivalent to integrated Doppler, in order to eliminate the bias $k^*\lambda$. This is an acceptable procedure so long as the induced correlations are incorporated into the estimation procedure properly. Here, to avoid this correlation, the bias $k^*\lambda$ is included in the vector of estimable quantities. In the construction of the optimal filter, it will be assumed that the data are collected in navigation mode (i.e., the receiver is moving arbitrarily). Thus the prediction capability of ρ^* due to the unknown motion has an uncertainty very much larger than the assumed 1 m measurement noise. The parameter prediction model from sample (n − 1) to sample n is then given by the following matrix equation:

$$\mathbf{x}_n = \mathbf{\Phi}_{n-1}\,\mathbf{x}_{n-1} + \mathbf{w}_{n-1}, \tag{5}$$

where $\mathbf{x}_n^T = [\rho_n^*, s_n, k^*\lambda]$, and $\mathbf{\Phi}_{n-1}$ is the state transition matrix. The $\mathbf{w}_{n-1}$ is a vector representing the error (whose expectation is zero) in the prediction from sample (n − 1) to sample n, with $E\{\mathbf{w}_{n-1}\mathbf{w}_{n-1}^T\} = \mathbf{Q}_{n-1}$ representing the 3 x 3 covariance matrix associated with this predictor error.

The state transition matrix, $\mathbf{\Phi}$, is defined as follows:

$$\mathbf{\Phi}_{n-1} = \begin{bmatrix} 1 & 0 & 0 \\ 0 & e^{-|\Delta t|/\tau} & 0 \\ 0 & 0 & 1 \end{bmatrix}$$

where τ is a constant and Δt is the time interval $t_n - t_{n-1}$. It will be shown that the ρ^* corrector relies entirely on the measurements since no knowledge of receiver motion is used. The change in ionosphere assumes an exponential decay model, with τ the correlation time. The model used here is rather insensitive to the value of τ since the driving noise component will dominate. This model is chosen really to enforce the concept of a zero mean. That is, if a large time interval is encountered (say, several hours), then the best predictor of the ionospheric error is zero, regardless of any knowledge at the previous time. The predictor of $k^*\lambda$ is without additional error since this quantity is truly a bias if no loss of lock on the carrier tracker occurs. Loss of lock situations can be accommodated by supplying a large value in the corresponding location of the driving noise matrix.

Letting the 3 x 3 matrix $P_{n-1}(+)$ be the previous updated covariance matrix of the estimate of ρ^*, s, and $k^*\lambda$, then the predicted state covariance matrix is as follows:

$$\mathbf{P}_n(-) = \mathbf{\Phi}_{n-1}\,\mathbf{P}_{n-1}(+)\,\mathbf{\Phi}^T_{n-1} + \mathbf{Q}_{n-1}, \tag{6}$$

The very uncertain prediction of ρ^* is realized by setting the (1,1) element of $\mathbf{Q}$ to infinity. As long as tracking of the satellite signals is maintained (no loss of lock), there is no additional error in the prediction of $k^*\lambda$ since it is a constant. As stated earlier, the uncertainty in the prediction of s is influenced mainly by the (2,2) element of $\mathbf{Q}$. Here it is assumed that the uncertainty in the ionospheric model prediction will increase linearly with prediction time, and that the (2,2) element will have the form $(Q_s\Delta t)^2$, where Q_s is a constant, and Δt is the same time interval discussed earlier $(t_n - t_{n-1})$. The other elements of $\mathbf{Q}$ are zero. For very quiet periods, a small value of Q_s can be chosen, e.g., 0.25 m/min. Periods of moderate to active solar activity require the use of a larger value, e.g., 1.0 m/min. A Q_s value of zero corresponds to a filter which assumes that the integrated Doppler, or equivalent carrier phase differences in time, yields very accurate measures of range change. Equation (6) evaluates to:

$$\mathbf{P}_n(-) = \mathbf{\Phi}_{n-1}\mathbf{P}_{n-1}(+)\mathbf{\Phi}^T_{n-1} + \begin{bmatrix} \infty & 0 & 0 \\ 0 & (Q_s\Delta t)^2 & 0 \\ 0 & 0 & 0 \end{bmatrix} \tag{7}$$

The inverse of $\mathbf{P}_n(-)$ is required in the filter equations. Assuming all the elements of $\mathbf{P}_n(-)$ to be finite, inverting, and then taking the limit as the (1,1) element goes to infinity, the following can be shown:

$$\mathbf{P}_n^{-1}(-) = \begin{bmatrix} \infty & p_{12} & p_{13} \\ p_{21} & p_{22} & p_{23} \\ p_{31} & p_{32} & p_{33} \end{bmatrix}^{-1} = \begin{bmatrix} 0 & 0 & 0 \\ 0 & \multicolumn{2}{c}{\multirow{2}{*}{$\begin{bmatrix} p_{22} & p_{23} \\ p_{32} & p_{33} \end{bmatrix}^{-1}$}} \\ 0 & & \end{bmatrix}$$

more preliminary step is required to generate a model of the measurements in terms of the parameters ρ^*, s, and $k^*\lambda$. After examination of equations (4a) and (4b), the following equivalent measurement model is constructed:

$$\mathbf{z}_n = \mathbf{H}_n\mathbf{x}_n + \mathbf{v}_n \tag{8}$$

with $\mathbf{H} = \begin{bmatrix} 1 & 1 & 0 \\ 1 & -1 & 1 \end{bmatrix}$.

IONOSPHERIC BEHAVIOR

In [5], it is stated that the GPS broadcast model evaluation will normally be good to 50 percent. Typical corrections are in the neighborhood of 4–12 m. However, this broadcast model generally cannot be expected to accommodate the very large excursions during intense solar activity. It is also stated in [5] that while the correlation distance averaged over several hours is of the order of 1000 mi, at an instant this correlation distance will be very much smaller, of the order of 100 km, which is common during periods of traveling ionospheric disturbances (TIDs). Here a conservative approach will be taken that errors in the ionospheric corrections do not completely cancel over distances where pseudoranges will be used to infer differential coordinates. It is understood that precision could suffer during periods of similar ionospheric behavior, but it is hoped that unpredictable excursions over local areas will not invalidate the results of the filtered pseudoranges. Thus typical values for the uncertainty in the ionospheric correction will be in the range of 2–6 m.

In 1983, when the first single-frequency (noncode) phase recording receivers were introduced, it was found that they could recover baselines good to one part per million (10^{-6}) of baseline length. However, now that we are in the middle of the peak of the 11 year solar cycle, baseline recoveries from single-frequency receiver data generally are not at the 1983 precision. Recently, [6] reported on the results of an experiment conducted in November 1988, designed to test the spatial variation in measurements collected at the earth's surface due to the ionosphere. The "Nautilus" network, located in Texas, consisted of 12 GPS receivers placed in an exponential spiral, covering about 300 km. Spatial ionospheric variabilities equal to 10^{-5} times baseline length were found. This means there could be 1 m differences for every 100 km of receiver separation for the case of differential positioning.

During December 1988, researchers from the Ohio State University Byrd Polar Research Center collected GPS data in Antarctica with two Texas Instruments TI 4100 receivers over a baseline of 9 km. Since the TI 4100 is a dual-frequency receiver, the precise phases were used to analyze the spatial difference between the ionospheric changes in time over this 9 km baseline. Figure 1 shows the observed ionospheric drift obtained from dual-frequency carrier phase measurements at stations 90 and 91 for Satellite 12 for an 80 min period (L1 frequency). One first notices the similarity, which is expected since the receivers are fairly close. The difference of these curves is given in Figure 2. Here one sees clearly that even at distances of 9 km, at times there are variations of 0.5 m (55 ppm)! Variations at the 10 cm level are common (10 ppm), which agrees with the result in [6]. Data from two satellites on two days (351, 352) were analyzed. Day 352 was quieter. Table 1 gives the RMS scatter from the two satellites on each day. Values ranging from 2 to 20 ppm are observed. Thus, to obtain optimal results, some information regarding this error must be included in both absolute and differential positioning algorithms using data from single-frequency receivers.

THE FILTER

With equations (3)–(8), the filter can be generated. Because of the presence of infinity in the predicted state covariance in equation (7), the Bayes sequential

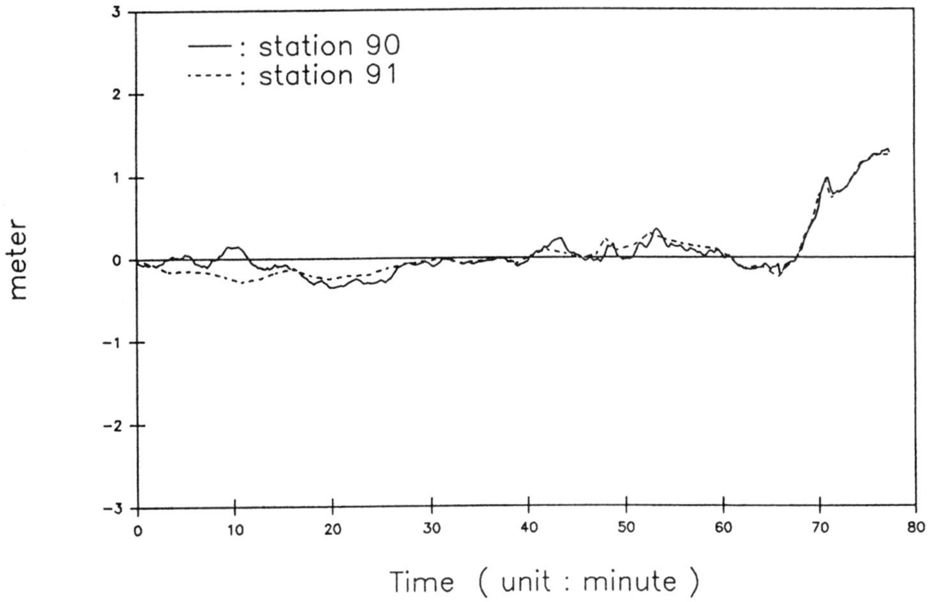

Fig. 1—Change in Ionospheric Effect on Phase Ranges; Day 351, 1988 (T14100 Receivers in Antarctica—Satellite #12)

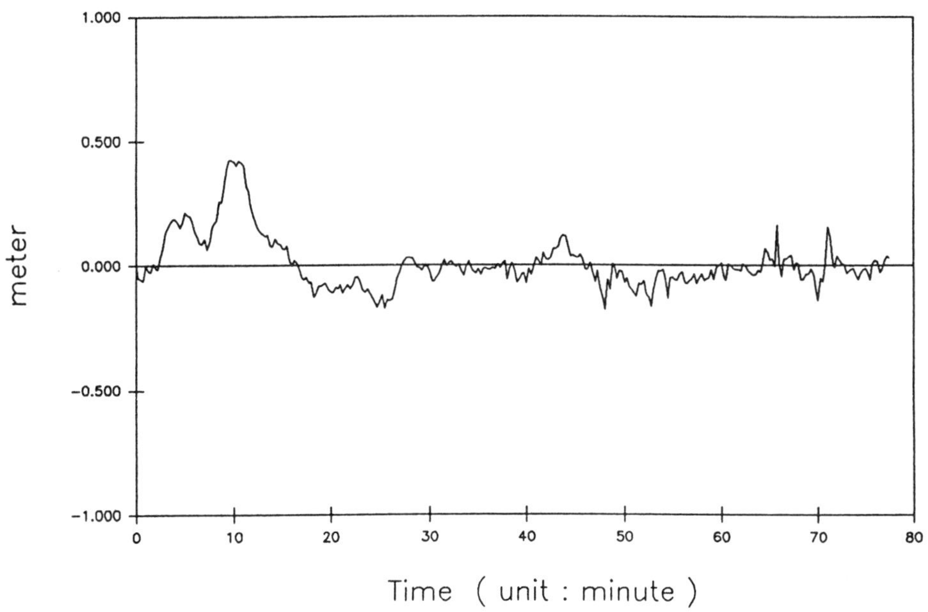

Fig. 2—Intersite Difference in Ionospheric Effect on Phase Ranges

Table 1—RMS Scatter of Ionospheric Contribution to L_1 Phase Range

Day of Year (1988)	Satellite 11 (m/ppm)	Satellite 12 (m/ppm)
351	.172/18.7	.181/19.6
352	.069/7.5	.024/2.6

formulation is used here. For completeness, the filter equations are given as follows [7]:

$$\mathbf{x}_n(-) = \mathbf{\Phi}_{n-1}\,\mathbf{x}_{n-1}(+) + \mathbf{w}_{n-1} \tag{9a}$$

$$\mathbf{P}_n(-) = \mathbf{\Phi}_{n-1}\,\mathbf{P}_{n-1}(+)\,\mathbf{\Phi}^T_{n-1} + \mathbf{Q}_{n-1} \tag{9b}$$

$$\mathbf{P}_n^{-1}(+) = \mathbf{P}_n^{-1}(-) + \mathbf{H}_n^T\,\mathbf{R}_n^{-1}\,\mathbf{H}_n \tag{9c}$$

$$\mathbf{K}_n = \mathbf{P}_n(+)\,\mathbf{H}_n^T\,\mathbf{R}_n^{-1} \tag{9d}$$

$$\mathbf{x}_n(+) = \mathbf{x}_n(-) + \mathbf{K}_n[\mathbf{z}_n - \mathbf{H}_n\,\mathbf{x}_n(-)] \tag{9e}$$

A computer program was written to perform the filtering. Since filtered pseudoranges are the goal of this investigation, the results of the filter gain values for pseudoranges are presented in Figure 3. The figure shows the filter gains computed for the phase measurements with four different values of Q_s. Since the filter is unbiased, filter gains for pseudorange measurements are equal to the phase gains subtracted from 1.0. It is clear that no information about any state variable is available prior to the first data sample. Thus it is reasonable that initally, the pseudorange measurements will contribute 100 percent to the estimate of ρ^* (or equivalently, phase measurements will contribute nothing at the first epoch to the filtered pseudorange). However, as knowledge of $k^*\lambda$ and s improves, the phase measures will then contribute to estimates of ρ^*. From Figure 3 it is seen that after approximately 30 data samples, the gain coefficients reach steady state values. Here 0.3 m and 1 mm were the assumed uncertainties for the pseudorange and phase range noise, respectively. Figure 4 shows the results of the ionosphere estimates for different values of Q_s. These results can be compared with the actual ionosphere change as measured by the available dual-frequency phase measurements. Values of Q_s greater than or equal to 0.5 m/min seem to do a very good job of tracking the ionospheric contribution without any detrimental lag. Figures 5 and 6 show the standard deviations of the ambiguities and pseudoranges. Chi square values of actual single station data filtering are given in Figures 7 and 8. For $Q_s = 4$ m/min and for $Q_s = 0$, Figure 7 shows that the chi square values represent both too tight a model ($Q_s = 0$) and too loose modeling. $Q_s = 0.5$ m/min (Figure 8) gives much more desirable values. Values of 0.5 are considered ideal for the data analyzed here since there is a good balance between measurement noise and uncertainties in the predicted states.

Three points are obvious about the steady state characteristics of the filter when the ionospheric correction error rate is uncertain at, say, 2 m/min or larger. The first is that since the pseudorange and phase range are averaged, the ionospheric contribution is then canceled. In fact there is no need even to

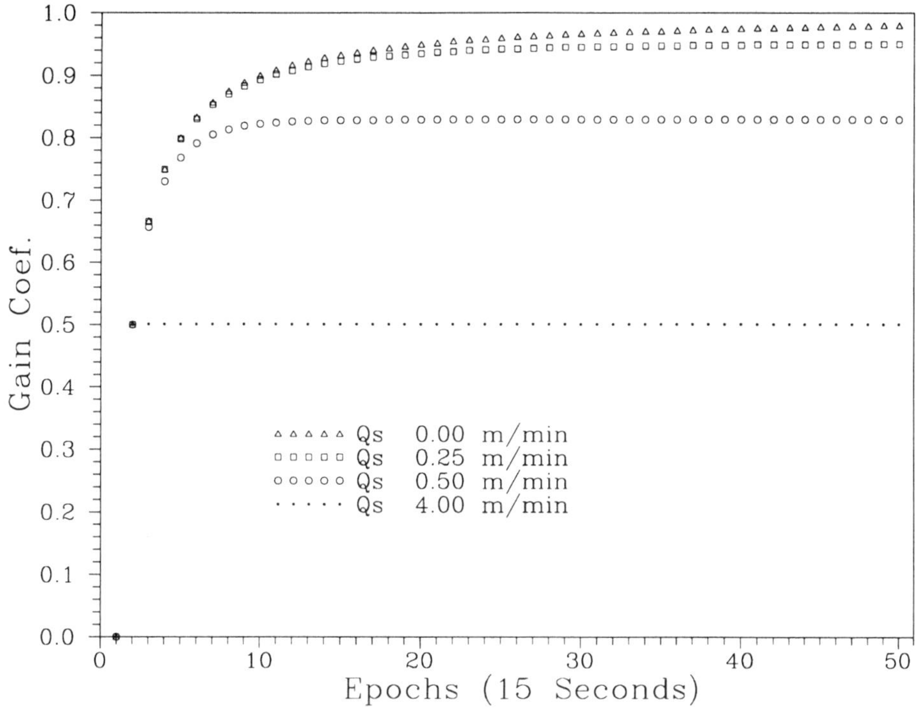

Fig. 3—Phase Weights Used to Generate Filtered Pseudorange

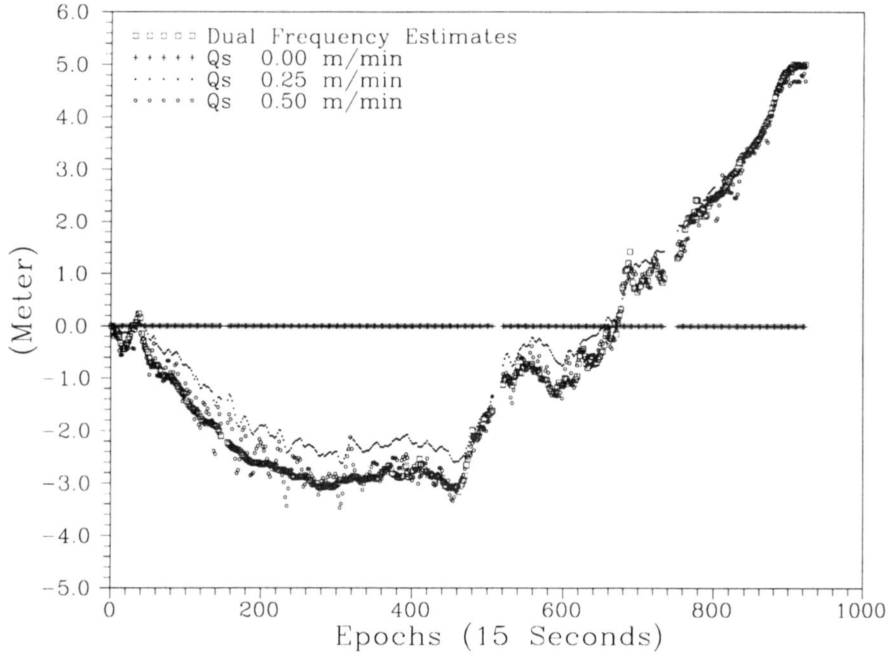

Fig. 4—Estimated Ionospheric Contributions

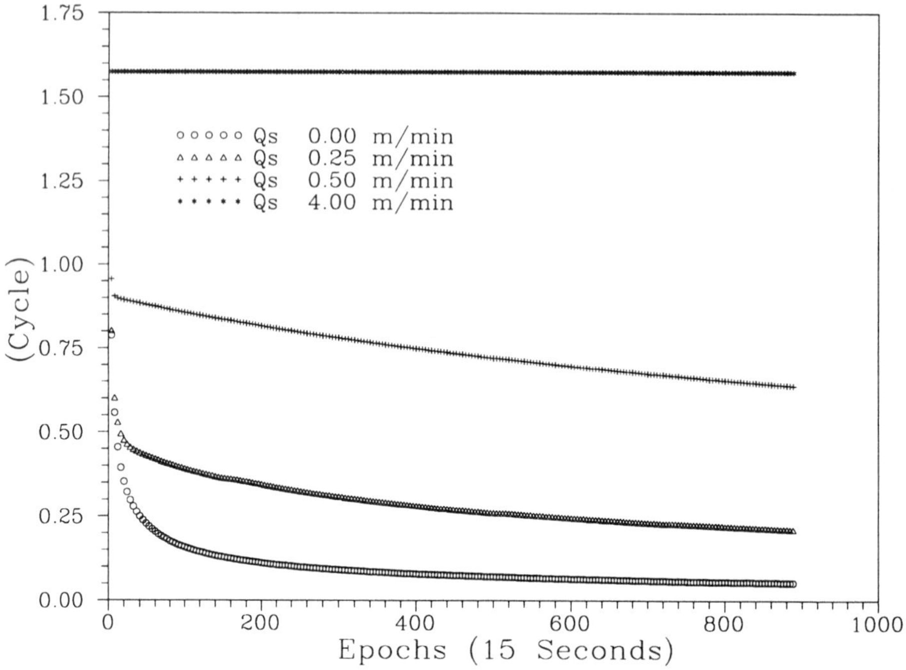

Fig. 5—Ambiguity Standard Deviations

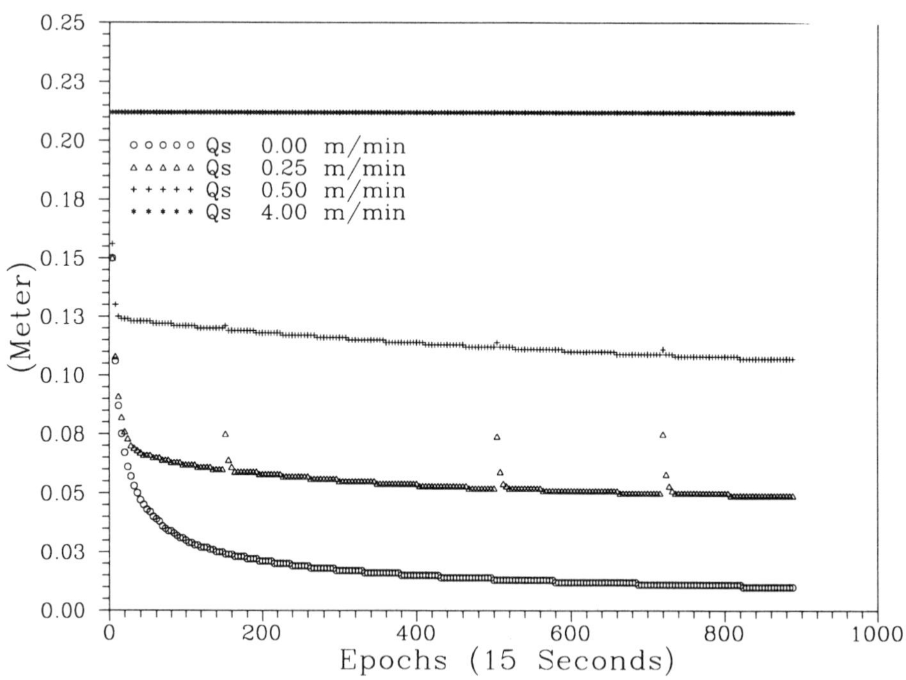

Fig. 6—Pseudorange Standard Deviations

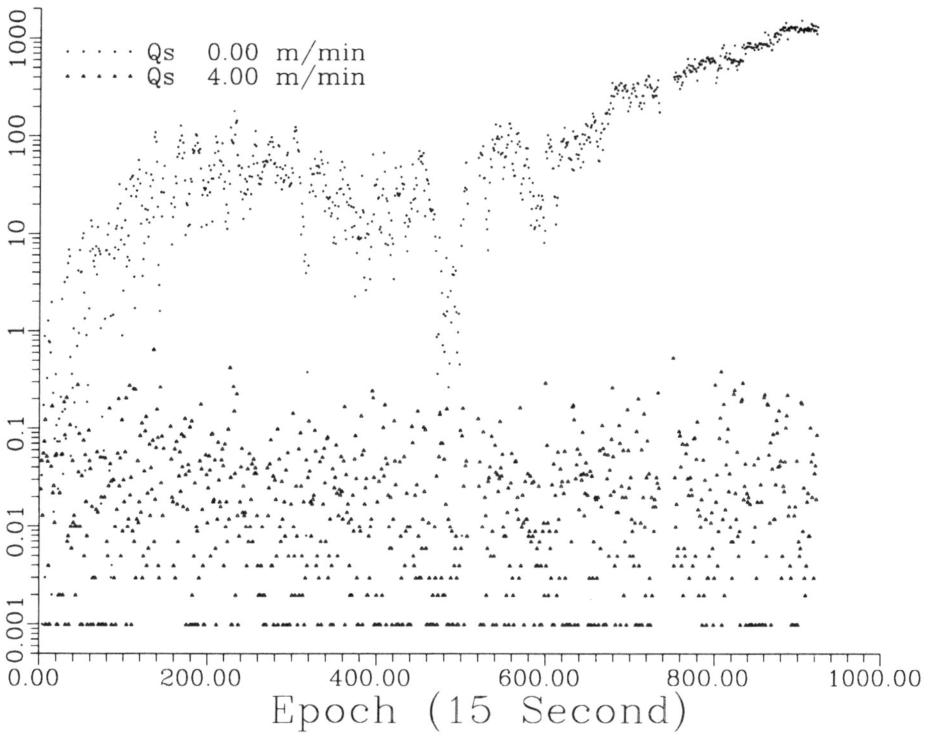

Fig. 7—Epoch Chi Square Values (Q_s = 0.00 m/min and 4.00 m/min)

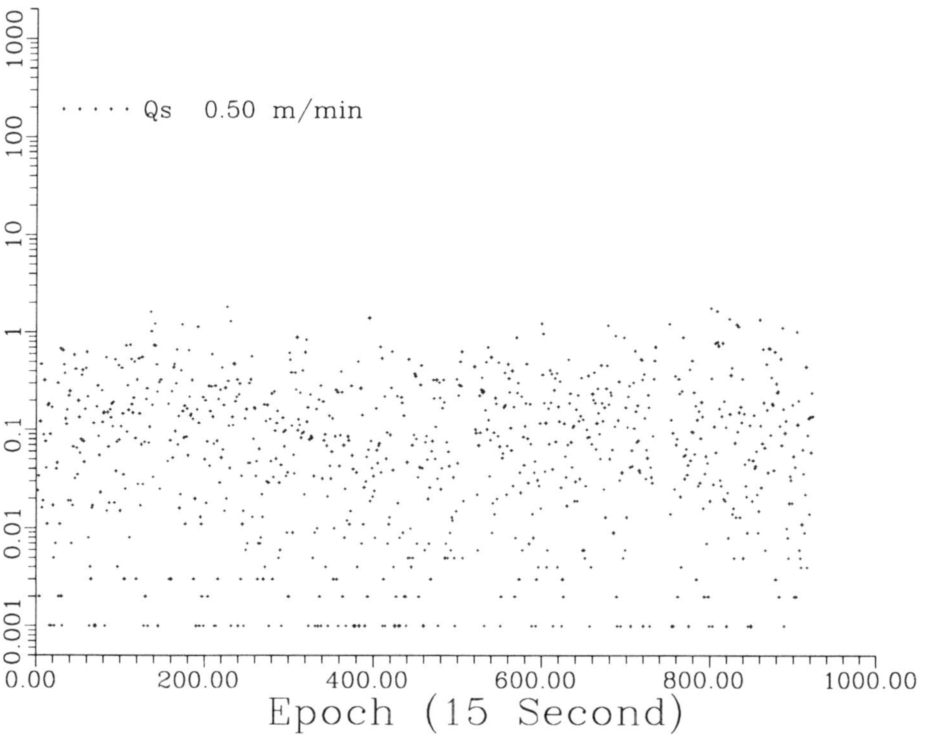

Fig. 8—Epoch Chi Square Values (Q_s = 0.50 m/min)

make the correction. This is very desirable since the ionospheric contribution to the measurements is the least predictable element. Second, if these steady state coefficients are used, then the best one can do is to reduce the noise of the pseudorange by a factor of 1/2. Claims of reducing differential pseudorange noise to 10 cm in 2 min in a single-frequency receiver obviously consider the ionospheric contributions to cancel [2]. And there surely are times when one could witness such cancellation of error. However, if TIDs should happen to cause anomalous behavior at one receiver but not similar behavior at the other, the resulting error would be approximately the position dilution of precision (PDOP) times the magnitude of this discrepancy. This magnitude could easily be of the order of a few meters. Multiply this by a PDOP of, say, 5 m/m, and one could experience a relative error of many meters indeed.

The emphasis here has been on the real-time filtering problem in the presence of an unknown and unpredictable ionosphere. However, there will be times, especially in post-mission mode, when additional knowledge of the ionospheric activity exists. If this additional information leads one to believe that there is more complete cancellation of the ionospheric contributions, then the Q_s values in the driving noise covariance matrix Q can take on reduced values. Calculating statistics on the prediction of a measurement involving only the s and $k^*\lambda$ could prove to be a very useful indicator of either overmodeling or undermodeling. One possible combination is found by subtracting (4b) from (4a):

$$R_n - \phi_n = 2s_n - k^*\lambda - v_{\phi_n} + v_{R_n} \tag{10}$$

Since (10) is void of those filter elements which are considered unpredictable, one can apply the following logic. If the squares of predicted residuals are consistently very small relative to the associated variances, the chosen Q_s is probably too large. On the other hand, if the squares of predicted residuals are large relative to the associated variances, then the value of Q_s is probably too small; subsequent large values will verify this assumption. Quantitive descriptors have been discussed in [8]. With such an approach, an adaptive scheme could prove to be very useful in processing the observed code and carrier measurements in a truly optimum manner.

OPTIMAL SMOOTHING

When one does not require estimates of pseudorange in real time, an optimal smoothing algorithm should be used. There are two ways to achieve such an optimal smoother. One is to set up the equations in batch mode, where only one estimate of $k\lambda$ is made for all samples with uninterrupted phase signals (no loss of lock during tracking). One can also use the same sequential filtering equations presented here when the outputs of the forward and backward filter are combined in an optimal matter. For the case of smoothing, a batch approach is more efficient. One will recognize this system of equations as being very similar to that of the pass-by-pass estimate of the frequency offset in Doppler positioning or to the optimal estimates of position from GPS pseudoranges when the antenna is not moving. The key element here is to realize that the real-time and postprocessing modes require different algorithms if optimal solutions are desired. It is conventional to refer to the real-time and postprocessing algorithms as filters and smoothers, respectively.

CONCLUSIONS

It is clear that in the absolute positioning mode, the optimal filter of data from single-frequency receivers must account for the contribution of the ionosphere. What requires some decision on the part of the analyst is the extent to which the error in the ionospheric correction contaminates differential positioning measurements. In steady state, only by averaging pseudoranges and phase ranges can one protect oneself completely, with no lag, against spatially unpredictable ionospheric behavior with only single-frequency data. It has been shown here that spatial differences of 20 cm and larger have been experienced over distances of only 10 km. One will often use differential positioning techniques over hundreds of kilometers, which could exacerbate the problem. The potential for contamination of differential position estimates is equal to the PDOP times what is not eliminated through differencing. During a severe ionospheric "storm," this could amount to several meters. The alternatives are to use dual-frequency receivers or average pseudorange and phase range as presented here. Considering the potential damage that unpredictable ionospheric activity can cause, choosing the optional adaptive filter should be considered seriously. Implementation is straightforward. Loss-of-lock situations are easily accommodated by setting the variance of the $k\lambda$ estimate to infinity, which is equivalent to restarting the estimator. If one chooses to assume that the ionospheric uncertainty is significantly smaller in differential positioning, then the three-state estimator algorithm with smaller Q_s values is still optimal for this case, with no further ramping/restarting procedure required to accommodate the oppositely signed contributions of the ionosphere in the phase range and pseudorange measurements. An adaptive scheme should work equally well in the differential mode. No restart procedure should be required, except for situations involving loss of lock.

ACKNOWLEDGMENT

I wish to acknowledge the funding supporting this effort from the NASA Commercial Center for Development of Space at the Ohio State University Center for Mapping and the Departments of Transportation of 38 states through the Federal Highway Administration. Discussions with Peter Loomis, Burkhard Schaffrin, H.-J. Euler, and Y. H. Tseng were very beneficial in formulating the ideas presented here.

REFERENCES

1. Gurtner, W., Mader, G. and MacArthur, D., *A Common Exchange Format for GPS Data*, Proceedings of the 5th International Geodetic Symposium of Satellite Positioning, Las Cruces, NM, March 13–17, 1989, pp. 920–31.
2. Ashjaee, J., Lorentz, R., Sutherland, R., Dutilloy, J., Minagio, J., Abtahi, R., Eichner, J., Kosmalska, J., and Helkey, R., *New Developments and Ashtech M-XII*, Proceedings of ION GPS-89, The Second International Technical Meeting of the Satellite Division of The Institute of Navigation, Colorado Springs, CO, September 27–29, 1989, pp. 195–98.
3. Hatch, R., *The Synergism of GPS Code and Carrier Measurements*, Proceedings of the Third International Geodetic Symposium on Satellite Doppler Positioning, Las Cruces, NM, February 8–12, 1982, pp. 1213–31.

4. Loomis, P., Kremer, G., and Reynolds, J., *Correction Algorithms for Differential GPS Reference*, NAVIGATION, Journal of The Institute of Navigation, Vol. 36, No. 2, Spring, 1989, pp. 179–93.
5. Klobuchar, J., *Ionospheric Time Delay Corrections for Advanced Satellite Ranging Systems*, AGARD Conference Proceedings No. 209 on Propagation Limitations of Navigation and Positioning Systems, NATO, Istanbul, October 20–22, 1976.
6. Counselman, C., *Ambiguity Bootstrapping to Determine GPS Orbits and Baselines*, Second Symposium on GPS Applications in Space, Air Force Geophysics Laboratory, October 10, 1989 [abstract].
7. Gelb, A. (editor), *Applied Optimal Estimation*, M.I.T. Press, Cambridge, MA, and London, England, 1974.
8. Teunissen, P. J. G. and Salzmann, M. A., *A Recursive Slippage Test for Use in State-Space Filtering*, Manuscripta Geodaetica, Vol. 14, No. 6, 1989, pp. 383–90.

NAVIGATION: *Journal of The Institute of Navigation*
Vol. 38, No. 1, Spring 1991
Printed in U.S.A.

Kinematic GPS for Differential Positioning: Resolving Integer Ambiguities on the Fly

PATRICK Y. C. HWANG
Rockwell International Corporation, Cedar Rapids, Iowa
Received October 1990
Revised January 1991

ABSTRACT

In kinematic GPS, the initial integer ambiguity must be resolved either by a static survey over time, or by instant calibration with a known baseline or an antenna exchange. While these standard methods require maintaining a baseline stationary to an earth-fixed reference frame during the initialization, there are situations when at least one of the receivers may be constantly in motion. This paper proposes two ideas for adapting standard kinematic techniques to situations that do not naturally allow for the constraint of a fixed baseline. The first calls for extracting the information needed to resolve the integer ambiguity from the very data collected while the kinematic survey is in progress. The second idea addresses the use of the antenna exchange technique for mobile platforms where the original locations of the antennas are not likely to remain stationary during the physical exchange. Both ideas count on information from additional measurements to augment their respective measurement models.

INTRODUCTION

For the past decade of GPS development, significant advances made in receiver and systems technologies have improved the practicality and applicability of carrier phase methodologies. Today, these methodologies show potential usefulness beyond their surveying roots. Several key breakthroughs were especially significant in the evolution of the contributing technologies, particularly the conception of kinematic surveying [1].

In kinematic GPS surveying, it is generally recognized that, regardless of how the survey is actually conducted, the initial integer ambiguity must be resolved by choosing any one of three standard methods, depending on the circumstances constraining the survey: (1) performing a static survey, (2) using the antenna exchange technique, or (3) calibrating with a known baseline. These methods all involve the special need to maintain two fixed points for a baseline. However, if at least one of the receivers is located out at sea, in the air, or in space, then a baseline that is stationary with respect to an earth-fixed frame of reference cannot be easily established.

In exploring new concepts involving carrier phase measurements in mid-1989, the author came up with the notion that, with redundancy from the use of measurements from more than four satellites, a kinematic survey need not start out with the two antennas being stationary. Although this idea was conceived of independently, it was later discovered that the same idea had, in fact, already been proposed in a paper delivered earlier that year [2]. Nevertheless, there is enough of a difference in the two viewpoints to justify an independent reproposing of this idea, which we shall call kinematic on the fly, a term admittedly more descriptive of the problem than of the solution. To complete the initialization process, the viewpoint in [2] requires what amounts to a complete resolution of all the unknown variables. The proposed model in this paper takes on a somewhat more optimistic outlook in requiring only a partial resolution.

A second idea put forth in this paper generalizes a kinematic-related concept commonly called the antenna exchange (or swap) for use on a moving platform. The supplemental information needed to meet the increase in complexity over the original antenna exchange model is derived not from extra satellites, but with an extra antenna.

To facilitate the main discussion later on, a tutorial section follows in the next section, where the analytical tools used in conjunction with explaining basic GPS surveying concepts are introduced. This, in turn, is followed by formulation of the two ideas mentioned above.

PERSPECTIVE ON CARRIER PHASE MEASUREMENT MODELS

To provide some background in rudimentary concepts of kinematic surveying, we shall look at various GPS carrier phase surveying models. Analyses of these models will concentrate on their solvability, with the concept of dilution of precision (DOP) used to determine their quality. The DOP factor gives a simple interpretation of how much one unit of measurement error contributes to the derived position solution error for a given situation. Naturally, the smaller the DOP, the better the quality of the solution. For illustrative purposes, the examples adopted are one-dimensional in position.

Static Surveying

In the GPS static surveying model shown in Figure 1, two stationary receivers located at the ends of a baseline are required to track and measure the phase of the GPS carrier signal arriving at their respective antenna locations. It is presumed that all changes in the carrier phase are exactly accounted for from the moment tracking begins when the first measurement is made.

$$\phi_A - \phi_B = \Delta\phi = \Delta x \cos\theta_0$$

$$\Delta\phi = \Delta\phi_0 - N = \Delta x \cos\theta_0$$

$$\Delta\phi_0 = \Delta x \cos\theta_0 + N \tag{1}$$

where ϕ_A is phase at A, ϕ_B is phase at B, $\Delta\phi$ is phase difference between locations A and B, θ_0 is initial satellite elevation angle, Δx is baseline separation between A and B (in units of wavelength), $\Delta\phi_0$ is initial measurement, and N is initial integer ambiguity (in number of whole wavelength cycles).

At the very first instance of achieving carrier track, the true relationship (i.e., the total phase delay) between the phase measurements seen at both receiver locations is unknown because of the nature of oscillatory signals. The

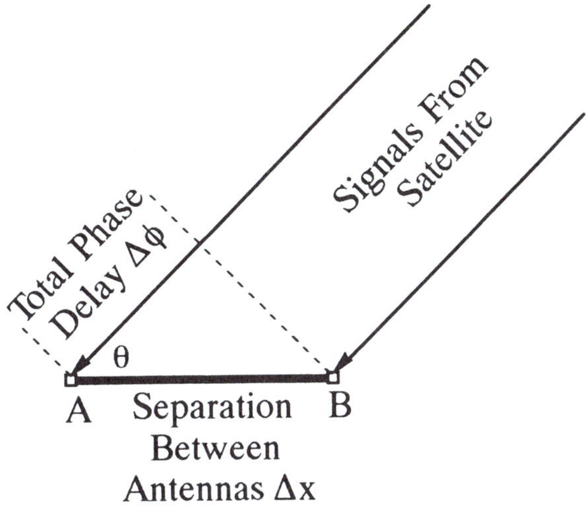

Fig. 1—Static Surveying

unobservable portion of this phase delay is, however, known to be some multiple of a whole cycle. This quantity representing the number of whole cycles needs to be found in the problem commonly referred to as the *initial integer ambiguity*.

For static surveying, the resolution of the initial integer ambiguity has to be accomplished by observing phase measurements over a period of time. It is clear from equation (1) that the two unknowns Δx and N cannot be solved with one measurement equation at just a single instant of time. With a second measurement, $\Delta\phi_1$, made presumably at a different time, t_1, the measurement situation becomes two-dimensional in nature.

$$\begin{bmatrix} \Delta\phi_0 \\ \Delta\phi_1 \end{bmatrix} = \begin{bmatrix} \cos\theta_0 & 1 \\ \cos\theta_1 & 1 \end{bmatrix} \begin{bmatrix} \Delta x \\ N \end{bmatrix} \tag{2}$$

A solution exists for $[\Delta x \ N]^T$ if, and only if, the 2×2-coefficient matrix containing the geometric parameters is invertible. And if that is the case, the DOP factor is used to provide an indication of the quality of the solution. The DOP determines the magnification factor of the measurement noise that is translated into the derived solution. These conditions constitute the central notion of observability that is widely used in control and estimation theory. The DOP factor for this situation is obtained by generating the variance of the solution vector with respect to a one-unit error variance in the measurement.

$$\begin{aligned} \mathrm{Var}\begin{bmatrix} \Delta x \\ N \end{bmatrix} &= \frac{1}{D}\begin{bmatrix} 1 & -1 \\ -\cos\theta_1 & \cos\theta_0 \end{bmatrix} \begin{bmatrix} 1 & 0 \\ 0 & 1 \end{bmatrix} \begin{bmatrix} 1 & -\cos\theta_1 \\ -1 & \cos\theta_0 \end{bmatrix} \frac{1}{D} \\ &= \frac{1}{D^2}\begin{bmatrix} 2 & -(\cos\theta_0 + \cos\theta_1) \\ -(\cos\theta_0 + \cos\theta_1) & \cos^2\theta_0 + \cos^2\theta_1 \end{bmatrix} \end{aligned} \tag{3}$$

where the determinant $D = \cos\theta_0 - \cos\theta_1$. Note here that, being related to the inverse, the determinant is also a good indicator of the solution's quality.

From equation (3), the DOP factor associated with N is

$$\sqrt{(\cos^2\theta_0 + \cos^2\theta_1)} / |\cos\theta_0 - \cos\theta_1|,$$

which suggests two things: (1) the rate of change of the geometry for a fixed time interval determines the observability of the measurement situation; and (2) hypothetically, if GPS were a geostationary system of satellites, cosθ in this problem would be a constant, and this problem would have no solution.

In a real-life three-dimensional GPS static survey, the amount of time necessary to obtain a solution is typically on the order of 20 to 30 min for short (less than 1 km) baselines [1]. The observation time required increases with baseline length as a result of such unmodeled error contributions as atmospheric delay decorrelation between the two paths, and differences in the multipath seen at the two antenna locations.

Kinematic Surveying

As an extension to static surveying, kinematic surveying allows at least one of the antennas to roam from point to point while keeping continuous track of the carrier signal. Continuity in the measured carrier phase profile provides the user with an exact history of position changes of the roving antenna since leaving its initial location. The results obtained from this type of surveying operation can be very accurate (subdecimeter level), provided that some sort of initialization procedure to resolve the initial integer ambiguity had already been made before the roving antenna was moved. In Figure 2, the static example of Figure 1 is extended to illustrate the kinematic model. For the additional kinematic motion, we use a new variable, δx, to denote the position increment from the initial. By augmenting equation (1) to include this motion, we obtain:

$$\begin{aligned} \Delta\phi + \delta\phi &= (\Delta x + \delta x)\cos\theta \\ \Delta\phi_0 - N + \delta\phi &= \Delta x \cos\theta + \delta x \cos\theta \\ \Delta\phi_0 + \delta\phi &= \delta x \cos\theta + \Delta x \cos\theta + N \end{aligned} \qquad (4)$$

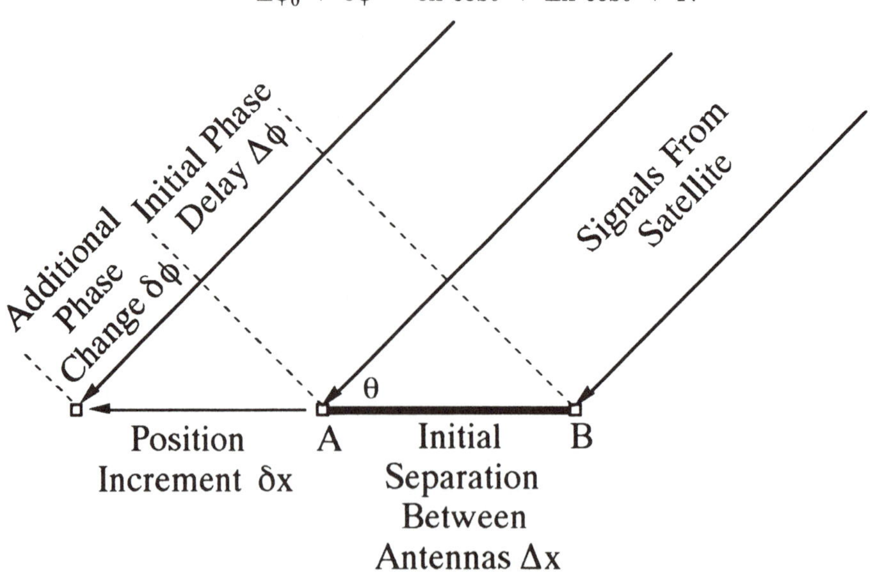

Fig. 2— Kinematic Surveying

where $\delta\phi$ is incremental phase change as a result of motion, and δx is position change.

At this point, the variables Δx and N should already have been solved from the initialization. This leaves δx, which can be solved instantaneously from the single measurement $(\Delta\phi_0 + \delta\phi)$. The implication here is that beyond the process of initialization, as long as continuous carrier tracking is maintained, the position derived by kinematic means is obtained in a similar way and under the same observability conditions as conventional positioning using pseudoranges. In part, this means that the accuracy of a kinematic solution is dictated by the standard position dilution of precision (PDOP) measure used for GPS code positioning. It should be obvious that the same also applies to static surveying when solving for Δx after completing its initialization (i.e., with N solved). To keep the notation simple, the geometric connection term, $\cos\theta$, is shown to be the same for equations (1) and (4) although in general, θ is not required to be constant in time. Certainly, for the case of GPS, the geometry will always be changing.

There are several ways to accomplish the initialization of a kinematic survey. The most obvious and perhaps least desirable is to run a time-consuming static survey. Once one static survey has been made, the same baseline can, of course, be reused to initialize other kinematic surveys as well, if operational circumstances permit. We note here that knowledge of the baseline antenna separation implies knowledge of the total phase delay, or equivalently, the initial integer ambiguity.

The most intriguing alternative among initialization techniques, though, is one called the antenna exchange (or swap), first introduced in 1985 [3]. This technique uses as its principal idea the movement of the antennas to help speed up resolution of the initial integer ambiguity. It suggests that by moving one antenna to the location of the other, the total phase delay can be solved just as effectively as waiting for the satellites to "move appreciably," as is the case for a static survey. However, since there is no way to merge physically two antennas that need to be occupying exactly the same location at the end of the antenna transfer, the next best thing to do is to march the other antenna off to the location vacated by the first one. This completes "the exchange." It can be seen from Figure 3 that the antenna exchange can also be interpreted as a kinematic-type movement of one antenna by the amount $\delta x = -2\Delta x$, while the other is kept stationary. As was the case in equation (1), we have two

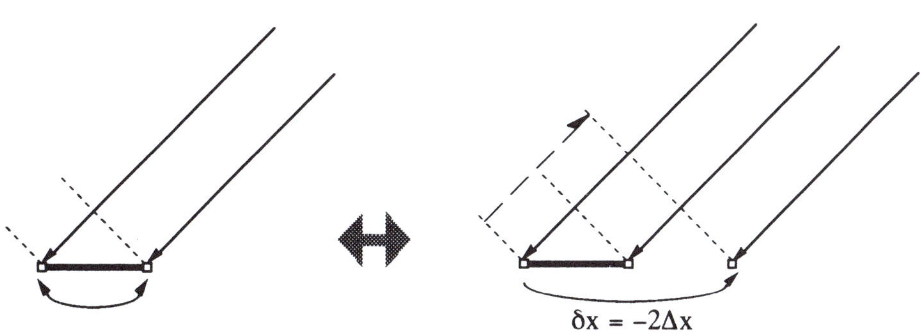

Fig. 3—Antenna Exchange—Two Equivalent Representations

variables to solve for; therefore two measurements are needed. These measurements need to be made once each before and after the exchange of the antennas.

Before exchange:

$$\Delta\phi_0 = \Delta x \cos\theta_0 + N$$

After exchange:

$$\begin{aligned}\Delta\phi_0 + \delta\phi &= (\Delta x + \delta x)\cos\theta_1 + N \\ &= (\Delta x - 2\Delta x)\cos\theta_1 + N \\ &= -\Delta x \cos\theta_1 + N\end{aligned}$$

where $\delta x = -2\Delta x$. The two simultaneous equations can be arranged into the following matrix equation:

$$\begin{bmatrix} \Delta\phi_0 \\ \Delta\phi_0 + \delta\phi \end{bmatrix} = \begin{bmatrix} \cos\theta_0 & 1 \\ -\cos\theta_1 & 1 \end{bmatrix} \begin{bmatrix} \Delta x \\ N \end{bmatrix} \tag{5}$$

A solution for equation (5) exists by the simple argument that its coefficient matrix is invertible, provided that the matrix determinant $\cos\theta_0 + \cos\theta_1 \neq 0$.

$$\begin{bmatrix} \Delta x \\ N \end{bmatrix} = \begin{bmatrix} \dfrac{1}{\cos\theta_0 + \cos\theta_1} & \dfrac{-1}{\cos\theta_0 + \cos\theta_1} \\ \dfrac{\cos\theta_1}{\cos\theta_0 + \cos\theta_1} & \dfrac{\cos\theta_0}{\cos\theta_0 + \cos\theta_1} \end{bmatrix} \begin{bmatrix} \Delta\phi_0 \\ \Delta\phi_0 + \delta\phi \end{bmatrix}$$

If the exchange is carried out quickly enough such that $\theta_1 \approx \theta_0$, we will then have

$$\begin{bmatrix} \Delta x \\ N \end{bmatrix} \approx \begin{bmatrix} \dfrac{1}{2\cos\theta_0} & \dfrac{-1}{2\cos\theta_0} \\ \dfrac{1}{2} & \dfrac{1}{2} \end{bmatrix} \begin{bmatrix} \Delta\phi_0 \\ \Delta\phi_0 + \delta\phi \end{bmatrix}$$

The DOP factor for this measurement situation is obtained by setting the phase noise covariance to a unit matrix. The solution variance then becomes

$$\mathrm{Var}\begin{bmatrix} \Delta x \\ N \end{bmatrix} \approx \begin{bmatrix} \dfrac{1}{2\cos^2\theta_0} & 0 \\ 0 & \dfrac{1}{2} \end{bmatrix} \tag{6}$$

Equation (6) points out that the DOP associated with the initial integer N is $\sqrt{1/2}$. In effect, the initialization with the antenna exchange method can be totally independent of *changes* in satellite geometry.

While the antenna exchange method provides the fast way to resolve the initial integer ambiguity, the two antennas must be close to one another at some point of a kinematic survey for this to be practical. Where antenna separations are prohibitively large, the conventional static method is still widely resorted to.

Generalization to Real-Life GPS Models

The simple models used above to convey the concepts of carrier phase surveying have also been useful in deriving insights into the limitations of these concepts. In transferring these insights to real-life GPS situations, several issues must be taken into consideration. The discussion in this section covers the issues of the benefits of higher sampling rates, raising of the spatial dimensionality of the model to three dimensions and accommodation of receiver timing errors, and extension of the observability criteria introduced with the one-dimensional models to higher dimensions.

In the above analyses, where the change in geometry over a period of time is a key ingredient of the solution, we have kept the illustrative models to the bare essentials by using only measurements made at two time points. These two time points, therefore, represent the start and the end of a time interval during which the rate of change in the satellite geometry dictates the speed of the solution's convergence. In an actual situation, though, measurements sampled in between the two boundaries of this time interval also contribute useful information. Thus, the higher the measurement sampling rate, the faster the initialization can be resolved. However, when successive measurements become correlated, such improvements with higher sampling rates gradually become negated.

To convert from one to three spatial dimensions, the number of position-related variables need only be trebled, and the number of satellites needed to solve this expanded set of variables will have to be proportionately increased. For the GPS situation, there is also a timing error in the receivers that must be accounted for. It is seldom convenient or necessary to deal with the timing error in a carrier phase surveying model only because this error is inestimable.

To see this, we again resort to a simplified illustration. A third variable, Δt, is introduced into equation (1) to give

$$\Delta\phi_0 = \Delta x \cos\theta + N + \Delta t \tag{7}$$

To solve this equation, we need more measurements. One way to satisfy this need may be to introduce a second satellite, which then adds another initial integer variable. Another way is to add a third measurement from the original lone satellite at a different time. In either case, or even when the two methods are combined, it can be shown with the kind of analysis used before that the sets of variables formulated cannot be resolved. Heuristically, this is due to the inability of the measurement situation to distinguish between the initial integer ambiguity, N, and the timing error, Δt, both of which reside in the range (measurement) space of the model.

Hence, to retain the form of the more insightful models given by equations (2), (4), and (5), we eliminate the timing error from consideration altogether. This can be realized by forming the so-called double difference. Consider two single-difference measurements (as in equation (7)), the first associated with satellite i and the second with satellite j:

$$\Delta\phi_0^{(i)} = \Delta x \cos^{(i)} + N^{(i)} + \Delta t$$

$$\Delta\phi_0^{(j)} = \Delta x \cos^{(j)} + N^{(j)} + \Delta t$$

A double-difference measurement can be formed by combining the two measurements above in such a way as to eliminate the common term Δt:

$$\Delta\phi_0^{(i)} - \Delta\phi_0^{(j)} = \Delta x (\cos^{(i)} - \cos\theta^{(j)}) + N^{(i)} - N^{(j)}$$

$$= \Delta x (\cos\theta^{(i)} - \cos\theta^{(j)}) + N^{(i,j)}$$

Assuming that the phase measurements are made from both satellites at the two antennas nearly simultaneously, the timing error conveniently cancels out in the double-difference measurement. In so doing, the initial integer $N^{(i,j)}$ is clearly a difference between a pair of integers. The number of independent double-difference measurements will always turn out to be one less than the total number of satellites being tracked.

Finally, the use of the coefficient-matrix determinant and the DOP factors as indicators of the solution's accuracy is also valid in three-dimensional situations. A DOP measure of solution accuracy can be derived for each of the different measurement situations. As pointed out before, for kinematic positioning beyond initialization, the DOP measure is identical to the PDOP from conventional GPS code positioning.

KINEMATIC ON THE FLY

A kinematic survey has traditionally been made up of two parts, the initialization and then the surveying. When dealing with postprocessed data, the order of the two is unimportant. Sometimes, for redundancy, the initialization procedure is even carried out twice, once before and then once after the survey for verification of the initial integer ambiguity. In real time, the initialization has to be performed and completed first in order for the surveying portion that follows to be meaningful. To reiterate, the choice of initialization methods varies from running a time-consuming static survey, to instant techniques such as calibration by using an already-surveyed baseline, or performing an antenna exchange. All of these alternatives carry the obvious limitation of requiring a baseline that is fixed to the GPS reference frame, at least for the duration of the initialization.

Nonstatic Initialization

To combine the two parts of the kinematic problem into one, let us refer to equation (4) describing kinematic surveying, except that now, Δx and N are yet unsolved. Clearly, we have three variables that can be solved by piling on more measurements, but not from the same satellite over time. Such a measurement situation would be unsolvable because every measurement added at a different time point would also add a different δx variable, so there would never be enough measurements to deal with the set of variables created. However, by adding another satellite to the measurement set, there is enough additional information to overcome the previously underdetermined condition.

$$\begin{bmatrix} \Delta\phi_0^{(1)} \\ \Delta\phi_1^{(1)} \\ \Delta\phi_0^{(2)} \\ \Delta\phi_1^{(2)} \end{bmatrix} = \begin{bmatrix} \cos\theta_0^{(1)} & 0 & 1 & 0 \\ \cos\theta_1^{(1)} & \cos\theta_1^{(1)} & 1 & 0 \\ \cos\theta_0^{(2)} & 0 & 0 & 1 \\ \cos\theta_1^{(2)} & \cos\theta_1^{(2)} & 0 & 1 \end{bmatrix} \begin{bmatrix} \Delta x \\ \delta x \\ N^{(1)} \\ N^{(2)} \end{bmatrix} \tag{8}$$

The parenthetical superscript assigned to a variable denotes the index of the satellite associated with it. The determinant of the coefficient matrix in equation (8) is $\cos\theta_0^{(1)} \cos\theta_1^{(2)} - \cos\theta_1^{(1)} \cos\theta_0^{(2)}$.

As we might have expected intuitively, the determinant is maximal if the temporal change between $\cos\theta_0$ and $\cos\theta_1$ and the spatial (satellite) separation between $\cos\theta^{(1)}$ and $\cos\theta^{(2)}$ are large.

Three-Dimensional Model

The one-dimensional example of equation (8) serves to illustrate the fundamental principles involved. This will now be generalized to three dimensions, where we have three initial position and three incremental position variables to begin with. In addition, we also have to add one initial integer variable for every satellite pair (double difference). In all, we need a total of six satellite pairs to provide twelve double-difference measurements (at two time points) to match the variables generated: three initial positions, three incremental positions, and six initial integers. This means a total of seven satellites are required to satisfy this model.

Although the five- and six-satellite models are also solvable, they have significantly poorer observability because they rely on redundancy in measurements made over more than two time points to make up for the lack of satellites. And when a model relies on measurements made at more than two time points, it is essentially attempting to derive information not just from the first-order, but higher-order changes in phase as well. Such changes are comparatively slow to evolve. This is not unlike using fewer than four satellites for a static survey. It is noteworthy here that to obtain results comparable to those of a static survey made with four satellites, a kinematic-on-the-fly survey requires at least seven satellites.

Kalman Filter

Based on the above foundations, a three-dimensional GPS measurement model can be formalized for any suitable estimator. In this paper, we choose the Kalman filter as the estimator for analytical purposes.

$$\begin{bmatrix} \Delta\phi^{(1)} - \Delta\phi^{(2)} \\ \Delta\phi^{(2)} - \Delta\phi^{(3)} \\ \Delta\phi^{(3)} - \Delta\phi^{(4)} \\ \Delta\phi^{(4)} - \Delta\phi^{(5)} \\ \Delta\phi^{(5)} - \Delta\phi^{(6)} \\ \Delta\phi^{(6)} - \Delta\phi^{(7)} \end{bmatrix} = \begin{bmatrix} h_x^{(1)} - h_x^{(2)} & h_y^{(1)} - h_y^{(2)} & h_z^{(1)} - h_z^{(2)} & 1\,0\,0\,0\,0\,0 \\ h_x^{(2)} - h_x^{(3)} & h_y^{(2)} - h_y^{(3)} & h_z^{(2)} - h_z^{(3)} & 0\,1\,0\,0\,0\,0 \\ h_x^{(3)} - h_x^{(4)} & h_y^{(3)} - h_y^{(4)} & h_z^{(3)} - h_z^{(4)} & 0\,0\,1\,0\,0\,0 \\ h_x^{(4)} - h_x^{(5)} & h_y^{(4)} - h_y^{(5)} & h_z^{(4)} - h_z^{(5)} & 0\,0\,0\,1\,0\,0 \\ h_x^{(5)} - h_x^{(6)} & h_y^{(5)} - h_y^{(6)} & h_z^{(5)} - h_z^{(6)} & 0\,0\,0\,0\,1\,0 \\ h_x^{(6)} - h_x^{(7)} & h_y^{(6)} - h_y^{(7)} & h_z^{(6)} - h_z^{(7)} & 0\,0\,0\,0\,0\,1 \end{bmatrix} \begin{bmatrix} \Delta x \\ \Delta y \\ \Delta z \\ N^{(1,2)} \\ N^{(2,3)} \\ N^{(3,4)} \\ N^{(4,5)} \\ N^{(5,6)} \\ N^{(6,7)} \end{bmatrix} \tag{9}$$

where $\Delta\phi^{(i)}$ is the single-difference measurement from satellite i between the two antennas, and $[h_x^{(i)} h_y^{(i)} h_z^{(i)}]^T$ is the unit direction vector to satellite i.

$N^{(i,j)}$ denotes the initial integer variable formed from the difference between integers associated with single-difference measurements from satellites i and j.

In equation (9), the first three components of the state vector are positions, and the remaining are initial integers. In the Kalman filter formulation, the initial position and incremental position variables for one dimension, Δx and δx used in equation (8), are combined into one state that is allowed to random walk. Hence, the random process model for the position states is made up of three independent random walk components, while the remaining six integer states are treated as random constants.

Implementation of the standard Kalman filter algorithms with the parameters defined above was straightforward [4]. The measurement noise variance was chosen to be $(0.002\ \text{m})^2$, and the process noise variance of the position states was $(100\ \text{m})^2$, with a time step interval of 30 s. A simulation program was written to generate a typical satellite constellation, and to compute the relevant unit direction vectors. This was done to analyze the convergence of the filter covariance over time. Each plot in Figure 4 represents the square root of the variance associated with an initial integer state. Recall that each integer state corresponds to a satellite pair coupled as a result of double differencing. Results under equivalent conditions for a four-satellite static surveying case are given in Figure 5. An exact comparison between the two cases is meaningless because the measurement situations are different. For the chosen situations, though, the static results appear to converge faster than the seven-satellite kinematic results.

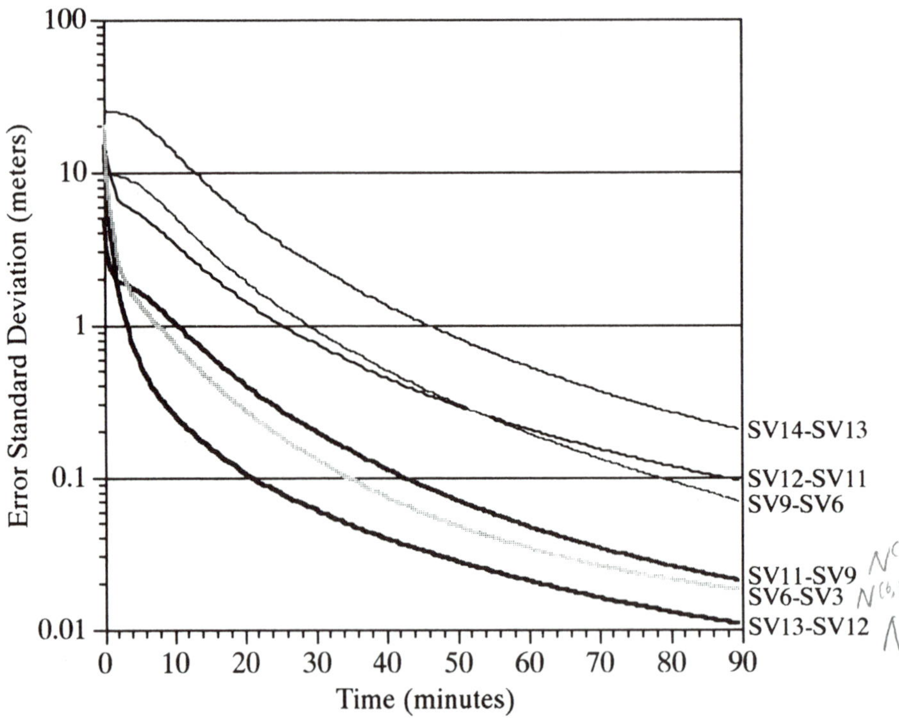

Fig. 4—Initial Integer Resolution: Kinematic on the Fly (seven satellites)

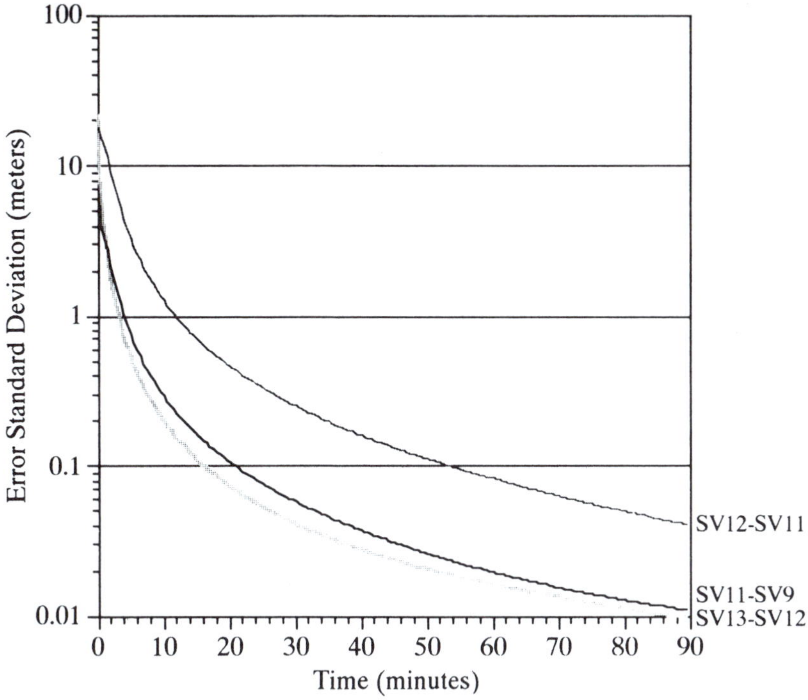

Fig. 5—Initial Integer Resolution: Static (four satellites)

The multiple plots of Figure 4 exhibit varying rates of convergence, which can be attributed to differences in satellite geometry. Although some of the integer estimates converge upon their solution faster than others, the total solution, including the position estimates, is not complete until all the integers have been resolved. Fortunately, it turns out that we do not need a total solution to complete the initialization. As a matter of fact, we need only three out of the six.

Based on the results of Figure 4, we shall choose to resolve the faster-converging double-difference initial integers $N^{(13,12)}$, $N^{(11,9)}$, and $N^{(6,3)}$ first. Then, by selecting the appropriate three rows from equation (9) and rearranging them, we can form a new 3-tuple vector equation to solve the position vector $[\Delta x \ \Delta y \ \Delta z]^T$.

$$\begin{bmatrix}\Delta x\\ \Delta y\\ \Delta z\end{bmatrix} = \begin{bmatrix} h_x^{(13)} - h_x^{(12)} & h_y^{(13)} - h_y^{(12)} & h_z^{(13)} - h_z^{(12)} \\ h_x^{(11)} - h_x^{(9)} & h_y^{(11)} - h_y^{(9)} & h_z^{(11)} - h_z^{(9)} \\ h_x^{(6)} - h_x^{(3)} & h_y^{(6)} - h_y^{(3)} & h_z^{(6)} - h_z^{(3)} \end{bmatrix}^{1} \begin{bmatrix} \Delta\phi^{(13)} - \Delta\phi^{(12)} - N^{(13,12)} \\ \Delta\phi^{(11)} - \Delta\phi^{(9)} - N^{(11,9)} \\ \Delta\phi^{(6)} - \Delta\phi^{(3)} - N^{(6,3)} \end{bmatrix}$$

This measurement situation involves six satellites. Fortunately, further simplification is possible before proceeding. The solution obtained for $[\Delta x \ \Delta y \ \Delta z]^T$ can be used to resolve the remaining initial integer ambiguities, $N^{(14,13)}$, $N^{(12,11)}$, and $N^{(9,6)}$, by extracting and modifying selected rows of equation (9):

$$\begin{bmatrix} N^{(14,13)} \\ N^{(12,11)} \\ N^{(9,6)} \end{bmatrix} = \begin{bmatrix} \Delta\phi^{(14)} - \Delta\phi^{(13)} \\ \Delta\phi^{(12)} - \Delta\phi^{(11)} \\ \Delta\phi^{(9)} - \Delta\phi^{(6)} \end{bmatrix} - \begin{bmatrix} h_x^{(14)} - h_x^{(13)} & h_y^{(14)} - h_y^{(13)} & h_z^{(14)} - h_z^{(13)} \\ h_x^{(12)} - h_x^{(11)} & h_y^{(12)} - h_y^{(11)} & h_z^{(12)} - h_z^{(11)} \\ h_x^{(9)} - h_x^{(6)} & h_y^{(9)} - h_y^{(6)} & h_z^{(9)} - h_z^{(6)} \end{bmatrix} \begin{bmatrix}\Delta x\\ \Delta y\\ \Delta z\end{bmatrix}$$

Now we can proceed with a reduced-order model given by equation (10) that requires only four satellites (a, b, c, d), which, in practice, should be selected on the basis of their PDOP.

$$\begin{bmatrix} \Delta\phi^{(a)} - \Delta\phi^{(b)} - N^{(a,b)} \\ \Delta\phi^{(b)} - \Delta\phi^{(c)} - N^{(b,c)} \\ \Delta\phi^{(c)} - \Delta\phi^{(d)} - N^{(c,d)} \end{bmatrix} = \begin{bmatrix} h_x^{(a)} - h_x^{(b)} & h_y^{(a)} - h_y^{(b)} & h_z^{(a)} - h_z^{(b)} \\ h_x^{(b)} - h_x^{(c)} & h_y^{(b)} - h_y^{(c)} & h_z^{(b)} - h_z^{(c)} \\ h_x^{(c)} - h_x^{(d)} & h_y^{(c)} - h_y^{(d)} & h_z^{(c)} - h_z^{(d)} \end{bmatrix} \begin{bmatrix} \Delta x \\ \Delta y \\ \Delta z \end{bmatrix} \tag{10}$$

The error covariance matrix furnished by the Kalman filter computations gives an accurate account of the statistics associated with the estimates if the model accurately reflects the true situation. This information can be used to determine the convergence criteria for terminating the initialization. It should be noted here that in a practical implementation, the initialization process can be further enhanced by imposing constraints on the value of the initial integer variables to be, in fact, integers. This additional piece of information can help speed up the search for the correct solution, although it should be no surprise that this benefit does not come without a price. The imposition of an integer constraint makes this a highly nonlinear optimization problem. Such a situation is less forgiving when the solution is incorrect if the noise is not exactly that which is modeled. In other words, if the integer resolution is erroneous, the error may end up being quite large. Caution must be exercised in trying to squeeze out too much information with integer constraints to speed up the rate of convergence at the expense of the robustness and reliability of the solution, unless the noise model is very well known.

Comparing Viewpoints

The formulation in [2] for the kinematic-on-the-fly problem also uses a Kalman filter for an analytical description. It treats the initial integers, however, as single-differenced integers, each associated with one satellite in forming a single-difference measurement between the two antennas. Because of the unsolvability of a GPS model that includes a timing error, as described previously in conjunction with equation (7), the single-differenced integers are also not observable as a consequence; only double-differenced integers (pairs) are. As a result, the other formulation had to depend on the covariance of the position variables as a measure of convergence, which by themselves represent the total convergence of the entire solution.

As pointed out previously, though, we require only a partial convergence of three integer pairs to begin the survey. When we take into consideration the good as well as the poor geometries that we have to contend with, clearly it will take longer to resolve all six integer pairs than a subset of the fastest three. For this reason, the viewpoint presented in this paper is somewhat more optimistic than that of [2] for obtaining a solution.

ANTENNA EXCHANGE ON A MOVING PLATFORM

The original antenna exchange technique calls for a mutual exchange of antennas between their fixed locations. Since the exchange is a physical one, the baseline used between the antennas needs to remain unchanged during the time of exchange. This is not a problem with surveys done on solid ground. However, if the antennas are in constant motion, there is no way to determine their original locations at the end of an exchange. Still, if the spatial relationship between the antennas during the exchange is either known or maintained, it is possible to expand the use of the antenna exchange technique to a moving platform.

Moving Platform Initialization

The initialization technique proposed here calls for a third antenna to be included in an array. This array must remain rigidly fixed to a platform that can be moving, such as a ship's deck. In the array, the antennas are collinear, and spaced by known distances. For convenience, the formulation pursued here assumes equal spacing between the antennas.

Carrier phase measurements are made at all three antennas. Two sets of double-difference measurements can be made by forming two antenna pairings; the third pairing is redundant. Then, by mutually exchanging two of the antennas while leaving the third antenna fixed, all initial integer ambiguities involved are instantly resolved (Figure 6). To see this, let us again resort to the kind of solvability analysis performed before. This time, a two-dimensional example is used instead of a one-dimensional example.

$$\begin{bmatrix} \Delta\phi_{01}^{(1)-} \\ \Delta\phi_{01}^{(2)-} \\ \Delta\phi_{21}^{(1)-} \\ \Delta\phi_{21}^{(2)-} \\ \Delta\phi_{01}^{(1)+} \\ \Delta\phi_{01}^{(2)+} \\ \Delta\phi_{21}^{(1)+} \\ \Delta\phi_{21}^{(2)+} \end{bmatrix} = \begin{bmatrix} h^{-}{}_{x}^{(1)} & h^{-}{}_{y}^{(1)} & 0 & 0 & 1\,0\,0\,0 \\ h^{-}{}_{x}^{(2)} & h^{-}{}_{y}^{(2)} & 0 & 0 & 0\,1\,0\,0 \\ 2h^{-}{}_{x}^{(1)} & 2h^{-}{}_{y}^{(1)} & 0 & 0 & 0\,0\,1\,0 \\ 2h^{-}{}_{x}^{(2)} & 2h^{-}{}_{y}^{(2)} & 0 & 0 & 0\,0\,0\,1 \\ -h^{+}{}_{x}^{(1)} & -h^{+}{}_{y}^{(1)} & -h^{+}{}_{x}^{(1)} & -h^{+}{}_{y}^{(1)} & 1\,0\,0\,0 \\ -h^{+}{}_{x}^{(2)} & -h^{+}{}_{y}^{(2)} & -h^{+}{}_{x}^{(2)} & -h^{+}{}_{y}^{(2)} & 0\,1\,0\,0 \\ h^{+}{}_{x}^{(1)} & h^{+}{}_{y}^{(1)} & h^{+}{}_{x}^{(1)} & h^{+}{}_{y}^{(1)} & 0\,0\,1\,0 \\ h^{+}{}_{x}^{(2)} & h^{+}{}_{y}^{(2)} & h^{+}{}_{x}^{(2)} & h^{+}{}_{y}^{(2)} & 0\,0\,0\,1 \end{bmatrix} \begin{bmatrix} \Delta x \\ \Delta y \\ \delta x \\ \delta y \\ N_{01}^{(1)} \\ N_{01}^{(2)} \\ N_{21}^{(1)} \\ N_{21}^{(2)} \end{bmatrix} \qquad (11)$$

The superscripts − and + denote time instants before and after the antenna exchange, respectively. The subscripts 01 and 21 refer to the pairings between antennas 0 and 1, and antennas 2 and 1. Without writing out the entire inverse of the coefficient matrix, the partial solution is given below for just the initial integers.

$$\begin{bmatrix} N_{01}^{(1)} \\ N_{01}^{(2)} \\ N_{21}^{(1)} \\ N_{21}^{(2)} \end{bmatrix} = \begin{bmatrix} 2/3 & 0 & -1/3 & 0 & 1/3 & 0 & 1/3 & 0 \\ 0 & 2/3 & 0 & -1/3 & 0 & 1/3 & 0 & 1/3 \\ -2/3 & 0 & 1/3 & 0 & 2/3 & 0 & 2/3 & 0 \\ 0 & -2/3 & 0 & 1/3 & 0 & 2/3 & 0 & 2/3 \end{bmatrix} \begin{bmatrix} \Delta\phi_{01}^{(1)-} \\ \Delta\phi_{01}^{(2)-} \\ \Delta\phi_{21}^{(1)-} \\ \Delta\phi_{21}^{(2)-} \\ \Delta\phi_{01}^{(1)+} \\ \Delta\phi_{01}^{(2)+} \\ \Delta\phi_{21}^{(1)+} \\ \Delta\phi_{21}^{(2)+} \end{bmatrix} \qquad (12)$$

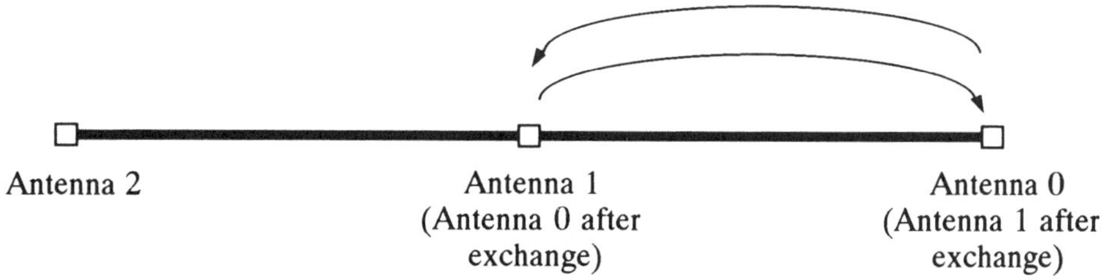

Fig. 6—Moving Platform Antenna Exchange Array

As in the one-dimensional case (see equation (6)), the integer solution is totally independent of satellite geometry. Rather, the nonzero numbers in the linear connection matrix reflect the proportion of the spacing among the three antennas in the array.

Three-Dimensional Model

To extend the above example to a GPS model, we need to include variables for one more dimension of position. Correspondingly, for the measurement set, two more satellites are added, for a total of four. The four satellites provide three double differences for each of the two antenna pairs. Thus, by making measurements before and after the antenna exchange, we have a total of twelve measurements needed to solve the twelve variables: three initial positions, three incremental positions, and six initial integers.

Kalman Filter

When generating a measurement model for a Kalman filter based on the above formulation, we can again combine the initial position and incremental position variables, just as was done for the kinematic-on-the-fly situation. The first three components of the state vector are positions, which are modeled as independent random walk, and the remainder are initial integers, treated as random constants.

Before exchange:

$$\begin{bmatrix} \Delta\phi_{01}^{(1)} - \Delta\phi_{01}^{(2)} \\ \Delta\phi_{01}^{(2)} - \Delta\phi_{01}^{(3)} \\ \Delta\phi_{01}^{(3)} - \Delta\phi_{01}^{(4)} \\ \Delta\phi_{21}^{(1)} - \Delta\phi_{21}^{(2)} \\ \Delta\phi_{21}^{(2)} - \Delta\phi_{21}^{(3)} \\ \Delta\phi_{21}^{(3)} - \Delta\phi_{21}^{(4)} \end{bmatrix} = \begin{bmatrix} h_x^{(1)} - h_x^{(2)} & h_y^{(1)} - h_y^{(2)} & h_z^{(1)} - h_z^{(2)} & 1\,0\,0\,0\,0\,0 \\ h_x^{(2)} - h_x^{(3)} & h_y^{(2)} - h_y^{(3)} & h_z^{(2)} - h_z^{(3)} & 0\,1\,0\,0\,0\,0 \\ h_x^{(3)} - h_x^{(4)} & h_y^{(3)} - h_y^{(2)} & h_z^{(3)} - h_z^{(4)} & 0\,0\,1\,0\,0\,0 \\ 2(h_x^{(1)} - h_x^{(2)}) & 2(h_y^{(1)} - h_y^{(2)}) & 2(h_z^{(1)} - h_z^{(2)}) & 0\,0\,0\,1\,0\,0 \\ 2(h_x^{(2)} - h_x^{(3)}) & 2(h_y^{(2)} - h_y^{(3)}) & 2(h_z^{(2)} - h_z^{(3)}) & 0\,0\,0\,0\,1\,0 \\ 2(h_x^{(3)} - h_x^{(4)}) & 2(h_y^{(3)} - h_y^{(4)}) & 2(h_z^{(3)} - h_z^{(4)}) & 0\,0\,0\,0\,0\,1 \end{bmatrix} \begin{bmatrix} \Delta x \\ \Delta y \\ \Delta z \\ N_{01}^{(1,2)} \\ N_{01}^{(2,3)} \\ N_{01}^{(3,4)} \\ N_{21}^{(1,2)} \\ N_{21}^{(2,3)} \\ N_{21}^{(3,4)} \end{bmatrix} \quad (13)$$

After exchange:

$$\begin{bmatrix} \Delta\phi_{01}^{(1)} - \Delta\phi_{01}^{(2)} \\ \Delta\phi_{01}^{(2)} - \Delta\phi_{01}^{(3)} \\ \Delta\phi_{01}^{(3)} - \Delta\phi_{01}^{(4)} \\ \Delta\phi_{21}^{(1)} - \Delta\phi_{21}^{(2)} \\ \Delta\phi_{21}^{(2)} - \Delta\phi_{21}^{(3)} \\ \Delta\phi_{21}^{(3)} - \Delta\phi_{21}^{(4)} \end{bmatrix} = \begin{bmatrix} -(h_x^{(1)} - h_x^{(2)}) & -(h_y^{(1)} - h_y^{(2)}) & -(h_z^{(1)} - h_z^{(2)}) & 1\,0\,0\,0\,0\,0 \\ -(h_x^{(2)} - h_x^{(3)}) & -(h_y^{(2)} - h_y^{(3)}) & -(h_z^{(2)} - h_z^{(3)}) & 0\,1\,0\,0\,0\,0 \\ -(h_x^{(3)} - h_x^{(4)}) & -(h_y^{(3)} - h_y^{(4)}) & -(h_z^{(3)} - h_z^{(4)}) & 0\,0\,1\,0\,0\,0 \\ h_x^{(1)} - h_x^{(2)} & h_y^{(1)} - h_y^{(2)} & h_z^{(1)} - h_z^{(2)} & 0\,0\,0\,1\,0\,0 \\ h_x^{(2)} - h_x^{(3)} & h_y^{(2)} - h_y^{(3)} & h_z^{(2)} - h_z^{(3)} & 0\,0\,0\,0\,1\,0 \\ h_x^{(3)} - h_x^{(4)} & h_y^{(3)} - h_y^{(4)} & h_z^{(3)} - h_z^{(4)} & 0\,0\,0\,0\,0\,1 \end{bmatrix} \begin{bmatrix} \Delta x \\ \Delta y \\ \Delta z \\ N_{01}^{(1,2)} \\ N_{01}^{(2,3)} \\ N_{01}^{(2,3)} \\ N_{21}^{(1,2)} \\ N_{21}^{(2,3)} \\ N_{21}^{(3,4)} \end{bmatrix} \quad (14)$$

where $\Delta\phi_{m,n}^{(i)}$ is the single-difference measurement from satellite i between the two antennas m and n, and $[h_x^{(i)} h_y^{(i)} h_z^{(i)}]^T$ is the unit direction vector to satellite i.

During the antenna exchange, no measurements are processed by the filter. Over this duration of time, the measurement situation is undefined, being covered neither by equation (13) nor equation (14).

CONCLUSIONS

This paper has presented ideas that may be useful in extending kinematic techniques to application areas beyond terrestrial surveying. By exploiting the redundancy available in more than four satellite measurements, the requirement for static initialization in kinematic surveying can be discarded. This initialization, which resolves the initial integer ambiguity, can instead be performed while the roving receiver is in motion. Although there are time savings that may be derived from this, the real benefit of such flexibility lies in its real-time applicability to vehicles that are constantly in motion.

To further expand the utility of kinematic techniques to such situations, an augmentation of the original antenna exchange technique was proposed for moving baselines. This technique uses a collinear three-antenna array to initialize a kinematic survey from any mobile platform, such as the deck of a ship, or a moving terrain vehicle. A potential application for this initialization technique involves the relative position determination of an element which, originally located on the mobile platform, is then dispatched from the platform. One example of this element would be a series of buoys with hydrophones deployed and towed by a seismic surveying ship at sea. Another example could be a reference target, dropped by parachute, for camera orientation in a photogrammetric survey where no such fixed references on the ground are available or are within the camera's field of view.

By extending the applicability of kinematic techniques to positioning and navigational settings, the overall usefulness of GPS is further enhanced. As we near the dawn of full GPS operation, the gradual uncovering of the system's true potential appears still to be incomplete. Meanwhile, its progress continues to astound.

ACKNOWLEDGMENTS

The author would like to thank Dr. Peter Loomis of Trimble Navigation for bringing to his attention the paper that originally proposed the redundant measurement solution to the kinematic-on-the-fly problem [2].

Based on a paper presented at IEEE Plans-90, Las Vegas, NV, March 1990.

REFERENCES

1. Remondi, B. W., *Kinematic and Pseudo-Kinematic GPS,* Proceedings of the Satellite Division of The Institute of Navigation First International Technical Meeting, Colorado Springs, CO, September 21–23, 1988.
2. Loomis, P. V. W., *A Kinematic GPS Double Differencing Algorithm,* Proceedings of the Fifth International Geodetic Symposium on Satellite Positioning, New Mexico State University, Las Cruces, NM, March 13–17, 1989.
3. Remondi, B. W., *Performing Centimeter-Level Surveys in Seconds with GPS Carrier Phase: Initial Results,* NAVIGATION, Journal of The Institute of Navigation, Vol. 32, No. 4, Winter 1985–86.
4. Brown, R. G., *Introduction to Random Signal Analysis and Kalman Filtering,* New York, Wiley, 1983.

NAVIGATION: *Journal of The Institute of Navigation*
Vol. 38, No. 2, Summer 1991
Printed in U.S.A.

High-Precision Real-Time GPS Positioning Concepts: Modeling and Results

NICHOLAS C. TALBOT

Royal Melbourne Institute of Technology, Melbourne, Australia

Received September 1990
Revised January 1991

ABSTRACT

GPS has been used successfully for a wide range of high-precision surveying applications. Most precise positioning results have been achieved through post-processing of data. This denies the use of GPS for those surveying and navigation applications which require on-line position computation.

A Kalman filtering scheme is presented which is designed to perform high-integrity, real-time GPS relative positioning in static and kinematic modes to centimeter-level accuracy. A continuous-kinematic data set for a train has been processed as an example to explain the operation of the system and illustrate achievable accuracy. Results of the test proved that the repeatability of vertical positions for a train traveling at 120 km/h was better than 5 cm.

INTRODUCTION

The classical approach to precise GPS positioning is the static technique, in which receivers simultaneously track the phase of the GPS carrier signals to four or more satellites for 1–2 h. This typically yields centimeter-level differential positioning accuracy [1].

The precise kinematic GPS technique offers a significant improvement in acquisition efficiency as compared with static positioning [2]. Kinematic positioning relies on a rapid initialization procedure such as an antenna swap, a known baseline occupation, or a static baseline observation to estimate the integer phase ambiguities [3]. Once the phase ambiguities have been resolved and continuous carrier phase tracking maintained to at least four satellites, the position of a roving receiver can be computed relative to a fixed reference receiver at each measurement epoch.

Variations on the kinematic technique include the stop-go method, where points of interest are occupied for 15 s to 1–2 min, but the trajectory of the receiver during transportation between marks is not used explicitly for processing [4]. The semikinematic method uses a Kalman filter to propagate the trajectory of the roving receiver between station marks and batch least squares during static mark occupation [5].

The limitation of precise GPS postprocessing schemes is the inability to use the system for purposes that require on-line computation and the inability to detect cycle slips correctly as they occur. With the provision of a data link between the reference and roving receivers, it is possible to process data in real time and still achieve accuracies equivalent to those provided by postprocessing [6]. This greatly broadens the applicability of the system to a wide range of precise surveying and navigation applications, such as land subdivision and hydrographic surveying [7]. The proposed real-time positioning system is shown in Figure 1.

The technique outlined below presents a new method for processing GPS phase data in real time for short-baseline (<15 km) estimation. It is based on a multiconfiguration Kalman filtering scheme that uses double-difference phase data to estimate centimeter-level differential positions. The integrity of the computed results is of paramount importance; hence the scheme incorporates several checks on the positioning filter performance. Although no real-time tests have been attempted thus far, the operation and results of the scheme are outlined for a continuous kinematic survey conducted on a moving train at Pueblo, Colorado [8].

GPS PHASE MEASUREMENTS

Measurement Model

The double-difference GPS phase observable is useful for estimating the position of a roving receiver relative to a fixed reference. The simplified double-difference phase measurement model is given by the following expression:

$$\nabla\Delta\phi_{12}^{ij} = \nabla\Delta\rho_{12}^{ij} + \nabla\Delta N_{12}^{ij} + v \tag{1}$$

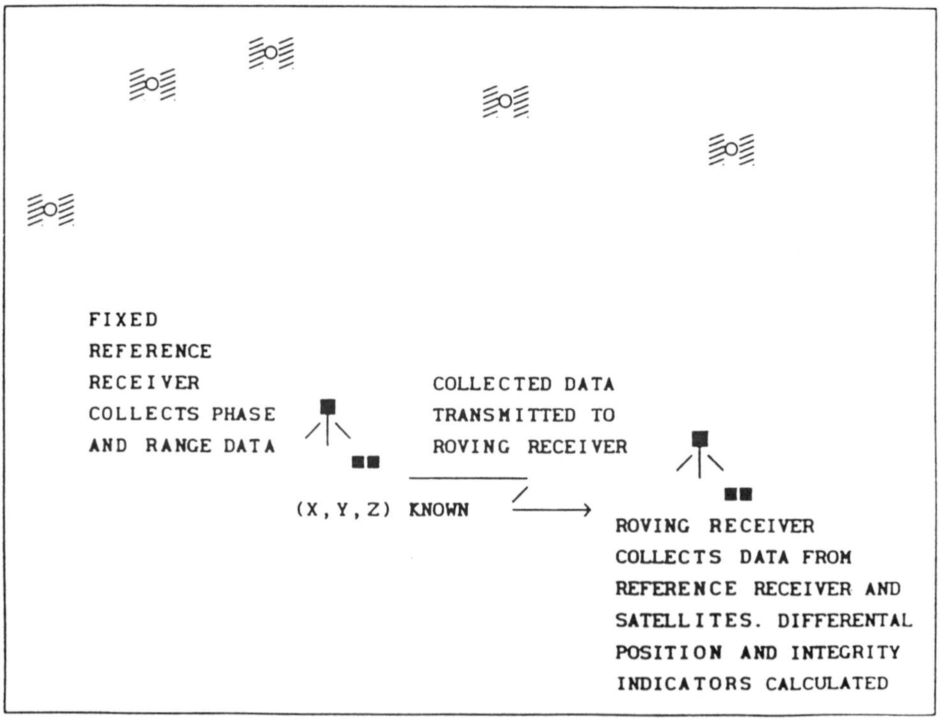

Fig. 1—Real-Time Differential Positioning And Integrity Monitoring Configuration

where $\nabla\Delta\phi_{12}^{ij}$ is a double-difference phase observation, which is subject to a residual or observation noise term, v. The double-difference phase measurement is formed by combining phase data from reference and roving receivers to satellites i and j:

$$\nabla\Delta\phi_{12}^{ij} = \phi_1^i - \phi_1^j - \phi_2^i + \phi_2^j \tag{2}$$

Subscript $_1$ identifies the reference receiver and $_2$ the roving receiver.

For real-time operation, the phase data from the reference receiver (ϕ_1^i and ϕ_1^j) is transmitted to the roving receiver(s), whereas for postprocessing, the data is stored in the field and brought together after the mission.

$\nabla\Delta\rho_{12}^{ij}$ is the double-differenced range term, which is a function of the unknown roving receiver WGS84 coordinates (X_2, Y_2, Z_2), known reference receiver coordinates (X_1, Y_1, Z_1), and satellite coordinates evaluated from the broadcast satellite ephemeris:

$$\nabla\Delta\rho_{12}^{ij} = \rho_1^i - \rho_1^j - \rho_2^i + \rho_2^j \tag{3}$$

For satellite i and receiver 1, the undifferenced range term is given by:

$$\rho_1^i = \left[(X^i - X_1)^2 + (Y^i - Y_1)^2 + (Z^i - Z_1)^2\right]^{1/2} \tag{4}$$

with like expressions for satellite j and receiver 2.

The double-differenced phase ambiguity term $\nabla\Delta N_{12}^{ij}$ is included in the double-difference measurement model to recognize the inability of the GPS receiver to determine the whole number of carrier wavelengths between the antenna and satellite at the start of phase tracking [1]. There are (s − 1) unique double-difference ambiguity terms for s observed satellites. In theory, the double-difference phase ambiguities represent integer values ($\nabla\Delta N_{12}^{ij}$) given by the following combination of the raw phase ambiguities:

$$\nabla\Delta N_{12}^{ij} = N_1^i - N_1^j - N_2^i + N_2^j \tag{5}$$

Effects of atmospheric, satellite ephemeris, and residual clock errors have been ignored in equation (1), since the processing scheme is aimed only at short baselines (<15 km). These errors and small random measurement errors are absorbed in the residual term v shown in equation (1). A more detailed discussion of the double-difference measurement model for static and kinematic positioning can be found in [1] and [9].

The double-difference observable is highly suited to static and kinematic positioning because of its relatively simple form. This is due to the absence of clock terms, and most importantly, the integer nature of the ambiguity terms is preserved [9].

Cycle Slips

Cycle slips are the limiting factor in GPS carrier phase positioning. A cycle slip occurs when tracking of the incoming carrier signal is interrupted. This causes an integer jump in the carrier phase data once tracking is regained [1]. Signal path obstruction, rapid acceleration, and high atmospheric disturbances are some of the possible causes of cycle slips. To achieve centimeter-level position accuracy, cycle slips must be correctly identified and repaired.

When both reference and roving receivers are stationary, cycle slips appear as integer steps in the double-difference residuals. Repair is achieved by applying the integer jump caused by the slip to the corresponding double-difference phase ambiguity term of the offending satellite combination. Cycle slips are automatically identified during the Doppler integration process in several surveying GPS units, for example, the Trimble 4000 series. The motion of the roving receiver during kinematic operation makes cycle slip identification far more difficult than in the static case. Analysis of the measurement residuals offers another method for cycle slip identification, covered below.

Once a cycle slip has been identified, it is possible to solve for the integer jump in the errant satellite by using the position calculated from the remaining double-difference measurements. At least three double differences are required to fix the position of the roving receiver, although four are advisable to ensure correct reinitialization after cycle slips have been resolved.

KALMAN FILTERING

Filter Equations

Kalman filtering techniques are well suited to real-time positioning [10]. First, Kalman filtering enables time variation of the position parameters to be modeled and included in the estimation process. Unlike batch least squares, where all data must be collected and postprocessed, Kalman filtering allows the double-difference phase measurements to be processed as they become available. Minimal computer storage is needed for Kalman filtering since the equations are recursive (see the equations in Table 1 and [11]).

Table 1—Summary of Discrete Kalman Filter Equations and Definitions (The notation has been adapted from [12].)

Equation	No.	Step
$\bar{\mathbf{x}}_k = \mathbf{\Phi}\, \hat{\mathbf{x}}_{k-1}$	(6.1)	Prediction
$\mathbf{Q}_{\bar{x}_k\bar{x}_k} = \mathbf{\Phi}\, \mathbf{Q}_{\hat{x}_{k-1}\hat{x}_{k-1}}\, \mathbf{\Phi}^T + \mathbf{Q}_{w_k w_k}$	(6.2)	
$\hat{\mathbf{x}}_k = \bar{\mathbf{x}}_k + \mathbf{K}_k \left[\mathbf{l}_k - \mathbf{A}_k \bar{\mathbf{x}}_k \right]$	(6.3)	Measurement Update
$\mathbf{Q}_{\hat{x}_k\hat{x}_k} = \left[\mathbf{I} - \mathbf{K}_k \mathbf{A}_k \right] \mathbf{Q}_{\bar{x}_k\bar{x}_k}$	(6.4)	
$\mathbf{K}_k = \mathbf{Q}_{\bar{x}_k\bar{x}_k}\, \mathbf{A}_k^T \left[\mathbf{A}_k \mathbf{Q}_{\bar{x}_k\bar{x}_k}\, \mathbf{A}^T + \mathbf{Q}_{v_k v_k} \right]^{-1}$	(6.5)	

$\bar{\mathbf{x}}_k$, $\hat{\mathbf{x}}_k$	Predicted and filtered state vectors
$\mathbf{\Phi}$	State transition matrix
$\mathbf{Q}_{\bar{x}_k\bar{x}_k}$, $\mathbf{Q}_{\hat{x}_k\hat{x}_k}$	Covariances of the predicted and filtered states
$\mathbf{Q}_{w_k w_k}$	System-driving noise covariance
$\mathbf{l}_k$	Measurement vector
$\mathbf{A}_k$	Measurement design matrix
$\mathbf{K}_k$	Kalman gain matrix
$\mathbf{Q}_{v_k v_k}$	Measurement noise covariance

A Kalman filter state description that is general enough to handle static and kinematic positioning modes is the constant velocity model [13]:

$$\mathbf{x} = \left[\begin{array}{c|c|c} X_2\, Y_2\, Z_2 & \dot{X}_2\, \dot{Y}_2\, \dot{Z}_2 & \nabla\Delta N^1\, \nabla\Delta N^2 \ldots \nabla\Delta N^{s-1} \end{array} \right] = \left[\begin{array}{c|c|c} \mathbf{c} & \mathbf{d} & \mathbf{b} \end{array} \right] \tag{7}$$

where the partitioned state vector includes the coordinates of the roving receiver (X_2, Y_2, Z_2), (partition **c**); their first derivatives ($\dot{X}_2$, $\dot{Y}_2$, $\dot{Z}_2$), (partition **d**); and the double-difference phase ambiguities ($\nabla\Delta N^1\, \nabla\Delta N^2 \ldots \nabla\Delta N^{s-1}$), (partition **b**). The state transition matrix has the following form for kinematic positioning:

$$\mathbf{\Phi} = \left[\begin{array}{c|c|c} \mathbf{I} & \mathbf{I} \times \Delta t & 0 \\ \hline 0 & \mathbf{I} & 0 \\ \hline 0 & 0 & \mathbf{I} \end{array} \right] \begin{array}{c} \mathbf{c} \\ \mathbf{d} \\ \mathbf{b} \end{array} \quad \text{(columns: } \mathbf{c}, \mathbf{d}, \mathbf{b}\text{)} \tag{8}$$

where **I** is the identity matrix, and Δt is the measurement epoch interval. When the roving receiver is static, the state transition matrix degenerates to the identity. The elements of the initial state covariance matrix ($\mathbf{Q}_{\hat{x}_0\hat{x}_0}$) and system-driving noise covariance matrix ($\mathbf{Q}_{w_k w_k}$) are set according to the positioning scenario, which is described below.

The linearized double-difference residual equation is given by:

$$\tilde{\mathbf{v}}_k = \mathbf{A}_k \tilde{\mathbf{x}}_k + \mathbf{v}_k \tag{9}$$

The design matrix $\mathbf{A}_k$ contains measurement partials for the roving receiver coordinates and double-difference phase ambiguity terms. The predicted residual vector $\tilde{\mathbf{v}}_k$ contains observed minus computed double-difference phase measurements. Both the measurement partials and the computed double-difference phase terms are evaluated at the predicted state $\tilde{\mathbf{x}}_k$. The vector $\mathbf{v}_k$ represents measurement noise or residual terms. The variances of all double differences are assumed equal, and the measurement noise covariance is given a diagonal form:

$$\mathbf{Q}_{v_k v_k} = 0.0025 \text{ cycles}^2 \times \mathbf{I} \tag{10}$$

Integrity Indicators

The success of the Kalman filter implementation rests on the proper modeling of the time variation of the state vector terms through the kinematic model and the appropriate measurement modeling. To test the on-line performance of the Kalman filtering system, the Local-Overall Model (LOM) test and the Global-Overall Model (GOM) test are used; their derivation and interpretation can be found in [14].

LOM is computed at each epoch by forming the quadratic product of the observed minus computed double-difference residuals, $\tilde{\mathbf{v}}_k$, normalized by their inverse covariance:

$$\mathcal{T}_k = \tilde{\mathbf{v}}_k^T \left[\mathbf{Q}_{\tilde{v}_k\tilde{v}_k} \right]^{-1} \tilde{\mathbf{v}}_k \tag{11}$$

The predicted residual covariance is readily obtained from matrices that are usually computed as part of the filtering process:

$$\mathbf{Q}_{\tilde{v}_k\tilde{v}_k} = \mathbf{Q}_{v_kv_k} + \mathbf{A}_k\mathbf{Q}_{x^-_k\tilde{x}_k}\mathbf{A}_k^T \tag{12}$$

An error in specifying the correct kinematic or measurement model is suspected if:

$$\mathcal{T}_k \geq \chi^2_\alpha(m_k, 0) \tag{13}$$

where $\chi^2_\alpha(m_k, 0)$ is the upper α probability point of the chi-square distribution, where the mean is zero and m_k is the number of measurements observed at epoch k, or the number of degrees of freedom.

The LOM test is very effective for detecting unannounced cycle slips in the data which materialize as abrupt jumps in the predicted double differences. Identification of the actual measurement that caused the condition of equation (13) is complex for kinematic positioning, particularly when there are multiple slippages at a given epoch [7].

The weighted sum of the LOM statistics, termed the GOM test-statistic, can be used for detecting global deficiencies in the measurement and kinematic models [14]:

$$T_k = \frac{\sum_{i=1}^{k} \tilde{\mathbf{v}}_i^T \mathbf{Q}_{\tilde{v}_i\tilde{v}_i}^{-1} \tilde{\mathbf{v}}_i}{\sum_{i=1}^{k} m_i} = \frac{\sum_{i=1}^{k} m_i \mathcal{T}_i}{\sum_{i=1}^{k} m_i} \tag{14}$$

A misspecification is suspected if

$$T_k \geq \chi^2_\alpha \left(\sum_{i=1}^{k} m_i, 0 \right) \tag{15}$$

Although a specific cause of a GOM failure is difficult to identify, possible causes in static double-difference phase data processing include small cycle slips, improper ambiguity initialization, uncorrelated atmospheric errors, measurement multipath, and unusually large ephemeris errors (which were not included in the working Kalman filter models).

Filtered State Covariance Trace

A conservative estimate of the precision of the remote station position can be computed from the filtered state covariance matrix at each measurement epoch in the following manner:

$$\sigma_c = \sqrt{(\sigma_X^2 + \sigma_Y^2 + \sigma_Z^2)} \tag{16}$$

where σ_c is an RMS error representation of the grouped remote station coordinate parameters, and the filtered state covariance matrix is taken as:

$$Q_{\hat{x}_k\hat{x}_k} = \left[\begin{array}{c|c|c} \begin{matrix} \sigma_X^2 & \sigma_{XY} & \sigma_{XZ} \\ \sigma_{YX} & \sigma_Y^2 & \sigma_{YZ} \\ \sigma_{ZX} & \sigma_{ZY} & \sigma_Z^2 \end{matrix} & \boldsymbol{\sigma}_{cd} & \boldsymbol{\sigma}_{cb} \\ \hline \boldsymbol{\sigma}_{dc} & \boldsymbol{\sigma}_d^2 & \boldsymbol{\sigma}_{db} \\ \hline \boldsymbol{\sigma}_{bc} & \boldsymbol{\sigma}_{bd} & \boldsymbol{\sigma}_b^2 \end{array}\right] \tag{17}$$

For kinematic differential positioning, the magnitudes of the covariance elements are dependent primarily on satellite geometry and the specified double-difference phase measurement variance.

POSITIONING SCENARIO CONCEPT

A positioning scenario concept is implemented to accommodate the operation of the filtering system in static and kinematic positioning modes [7]. A change in the positioning scenario is triggered by the user to reconfigure the state vector terms and the system-driving noise covariance elements. For example, to commence a kinematic survey, the user may select the static_bias_solution scenario. In this case, coordinates of the occupied station must be entered, and the filter is configured to estimate only the double-difference phase ambiguity terms. The filter automatically identifies the integer phase ambiguities and constrains them to their integer values. After ambiguity resolution, the user can select the kinematic scenario and move the roving receiver. During the kinematic scenario, the trajectory of the vehicle is projected and updated with the double-difference measurements. At the stop points in a stop-go kinematic survey, the static_coordinate_solution scenario is selected, and the filter is reconfigured to provide a sequential solution for the remote receiver coordinate terms only.

The advantage of the real-time processing scheme is the ability to detect cycle slips and gauge whether reinitialization of the survey is necessary. The operation of each positioning scenario is detailed below.

Static Scenario

The static scenario incorporates the solution of both $\nabla\Delta N$ terms and coordinate parameters, while the velocity terms are identically constrained to zero (see Table 2). This implies that the coordinates and ambiguities are modeled as constants and leads to their sequential adjustment. The filter automatically identifies and constrains the phase ambiguity terms to integer values as soon as their respective formal precisions drop below 0.0225 cycles^2 and are less than 0.15 cycles from an integer. This technique is termed sequential ambiguity resolution and leads to accelerated baseline recovery [15]. Data collection can be stopped as soon as the precision of the coordinates reaches an acceptable level (1–5 cm), which for short baselines with at least five tracked satellites is usually achieved in 15–30 min.

Table 2—Static Scenario Filter Parameter Summary

Initial State Covariance			System Driving Noise Covariance		
c	d	b	c	d	b
10			10^{-20}		
	10^{-2}			10^{-20}	
		10^{20}			10^{-20}

Initial State Vector

c = Pseudoranged Position
d = 0
b = 0

Static Bias Solution Scenario

Double-difference phase ambiguity initialization can be performed with either the static scenario or the static_bias_solution scenario. The latter is used when a known baseline is occupied. The coordinate partition of the state vector must be initialized with a priori estimates of the roving station coordinates relative to the reference receiver and accurate to ± 5 cm. As with the static scenario, the phase ambiguities ($\nabla\Delta N^i$) are automatically constrained once their formal precisions reach 0.0225 cycles² and their values are sufficiently close to integers. The filter parameters used for the static_bias_solution are shown in Table 3.

Since only phase ambiguity terms are estimated, the measurement design matrix includes only ambiguity partials. Only two to three measurement epochs are required to estimate and constrain the ambiguities.

Kinematic Scenario

Once the ambiguities have been constrained, the position and velocity of the roving receiver can be evaluated. The system-driving noise covariance for the kinematic scenario is given by:

$$Q_{w_k w_k} = \begin{bmatrix} Iq\Delta t^3/3 & Iq\Delta t^2/2 & 0 & c \\ Iq\Delta t^2/2 & Iq\Delta t & 0 & d \\ 0 & 0 & 0 & b \\ c & d & b & \end{bmatrix} \tag{18}$$

Table 3—Static_Bias_Solution Filter Parameter Summary

Initial State Covariance			System-Driving Noise Covariance		
c	d	b	c	d	b
10^{-20}			10^{-20}		
	10^{-20}			10^{-20}	
		10^{20}			10^{-20}

Initial State Vector

c = Accurate Coordinate Values
d = 0
b = 0

where the coordinate states are uncoupled, and the system-driving noise is assumed to enter through random accelerations. The system-driving noise values for each component are assumed to have equal spectral density q. The selection of the system-driving noise spectral density is made by the user according to the expected dynamics of the roving receiver platform. Three levels of dynamics are available: kinematic_Level_1 for back-pack GPS; kinematic_Level_2 for medium dynamics, such as ship mounting or land vehicles; and kinematic_Level_3 for high dynamics for aircraft. The filter parameters for the kinematic solution are shown in Table 4.

Static Coordinate Solution Scenario

The static_coordinate_solution is used for stop-go kinematic work and can be invoked after the initial phase ambiguities have been constrained. The filter parameters for the static_coordinate_solution scenario are shown in Table 5.

The kinematic model specification enforces a sequential adjustment of the state coordinate parameters. Only one to two measurement epochs are needed to determine precisely the coordinates of the ground mark, provided reasonable satellite geometry exists.

COMPUTER IMPLEMENTATION

The filtering scheme—KRONOS (**K**alman **R**eduction **O**f **N**AVSTAR **O**rbiting **S**atellites)—was developed on a Toshiba T1600 IBM AT compatible laptop computer, using the Turbo Pascal V5 development environment [16].

Table 4—Kinematic Scenario Filter Parameter Summary

Initial State Covariance			System Driving Noise Covariance		
c	d	b	c	d	b
10^2			$q\Delta t^3/3$	$q\Delta t^2/2$	
	1		$q\Delta t^2/2$	$q\Delta t$	
		10^{-20}			10^{-20}

Initial State Vector

c = Coordinates from Previous Scenario
d = 0
b = Constrained Ambiguity Terms

Table 5—Static_Coordinate_Solution Scenario Filter Parameter Summary

Initial State Covariance			System-Driving Noise Covariance		
c	d	b	c	d	b
Set from Previous Scenario			10^{-20}		
	10^{-20}			10^{-20}	
		10^{-20}			10^{-20}

Initial State Vector

c = Predicted State
d = 0
b = Constrained Ambiguity Terms

Although the current system has only been tested off-line, it has been designed specifically to operate in an on-line environment in conjunction with a radio data link. Presently, data collected at reference and roving receivers are preprocessed, and then merged into a consolidated virtual array file that allows unlimited data handling.

No user interaction is required during the process run once the input data files have been specified. All scenario changes are embedded in the consolidated data file rather than being accepted interactively, which would occur on-line.

FULL KINEMATIC PUEBLO TRAIN EXAMPLE

A unique high-dynamic test was conducted at the American Association of Railroads' Transportation Test Center near Pueblo, Colorado, in October 1989. Several GPS receivers were mounted on a railcar and driven at speeds up to 125 km/h around a 23.2 km section of the test track [8]. Figure 2 shows a plan view of the track and Figure 3 a speed profile for the first loop.

The Trimble 4000SST data obtained from the test had characteristics listed in Table 6. Even though dual-frequency phase data was available for investigation, the study concerned only the L1 phase data.

The survey started with a short static baseline observation for resolution of the initial phase ambiguity terms. The train was then driven around the track and stopped for a short time near the initialization point. Several passes were made around the track; however, the following analysis is limited to only the first two. A log of the significant event times for the data is shown in Table 7.

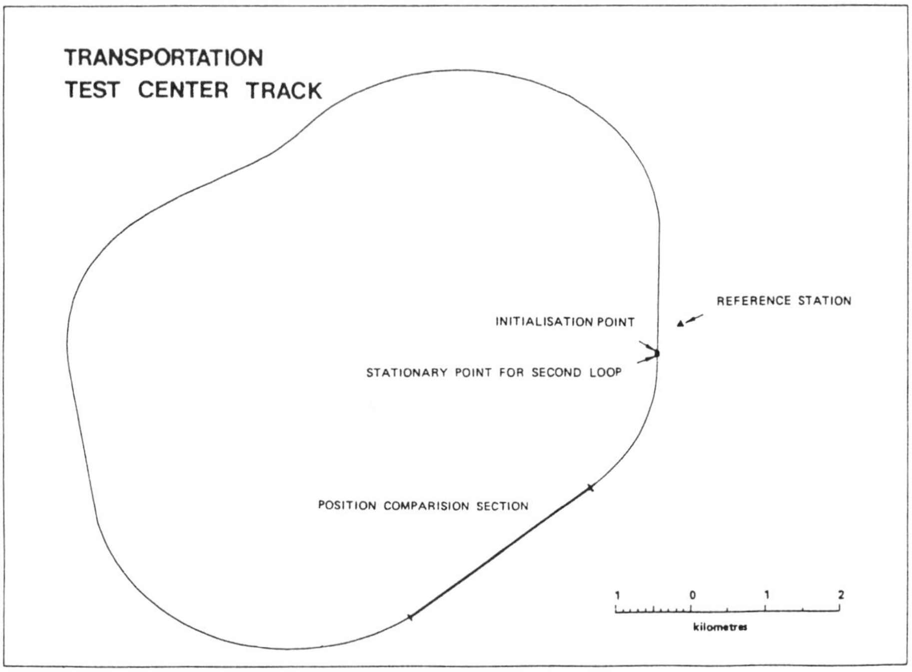

Fig. 2—Pueblo Transportation Test Center

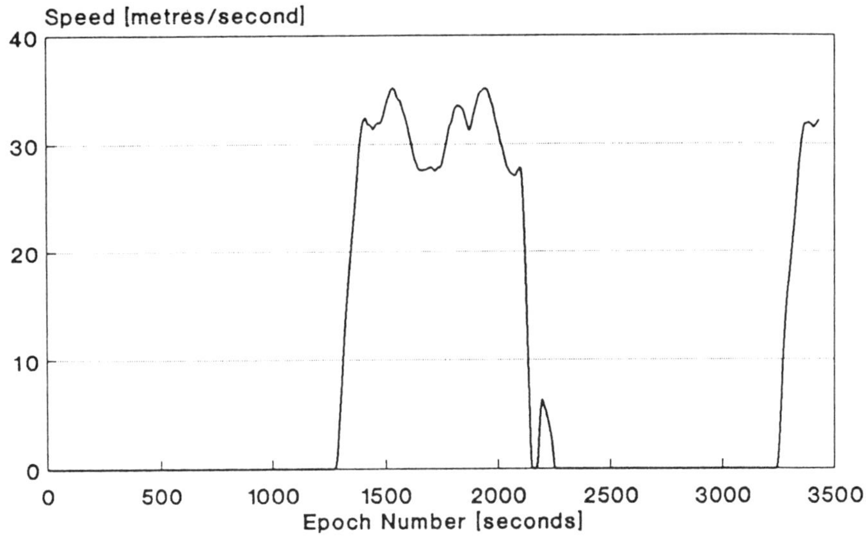

Fig. 3—Speed Profile for the Train

Table 6—Train Data Characteristics

Day of Year: 299 (October 26, 1989)
Observation Time Span: 2 h, 47 min
Data Update Rate: 1 s
Observed Satellites: 3, 6, 9, 11, 12, 13, 14 (PRN Numbers)
Data Type: L1 C/A-Code and Continuous Phase
L2 Codeless Continuous Phase

Table 7—Significant Positioning Events in the Kinematic Train Survey

Epoch (s)	Event	Scenario
0–1277	Roving Receiver Static for Ambiguity Resolution	Static
1278–2146	Train Driven Around Track (Loop 1)	
2147–2169	Train Stationary	
2170–2245	Train Reversed Back Towards Initialization Point	Kinematic
2246–3240	Train Stationary Near Initialization Point	
3241	Train Moved off on Loop 2	

The kinematic sections of the data were processed using the kinematic_Level_2 scenario, which corresponds to a driving noise spectral density of 0.20 m^2/s^3.

RESULTS

All the double-difference phase ambiguity estimates were identified and constrained to integers in less than 750 epochs (12.5 min) using the sequential ambiguity resolution strategy. The real-value estimates of the double-difference phase ambiguity terms just prior to their formal precision convergence below the critical level of 0.0225 cycles^2 are shown in Table 8.

The closeness of the ambiguity terms to integers confirms that they were correctly identified. The presence of a data link would be a clear advantage during the initialization procedure. Following ambiguity resolution at epoch 750, the roving receiver could have embarked immediately on the kinematic stage of the survey. Since no data link was used during data collection, the roving receiver remained at the initialization point for a conservative 1277 epochs.

To judge the precision of the system, a comparison of the kinematic vertical position results for a 3.2 km straight section of track, highlighted in Figure 2, was made for the two passes. The filtered positions from both passes were transferred into the LISCAD™ [17] surveying computations package for graphical and numerical analysis. The longitudinal section shown in Figure 4 portrays the vertical offsets of the results of the second pass relative to the first. It should be noted that a simple straight-line fit has been made to the first pass fixes for purposes of offset computation. This assumption causes an apparent bias in the vertical offset differences between chainages 1130 and 2820 m, where the track undergoes a vertical curve. Nevertheless, the vertical offsets indicate excellent agreement between loops 1 and 2. The maximum offset is 0.044 m, with most offsets being less than 0.03 m.

Satellite 14 rose at epoch 3248, and its phase ambiguity value was automatically initialized using the filtered position fix of the other satellites. As shown in Figure 4, lock was lost on satellite 14 during the first 50 epochs of phase tracking. The slip was successfully identified by the receiver and repaired during processing.

The excellent comparison between solutions must be tempered with the fact that five satellites were tracked for the duration of the comparison, and satellite geometry was strong. The strength of satellite geometry can be seen from the graph of the filtered state precision (σ_c) versus time in Figure 5, which has values less than 0.05 m.

During epochs 2246–3240, when the train was stationary near the initialization point, the static_coordinate_solution was not invoked. Instead, the kinematic scenario was used to provide epoch-by-epoch solutions. This enabled an assessment of the internal accuracy of the position fixes. The variance of the position fix errors about the mean was found to be 0.005 m horizontally and 0.009 m vertically.

Table 8—Estimated Double-Difference Phase Ambiguity Terms Just Prior to Their Constraint

Ambiguity Term PRN6–PRNi	Estimated Value
6–9	218571.026
6–11	139542.987
6–12	395190.972
6–13	331385.051

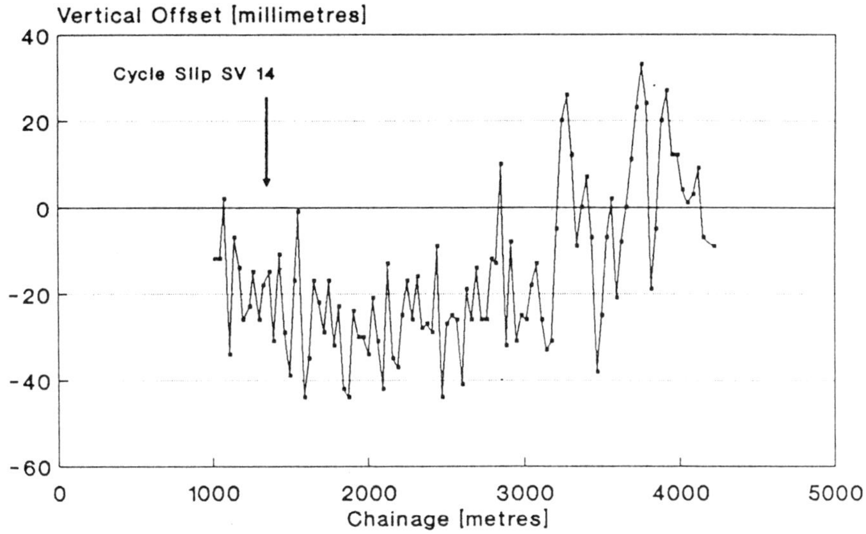

Fig. 4—Longitudinal Section of Vertical Offsets Between Position Fixes of Loops 1 and 2

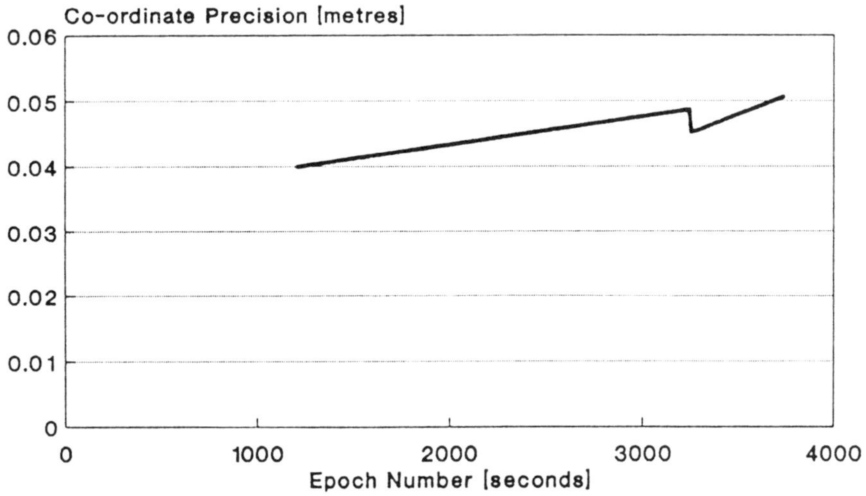

Fig. 5—Precision of the Remote Receiver

The following conclusions can be drawn from the graph of the LOM statistics versus time shown in Figure 6:

1) The critical chi-square value of the LOM test was not exceeded at any point in the data, based on a 95 percent confidence level. This confirms that the specified kinematic and measurement models are satisfactory.

2) When the train was stationary between epochs 2246 and 3240, the predicted residuals contained measurement errors only. This caused the LOM values to be far smaller than expected. Conversely, during periods of acceleration and deceleration, the magnitude of the predicted residuals

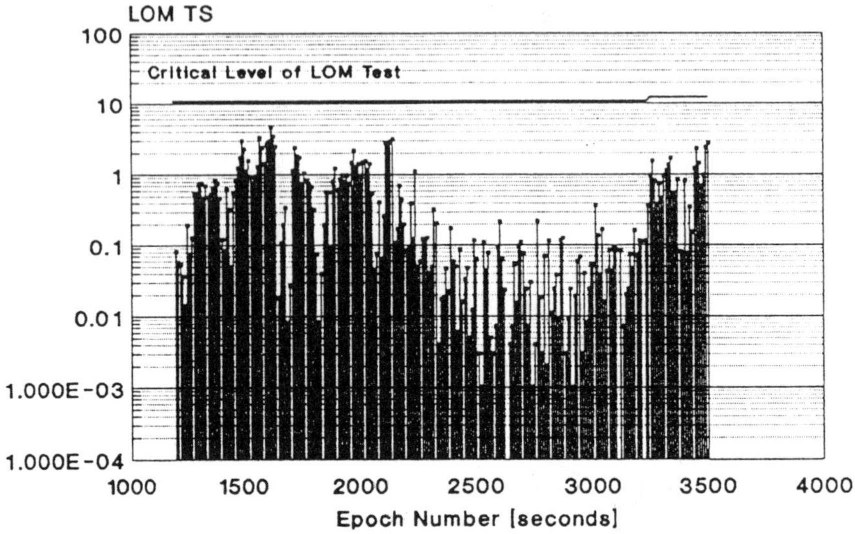

Fig. 6—Local-Overall Model Statistics

was governed predominantly by the state prediction errors inherent in the constant velocity model.

CONCLUSION

The concept of recursive differential GPS carrier phase processing has been verified by processing the GPS data previously collected for a continuous kinematic train experiment. The processing scheme is based around a Kalman filter algorithm that accepts user-input positioning scenario changes to reconfigure the filter, and thus handle static and kinematic modes of operation.

Comparison of 100 position fixes computed from repeated loops of a train track shows differences in the vertical component of less than 5 cm. This was achieved for a train traveling at 120 km/h and favorable five-satellite geometry with infrequent cycle slips.

The ability to use GPS for high-precision real-time relative navigation is a major benefit of the proposed scheme. Improved efficiency in field data acquisition over existing methods is an added strength of real-time processing. A static survey can be stopped once coordinate precision reaches an acceptable level.

The inclusion of the Local-Overall Model (LOM) and Global-Overall Model (GOM) statistics to detect cycle slips and verify correct filter performance is advised for a practical implementation. Development of an operational real-time system is proceeding.

ACKNOWLEDGMENTS

The author expresses thanks to Peter Griffioen and the technical staff at Trimble Navigation for the supply of the train data and its preprocessing. Craig Hill, Rod Allan, and Renado Schnitzer are also acknowledged for their assistance in generating plots of the data and the presentation of the results.

REFERENCES

1. Remondi, B. W., *Using the Global Positioning System (GPS) Phase Observable for Relative Geodesy: Modeling, Processing and Results*, NOAA, Rockville, MD. Reprint of doctoral dissertation, Center for Space Research, University of Texas, Austin, 1984.
2. Mooyman, K., and Quirion, C. A., *High Production Kinematic GPS Surveying*, American Congress on Surveying and Mapping, Surveying and Mapping Journal, Vol. 49, No. 3, September 1989.
3. Remondi, B. W., *Performing Centimeter-Level Surveys in Seconds with GPS Carrier Phase: Initial Results*, NAVIGATION, Journal of The Institute of Navigation, Vol. 32, No. 4, Winter 1985, pp. 386–400.
4. Quirion, C. A., *Automated Kinematic GPS Post-Processing for High Production Surveys*, Presented at the Fifth International Geodetic Symposium on Satellite Positioning, Las Cruces, NM, March 13–17, 1989.
5. Cannon, M. E., *High-Accuracy GPS Semikinematic Positioning: Modeling and Results*, NAVIGATION, Journal of The Institute of Navigation, Vol. 37, No. 1, Spring 1990, pp. 53–64.
6. Ferguson, K. E., and Veatch, E. R., *Centimeter Level Surveying in Real-Time*, Second International Symposium on Positioning with GPS, Ottawa, Canada, September 3–7, 1990.
7. Talbot, N. C. *The Effects of GPS Signal Degradation upon Differential Positioning in Static and Kinematic Modes*. Doctoral dissertation in preparation, Department of Land Information, RMIT, Melbourne, Australia, 1990.
8. Gibbons, G. (Ed.), *Long Base Line Kinematic GPS Tested on Rail Car*, GPS World, May/June 1990, pp. 16–18.
9. Leick, A., *GPS Satellite Surveying*, New York: John Wiley & Sons, 1990.
10. Sorenson, H. W. (Ed.), *Kalman Filtering: Theory and Application*, New York: IEEE Press, 1985.
11. Gelb, A. (Ed.), *Applied Optimal Estimation*, Cambridge, MA: MIT Press, 1974.
12. Merminod, B., *The Use of Kalman Filters in GPS Navigation*, UNISURV S-35, University of New South Wales, School of Surveying, Sydney, Australia, 1989, p. ix.
13. Schwarz, K. P., Cannon, M. E., and Wong, R. V. C., *A Comparison of GPS Kinematic Models for the Determination of Position and Velocity Along a Trajectory*, Manuscripta Geodaetica, Vol. 14, 1989, pp. 345–53.
14. Teunissen, P. J. G. and Salzmann, M. A., *A Recursive Slippage Test for Use in State-Space Filtering*, Manuscripta Geodaetica, Vol. 14, No. 6, 1989, pp. 383–90.
15. Talbot, N. C., *Sequential Phase Ambiguity Resolution for Real-Time Static Differential GPS Positioning*, Paper submitted to Manuscripta Geodaetica, September 1990.
16. Borland, *Turbo Pascal Version 5 Reference Guide*, Scotts Valley, USA, Borland International, 1988.
17. LISTECH Pty., Ltd., *LISCAD Reference Manual*, Melbourne, Australia, 1990.

NAVIGATION: *Journal of The Institute of Navigation*
Vol. 35, No. 2, Summer 1988
Printed in U.S.A.

High-Accuracy Kinematic Positioning by GPS-INS

R. V. C. WONG, K. P. SCHWARZ, and M. E. CANNON

The University of Calgary, Calgary, Alberta, Canada
Received October 1987
Revised February 1988

ABSTRACT

Recent results achieved with relative GPS positioning techniques indicate that accuracies at the meter level are possible in land vehicle mode if the cycle slip problem can be minimized. One of the possible solutions to the problem is the integration of GPS and inertial data. Results from inertial surveys show that, with regular, accurate coordinate or range updates, an INS will give velocity estimates at the cm/s level [1]. By integrating differential GPS measurements with an INS, the effect of cycle slips over short intervals may be eliminated from the positioning results. A Kalman filter-smoother to handle this problem has been developed. It integrates differential range and phase measurements with data from an inertial navigation system. The optimal backward smoother improves the filter estimates for periods of poor geometry and multiple cycle slips. The package has been tested with GPS and INS data from a baseline survey near Calgary, Canada. Results of the test show that sub-meter kinematic positioning accuracies and cm/s velocity accuracies are achievable with an integrated GPS-INS.

INTRODUCTION

In recent years, relative GPS positioning techniques have been used to obtain highly accurate results in navigation and kinematic positioning. By combining differential pseudorange and carrier phase measurements from two P-code multiplexing or multichannel GPS receivers, positioning accuracies of 1–2 m (1σ) are achievable in land vehicle mode, provided the problems caused by multiple cycle slips can be solved or minimized [2]. Increasing the receiver tracking bandwidth addresses only part of the problem [3]. It reduces the occurrence of cycle slips due to receiver dynamics, but it cannot eliminate cycle slips that occur when the satellite signals are blocked by surrounding objects or lost because of low satellite elevations. It also results in a loss of positioning accuracy.

The solution of the cycle slips problem is usually done in two steps. The first step is the detection of the cycle slips. It can be accomplished by checking the misclosures of the measurements against their estimated variances, or by comparing the L1 measurements with the L2 ones if dual frequency data are available. The second step is the estimation and removal of the effects of the cycle slips. It has been suggested that the number of cycle slips may be estimated by a Kalman filter, which keeps track of the rate of change of the carrier phase [4]. Such a method requires an accurate estimate of the position of the receiver, which is often not available because accuracies deteriorate very rapidly during long periods of multiple or continuous cycle slip. Another method for minimizing the cycle slip problem is the use of coordinate and velocity data from an INS to maintain the accuracy of the GPS receiver over the period when cycle slips occur. This method, which in this paper is referred to as integrated GPS-INS, can be used to predict the coordinates and velocity of the navigation solution during periods of cycle slips at about the same level of accuracy as before the loss of cycle counts. The integrated GPS-INS will also estimate the roll, pitch, and azimuth of the system, which is required in many applications. Compared to a GPS-only approach, the cost of the integrated system would be 20 to 50 percent higher. The integration is a trade-off between cost on the one hand and performance, versatility, accuracy, and reliability on the other.

Among the different estimation methods available for the integration of INS data with measurements from GPS receivers, the most commonly used is the Kalman filter approach. This method allows users to predict and update the error states of the INS and GPS receiver with the range and phase measurements from the receiver in real time or post-mission. Results of integrating a single receiver with the INS, using either pseudorange or phase data, show that this method is not suited for high-accuracy applications (e.g., [1, 5]). The use of differential techniques involving two receivers is therefore indicated. By combining pseudorange and carrier phase measurements, the accuracies of the integrated GPS-INS can be further enhanced. In this case, the phase measurements are added to the pseudoranges to form the more precise biased-ranges, which can then be used to update the error states of the GPS-INS in the same manner as the pseudoranges.

Research into the use of biased-ranges to aid the INS was conducted at the University of Calgary to determine the accuracy of a land-mode integrated GPS-INS. A Kalman filter-smoother has been developed to integrate the pseudorange and phase measurements with the raw position and velocity data from an INS to produce precise coordinates and velocities of the vehicle. The filter-smoother was tested with data gathered with two TI4100 GPS receivers and the Ferranti Inertial Land Surveyor (FILS) over an L-shaped baseline 60 km east of Calgary. This paper presents the results of the error analysis of the baseline test.

Although the filter-smoother was developed for land vehicle navigation, it can also be modified for applications in the marine and airborne environments. The basic error behavior of the integrated GPS-INS remains the same at sea or in the air; however, because of the differences in dynamics, the cycle slip problems may become more or less severe, depending on the application and the environment involved.

PRINCIPLE OF OPERATION

The basic components of an integrated GPS-INS are an INS, one or two GPS satellite receivers, and a computer. An INS can determine the acceleration and angular rotation experienced by the system at a very high rate (> 20 Hz). These data are then integrated, in real time, into raw coordinates and velocities of the INS. Because of the error sources in the INS, the coordinate and velocity data from the system contain time-dependent errors that perturb the position and velocity in a sinusoidal manner. The periods of these errors are functions of the Schuler frequency and the earth rate of rotation. The error behavior of an INS is well documented in [6] and can be estimated with a discrete Kalman filter [7] if accurate external control measurements are available to update the error states of the INS. The type of INS used will affect the error behavior to some extent. The INS used in the baseline test is a gimballed local-level system. Plans are in progress to integrate a strapdown ring-laser-gyro inertial reference unit with two Trimble 4000S GPS receivers to produce a less expensive integrated GPS-INS. Any time-tagged instantaneous measurement that can be expressed as a linearized function of the error states of the INS may be used to update estimates of the Kalman filter. In the case of integrated GPS-INS, such measurements are provided by the GPS receiver in the form of either pseudoranges or biased-ranges derived from the carrier phase data. Here, the GPS carrier phase data are the sum of the integrated Doppler count, i.e., the integer portion, and the fractional phase measurement.

The change in range to a satellite between two epochs can be computed from the phase data by the simple expression

$$\Delta\bar{\rho}_{k+1} = c\,\{\Delta N_{k,k+1} + \Delta\Phi_{k,k+1}\}/f \tag{1}$$

where

$\Delta\bar{\rho}_{k+1}$ = the change of range between epochs k and k + 1,
c = the speed of light,
$\Delta N_{k,k+1}$ = the integrated Doppler count between epochs k and k + 1,
$\Delta\Phi_{k,k+1}$ = the difference in the fractional phase measurements at k and k + 1, and
f = the carrier frequency.

Since fractional phase measurements can be accurately observed to better than 5 cm, the delta range is a more precise form of measurement than the pseudorange. The biased-ranges used in this research are constructed by adding the pseudoranges, ρ_o, at a reference epoch to the range changes, $\Delta\bar{\rho}_i$, measured at subsequent epochs, i.e.,

$$\bar{\rho}_k = \rho_o + \sum_{i=1}^{k} \Delta\bar{\rho}_i. \tag{2}$$

The term "biased-range" has been used to indicate that $\bar{\rho}_k$ contains a bias, which is the sum of the receiver clock error and the measurement error of the initial pseudorange (ρ_o).

The mathematical model for the biased-range is

$$\bar{\rho}_k = r_k + \delta t_k + e_\rho + e_{\Delta\rho}. \tag{3}$$

In equations (2) and (3),

$\bar{\rho}_k$ = the biased-range at epoch k,
ρ_o = the pseudorange at reference epoch o,
$\Delta\bar{\rho}_i$ = the change in the biased-range between epoch i-1 and i,
r_k = the true range at epoch k,
δt_k = the receiver clock error at epoch k,

e_ρ = the pseudorange error at epoch o, including both systematic and stochastic errors, and

$e_{\Delta\rho}$ = the mean error of all $\Delta\rho$ measured from o to k, including measurement noise, multipath errors, and residual atmospheric and ionospheric effects.

Since the errors, e_ρ, in the pseudoranges at epoch zero can be modeled as constant biases, the measurement accuracy of the biased-ranges obtained after the reference epoch is equal to that of the range changes or delta-ranges, $\Delta\bar{\rho}_i$. In the following, biased-ranges are used to update the error states of the integrated GPS-INS except for the periods of multiple cycle slips, when pseudoranges are used to control the growth of errors in the filter estimates. Figure 1 shows the relationship between the GPS measurements and the integrated GPS-INS. In this research, cycle slip detection is accomplished by comparing the L1 and L2 delta ranges. Details of the procedure are given in the Appendix.

In the differential mode, the integrated GPS-INS utilizes the GPS biased-ranges from two satellite receivers to update its error states. The ranges observed in the stationary receiver at a known location are subtracted from the ranges to the same satellites gathered by the moving receiver to produce differential ranges that contain only small parts of the errors common to both receivers. The common characteristics of these errors in the measurements, such as satellite ephemeris errors, and ionospheric and tropospheric delays, diminish with increasing distance between the two receivers. The disadvantage of the differential mode is that at present, because of the lack of real-time data links between the stationary and moving receivers, it can be applied only in post-mission processing.

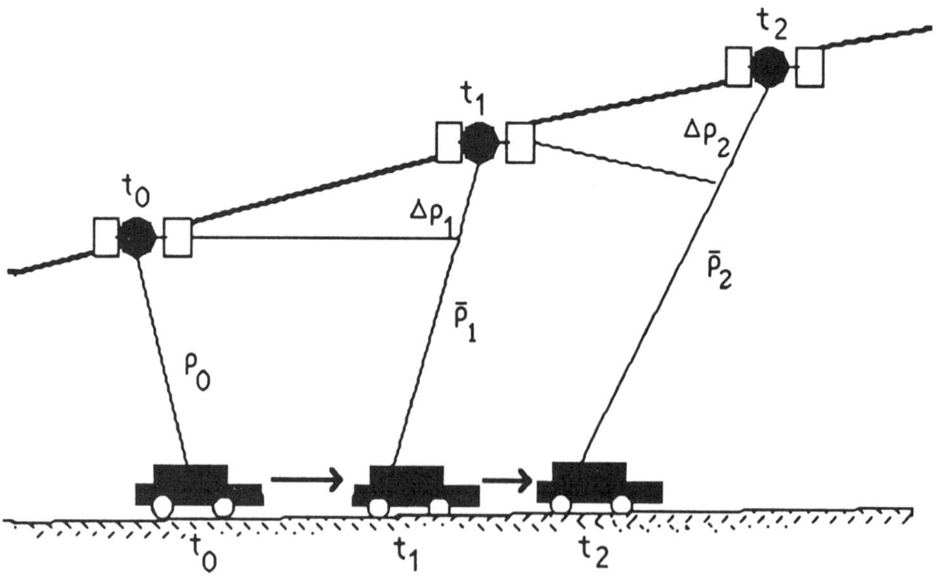

Fig. 1—Relationship Between GPS Measurements and the GPS-INS

KALMAN FILTER-SMOOTHER

The formulae for discrete Kalman filtering and optimal smoothing are well documented, for example, in [8] and [9]. Their applications in inertial surveying and in integration of GPS signals with INS have been described in [10] and [11]. In this research, a Kalman filter with 14–18 error states was developed to integrate one or two GPS receivers with an INS. These error states are

$$\underline{\mathbf{X}} = [\epsilon_E, \epsilon_N, \epsilon_U, \delta\phi, \delta\lambda, \delta\dot{\phi}, \delta\dot{\lambda}, \delta h, \delta\dot{h}, d_E, d_N, d_U, \delta t, \delta\dot{t}, e_{\rho 1}, e_{\rho 2}, e_{\rho 3}, e_{\rho 4}], \tag{4}$$

where

$\epsilon_E, \epsilon_N, \epsilon_U$ = the misalignment of the INS east, north, and vertical axes
$\delta\phi, \delta\lambda, \delta h$ = the latitude, longitude, and height errors
$\delta\dot{\phi}, \delta\dot{\lambda}, \delta\dot{h}$ = the velocity errors in the ellipsoidal coordinates,
d_E, d_N, d_U = the INS gyro drift rates,
δt = the receiver clock bias,
$\delta\dot{t}$ = the receiver clock drift rate, and
$e_{\rho i}$ = the bias in the biased-range to satellite i.

The first 12 error states in equation (4) are the basic errors associated with an INS. The propagation of the error states is done with a transition matrix, numerically derived from a dynamics matrix similar to the one given in [12]. The receiver clock bias is treated as a constant quantity in the filter, and the clock drift rate is modeled as a first-order Markov process. The clock drift rates in the receivers used in the baseline survey were found to be noisy and growing with time; therefore the correlation time was chosen to be 20 h, and the initial variance of the Markov process was estimated to be approximately 2.5 m/s. The biases in the biased-ranges are constant quantities. Whenever a cycle slip occurs in the carrier phase measurements to a satellite, the estimated bias becomes invalid and must be reset to zero. The new value for the bias must then be estimated with the next complete set of measured pseudoranges and delta-ranges from the satellite. In order to ensure continuity in the error propagation, the biases of the biased-ranges are modeled by another first-order Markov process, with a long correlation time (> 55 h) between cycle slips. The correlation time is reset to 1 s during the periods of cycle slips. Pseudoranges measured in the periods of cycle slips are used to limit the degradation of the accuracy of the integrated GPS-INS. In this error analysis, the estimated values of the standard deviations of the observables are 4.0 m for the differential pseudoranges and 0.2 m for the biased-ranges. Although these values seem large compared to the ones given by other authors, they agree best with the results of the analysis of actual measurements.

The Kalman filter estimates for the periods of cycle slips and poor geometry can be improved by means of optimal smoothing. The optimal smoother used in this case is the modified Bryson-Frazier smoother [9]. The data required by the smoother are, in general, the state vector, its covariance matrix, the transition matrix, the observations, and their variances estimated by the Kalman filter. The results after smoothing represent a solution that uses all the observations obtained during the entire survey or navigation period.

BASELINE TEST AND ANALYSIS OF RESULTS

The Kalman filter-smoother was developed at the Department of Surveying Engineering of the University of Calgary for the integration of differential GPS measurements with an INS. As noted earlier, the package was tested with the data surveyed over an L-shaped baseline 60 km east of Calgary, in the summer of 1985. The survey was done with two TI4100 GPS receivers in P-code mode and a Ferranti Inertial Land Surveyor (FILS). The FILS is a gimballed inertial survey system of the local level type. A detailed description of the FILS can be found in [13]. One TI4100 was placed at a known station in Calgary, and the other was mounted in a truck together with the FILS. A computer was used to time-tag the inertial survey system outputs by recording the FILS time of the 1 Hz pulse output of the moving GPS receiver. The data supplied by the FILS are raw coordinates, velocities, attitudes, and time. The outputs from the TI4100 are the pseudoranges, carrier phase measurements, broadcast ephemerides, and GPS time.

The L-shaped baseline, as shown in Figure 2, is a 60 km traverse located approximately 60 km east of Calgary. The control points on the baseline are 4–6 km apart. The coordinates of the control points were surveyed by conventional surveying techniques to 0.3 m (1σ). The vehicle carrying the GPS receiver and the FILS were stopped every 4 min in order to record the velocity update data, which were later used to provide control information for the baseline tests. The entire survey can be divided into a forward and backward run of the traverse. The average speed of the forward run was about 60 km/h, and of the backward run 110 km/h. In the first half of the forward run, the geometry of the satellites was very good, but it rapidly deteriorated at the end of the run. Because of the low elevation of some of the satellites, that period was also plagued by multiple cycle slips. The geometry improved slightly during the backward run, with the HDOP number fluctuating around 2.0 while the VDOP number remained at 5.0. Throughout the survey, the geometry permitted a more accurate determination of the horizontal coordinates than of the height for the integrated GPS-INS. The dilution-of-precision numbers of the test are plotted in Figure 3. In this survey, the starting and end points of the traverse are considered known; therefore coordinate updates can be performed at the end of the forward and backward runs.

Because of the lack of accurate external kinematic control, the control coordinates and velocities for the baseline test were obtained by estimating, in a separate 12-state Kalman filter-smoother, the FILS error states from coordinate updates and regular zero velocity update measurements. The 12 error states of this Kalman filter are the same as the first 12 states given in equation (4). The use of such a Kalman filter-smoother in establishing control points along a traverse is shown in [7]. The main difference between the 12-state Kalman filter-smoother and the 18-state one used in the integrated GPS-INS is the use of 6 GPS-related states in the latter. The difference in the position estimation is that the former uses the zero velocity and known coordinates of the traverse points for update, while the latter uses the biased-ranges and pseudoranges for

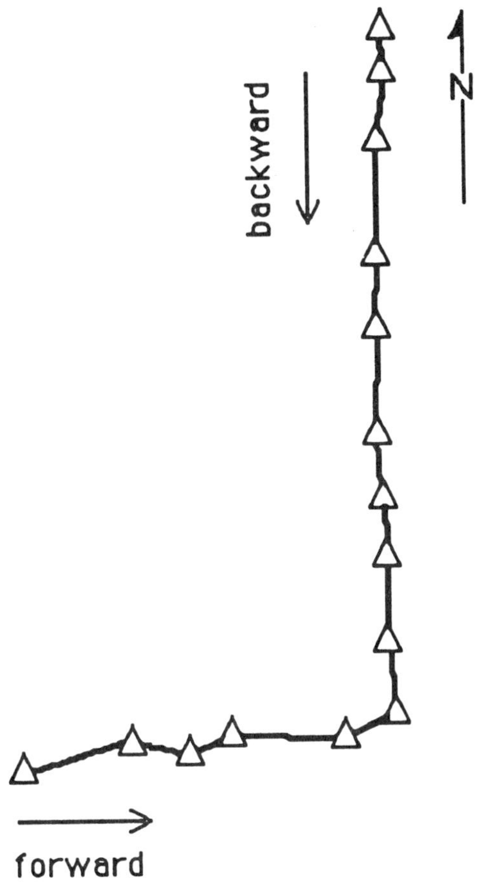

Fig. 2—L-Shaped Baseline

this purpose. Since the control points are close to each other and frequent zero velocity updates (once every 2 km) were used, the accuracies of the coordinates and velocities were estimated to be 0.5 m and 0.005 m/s, respectively. In this baseline test, the control data were generated at 30 s intervals for comparison with the output of the integrated GPS-INS.

The results of the comparison between the integrated GPS-INS and the control coordinates show that the accuracy of the horizontal coordinates is better than that of the height. This is mainly due to the poorer geometry for height determination during the baseline survey. The RMS calculated from the differences between the estimated coordinates and the control coordinates, along with the corresponding average velocity differences, are tabulated in Table 1 for two cases: the INS/pseudorange integration and the INS/biased-range integration. When differential pseudoranges are used to update the integrated GPS-INS, the latitude and longitude are determined accurately to better than 4 m by Kalman filtering, and their accuracy improves to better than 3 m after smoothing. The filtered height accuracy is below the 10 m level, and it is lowered to 5 m after smoothing. As an example of the general improvement in the accuracy of the coordinates after smoothing, the filtered and smoothed

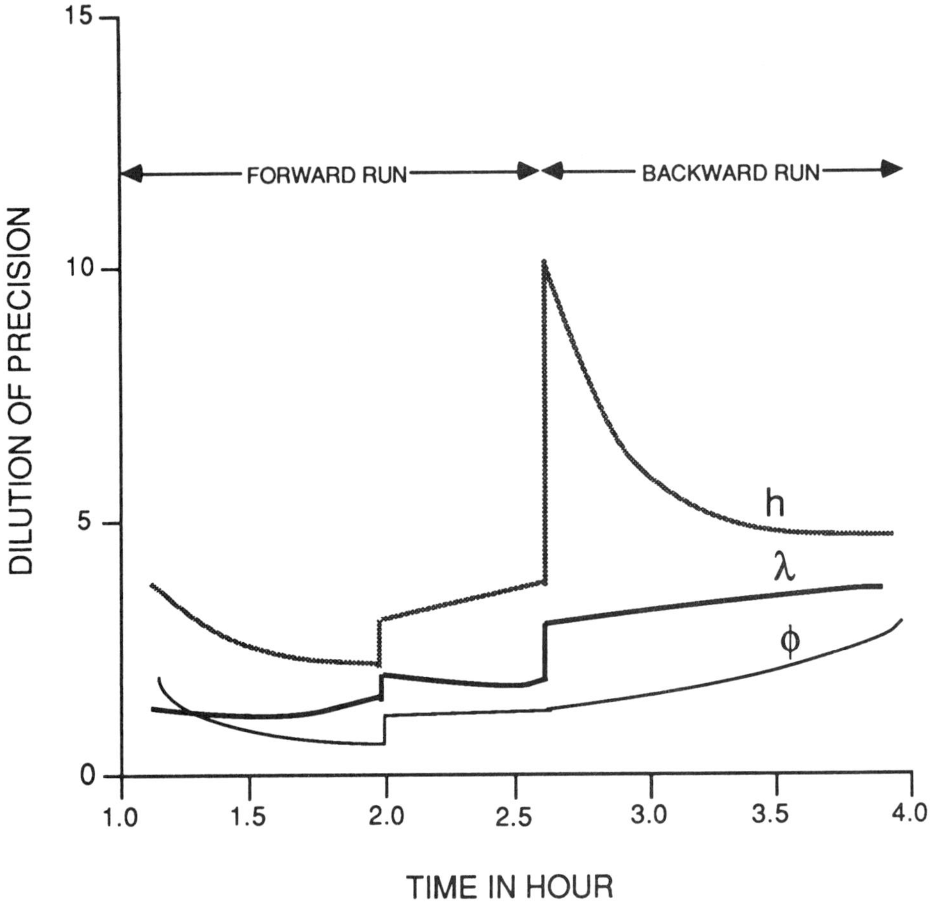

Fig. 3—Dilution-of-Precision Numbers (ϕ = lat., λ = long., h = hgt.)

Table 1—RMS Coordinate and Velocity Differences (1σ)

Update	Estimation	$\Delta\phi$	$\Delta\lambda$ (m)	Δh	ΔV_N	ΔV_E (cm/s)	ΔV_U
Pseudo-	Filtered	3.6	3.4	9.0	3.7	3.4	5.9
range	Smoothed	2.7	1.7	5.0	1.8	1.3	1.1
Biased-	Filtered	2.4	1.1	2.8	2.9	2.3	2.1
range	Smoothed	0.8	0.7	1.3	1.5	1.0	0.3

latitude differences are plotted in Figures 4 and 5. They show not only a general reduction in the size of the differences, but also a smoother error behavior.

When the more precise biased-ranges were used to update the integrated GPS-INS, the estimated coordinates, especially the height, became more accurate. Results in Table 1 indicate that an accuracy of about 2 m (1σ) is achievable after filtering, and that it improves to 1 m (1σ) after smoothing. It should be noted that the RMS values for the horizontal coordinates of 0.7 m and 0.8 m are close to the estimated accuracy of the control coordinates, which is about

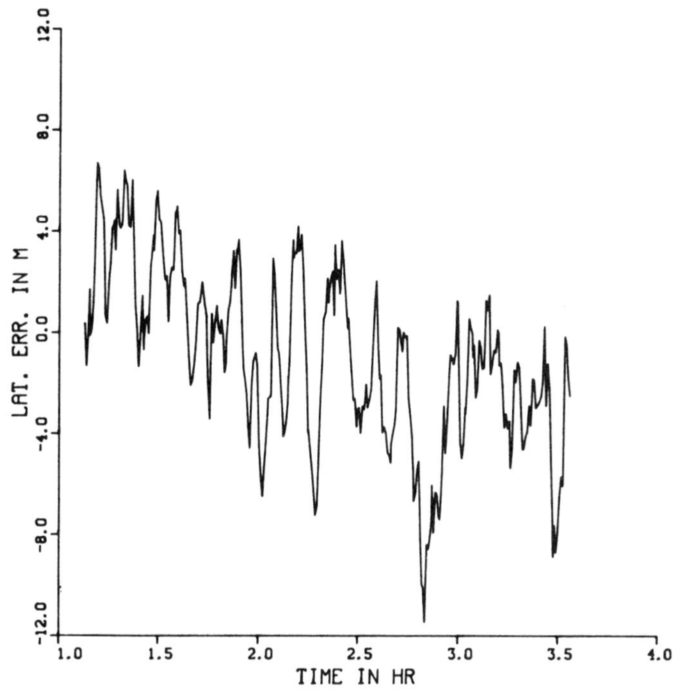

Fig. 4—Filtered Latitude Errors for Differential Pseudorange Updates

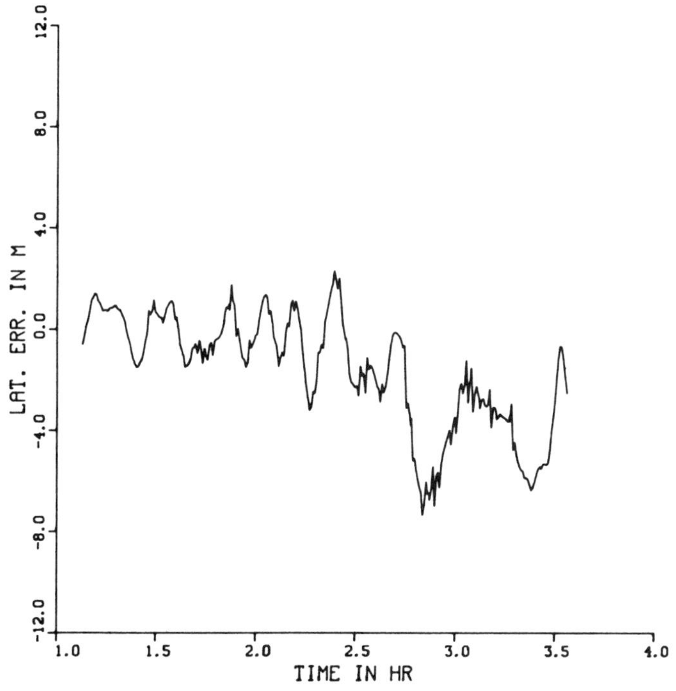

Fig. 5—Smoothed Latitude Errors for Differential Pseudorange Updates

0.5 m. Thus, the actual RMS values are closer to 0.5 m (1σ). Again, as an example, the latitude differences are plotted in Figures 6 and 7 to show the general improvement in the accuracy of the coordinates. In the first half-hour, when the geometry was good, the filtered coordinates were estimated to better than 1 m, and the coordinate error profiles became very gentle after optimal smoothing. The larger errors found in the smoothed coordinates between 2.3 and 2.5 h, when the survey was seriously affected by prolonged outages, poor geometry, and multiple cycle slips, indicate that the integrated GPS-INS cannot completely eliminate the effects of these problems.

Comparisons of the filtered and smoothed coordinates with the coordinates of the known points on the L-shaped baseline indicate that the errors of the estimates of the Kalman filter-smoother for these points are consistent with the general trends presented in Figures 4–7.

The velocity differences in Table 1 show that the accuracy of the filtered velocity is below 7 cm/s when differential pseudoranges are used, and it becomes less than 3 cm/s if biased-ranges are available for measurement updates. In both cases, the velocity accuracy is improved to the 1 cm/s level after smoothing.

CONCLUSIONS

Differential pseudoranges and phase data from two TI4100 GPS receivers were used to update the error states of an integrated GPS-INS. An 18-state Kalman filter-smoother developed at the University of Calgary was used to integrate the GPS receivers with a Ferranti Inertial Land Surveyor. The pack-

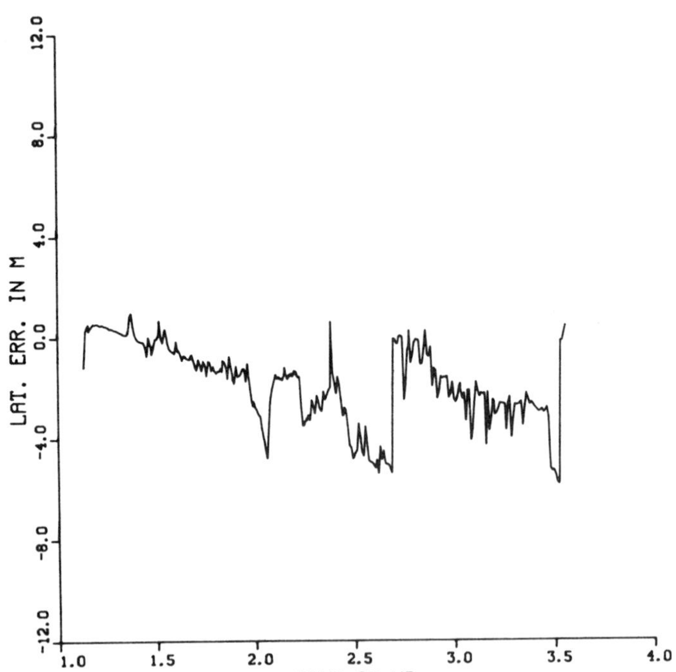

Fig. 6—Filtered Latitude Errors for Differential Biased-Range Updates

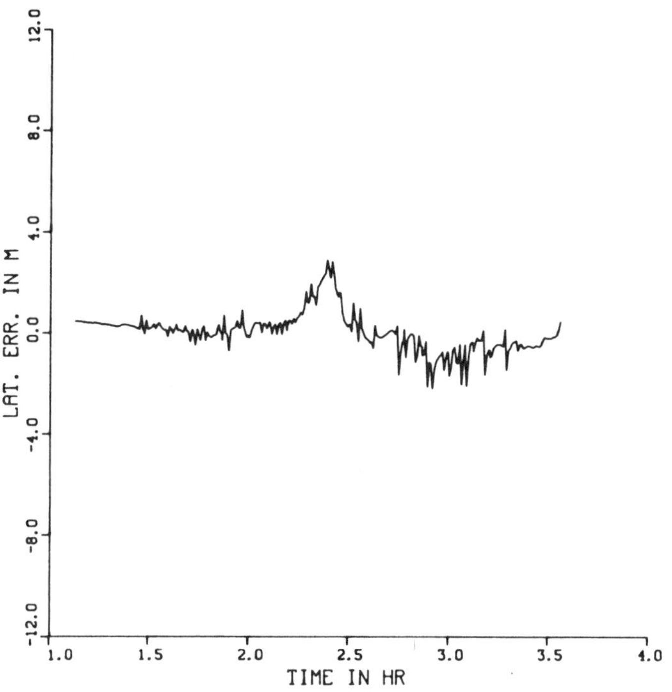

Fig. 7—Smoothed Latitude Errors for Differential Biased-Range Updates

age was tested with the data from a survey over an L-shaped baseline 60 km east of Calgary. Error analysis of the results of the test indicates that, in differential land-vehicle mode, coordinate accuracies at the 2 m level can be achieved after filtering if biased-ranges constructed from phase data are available to update the integrated GPS-INS. The accuracy may be enhanced to the 1 m level by an optimal smoother in post-mission processing if the survey ends with reasonably good geometry or at a known point. The velocity errors of the integrated system are, in general, below 3 cm/s after filtering, and they can be improved to the 1 cm/s level by using optimal smoothing. Results of the test also show that most of the effects of prolonged outages, poor geometry, and multiple cycle slips can be eliminated by integrating the two systems.

ACKNOWLEDGMENT

Nortech Surveys (Canada) Ltd. is acknowledged for their assistance in the preparation and decoding of the raw data used in the baseline tests.

REFERENCES

1. Schwarz, K. P., Wong, R. V. C., Hagglund, J., and Lachapelle, G., *Marine Positioning with a GPS-Aided Inertial Navigation System,* Proceedings of the First National Technical Meeting of the Institute of Navigation, San Diego, CA, January 1984.
2. Cannon, M. E., Schwarz, K. P., and Wong, R. V. C., *Kinematic Positioning with GPS—An Analysis of Road Tests,* Proceedings of the Fourth International Geodetic Symposium on Satellite Positioning, Austin, TX, 1986.

3. Remondi, B. W., *Performing Centimeter-Level Survey in Seconds with GPS Carrier Phase: Initial Results,* NOAA Technical Memorandum NOS NGS-43, U.S. Department of Commerce, Rockville, MD, 1985.
4. Goad, C., *Precise Positioning with the Global Positioning System,* Proceedings of the Third International Symposium on Inertial Technology for Surveying and Geodesy, Vol. 2, Banff, Canada, 1985.
5. Tearsly, S. P., *Flight Test Results of an Integrated GPS and Strapdown Inertial System,* Proceedings of the IEEE Position Location and Navigation Symposium 86, Las Vegas, NV, 1986.
6. Britting, K. R., *Inertial Navigation System Analysis,* Wiley-Interscience, NY, 1971.
7. Wong, R. V. C., *A Kalman Filter-Smoother for an Inertial Survey System of Local Level Type,* Masters Thesis, Publication # 20001, Department of Surveying Engineering, University of Calgary, Calgary, Canada, 1982.
8. Gelb, A., *Applied Optimal Estimation,* The M.I.T. Press, Fourth Printing, MA, 1978.
9. Bierman, G. J., *Fixed Interval Smoothing with Discrete Measurements,* International Journal of Control, Vol. 18, No. 1, 1973.
10. Schwarz, K. P., *Error Propagation in Inertial Positioning,* The Canadian Surveyor, Vol. 34, No. 3, September, 1980.
11. Wong, R. V. C., and Schwarz, K. P., *Dynamic Positioning with an Integrated GPS-INS: Formulae and Baseline Tests,* Technical Report #30003, Department of Surveying Engineering, University of Calgary, Calgary, Canada, 1983.
12. Schmidt, G. T., *Strapdown Inertial System—Theory and Applications—Introduction and Overview,* AGARD Lecture Series No. 95, U.S., 1978.
13. Hagglund, J., *The Ferranti Inertial Land Surveying System (FILS) as Part of an Integrated Navigation and Positioning System,* Masters Thesis, Publication #20020, Department of Surveying Engineering, University of Calgary, Calgary, Canada, 1987.

APPENDIX

The following algorithm is used in this research to detect the occurence of a cycle slip.

From the stationary GPS satellite receiver, we have

$$\theta_1 = \Delta\bar{\rho}_1(L1) - \Delta\bar{\rho}_1(L2), \tag{A-1}$$

and, from the moving receiver,

$$\theta_2 = \Delta\bar{\rho}_2(L1) - \Delta\bar{\rho}_2(L2), \tag{A-2}$$

where $\Delta\bar{\rho}(L1)$ and $\Delta\bar{\rho}(L2)$ denote the delta ranges determined from L1 and L2 frequencies, respectively, and the subscripts 1 and 2 denote the stationary and moving receivers. Usually, $\Delta\bar{\rho}(L1)$ should be very close to $\Delta\bar{\rho}(L2)$, the only difference being a differential ionospheric correction, which is frequency-dependent. If a cycle slip occurs, θ_1 and θ_2 show large values. Since cycle slips of 0.5 cycle are possible with certain receivers, the minimum change to be expected from a cycle slip is 0.1 m. Thus, cycle slip is assumed to have occurred when

$$|\theta_1| \text{ or } |\theta_2| > 0.1 \text{ m} \tag{A-3}$$

or

$$|\theta_1 - \theta_2| > 0.1 \text{ m}. \tag{A-4}$$

Another check on the validity of the data is also performed before measurement update by comparing the misclosure

$$w = \mathbf{Hx} - y \tag{A-5}$$

with its estimated variance

$$\sigma_w^2 = (\mathbf{HC_xH^T} + \sigma_y^2), \tag{A-6}$$

where $\mathbf{H}$ is the design matrix, y is the measurement, σ_y^2 is the variance of the measurement, and $\mathbf{C_x}$ is the variance-covariance matrix of the state vector $\mathbf{x}$. The measurement would be rejected if

$$|w| > 4.0\ \sigma_w. \tag{A-7}$$

If the measurement is a biased-range, then a cycle slip is assumed to have occurred.

NAVIGATION: *Journal of The Institute of Navigation*
Vol. 37, No. 1, Spring 1990
Printed in U.S.A.

High-Accuracy GPS Semikinematic Positioning: Modeling and Results

M. ELIZABETH CANNON

The University of Calgary, Calgary, Alberta, Canada
Received September 1989
Revised November 1989

ABSTRACT

The concept of GPS semikinematic positioning is reviewed, with an emphasis on accuracies at the centimeter level. Details of the mathematical model used herein are given. A two-step approach combines a Kalman filter to process kinematic carrier phase, phase rate, and pseudorange data, while a batch least-squares adjustment processes static carrier phase data. A semikinematic test performed over a well-controlled traverse in a mountainous region near Calgary is used to assess the performance of the algorithm. Several runs of the traverse were made with the station occupation time at 2–3 min. The analysis shows that centimeter-level accuracies can be achieved when five satellites are observed in good geometery, even when frequent cycle slips occur. An accuracy improvement technique of reverse processing is also investigated. Recommendations for improving semi-kinematic surveying are made.

INTRODUCTION

GPS semikinematic positioning is a procedure in which stations along a traverse are occupied for a short time, e.g. 2–4 min, and then the receiver is moved to the next station along the traverse. The name semikinematic refers to the two dynamic states of the receiver; part of the time is spent static on a station, while the rest of the time the receiver is in pure kinematic mode en route to the next station. This method is an extension of conventional GPS relative static surveying, which normally requires 1–2 h of data per station to obtain centimeter accuracy. The goal of semikinematic surveying is to maintain the centimeter-level accuracy while reducing the site occupation time from hours to minutes. By achieving this goal, dramatic cost and time savings can be made on a precise GPS survey. One application of the semikinematic technique is for densification of control points in geodetic networks.

A number of semikinematic tests have been performed to date, for example, those described in [1–6]. Centimeter accuracies have been achieved under varying conditions, ranging from very short baselines, for example, as described in [5 and 6], to baselines of up to 20 km [2]. Most of the processing techniques do not use the kinematic GPS data collected between the static stations. A method is presented in this paper which uses both the static and kinematic data in a two-step processing technique. A least-squares batch adjustment is used to process the static data, while a Kalman filter is implemented on the kinematic data. Also presented is a procedure of reverse-processing for accuracy improvement over semikinematic runs with accuracy degradation due to cycle slips.

MATHEMATICAL MODEL

Two methods are used for postprocessing of GPS data. A batch least-squares adjustment of double-differenced carrier phase is used to process the 2–3 min of static data, while a Kalman filter is implemented during the kinematic sections between static points. Data used in the Kalman filter include carrier phase data, as well as phase rate and pseudoranges. All measurements input to the Kalman filter are processed using double differencing. The advantage of using this approach as opposed to processing only static data is that no information is neglected. By processing kinematic data, information on cycle slip occurrence is gained. New carrier phase ambiguities can then be estimated during transit to the next station.

Static Data Processing Model

The GPS carrier phase data collected at static stations were processed using a batch least-squares adjustment. Details of the double-difference method are not given here, but can be found in [7]. The basic observation equation between two receivers and two satellites is [8]:

$$\nabla\Delta\phi = \nabla\Delta\rho + \lambda\nabla\Delta N - \nabla\Delta d_{ion} + \nabla\Delta d_{trop} + \nabla\Delta d_{\rho} + \epsilon \qquad (1)$$

where $\nabla\Delta$ denotes double difference, ϕ is the carrier phase observation (m), ρ is the range between a satellite and receiver (m), λ is the carrier phase wavelength (m), N is the carrier phase integer ambiguity (cycles), d_{ion} is the ionospheric correction (m), d_{trop} is the tropospheric correction (m), d_{ρ} is the orbital error (m), and ϵ is the measurement noise (m). Ionospheric and tropospheric corrections are not applied to the observations since the baselines are less than 10 km.

At the first station of the traverse, the baseline, i.e., the vector between the differential GPS monitor and remote receivers, was held fixed in order to estimate the carrier phase integer ambiguities. This assumes that the baseline is known a priori. If this is not the case, techniques such as antenna swapping [9] or a conventional static solution (i.e., longer observation span) could be implemented. After this first static solution is performed, the estimated remote receiver coordinates from the adjustment, along with their variances, are entered as a priori information into the Kalman filter. The integer ambiguities are stored, but are generally not estimated in the Kalman filter, as discussed in the next section.

In processing subsequent stations along the traverse, the ambiguities are estimated only if a cycle slip has occurred since the last static point. If no cycle slips have occurred, the ambiguities are held fixed to their previous integer values, and only station coordinates are estimated. Knowledge of the integer ambiguities in semikinematic positioning is crucial for maintaining centimeter-level accuracy, since 2–3 min of static data is then sufficient to determine precise station coordinates. In conventional GPS static differential surveys, a longer data span is necessary in order to accumulate sufficient geometric information to resolve the integer ambiguity parameters.

If cycle slips occur during the kinematic section, new carrier phase integer ambiguities for those satellites affected are estimated in the next static least-squares adjustment. In order to constrain the ambiguities somewhat, a priori information on the ambiguities is taken from the Kalman filter results. Clearly, the more accurate the a priori information input to the least-squares adjustment, the stronger the possibility of recovering the new carrier phase integer ambiguity.

Kinematic Data Processing Model

A Kalman filter is implemented to process the kinematic GPS data collected between stations. The underlying theory can be found in [10]. Measurements used to update the Kalman filter are differential carrier phase, phase rates, and pseudoranges. Phase rates are included to improve velocity determination, while pseudoranges assist in bounding the error during periods of multiple cycle slips. Both these measurement types improve the filter performance, especially the phase rates since they contribute to the modeling of the vehicle dynamics. All measurements are processed in double-difference mode. Standard deviations of 0.5 cm, 5 cm/s, and 4 m are applied to the double-differenced carrier phase, phase rates, and pseudoranges, respectively.

The state vector, $\mathbf{x}$, of the filter is

$$\mathbf{x} = \{\delta\phi, \delta\lambda, \delta h, \delta v_n, \delta v_e, \delta v_u\}^T \tag{2}$$

where $\delta\phi$ is the correction to latitude, $\delta\lambda$ is the correction to longitude, δh is the correction to height, δv_n is the correction to north velocity, δv_e is the correction to east velocity, and δv_u is the correction to up velocity.

Equation (2) gives the state vector if carrier phase ambiguities have been resolved to their integer values. In this case, there is no need for carrier phase ambiguity estimation. However, if cycle slips occur during a kinematic section, the state vector is augmented to estimate the ambiguities that have been affected. The augmented state vector is

$$\mathbf{x} = \{\delta\phi, \delta\lambda, \delta h, \delta v_n, \delta v_e, \delta v_u, \delta N_1, \ldots, \delta N_j\}^T \tag{3}$$

where δN is the magnitude of a satellite's carrier phase ambiguity, and j is the number of satellites that have had cycle slips. Cycle slips are detected by comparing the measured carrier phase measurement to the predicted measurement using the measured phase rate.

The Kalman filter prediction and update equations as given in [10] are listed below:

$$\text{Prediction: } \hat{\mathbf{x}}_{k+1}(-) = \mathbf{\Phi}_{k+1,k}\hat{\mathbf{x}}_k \tag{4a}$$

$$\mathbf{C}^x_{k+1}(-) = \mathbf{\Phi}_{k+1,k}\, \mathbf{C}^x_k \mathbf{\Phi}^T_{k+1,k} + \mathbf{C}^w_{k+1,k} \tag{4b}$$

$$\text{Update: } \hat{\mathbf{x}}(+) = \hat{\mathbf{x}}(-) + \mathbf{K}\ \{\mathbf{l} - \mathbf{A}\hat{\mathbf{x}}(-)\} \tag{4c}$$

$$\mathbf{C}^x(+) = \{\mathbf{I} - \mathbf{KA}\}\ \mathbf{C}^x(-) \tag{4d}$$

$$\mathbf{K} = \mathrm{C}^x(-)\mathbf{A}^T\{\mathbf{AC}^x(-)\mathbf{A}^T + \mathbf{C}^e\}^{-1} \tag{4e}$$

where $\hat{\mathbf{x}}$ is the estimated state vector, $\mathbf{\Phi}$ is the transition matrix, $\mathbf{C}^x$ is the state covariance matrix, $\mathbf{C}^w$ is theprocess noise matrix, $\mathbf{K}$ is the Kalman gain matrix, $\mathbf{l}$ is the observations, $\mathbf{A}$ is the design matrix, and $\mathbf{C}^e$ is the measurement noise matrix. The $(-)$ and $(+)$ represent estimates before and after measurement update, respectively.

Each kinematic section is initialized with position and covariance information from the double-difference solution. When a cycle slip occurs, the state vector is augmented by the carrier phase ambiguity term. At the next static station, this ambiguity is estimated in the least-squares double-difference adjustment. Ideally, using the a priori information from the Kalman filter, the new integer ambiguity is resolved. If the proper combination of integer ambiguities cannot be reliably resolved during the adjustment, the Kalman state vector is kept augmented for the next kinematic section.

This relationship of feeding information back and forth from the Kalman filter and least-squares adjustment exploits the use of all the data collected. Some semikinematic processing techniques, for example, those described in [3–5], do not process data between static points. However, as will be shown, use of the technique described above allows cycle slips to be identified when they occur and estimated during transit.

SEMIKINEMATIC TEST DESCRIPTION

Semikinematic tests were performed in August 1988, in the Kananaskis region, located approximately 80 km west of Calgary. An 8 km stretch of highway was selected as the test traverse. A control station was established every kilometer along the road using precise static GPS techniques. The resulting 8 points are known with an accuracy of 2–5 cm with respect to the monitor station, located less than 1 km from the first control point.

The Kananaskis region is a mountainous and densely forested area. This affected the length of the GPS coverage window since satellite masking was a major concern. During the 4 h window, three runs of the traverse were possible. Two days of GPS data was collected, for a total of six traverse runs. Five-channel, single-frequency Trimble 4000SX GPS receivers were used. The test schedule is outlined in Table 1.

As shown in Table 1, five satellites were visible for Runs 1 and 4, while only four satellites were tracked during the remaining runs. The satellite geometry was relatively good for the first and last runs of each day; however, for Runs 2 and 5, the PDOP temporarily exceeded 200.

Table 1—Semikinematic Test Schedule

Date	Run	Satellites Tracked (PRN Number)	Length of Run (min)
	1	6, 8, 11, 12, 13	59
Aug 17/88	2	3, 11, 12, 13	55
	3	3, 11, 12, 13	49
	4	6, 8, 11, 12, 13	72
Aug 23/88	5	3, 11, 12, 13	59
	6	3, 11, 12, 13	53

A tripod was set up on each control station before the test began, and each antenna height was measured. This was done to save time during the actual survey. At each stop, the antenna had to be mounted manually on the tripod. Since these tests were performed, a mobile lever has been developed which holds the antenna. Once mounted on a vehicle roof, the lever can be swung out so the antenna can be centered directly over the point.

The first station of the traverse was occupied for 10–15 min. While still logging GPS data, the vehicle was then driven at about 50 km/h to the next station. The second and subsequent stations were occupied for 2–3 min. Cycle slips were detected during kinematic sections of the traverse when the satellite signal was shaded because of the mountains and trees. Although it was apparent from the data-logging computer output that cycle slips were occurring, no additional procedures were invoked in the field to counteract their effect.

DISCUSSION OF RESULTS

The data were postprocessed using the program SEMIKIN, developed by the author at The University of Calgary. It implements the two-step model detailed above and runs automatically without operator intervention once an input file has been created. It was developed in FORTRAN on a VAX/VMS computer, but has since been installed on a Compaq 386 portable computer for field use. Future plans include translating the software to the C language.

Only four of the six semikinematic runs were processed since satellite masking was a severe problem during the second run of each day. The semikinematic results at each of the static stations were compared to the control coordinates independently determined for each station. Unfortunately, no control information was available during the kinematic sections of the traverse.

Forward Processing

Figures 1 through 4 show the comparisons with the control coordinates for the eight traverse points. The arrows (↑) on the figures show whether cycle

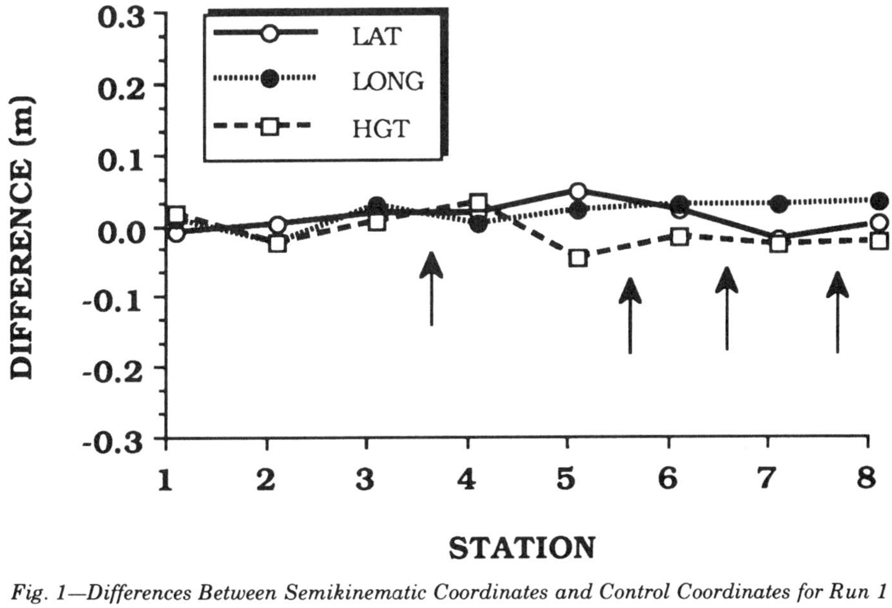

Fig. 1—Differences Between Semikinematic Coordinates and Control Coordinates for Run 1

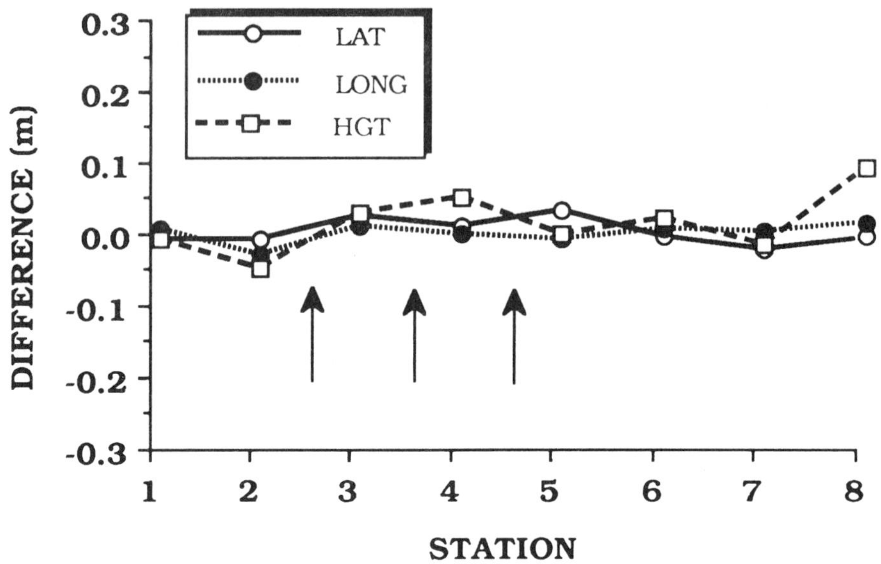

Fig. 2—Differences Between Semikinematic Coordinates and Control Coordinates for Run 4

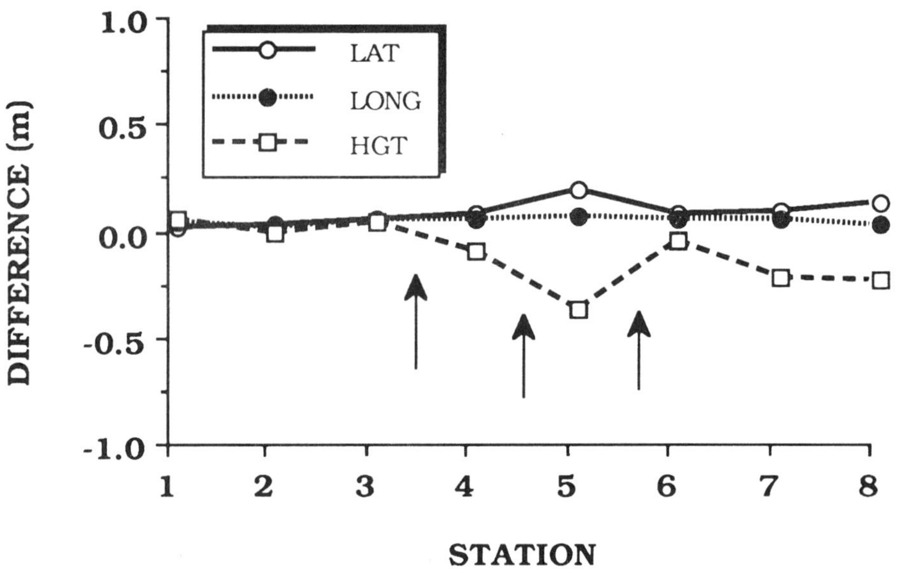

Fig. 3—Differences Between Semikinematic Coordinates and Control Coordinates for Run 3

slips were detected during a particular section. An arrow could represent an isolated cycle slip, or in some cases multiple cycle slips on more than one satellite simultaneously.

For Run 1, Figure 1 shows very good agreement between the semikinematic results and the control coordinates. Overall, the differences are a few centimeters, with maximum errors in latitude, longitude, and height being −3.6 cm, −4.0 cm, and −6.1 cm, respectively. Results of Run 4, plotted in Figure 2, also

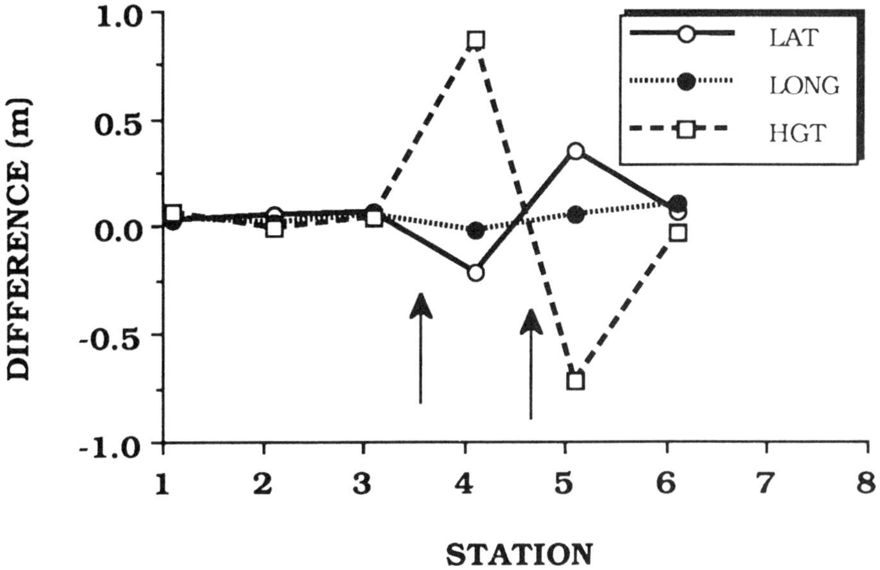

Fig. 4—Differences Between Semikinematic Coordinates and Control Coordinates for Run 6

show agreement on the order of a few centimeters. Maximum errors in this case are −3.7 cm, −2.4 cm, and −6.2 cm for latitude, longitude, and height, respectively. The height results are slightly worse than those for the other two dimensions; this may be due to residual atmospheric effects. Throughout both of these runs, five satellites were tracked. Even though cycle slips were detected in both runs, ambiguities could be resolved to their integer values at each static point. These results are an improvement over those reported in [1], in which ambiguities were not set to their integer values.

The results plotted in Figures 3 and 4 are for the two runs (3 and 6) in which only four satellites were tracked. Note the different scale for the vertical axis as compared to Figures 1 and 2. In Figure 3, the semikinematic accuracy for Run 3 is at the centimeter level until Station 4. A cycle slip was detected during the third kinematic section. After Station 3, the integer ambiguities could not be reliably resolved. This is because only four satellites were observed, and only 2–3 min of the static data was collected at each station. If the occupation time at each station were longer, say 5–7 min, the probability of resolving the integer ambiguities would be higher. The maximum differences for this run are 14.6 cm, −3.4 cm, and −41.7 cm for latitude, longitude, and height, respectively.

For the results of Run 6 in Figure 4, centimeter accuracy is obtained until Station 3, after which there is a severe degradation in accuracy. Again, after the cycle slips in the fourth kinematic section, the integer ambiguities could not be reliably resolved. Maximum differences in this case reach −26.9 cm, −6.2 cm, and 81.9 cm for latitude, longitude, and height, respectively. Processing was stopped after Station 6 on this run because of severe satellite shading problems.

Overall, the results are excellent for the runs in which five satellites were observed throughout. However, when cycle slips occurred during the four satellite runs, the integer ambiguities could not be reliably resolved after cycle slips. This is due to the lack of redundancy and will not occur once the full constellation is in place.

Improving Results by Reverse Processing

In order to improve results for Runs 3 and 6, the data was processed in reverse time, that is, starting at Station 8 (Station 6 in the case of Run 6) and ending at Station 1. Combining the estimated positions of forward processing and reverse processing should result in a better weighted position. This stems from the fact that the semikinematic coordinate accuracy is very good until cycle slips occur. If they occur during the middle of a run, the results of the remaining stations can be affected, as seen in Figures 3 and 4. Processing the data in reverse should at least improve the stations at the end of the forward run (i.e., the beginning of the reverse run). This technique was not applied to Runs 1 and 4 since the ambiguities were resolved to their integer values during the forward processing. However, processing a traverse in reverse gives a measure of the reliability of the results.

The same two-step static and kinematic methodology described earlier was used to process the data in reverse. The coordinates of Station 8 (Station 6 for Run 6) were held fixed to determine the initial "reverse" integer ambiguities. This assumes that the coordinates of the end station of the traverse are known, or that the traverse is closed at the first station. Alternatively, a conventional static survey could be performed at the last station of the traverse.

Results for Runs 3 and 6 processed in reverse are shown in Figures 5 and 6, respectively. Note that the results for Stations 6, 7, and 8 in Figure 5 are greatly improved as compared to Figure 3, as no cycle slips occurred after this time. Comparing Figures 5 and 6 to Figures 3 and 4, all the figures show that

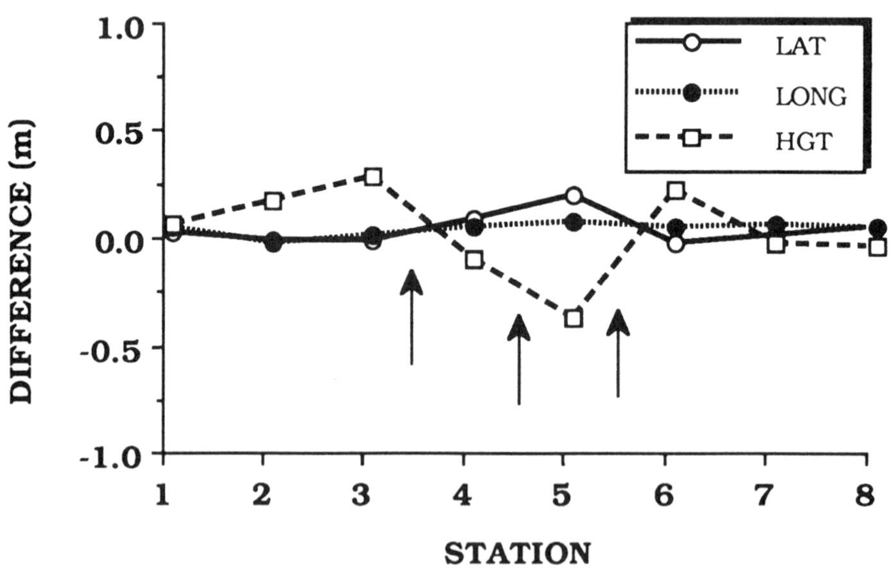

Fig. 5—Differences Between Semikinematic and Control Coordinates for Run 3 Processed from Station 8 to 1

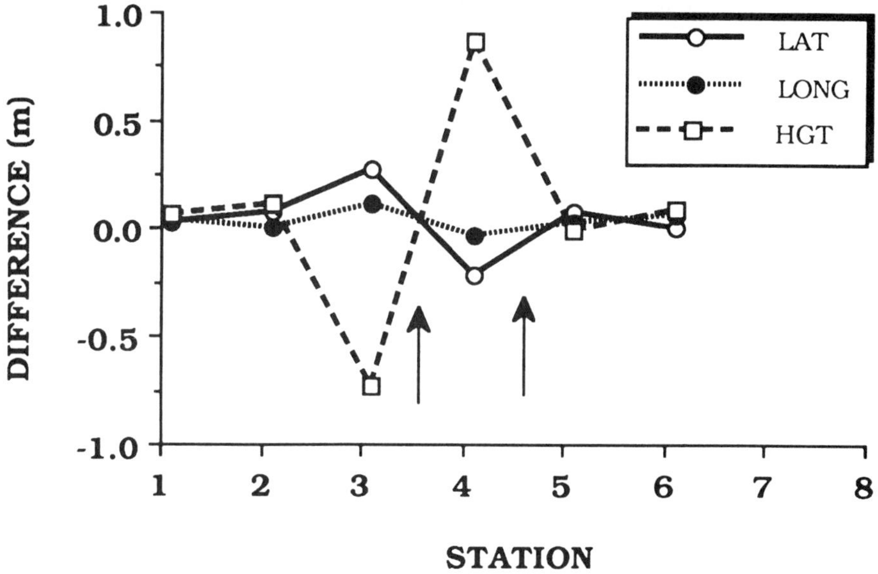

Fig. 6—Differences Between Semikinematic and Control Coordinates for Run 6 Processed from Station 6 to 1

accuracy degrades when cycle slips occur and integer ambiguities cannot be resolved.

The weighted means of Runs 3 and 6 were computed from the forward- and reverse-processed positions. A weighted mean value at a control point was computed as follows (using height as an example):

$$h_M = \frac{\frac{1}{\sigma_F^2}h_F + \frac{1}{\sigma_R^2}h_R}{\frac{1}{\sigma_F^2} + \frac{1}{\sigma_R^2}} \tag{5}$$

where h_M is the weighted mean height, h_F is the height from forward processing, σ_F^2 is the height variance from forward processing, h_R is the height from reverse processing, and σ_R^2 is the height variance from reverse processing.

Weighted mean results are shown in Figures 7 and 8. They are much better than either the forward or reverse runs in the sense that centimeter-level accuracy is achieved at most stations. As seen in the figures, the accuracy of stations in the middle of the runs is generally not improved since the effect of the cycle slips could not be reduced.

CONCLUSIONS

The effectiveness of the two-step approach for semikinematic processing has been demonstrated using a well-controlled traverse. The approach uses a batch least-squares adjustment at static stations, as well as a Kalman filter to process pure kinematic data between static points. Accuracies of a few centimeters have been achieved on runs with five-satellite visibility, even in the presence of numerous cycle slips.

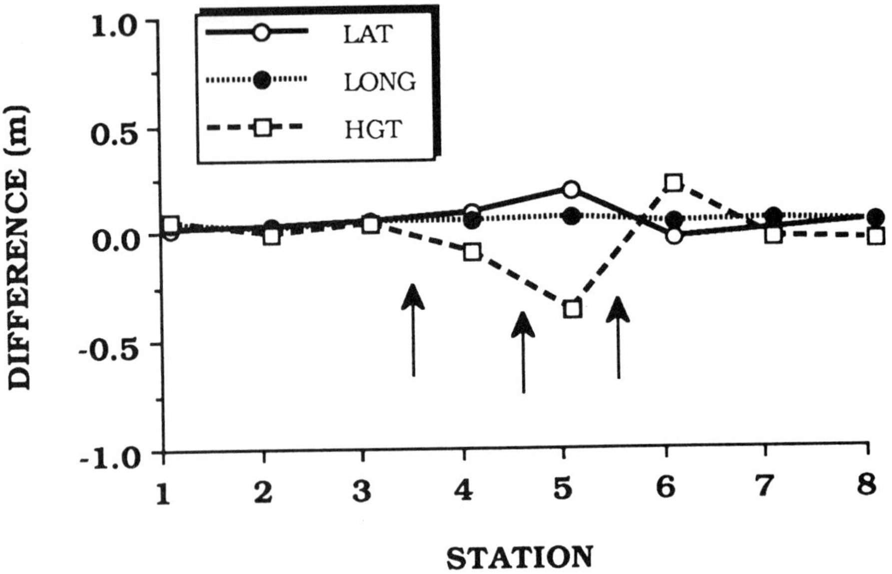

Fig. 7—Weighted Mean Differences Between Semikinematic and Control Coordinates for Run 3

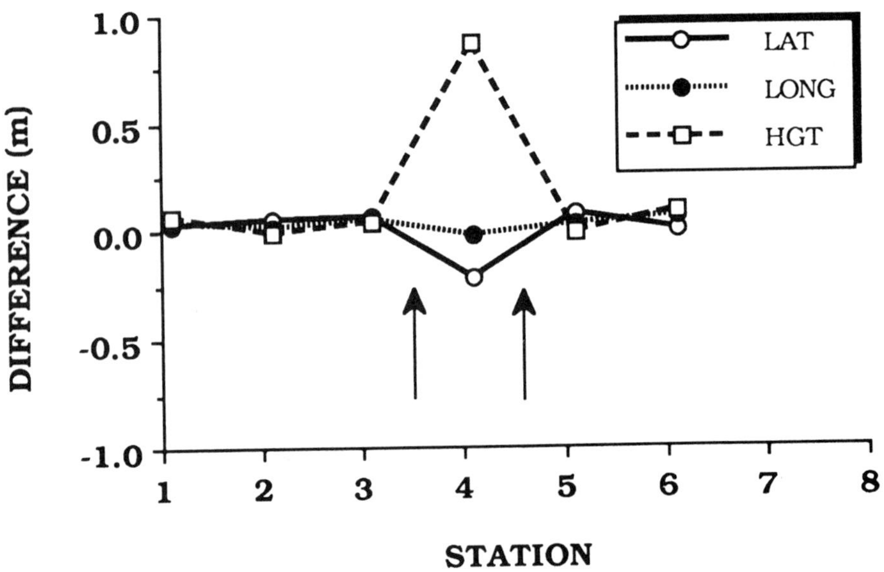

Fig. 8—Weighted Mean Differences Between Semikinematic and Control Coordinates for Run 6

On runs with only four satellites tracked, centimeter-level accuracies were not achieved after cycle slips occurred. However, the position accuracy was improved by processing the data in reverse and computing a weighted position from the forward and reverse positions. This technique is effective in improving accuracy near the ends of the traverse, but will not guarantee homogeneous accuracy at the centimeter level throughout the traverse. Another application of this technique is to measure the reliability of forward results; gross errors should be detectable by reverse processing.

For further semikinematic tests, it is clear that a minimum of five satellites should be tracked throughout a run. Also, at least 2–3 min of data should be collected at each station. Superior results would most likely be gained if longer time spans were observed, especially on runs with only four satellites visible.

The problem of four-satellite coverage will be obsolete when the full GPS constellation is operational. The visibility of six or seven satellites will greatly improve the reliability of centimeter-level accuracy throughout a semikinematic survey because of the increased redundancy. However, a further investigation into improved reliability measures and statistical testing procedures is required for semikinematic data sets collected under the full GPS constellation; such investigation will be extremely important for production semikinematic surveys.

ACKNOWLEDGMENTS

The author would like to thank graduate students at the Department of Surveying Engineering, The University of Calgary, for participating in the GPS data collection campaign in August 1988. Prof. K. P. Schwarz is also thanked for his support throughout the author's graduate studies.

Based on a paper presented at The Second International Technical Meeting of the Satellite Division of The Institute of Navigation, Colorado Springs, CO. This paper was the winner of the Satellite Division Student Paper Competition.

REFERENCES

1. Cannon, M. E., and Schwarz, K. P., *Semi-Kinematic Positioning Along a Well-Controlled Traverse,* Presented at the Fifth International Geodetic Symposium on Satellite Positioning, Las Cruces, NM, March 13–17, 1989.
2. McLellan, J. F., and Porter, T. R., *Semi-Kinematic GPS: Methods and Accuracies,* Presented at the Annual Meeting of the Canadian Institute of Surveying and Mapping, Halifax, Nova Scotia, June 6–9, 1989.
3. Minkel, LCDR, D. H., *Demonstration and Discussion of the Pseudo-Kinematic Method,* Presented at the Fifth International Geodetic Symposium on Satellite Positioning, Las Cruces, NM, March 13–17, 1989.
4. Quirion, C. A., *Automated Kinematic GPS Post-Processing for High Production Surveys,* Presented at the Fifth International Geodetic Symposium on Satellite Positioning, Las Cruces, NM, March 13–17, 1989.
5. Goad, C. C., *A Kinematic Survey of the Clinton Lake Dam,* Presented at the ASCE Specialty Conference GPS-88, Engineering Applications of GPS Satellite Surveying Technology, Nashville, TN, May 11–14, 1988.
6. Remondi, B. W., *Performing Centimeter-Level Surveys in Seconds with GPS Carrier Phase: Initial Results,* NAVIGATION, Journal of The Institute of Navigation, Vol. 32, No. 4, Winter 1985, pp. 386–400.
7. Remondi, B. W., *Using the Global Positioning System (GPS) Phase Observable for Relative Geodesy: Modeling, Processing, and Results,* CSR-84-2, Center for Space Research, The University of Texas at Austin, 1984.
8. Wells, D. E., Beck, N., Delikaraoglou, D., Kleusberg, A., Krakiwsky, E. J., Lachapelle, G., Langley, R. B., Nakiboglou, M., Schwarz, K. P., Tranquilla,

J. M., and Vaníček, P., *Guide to GPS Positioning,* Canadian GPS Associates, Fredericton, New Brunswick, 1986.

9. Hofmann-Wellenhof, B., and Remondi, B. W., *The Antenna Exchange: One Aspect of High Precision GPS Kinematic Survey,* Presented at the International GPS-Workshop, Darmstadt, Federal Republic of Germany, April 10–13, 1988.

10 Gelb, A., ed., *Applied Optimal Estimation,* M.I.T. Press, Cambridge, MA, 1974.

NAVIGATION: *Journal of The Institute of Navigation*
Vol. 38, No. 1, Spring 1991
Printed in U.S.A.

Pseudo-kinematic GPS Results Using the Ambiguity Function Method

BENJAMIN W. REMONDI
National Oceanic and Atmospheric Administration,
Rockville, Maryland
Received June 1990
Revised January 1991

ABSTRACT

Pseudo-kinematic (PK) positioning with GPS is a survey technique employed by the National Geodetic Survey. This technique, which could also be called broken static surveying (or intermittent static surveying), must not be confused with kinematic surveying. The kinematic survey method requires continuous carrier phase tracking while moving between survey monuments and places strong restrictions on the number of satellites without cycle slips. The PK method is a static technique whereby a geodetic monument is occupied for, typically, 1 to 5 min for two or more occupations separated substantially in time (e.g., 1 h). Tracking during the transitions between monuments is categorically not required, but is often done for practical reasons.

PK promises substantial productivity gains over classical static surveys and can be employed where the regular kinematic method is impractical. Furthermore, PK is safer than the regular static mode in that a constant antenna height is normally involved. These advantages are discussed. The paper defines the PK method; provides a brief history; considers alternative possible names; presents the mathematical and physical basis for the method, with emphasis on what to do about all the cycle slips between occupations separated by 1 h or more; and provides a number of examples.

INTRODUCTION

Pseudo-kinematic surveying (or intermittent static, broken static, or snapshot static surveying) was first conceived in the 1982–83 time frame [1]. The idea sprung from research related to the ambiguity function method (AFM) [2]. After months of processing first simulated data (1982) and later real data (1983), it became obvious that the effectiveness of the method depended primarily on geometric change; that is, the satellite-receiver geometry needs to change. It became clear that a few minutes of data at one time and a few minutes of data an hour later were nearly as good, geometrically, as the full hour of data. Thus the original conception of the pseudo-kinematic survey technique (hereafter PK) was associated with the AFM. Because the receivers at that time did not lend themselves to rapid survey techniques and because single-dfference, double-difference, and triple-difference processing methods proved superior to the AFM, PK remained no more than an interesting idea. PK and AFM were closely associated because the AFM was immune to cycle slips. If one were to revisit sites with period of 1 h (or more) between visits, fixing cycle slips could prove to be difficult with the standard least-squares postprocessing methods.

In 1983, PK was experienced for the first time in an unexpected way. A data set had two gaps of possibly 30 min or more. Fixing cycle slips over these gaps was possible, making it clear that the receiver could have been elsewhere during those outages. In 1984, the National Geodetic Survey (NGS) of the National Oceanic and Atmospheric Administration (NOAA) set up a test network in California for testing existing GPS receivers. Thus numerous monuments were established within tens of meters of each other, using NGS Macrometers(TM)[1]. In fact, each of these monuments was occupied for hours. I had considered demonstrating PK at that time, and only the practical inconveniences kept NGS from doing so. In retrospect, the inconveniences were relatively minor.

In 1985, PK processing was accomplished successfully for the first time. The success, however, was based on achieving a first estimate using a triple-difference least-squares (TLSQ) processing program. Using this first estimate, cycle slips were fixed, and definitive processing was performed with a double-difference least-squares (DLSQ) processing program.

During this period, it was obvious that cycle slip fixing based on the TLSQ solution was a limiting factor, and AFM or similar search techniques would be required to expand the method for general use. It is remarkable how well the TLSQ solution has worked. This is particularly true where satellite visibility is superior (e.g., Arizona). Obtaining a TLSQ solution accurate enough to fix cycle slips is limited by baseline length, duration of gap, and number of satellites. With these limitations in mind, one can employ PK safely and productively based on TLSQ and fixed DLSQ. In spite of reasonable success, PK was never intended to be used in this way and especially to be under restrictions of the duration of the tracking gap. PK was intended to be useful over the same baselines as "classical" static surveys (i.e., without gaps), where integers would be clearly determined using conventional DLSQ processing. PK can be used (potentially) up to 100 km or more with single-frequency data under ideal conditions where the error sources are minimal or can be minimized. Obviously, with dual-frequency processing, this could be extended considerably.

On a practical matter, PK is normally performed with fixed-height survey poles. NGS is considering performing classical static surveying with fixed-height survey poles as well. Up to now, most of the regular static surveys have been done with tripods.

THE PSEUDO-KINEMATIC METHOD

The PK method is a static survey method. This techique is often confused with the kinematic survey method which requires continuous carrier phase tracking while moving between survey monuments and has strong restrictions on the number of satellites without cycle slips. Pseudo-kinematic means false kinematic and hence not kinematic, but, rather, static. It is a static technique whereby a geodetic monument is occupied for, typically, 1 to 5 min for two or

[1]Macrometer is a trademark of Western Geophysical Division, Western Atlas International, Houston, Texas.

more occupations separated substantially in time (e.g., 1 h). Tracking during the transitions between marks is categorically not required, but is often done for practical reasons (e.g., to avoid having too many files and to achieve more rapid reacquisition of those satellites which were obscured en route). In fact, in PK mode, the rover receiver (the receiver that travels) may be turned off while traveling between monuments.

PK surveying promises substantial productivity gains over regular classical static methods, where the receiver stays at a mark for 1–2 h. Field productivity improvements can be tenfold for certain surveying activities. Postprocessing in PK mode can be fast for those cases where searches (to isolate the correct solution) are not needed and tends to be slow where searches are involved. Search techniques, while historically slow, are not as inherently slow as one might imagine. Even in sequential processing computers, search techniques can be made acceptably fast. In a parallel processing environment, they can be made almost arbitrarily fast.

The PK method can be distinguished from the single occupation rapid static methods, whereby a station is occupied for 5 min or less [3]. These methods are equally valid and will yield even more productivity gains. They require high-quality dual-frequency P-code data. They will be extremely effective in near-real-time processing. These methods may be less accurate than PK, however. In contrast, the PK method can be fully effective using just single-frequency carrier phase data from a C/A-code receiver.

The PK results presented in this paper are based on single-frequency carrier phase data from five or six GPS satellites, where two or more site visits are assumed. Although the cases presented herein assume the same satellites are tracked during the multiple visits, it should be clear that this is not required.

PSEUDO-KINEMATIC PROCESSING BASED ON TRIPLE-DIFFERENCE/DOUBLE-DIFFERENCE LEAST-SQUARES PROCESSING

In its most restricted form, PK postprocessing is accomplished as follows. One first obtains a TLSQ solution. With only, say, 2 min of carrier phase data (for each of two visits), this solution can be expected to be poor (e.g., 50 cm). (The measurement recording interval is not too critical; 10–20 s is satisfactory.) Next, one makes the cycle connection between visits based on the TLSQ solution. A correlated TLSQ solution will obviously be more accurate than a TLSQ solution that ignores the correlation of triple differences. Roughly speaking, the TLSQ solution must be good to 10–20 cm if the site visitations are 1 h apart. For example, suppose a site is initially occupied for 2 min, and suppose 30 min later it is occupied a second time for 2 min. The 4 min of data is processed in TLSQ mode. Assuming the solution is good to roughly 30 cm, the cycle slips (we are imagining, here, that the receiver has actually been turned off) normally are correctly fixed. Generally speaking, this is a short baseline (1–10 km) technique that requires good satellite visibility. Correct fixing of cycle slips is normally obvious under these assumptions. The TLSQ solution does not have to be as good if the time period between visits is shorter (e.g., 15 min). This, however, limits the full use of this method. If time gaps of 1–2 h and longer can be overcome, this will obviously result in higher productivity and greater accuracy. It should be clear that this restricted mode (i.e., TLSQ/

DLSQ) is not compatible with single-frequency receivers over long baselines and with 1–2 h time periods between visits. (This would require extremely good error models.)

A BRIEF DESCRIPTION OF THE AMBIGUITY FUNCTION METHOD (AFM)

Complete details on the AFM can be found in [2]. Only recent algorithmic advances to speed up the computations are missing from that document. The ambiguity function is simply:

$$f(\mathbf{r}) = \sum_{i=1}^{nepochs} \left| \sum_{k=1}^{nprns} e^{j\left(\phi^k - \frac{\rho^k(\mathbf{r})}{\lambda}\right)} \right| \tag{1}$$

where ϕ^k is the single-difference phase based on receivers 1 and 2 and satellite k, ρ^k is the difference of the two computed receiver-satellite ranges, and λ is the nominal carrier wavelength. Note that the phase argument is in units of cycles. Note also that integer ambiguities are absent from the model. This is because the phasor $\exp(j\phi)$ is identical to $\exp[j(\phi + N)]$, where N is an integer number of cycles. Note as well that the receiver epoch clock error is absent from the model. This is allowed since all the satellite phasors at an epoch would otherwise rotate by the same angle Ω, yielding the same magnitude of the sum of phasors at an epoch.

The AFM works as follows. Select the next candidate position **r** and compute f(**r**). Realize that f(**r**) is bounded by the number of single-difference measurements in that one single-difference measurement can increment f(**r**) by at most unity. Thus, for 200 single-difference measurements, f(**r**) can achieve a value not to exceed 200. The concept of the AFM is really very simple. The difficulty is in achieving computational speed. This has been achieved. Further speed enhancements are being investigated, and it seems possible that a goal of considering a million positions in 1 min is possible on existing fast ($12,000) desktop computers. Generally speaking, there is no current need to consider a million positions, and, for present applications, 125,000 positions seem completely adequate. I intend to document these findings in the future.

PSEUDO-KINEMATIC PROCESSING BASED ON THE AMBIGUITY FUNCTION METHOD

To get the full potential of this technique, we first obtain the TLSQ solution (correlated or uncorrelated). After this, we use a search technique. There are many possible techniques capable of performing this search. To keep this discussion focused, I shall limit it to the AFM.

Let us assume that standard deviations resulting from TLSQ have been normalized so as to be realistic and believable. The AFM then searches all space around the TLSQ solution (correlated or uncorrelated) inside a box whose sides are $4\sigma_x$, $4\sigma_y$, and $4\sigma_z$. Let us assume the box is divided into a grid of points separated in all directions by 0.25–0.33 cycles. For L1 processing this could be, for example, 5–6 cm. A typical search volume might be a cube with 2 m sides. Thus, for example, 41^3 or 68,921 positions would be considered. At each of these 68,921 points, all the relevant single-difference measurements will be evaluated. Although this seems to be a Herculean task, it can be performed on a fast ($12,000) desktop computer in less than 1 min (and on a regular 20 MHz 80386 PC in 2–4 min). A maximal search volume might be a 4 m × 4 m × 2 m box. This search technique will uncover those positions where the fractional phase measurements agree with the fractional phase based on the model. In such a search there is no such thing as a cycle slip in that only the fractional phase measurement is employed. A search will uncover numerous functional maxima. For short baselines and two visits, there will be a predominant peak and possibly one or more significant but lesser peaks (this assumes reasonable satellite visibility; a representative example would be five satellites on both visits and perhaps an additional satellite on one of the two visits). If a peak is clearly greater than other peaks, this will almost certainly be the correct peak. Where necessary, a candidate peak can be verified by double-difference processing (DLSQ). The candidate peak is first used to fix cycle slips. DLSQ processing follows. For short baselines, the float-DLSQ solution will achieve almost perfect integers if the cycle slips were fixed correctly (which will be achieved only at the correct peak).

RESULTS

In the following, the results of four experiments are discussed. In all cases the data were collected in preparation for PK postprocessing.

Experiment No. 1 (April 24, 1989)

The first experiment involved an airport survey performed in Apalachicola, FL. These data were collected consistent with kinematic processing (i.e., no cycle slips) and PK processing (i.e., two visits). Although there were no cycle slips in this data set, it is still possible to process the data while not benefiting from the fact that there were no cycle slips. In this survey, the reference receiver was situated at mark APAL for the duration of the test, with the exception of antenna swaps, which took only a few minutes to perform. The rover receiver started at station APAZ and then performed two antenna swaps with the reference receiver (consequently returning the reference receiver to APAL and the rover to APAZ). Then the rover occupied marks AP18, AP36, AP31, AP13, AP06, and AP24, finally returning to APAZ. This circuit was then repeated. With the exception of APAZ, all marks were occupied two times for 2 min each time. APAZ was occupied four times for 5 min each time. All marks were within 2 km of APAL.

These data were processed routinely in kinematic mode as part of NGS operations [4]. These kinematic solutions, though accurate to possibly 1 to 3 cm, will provide approximate truth for the PK processing even though PK processing would normally be more accurate than kinematic processing. The rationale for using a less accurate solution for truth is not accuracy, but rather verification that our solution is the correct one.

Table 1 comprises seven baselines with five entries for each baseline. The first entry is the triple-difference (TLSQ) solution. This solution did not use a weight matrix. A weight matrix would be helpful in reducing the dimension of the search box, possibly by a factor of 2 (and the volume by 8). This would be an important consideration in an operational setting, but was unimportant in isolating the correct position. Note that the TLSQ solution is relatively poor.

Table 1—Ambiguity Function Method Results from the April 24, 1989, Apalachicola, FL, Survey

Baseline	Coordinates (m) dx	dy	dz	Remarks
APAL-APAZ				
TLSQ	18.964	−498.574	−870.138	
	(10 cm)	(14 cm)	(7 cm)	resid = 0.011
DDX(TLSQ)	18.918	−498.504	−870.136	resid = 0.041
AF(TLSQ)	18.926	−498.496	−870.144	94%, 88%
DDX(AF)	18.918	−498.504	−870.136	resid = 0.041
Truth	18.922	−498.506	−870.132	approximate
APAL-AP06				
TLSQ	−1259.255	−866.540	−1316.832	
	(25 cm)	(31 cm)	(22 cm)	resid = 0.015
DDX(TLSQ)	−1259.337	−866.857	−1316.802	resid = 0.280 bad cycle fix
AF(TLSQ)	−1259.691	−866.184	−1316.800	93%
DDX(AF)	−1259.690	−866.190	−1316.793	resid = 0.020
Truth	−1259.690	−866.182	−1316.798	approximate
APAL-AP13				
TLSQ	−1051.322	−271.354	−312.845	
	(20 cm)	(24 cm)	(16 cm)	resid = 0.012
DDX(TLSQ)	−1051.308	−271.676	−312.832	resid = 0.146 bad cycle fix
AF(TLSQ)	−1051.288	−271.126	−312.921	93%
DDX(AF)	−1051.287	−271.122	−312.916	resid = 0.030
Truth	−1051.289	−271.130	−312.914	approximate
APAL-AP18				
TLSQ	55.464	27.096	38.879	
	(23 cm)	(26 cm)	(16 cm)	resid = 0.012
DDX(TLSQ)	55.670	26.837	38.907	resid = 0.016
AF(TLSQ)	55.668	26.841	38.907	94%, 85%
DDX(AF)	55.670	26.837	38.907	resid = 0.016
Truth	55.671	26.833	38.907	approximate
APAL-AP24				
TLSQ	94.780	−352.720	−626.084	
	(31 cm)	(45 cm)	(24 cm)	resid = 0.016
DDX(TLSQ)	94.677	−352.466	−626.003	resid = 0.020
AF(TLSQ)	94.670	−352.447	−626.018	94%, 88%
DDX(AF)	94.677	−352.466	−626.003	resid = 0.020
Truth	94.675	−352.460	−626.007	approximate
APAL-AP31				
TLSQ	121.658	−738.019	−1302.870	
	(17 cm)	(20 cm)	(13 cm)	resid = 0.010
DDX(TLSQ)	121.730	−738.090	−1302.889	resid = 0.023
AF(TLSQ)	121.736	−738.080	−1302.898	93%, 89%
DDX(AF)	121.730	−738.090	−1302.889	resid = 0.023
Truth	121.735	−738.090	−1302.890	approximate
APAL-AP36				
TLSQ	115.874	−765.752	−1350.151	
	(28 cm)	(31 cm)	(20 cm)	resid = 0.015
DDX(TLSQ)	115.937	−765.755	−1350.216	resid = 0.025
AF(TLSQ)	115.936	−765.738	−1350.230	93%, 87%
DDX(AF)	115.937	−765.755	−1350.216	resid = 0.025
Truth	115.938	−765.755	−1350.218	approximate

The one-sigma uncertainties associated with the TLSQ solution have been included in parentheses. The dimensions of the search box were approximately four times these values.

The next entry (DDX(TLSQ)) is the DLSQ solution based on using the TLSQ solution for cycle slip repair (i.e., DLSQ(TLSQ)). Notice that this failed for baselines APAL-AP06 and APAL-AP13. For the other baselines, this step was successful and provided a definitive solution. For the two cases that failed the TLSQ/DLSQ processing, cycle slips of magnitude 1 or 2 were wrongly introduced; this, naturally, was the cause of the failure. The AFM, of course, remedies this.

The next entry is the solution of the AFM search (i.e., AF(TLSQ)). Note that this step was always successful. The percentage represents how close f(**r**) came to its maximum theoretical peak (when multiple significant peaks were encountered, the second-highest peak is also noted). For a specific value of **r**, the percentage is computed by 100·f(**r**)/number of measurements.

The next-to-the-last entry (DDX(AF)) is the result achieved when the AFM solution is passed to fixed-DLSQ processing. In general, the fixed-DLSQ solution was neither better nor worse than the AFM solution.

The last entry is approximate truth. In this example, the survey marks were occupied two times for 2 min each time. The occupations were spaced approximately 50 min apart. Either five or six satellites were common to the two visits. In the case where five satellites were common, a sixth satellite was tracked during one of the two occupations. Measurements were recorded every 15 s.

Experiment No. 2 (February 21–22, 1990)

In this experiment, the PK method was an important part of the experiment design. The objective in this experiment was to measure the deflection of the vertical. The accuracy objective was 0.25 arcsec. Geodetic leveling was also performed at these sites. This test was performed at the Naval Surface Warfare Center (NSWC), Dahlgren, VA. Preliminary geodetic results were presented at the Fall 1990 American Geophysical Union (AGU) meeting in San Francisco [5].

Five receivers were involved. The reference receiver was situated at site MBRE on a tripod and never moved; the remaining four receivers were rovers. Within 2 km of MBRE, in a loop roughly centered about MBRE, were 12 survey marks, alternately designated blue and red. Two of the rover receivers were to occupy the six red marks driving a counter-clockwise route which would complete the loop eight times. Two of the rover receivers did the same with respect to the blue marks. Based on the previous experiment, only two visits were needed to determine the location of these marks to a couple of centimeters. Nevertheless, eight occupations were decided upon since we wanted to achieve the highest possible accuracy. There were numerous tall trees and buildings en route. In fact, there were so many obstructions that almost every receiver had cycle slips on almost every satellite during almost every transition between marks. In short, there were many dozens of cycle slips for each receiver (possibly more than 100). This presented no problem whatsoever; in fact, it was expected. The receiver could just as well have been turned off en route, but was not so as to avoid a proliferation of data files. Each occupation took 4 min, and data were collected every 10 s. Clearly this quantity of data was excessive. It should also be pointed out that the number of satellites tracked during any specific occupation varied from three to six. In the most favorable situations, the same six satellites were observed on consecutive occupations of a mark.

Finally, the red team completed its loop in approximately 48 min, whereas the blue team took approximately 55 min. It is not surprising, therefore, that the one instance in which the TLSQ solution was unable to fix cycle slips correctly was associated with one of the blue team's occupations. For a given accuracy of the TLSQ solution, the length of the time gap between revisits is bounded. In this case AFM was employed. This survey was performed using both L1 and L2 frequencies and on each of two consecutive evenings. It should be clear that once the solution was obtained for L1 processing for the first night, there was no need of TLSQ or AFM processing for the L2 solutions of the first night or for the L1 or L2 solutions of the second night.

Let us designate the reference mark by R, the blue marks by B1 through B6, and the red marks by R1 through R6. If, furthermore, we distinguish between the two blue rovers, we can designate the resulting solutions as R-B1I and R-B1II, and similarly for red. With this in mind, an important aspect of the experiment was to isolate solutions based on one rover (e.g., R-B1II, R-B2II, . . ., R-B6II). To eliminate any possibility of instrument bias, the differences, such as [(R-B1II) − (R-B4II)], were formed. This, in effect, differenced out the reference receiver and placed the same instrument at both ends of a baseline. The rover antennae were placed on top of *fixed height* survey poles so that the verticle height above the mark was constant. Of course we recognize that the phase center is still a function of the direction to a satellite, but this procedure should eliminate some amount of bias.

Because two blue teams and two red teams were employed, two frequencies observed, and the survey performed twice, eight solutions were obtained for each baseline. These eight solutions are treated as being independent even though there may be some correlation involved. Table 2 lists means (m) and standard deviations of solutions based on a given rover receiver after differencing out the reference receiver. The standard deviations in length (upper triangle) tend to be about 3 mm, and those in height (lower triangle) tend to be about 6 mm. Although a final statement cannot be made at this time, preliminary results indicate that the ensemble-means agree in height at the 2 to 3 mm level. This is based on the a priori truth of the deflection of the vertical at the test site. It is known to approximately 0.1 arcsec (equivalent to 1 mm over 2 km). On the other hand, baseline length truth was not available, except for MBRE to HERO, where there was 1 mm agreement. For these baseline lengths, 2 to 3 mm agreement is anticipated. Prior tests were conducted to verify that the survey rods employed could be set up, horizontally, at the 1 mm level. The vertical set-up error is essentially 0.0 mm.

Experiment No. 3 (February 23, 1990)

From a GPS standpoint, this experiment was similar to Experiment No. 2. Here a central site, CQCP, was encircled, more or less, by five other sites of interest. The entire experiment was restricted to a 200 m network. CQCP was the site of the reference receiver. Four rover receivers were used. Their antennas were mounted using forced-centering devices. This guaranteed that, on

Table 2—Lengths (m), Differential Heights (m), and Respective Standard Deviations (mm) from the February 21–22, 1990, Deflection of the Vertical Test (lengths are presented above and heights below diagonal)

RED	AAFR	DCKR	PBRR	PNDR	RUNR	WHYR
AAFR		1044.251	1494.963	275.401	1147.797	1852.769
		(1.2 mm)	(2.5 mm)	(2.4 mm)	(1.9 mm)	(1.6 mm)
DCKR	−0.734		2261.520	1260.200	837.213	2075.090
	(4.7 mm)		(2.8 mm)	(3.2 mm)	(2.8 mm)	(2.5 mm)
PBRR	0.509	1.243		1227.171	1725.680	1145.879
	(5.1 mm)	(4.9 mm)		(2.5 mm)	(2.7 mm)	(2.0 mm)
PNDR	−0.488	0.246	−0.997		1192.328	1690.531
	(6.0 mm)	(7.5 mm)	(6.2 mm)		(1.0 mm)	(2.0 mm)
RUNR	0.741	1.475	0.232	1.229		1264.401
	(3.9 mm)	(5.8 mm)	(7.4 mm)	(7.1 mm)		(2.5 mm)
WHYR	0.164	0.898	−0.344	0.653	−0.576	
	(4.5 mm)	(4.8 mm)	(5.3 mm)	(3.7 mm)	(4.9 mm)	

BLUE	AAFB	DOCK	HERO	PNDB	RUNB	SHL3
AAFB		1152.020	1621.348	210.650	1828.959	1692.221
		(1.7 mm)	(3.0 mm)	(2.1 mm)	(2.7 mm)	(1.9 mm)
DOCK	2.464		1277.366	1321.689	2020.425	2628.451
	(6.4 mm)		(4.2 mm)	(2.8 mm)	(2.8 mm)	(2.9 mm)
HERO	0.853	3.317		1631.803	927.280	2189.166
	(4.9 mm)	(6.8 mm)		(2.9 mm)	(2.5 mm)	(2.8 mm)
PNDB	−0.036	2.429	−0.889		1728.934	1483.097
	(4.0 mm)	(5.2 mm)	(4.7 mm)		(3.2 mm)	(3.0 mm)
RUNB	0.341	2.805	−0.512	0.377		1555.228
	(4.5 mm)	(5.9 mm)	(3.7 mm)	(4.5 mm)		(3.0 mm)
SHL3	−0.875	1.589	−1.729	−0.840	−1.217	
	(2.3 mm)	(5.9 mm)	(5.5 mm)	(3.5 mm)	(5.7 mm)	

reoccupations, the physical device would have no significant error (perhaps 0.1 mm). Thus errors were not due to mechanical set-up, but were due to phase center variations, multipath, refraction, and so forth. The aim of this was accuracy. The truth in this case is known at the 0.5 mm level by careful terrestrial survey methods using modern instrumentation. The five monuments were ASTW, CQD1, CQD2, CQD3, and CQD4, located at the NOAA/NGS Instrumentation and Equipment Section facility in Corbin, VA.

The test was designed as a PK test. Thus cycle slips between marks were not a factor, and the receivers could have been shut off between sites. Each receiver occupied a site for exactly 3 min and then moved clockwise to the next one. Exactly 3 min was allotted for travel time. In all cases, set-up was completed in less than the allotted 3 min. All four rover receivers completed the circuit exactly 10 times. Thus all baselines from CQCP to the remote sites were measured by the four receivers. Subsequently, the reference receiver was differenced out. Before presenting the results, it should be stated that the satellite visibility was identical to that of Experiment No. 2 in that the survey was performed on February 23, 1990, just one night after the NSWC deflection of the vertical test (February 21–22, 1990) and only 100 km away. Thus only three satellites were tracked during some occupations, and a maximum of six satellites were tracked during some occupations. Because only 30 min separated occupations of a given mark by a specified receiver (i.e., once per loop), a maximum of six satellites was common to as many as three occupations.

Table 3 gives results from the reference mark to the remote sites. In the left-hand column are the means associated with four L1 solutions and four L2 solutions. The next column contains the associated standard deviations (mm). The final column contains the truth. Table 3a includes the distance results and Table 3b the height results. The accuracy of the azimuth component can be assumed to be somewhere in between.

Notice that the distance agrees at the 1 mm level except for CQD4, where the disagreement is 3 mm. Remarkably, all heights agree to about 1 mm except for CQD4, where the discrepancy is 9.7 mm. These results lead one to believe these multiple measurement (i.e., averaging) methods can deliver 1 mm distances and possibly 2–3 mm heights over short (100 m) distances.

Table 4 contains the results of this same experiment where the reference site has been differenced out. Here we are trying to eliminate small instrumentation biases.

A linear combination of noisy quantities yields a noisier quantity. Thus if no instrument-specific biases existed, the differenced quantities would be expected to have larger standard deviations than those quantities prior to differencing. The fact that the standard deviations have actually been reduced should point to the removal of biases. Clearly each monument has its biases, and each instrument has its biases. Although there are monument height biases, it is the removal of instrumentation biases that has led to smaller standard deviations.

This experiment has one disadvantage over Experiment No. 2. In Experiment No. 2, we maintained exactly the same height above all remote marks since the same unadjustable rod was used by a given rover at all sites. At the Corbin facility, a platform was set up over each survey mark to accommodate the forced-centering devices. These platforms are semipermanent, and extreme

Table 3a—Distance Results from the February 23, 1990, Corbin, VA Test

Baseline	Mean (m)	Standard Deviation (mm)	Truth (m)
CQCP-ASTW	34.9458	1.1	34.945
CQCP-CQD1	95.4716	1.2	95.472
CQCP-CQD2	86.9092	1.1	86.910
CQCP-CQD3	79.7833	1.5	79.783
CQCP-CQD4	106.9900	2.1	106.987

Table 3b—Differential Height Results from the February 23, 1990, Corbin, VA Test

Baseline	Mean (m)	Standard Deviation (mm)	Truth (m)
CQCP-ASTW	−0.2297	5.4	−0.2292
CQCP-CQD1	1.6918	6.3	1.6924
CQCP-CQD2	0.7414	4.5	0.7427
CQCP-CQD3	−0.2945	4.9	−0.2964
CQCP-CQD4	0.6385	7.3	0.6482

Table 4 —Length (m) and Differential Height (m) Results from the February 23, 1990, Corbin, VA Test, Where the Reference Receiver Is Differenced Out (lengths are presented above and heights below diagonal)

	ASTW	CQD1	CQD2	CQD3	CQD4
ASTW		106.1009	121.8316	82.5642	72.0967
		(0.8 mm)	(0.9 mm)	(0.9 mm)	(1.7 mm)
Truth		106.102	121.832	82.563	72.095
CQD1	−1.9214		122.8378	175.2505	150.0037
	(2.6 mm)		(0.8 mm)	(0.6 mm)	(0.9 mm)
Truth	−1.9216		122.838	175.250	150.002
CQD2	−0.9711	0.9503		123.4383	193.8942
	(3.6 mm)	(3.1 mm)		(1.2 mm)	(1.5 mm)
Truth	−0.9719	0.9497		123.439	193.892
CQD3	0.0649	1.9863	1.0360		127.2370
	(1.7 mm)	(2.2 mm)	(2.5 mm)		(1.8 mm)
Truth	0.0672	1.9888	1.0391		127.234
CQD4	−0.8682	1.0533	0.1030	−0.9330	
	(5.3 mm)	(4.8 mm)	(4.9 mm)	(4.2 mm)	
Truth	−0.8774	1.0442	0.0945	−0.9446	

care was used in measuring these heights. We believe these errors are below 0.5 mm (in the previous experiment, the vertical set-up error was essentially 0.0 mm).

This test will be repeated. It is important to know what procedures are required to achieve 1 mm distances consistently and, possibly, 2 mm heights over short baselines (200 m). Experiments No. 2 and No. 3 should remove any question about the ability to measure distances and even ellipsoidal height differences to a few millimeters. The appropriate question, especially for short baselines (e.g., 3 km), is what procedures should be followed if one is to measure distances to, say, 2 mm and ellipsoidal height differences to, say, 5 mm. Clearly these accuracies can be achieved over short distances, especially where the elevation changes are small and the meteorology is favorable.

Experiment No. 4 (February 9, 1990)

This is the most important example with respect to supporting the theme of this paper.

A reference receiver was situated at S274, which is a National Geodetic Reference System first-order horizontal control station and benchmark. At a distance of approximately 40 km is Malden Municipal Airport, MO, where the primary site is MALD. Another reference receiver was situated at MALD for the duration of this test. A third receiver was a rover receiver. This rover began at site MAAZ and proceeded to sites MA04, MA13, MA18, MA22, MA31, and MA36, finally returning to MAAZ. This circuit was completed a total of four times. Each site was occupied for 5 min, although this length of stay was not required. Thus each loop took about 62 min, and the entire experiment took about 4.25 h. In this experiment, there was no a priori truth. For approximate truth, the entire data set for the baseline S274-MALD was used to determine this vector with sufficient accuracy. This experiment was conducted at night, when the ionospheric refraction was minimal. As a consequence, the integer ambiguities were obvious and determined to within 0.1 cycle in normal processing. This solution will provide truth in the sense that the same solution will be attempted with only a few minutes of data. With MALD approximately known, it was a simple matter to determine the other (local) seven airport marks. This was achieved using PK processing with the TLSQ and DLSQ solutions. This was easy to achieve since there were four occupations of 5 min each, and the baselines were short (e.g., less than 2 km). Thus the approximate vector truth was obtained from S274 to MALD, as well as the approximate vector truth from S274 to the other seven airport monuments. The satellites tracked during the four occupations varied from three or four to a maximum of six at any one time. Altogether, seven satellites were observed.

In the easy part of the experiment, the seven baseline vectors were determined from MALD to the other runway monuments using four occupations, three occupations, and finally two occupations of 5 min each using the AFM. Although not included in this report, the results are roughly the same when only 2 min occupations are assumed and the remainder of the data is simply ignored.

For the case where two occupations are assumed, either six satellites are common to two occupations, or there are at least five common satellites and an additional satellite that is not common. Thus the geometry is quite good, but not significantly better than will be commonly available when the GPS constellation is complete.

Table 5 lists the results of the four-, three-, and two-occupation studies for the local MALD airport processing. For each baseline vector, there are seven entries. The first line lists the TLSQ solution and realistic standard deviations for the four-occupation case (denoted 4V for four visits). This is followed by the AFM solution for the four-occupation case. The next two lines reflect the three-occupation case (denoted 3V). This is followed by the two-occupation case (denoted 2V), and finally the truth vector.

Each AFM solution includes the amplitude of the highest maximum in comparison with the theoretical maximum and also the second-highest peak. Notice that as the number of occupations is reduced, the first and second maxima are less distinct. What may be surprising is that even when the peaks become essentially the same, the correct solution can be isolated (normally) by DLSQ processing. The right peak will fix cycle slips correctly and go on to yield a fixed double-difference solution with small residuals. The wrong peak will not behave this way.

The AFM search volume is based on the sigmas from TLSQ. These sigmas are realistic and even conservative. The search took place over approximately four-sigma space. In fact, the correct solution was normally well within the search volume. Clearly, if a satisfactory peak had not been achieved, a larger search volume would have been used. In none of the cases presented here was an expanded search volume needed.

This processing was straightforward, and, as stated earlier, only 2 min occupations were generally required; this may be documented in an upcoming report. It is important to note that the AFM solution is accurate and generally within the accuracy expectations of the fixed double-difference solution.

The final and most important aspect of this processing was in determining the baseline vectors from S274 to the eight airport marks with four, three, and

Table 5—Ambiguity Function Method Results from the February 9, 1990, Malden Municipal Airport Test

Baseline		Coordinates (m) dx	dy	dz	Remarks
MALD-MA04					
TLSQ	4V:	416.974	196.616	264.707	
		(14 cm)	(17 cm)	(7 cm)	
AF (TLSQ)	4V:	417.166	196.608	264.749	97%, 79%
TLSQ	3V:	417.028	196.644	264.646	
		(16 cm)	(21 cm)	(8 cm)	
AF (TLSQ)	3V:	417.170	196.602	264.746	95%, 85%
TLSQ	2V:	416.936	196.656	264.701	
		(29 cm)	(38 cm)	(15 cm)	
AF(TLSQ)	2V:	417.170	196.606	264.743	94%, 89%
Truth		417.162	196.608	264.734	
MALD-MA13					
TLSQ	4V:	416.552	177.393	239.169	
		(17 cm)	(19 cm)	(9 cm)	
AF(TLSQ)	4V:	416.410	177.501	239.143	95%, 80%
TLSQ	3V:	416.584	177.425	239.206	
		(18 cm)	(25 cm)	(10 cm)	
AF(TLSQ)	3V:	416.410	177.491	239.141	95%, 85%
TLSQ	2V:	416.540	176.821	239.178	
		(30 cm)	(45 cm)	(19 cm)	
AF(TLSQ)	2V:	416.414	177.521	239.128	94%, 89%
Truth		416.412	177.509	239.126	
MALD-MA18					
TLSQ	4V:	41.990	−264.539	−356.865	
		(13 cm)	(13 cm)	(6 cm)	
AF(TLSQ)	4V:	41.956	−264.575	−356.865	96%, 78%
TLSQ	3V:	42.009	−264.568	−356.869	
		(13 cm)	(16 cm)	(7 cm)	
AF(TLSQ)	3V:	41.951	−264.558	−356.859	95%, 85%
TLSQ	2V:	42.008	−264.560	−356.864	
		(18 cm)	(22 cm)	(10 cm)	
AF(TLSQ)	2V:	41.957	−264.560	−356.864	94%, 90%
Truth		41.950	−264.556	−356.869	
MALD-MA22					
TLSQ	4V:	−672.718	−438.812	−591.041	
		(20 cm)	(20 cm)	(9 cm)	
AF(TLSQ)	4V:	−672.610	−438.602	−591.075	95%
TLSQ	3V:	−672.708	−438.859	−591.043	
		(19 cm)	(23 cm)	(11 cm)	
AF(TLSQ)	3V:	−672.611	−438.603	−591.075	95%, 83%
TLSQ	2V:	−672.700	−438.857	−591.104	
		(25 cm)	(31 cm)	(15 cm)	
AF(TLSQ)	2V:	−672.614	−438.584	−591.090	93%, 87%
Truth		−672.616	−438.593	−591.085	

Table 5—Ambiguity Function Method Results from the February 9, 1990, Malden Municipal Airport Test—Continued

Baseline		Coordinates (m) dx	dy	dz	Remarks
MALD-MA31					
		−649.181	826.538	1113.925	
TLSQ	4V:	(15 cm)	(19 cm)	(9 cm)	
AF(TLSQ)	4V:	−649.207	826.429	1113.975	95%, 79%
		−649.191	826.611	1113.882	
TLSQ	3V:	(17 cm)	(26 cm)	(11 cm)	
AF(TLSQ)	3V:	−649.207	826.419	1113.982	95%, 79%
		−649.224	826.716	1113.772	
TLSQ	2V:	(14 cm)	(29 cm)	(12 cm)	
AF(TLSQ)	2V:	−649.210	826.429	1113.970	93%, 85%
Truth		−649.209	826.434	1113.968	
MALD-MA36					
		58.384	645.367	869.596	
TLSQ	4V:	(15 cm)	(19 cm)	(9 cm)	
AF(TLSQ)	4V:	58.426	645.457	869.630	95%, 79%
		58.355	645.395	869.601	
TLSQ	3V:	(18 cm)	(26 cm)	(11 cm)	
AF(TLSQ)	3V:	58.421	645.453	869.627	95%, 78%
		58.324	645.288	869.699	
TLSQ	2V:	(32 cm)	(45 cm)	(19 cm)	
AF(TLSQ)	2V:	58.416	645.446	869.631	92%, 85%
Truth		58.419	645.465	869.614	92%, 85%
MALD-MAAZ					
		8.175	378.968	510.962	
TLSQ	4V:	(15 cm)	(16 cm)	(7 cm)	
AF(TLSQ)	4V:	8.209	379.160	510.970	95%, 83%
		8.139	379.045	510.994	
TLSQ	3V:	(14 cm)	(16 cm)	(8 cm)	
AF(TLSQ)	3V:	8.214	379.151	510.978	94%, 86%
		8.309	378.901	511.078	
TLSQ	2V:	(16 cm)	(21 cm)	(10 cm)	
AF(TLSQ)	2V:	8.215	379.165	510.963	92%, 88%
Truth		8.214	379.154	510.963	

finally two 5 min occupations. These baselines are approximately 40 km in length. Table 6 lists the results.

In the four-occupation case, the solution was extremely easy and obvious. The maxima were generally at the 90 percent level, whereas other maxima were 80 percent or so, and below. Nevertheless, the AFM solution was input into the DLSQ processing as a check, though not included in Table 6. Correct cycle slips, ambiguities that are nearly integers, and small residuals are all required. In all cases these were achieved, and secondary peaks failed. Notice again that ambiguity function solutions compared favorably with the truth.

In the three-occupation case, notice that the maximum peak and the secondary peak are somewhat closer. Nevertheless, the maximum peak in all cases was the correct one. In fact, in this experiment there was not much difference

Table 6—Ambiguity Function Method Results from the February 9, 1990, Malden Municipal Airport Test

Baseline		Coordinates (m)			Remarks
		dx	dy	dz	
S274-MALD					
TLSQ	4V:	38350.787	−1640.139	−2400.060	
		(18 cm)	(22 cm)	(10 cm)	
AF(TLSQ)	4V:	38350.737	−1640.213	−2400.144	87%
TLSQ	3V:	38350.814	−1640.294	−2400.068	
		(19 cm)	(29 cm)	(12 cm)	
AF(TLSQ)	3V:	38350.738	−1640.236	−2400.133	85%, 77%
TLSQ	2V:	38350.794	−1640.662	−2400.042	
		(18 cm)	(38 cm)	(15 cm)	
AF(TLSQ)	2V:	38350.734	−1640.238	−2400.138	88%, 87%
Truth		38350.740	−1640.212	−2400.137	
S274-MA04					
TLSQ	4V:	38768.140	−1444.383	−2134.994	
		(24 cm)	(28 cm)	(12 cm)	
AF(TLSQ)	4V:	38767.890	−1443.626	−2135.378	91%, 80%
TLSQ	3V:	38768.170	−1444.558	−2135.003	
		(26 cm)	(35 cm)	(14 cm)	
AF(TLSQ)	3V:	38767.894	−1443.635	−2135.371	90%, 83%
TLSQ	2V:	38768.535	−1444.780	−2135.099	
		(39 cm)	(52 cm)	(22 cm)	
AF(TLSQ)	2V:	38767.899	−1443.656	−2135.367	91%, 88%
Truth		38767.902	−1443.604	−2135.403	
S274-MA13					
TLSQ	4V:	38767.208	−1462.538	−2161.162	
		(24 cm)	(28 cm)	(12 cm)	
AF(TLSQ)	4V:	38767.150	−1462.737	−2160.980	91%, 77%
TLSQ	3V:	38767.184	−1462.636	−2161.140	
		(27 cm)	(35 cm)	(14 cm)	
AF(TLSQ)	3V:	38767.150	−1462.736	−2160.982	91%, 83%
TLSQ	2V:	38767.063	−1462.245	−2161.201	
		(36 cm)	(49 cm)	(22 cm)	
AF(TLSQ)	2V:	38767.151	−1462.761	−2160.969	91%, 88%
Truth		38767.152	−1462.703	−2161.011	
S274-MA18					
TLSQ	4V:	38392.730	−1904.452	−2756.826	
		(18 cm)	(19 cm)	(8 cm)	
AF(TLSQ)	4V:	38392.696	−1904.768	−2757.002	90%
TLSQ	3V:	38392.764	−1904.519	−2756.947	
		(19 cm)	(22 cm)	(10 cm)	
AF(TLSQ)	3V:	38392.698	−1904.767	−2757.007	90%, 81%
TLSQ	2V:	38392.603	−1904.252	−2756.919	
		(22 cm)	(26 cm)	(12 cm)	
AF(TLSQ)	2V:	38392.695	−1904.780	−2757.007	92%, 91%
Truth		38392.690	−1904.768	−2757.006	

Table 6—Ambiguity Function Method Results from the February 9, 1990, Malden Municipal Airport Test—Continued

Baseline		Coordinates (m) dx	dy	dz	Remarks
S274-MA22					
		37678.083	−2079.005	−2991.192	
TLSQ	4V:	(22 cm)	(23 cm)	(10 cm)	
AF(TLSQ)	4V:	37678.125	−2078.780	−2991.234	92%, 80%
		37678.112	−2078.860	−2991.035	
TLSQ	3V:	(21 cm)	(25 cm)	(12 cm)	
AF(TLSQ)	3V:	37678.120	−2078.768	−2991.235	91%, 81%
		37678.113	−2078.828	−2991.069	
TLSQ	2V:	(24 cm)	(31 cm)	(15 cm)	
AF(TLSQ)	2V:	37678.121	−2078.800	−2991.225	92%, 90%
Truth		37678.124	−2078.805	−2991.222	
S274-MA31					
		37701.584	−813.518	−1286.309	
TLSQ	4V:	(19 cm)	(23 cm)	(11 cm)	
AF(TLSQ)	4V:	37701.526	−813.778	−1286.175	88%, 80%
		37701.565	−813.570	−1286.305	
TLSQ	3V:	(21 cm)	(32 cm)	(13 cm)	
AF(TLSQ)	3V:	37701.523	−813.804	−1286.171	87%, 80%
		37701.586	−814.131	−1286.210	
TLSQ	2V:	(18 cm)	(38 cm)	(16 cm)	
AF(TLSQ)	2V:	37701.526	−813.791	−1286.182	90%, 88%
Truth		37701.531	−813.778	−1286.169	
S274-MA36					
		38409.424	−994.511	−1530.747	
TLSQ	4V:	(17 cm)	(22 cm)	(10 cm)	
AF(TLSQ)	4V:	38409.158	−994.761	−1530.513	91%, 81%
		38409.450	−994.563	−1530.769	
TLSQ	3V:	(20 cm)	(30 cm)	(12 cm)	
AF(TLSQ)	3V:	38409.158	−994.789	−1530.495	90%, 83%
		38409.558	−994.716	−1530.712	
TLSQ	2V:	(39 cm)	(46 cm)	(20 cm)	
AF(TLSQ)	2V:	38409.155	−994.792	−1530.500	91%, 88%
Truth		38409.159	−994.747	−1530.523	
S274-MAAZ					
		38358.876	−1261.565	−1889.100	
TLSQ	4V:	(18 cm)	(20 cm)	(9 cm)	
AF(TLSQ)	4V:	38358.937	−1261.064	−1889.183	87%
		38358.892	−1261.531	−1889.061	
TLSQ	3V:	(19 cm)	(22 cm)	(11 cm)	
AF(TLSQ)	3V:	38358.943	−1261.021	−1889.187	89%, 81%
		38358.997	−1261.760	−1889.079	
TLSQ	2V:	(22 cm)	(29 cm)	(14 cm)	
AF(TLSQ)	2V:	38358.945	−1261.032	−1889.183	90%, 88%
Truth		38358.954	−1261.058	−1889.174	

between the four-occupation case and the three-occupation case. As in the four-occupation case, all strong secondary peaks were input to DLSQ processing as a test along with the maximum. In all cases the AFM maxima passed these tests, whereas the other, lesser maxima did not.

In the two-occupation case, the primary and secondary maxima are nearly indistinguishable. Nevertheless, in all cases except one, the maximum was the correct solution. In that exceptional case, the AFM maximum failed the DLSQ step in that large residuals were encountered. The second peak passed the DLSQ test in that small residuals were achieved. It should be noted that the two-occupation case was barely strong enough to succeed, even though this was a favorable data set taken during the quiet portion of the day. In spite of this, these results are most encouraging:

1) They indicate that dual-frequency AFM processing over moderate baselines will be very effective.

2) Gaps of 2–3 h in metropolitan surveys (20 km) will be productive, efficient, and accurate, especially with a full GPS constellation.

3) As the author has previously stated, the results prove [1] that PK works, that two or three occupations of 2 to 3 min are adequate to determine baselines of moderate length, and that the GPS receiver can in fact be turned off between visitations.

OTHER USEFUL TECHNIQUES

An Alternative Search Method

Previously it was stated that there are alternative search techniques that can isolate the correct solution. Here I discuss one that I have occasionally used in my PK research. This method assumes two or more occupations, as its objective is to make the cycle connection between occupations.

The method is conceptually simple. After the TLSQ solution is used to fix cycle slips, the DLSQ solution follows. If double-difference integers are very poor, the user should assume the cycle slips were not fixed correctly, add a vector increment to the TLSQ solution, and attempt to fix cycle slips again. The DLSQ solution would then be attempted again.

In PK processing, this method can be effective in that only a few observations are processed, and success is known after just one iteration. In fact, success or failure is known within seconds. In effect one searches a box about the TLSQ solution, but uses these candidates in a different way than in ambiguity function processing. The search grid can be very crude if the time period between occupations is not too large. The longer the gap, the finer the search grid. A typical example would be for 30 min gaps. In this case, if the TLSQ solution does not repair cycle slips correctly, a 20 cm search grid could be appropriate. The granularity of the search grid can be computed easily based on the amount of integrated Doppler error one can tolerate over the period between occupations. Without providing specific examples, I shall simply state that the method works rather well in a variety of situations, but the AFM and other search methods, not given here, are more general.

The Tandem Pseudo-kinematic Method (TPK)

The PK method as described so far in this paper assumes that the rover receiver will occupy a given mark two or more times.

An extension to this, which I used originally out of necessity and later by design, is as follows. A hypothetical example is used to describe the method.

The U.S. Capital Beltway around Washington, D.C. is approximately 100 km in circumference. Suppose there were geodetic marks located approximately 1 km apart along the circumference. Suppose the objective were to determine the locations of these 100 marks to within a couple of centimeters with respect to a central geodetic monument. One could use static or a combination of static and kinematic methods (since there are many overpasses). Clearly PK would be more productive. Still more productive would be TPK, which is executed as follows.

In the above example, the reference receiver is situated in downtown Washington, D.C., near the center of the Beltway. Two or more rovers will be assumed. The first begins the loop at a certain time. Then, 1–2 h later, a second rover occupies the same marks as the first rover and in the same order. For increased accuracy and reliability, an optional third tandem rover could follow the second. During postprocessing of a given geodetic mark, the carrier phase data are used from any receiver that occupied that mark. Naturally, it would not matter very much that these data were collected by many receivers as opposed to the traditional single receiver. One savings over PK is that the rover would not have to waste time in returning to the original mark. Another factor is that the survey could be completed in less than half the time of a one-rover PK survey. (Of course, the total amount of person-hours would be nearly the same.) Planning and executing a TPK survey would be simple.

The Simultaneous Tandem Pseudo-kinematic Method (STPK)

This method is similar to the TPK method described above, except the tandem rovers occupy their different marks simultaneously, and a fixed reference receiver is optional. Such a survey could be accomplished either using a rigorous preschedule plan, or with the aid of cellular phones. This method differs in another subtle way. Whereas TPK above can be executed with either two or three rovers, this method is more effective with three (or more) rovers. Three simultaneous rovers provide two occupations of all circumferential neighboring vectors, allowing subcentimeter accuracies. Note that even two rovers, operating in STPK mode, can be effective (assuming enough satellites) if a priori vectors are determined with respect to a (fixed) reference receiver (situated at some central location). The combination of the radial vectors and the high-precision circumferential vectors should yield a powerful network combination.

It should be obvious that the planning and executing of an STPK survey could be tricky. An STPK survey must have simultaneity, uniform travel times between marks, and a reasonable time gap between when the tandem rovers occupy a given mark. Thus, while this technique could be useful, it may have limited application.

The circumferential survey example, of course, is only illustrative. One does not actually have to circumnavigate the reference receiver. Best results would normally be achieved if the rovers ultimately returned to their starting marks, although this is not strictly required.

SUMMARY

Four examples have been presented to substantiate earlier claims that PK is a potentially powerful static survey method where only two or three occupations for only 2–3 min are required, allowing the rover, at least in theory, to turn off the GPS receiver between visits.

The first example demonstrated that the PK method worked with two visits of 2 min each. A search method was needed for two cases since, for those cases, the TLSQ solution was not sufficiently accurate to fix cycle slips between occupations. Either the alternative search method, as described above, or the AFM easily solves this problem. The AFM is preferred because it is more general. The alternative search method will encounter problems when attempting to fix cycle slips over baselines of moderate lengths where gaps of 2–3 h are involved; the AFM will not encounter such limitations.

The second and third examples demonstrated that the PK method can yield accurate survey results and that loss of lock between occupations does not present a problem.

The fourth example was potentially the most informative. It demonstrated that the AFM allowed PK processing from short to moderate baseline lengths (i.e., 40 km). To some extent, this example proved that PK is not subject to working only some of the time as has sometimes been conjectured. It can be expected to work all the time if certain limitations are respected.

The PK method has evolved over recent years into a technique that is useful today and has a bright future. The AFM has a long history, and was studied a great deal in the early 1980s [2, 6]. In the author's opinion, PK was always intended to be combined with a search method to achieve its full effectiveness. The AFM is a good candidate for this now that it is fast enough to do complex searches in less than 1 min.

A less general but alternative search method was presented. Abbreviated as TPK and STPK, two variations of the pseudo-kinematic method were introduced.

CONCLUSIONS

The PK method can achieve its full effectiveness when combined with an appropriate search method. Although the AFM is not the only search method, it is fast and effective; in a parallel processing future, it could possibly be performed in less than 1 s. These search techniques readily extend to the determination of integer ambiguities on the fly in that one can search for the correct starting location in the same way one searches for the static vector solution. The reason for this is simple: the starting position is not known, but one does know, precisely, the change in position from the starting position to the position occupied later. One selects the trial solution which generates the trajectory yielding the best integers.

With a central reference site, the PK method based on an effective search technique is a method for performing rapid and accurate surveys on metropolitan scales (e.g., Washington, D.C., and vicinity to about 25 km). This will become clearer as the full GPS constellation is deployed and advances in post-processing software are developed.

Based on a paper presented at The Institute of Navigation Annual Meeting, Atlantic City, NJ, June 1990.

PRODUCT DISCLAIMER

Mention of a commercial company or product does not constitute an endorsement by the National Oceanic and Atmospheric Administration. Use for publicity or advertisement purposes of information from this publication concerning proprietary products or the tests of such products is not authorized.

REFERENCES

1. Remondi, B. W., *Kinematic and Pseudo-kinematic GPS,* Proceedings of the Satellite Division's International Technical Meeting, The Institute of Navigation, Colorado Springs, CO, September 19–23, 1988, 115–21.
2. Remondi, B. W. (Center for Space Research, The University of Texas at Austin, Austin Texas), *Using the Global Positioning System (GPS) Phase Observable for Relative Geodesy: Modeling, Processing, and Results,* Ph.D. dissertation, 360 pp., National Geodetic Information Center, NOAA, Rockville, MD, 1984.
3. Blewitt, G., *A New Tool for Dense Monitoring of Crustal Strain: GPS Rapid Static Surveying,* EOS Transactions, American Geophysical Union, Baltimore, MD, Vol. 71, No. 17, April 24, 1990.
4. Remondi, B. W., *Applying Kinematic GPS Techniques at Our Nation's Airports,* Proceedings of IEEE Position Location and Navigation Symposium 1990, The IEEE AES Society, Las Vegas, NV, March 20–23, 1990, 262–66.
5. Evans, A. G. and Remondi, B. W., *Gravity Deflections of the Vertical Using Pseudo-kinematic GPS and Leveling: Initial Test Results,* EOS Transactions, American Geophysical Union, San Francisco, CA, Vol. 71, No. 43, 1990, p. 1279.
6. Counselman, C. C. and Gourevitch, S. A., *Miniature Interferometer Terminals for Earth Surveying: Ambiguity and Multipath with the Global Positioning System,* IEEE Transactions on Geoscience and Remote Sensing, Vol. GE-19, No. 4, October 1981.

NAVIGATION: *Journal of The Institute of Navigation*
Vol. 35, No. 1, Spring 1988
Printed in U.S.A.

Shipborne GPS Kinematic Positioning for Hydrographic Applications

G. LACHAPELLE, W. FALKENBERG, J. HAGGLUND
Nortech Surveys (Canada, Inc.), Calgary, Canada
D. KINLYSIDE
New South Wales Department of Lands, Bathhurst, Australia
M. CASEY
Canadian Hydrographic Service, Ottawa, Canada
P. KIELLAND
Service Hydrographique du Canada, Quebec, Canada
H. BOUDREAU
Canadian Hydrographic Service, Dartmouth, Nova Scotia
Received August 1987
Revised December 1987

ABSTRACT

The objectives of the Canadian Hydrographic Service with respect to GPS utilization are reviewed. A shipborne experiment undertaken in November 1986 and designed to test techniques developed to fulfill some of these objectives is described. The primary purpose of the experiment was to assess the performance of differential GPS positioning on a 10 m survey launch. The techniques utilized and results obtained on the launch are described. Additional tests conducted on a 38 m vessel are summarized. TI4100 GPS receivers were used in various combinations of P code, C/A code, and different user dynamics modes. Carrier phase measurements were combined with pseudoranges to improve the accuracy. The reference positions were provided by two Atlas Polarfix systems and a Mini-Ranger Falcon 484 system. The external horizontal position accuracy of differential GPS was estimated to be 4.5 m at the 95 percent confidence level. Initial results obtained with a low cost attitude sensor are compared with GPS for both the launch and the 38 m vessel. The use of GPS to estimate heave motion is also discussed. Prospects of further improving GPS accuracy for shipborne positioning are mentioned in the conclusions.

INTRODUCTION

The Canadian Hydrographic Service (CHS) is responsible for a variety of hydrographic surveying tasks. These include shipborne applications on inland waters and the three oceans surrounding Canada. Shipborne positioning has mostly been conducted up to now with a variety of land-based radio positioning systems, either in a stand-alone mode or in combination with ship's log and gyro and the Navy Navigation Satellite System (Transit). A real time integrated navigation system, BIONAV, was successfully developed for this purpose [24]. In the early 1980s, initial studies were conducted to assess the potential of Global Positioning System (GPS) to complement and eventually replace the above positioning systems. The global, continuous, all-weather coverage of GPS, its high accuracy level and independence from logistically complex shore based transmitting stations, constitute major advantages when compared to other systems.

Initial shipborne trials conducted by CHS in cooperation with Sheltech Canada (now Nortech Surveys) and The University of New Brunswick in 1981/82 using a slow switching single channel STI5010 receiver yielded accuracies of 50-100 m [9]. During the 1982 campaign, the integration of the STI5010 unit with a FILSII inertial navigation system was tested under a joint project between Sheltech/Nortech and The University of Calgary, and accuracies of 15 m were obtained [25]. In 1983, further tests conducted with TI4100 multiplex receivers resulted in accuracies of 25 m in single point P-code mode and 10 m in differential mode [10]. No attempt was made at that time to combine carrier phase with pseudorange measurements (code) to improve accuracy. Nevertheless, these latter tests confirmed the potential of GPS to replace many, if not all, of the existing positioning systems for shipborne hydrographic operations. GPS was subsequently used operationally by CHS for specific tasks, such as Loran-C calibration [5].

Following the 1983 sea trials, CHS established specific objectives regarding GPS. The overall objective was the development of differential techniques for the cost effective use of GPS in hydrographic applications. These techniques were to be based on the combination of code and carrier measurements. The emphasis was to be on C/A code utilization, with P code being used to assist in the evaluation of C/A code performance. This objective naturally includes, but is not limited to, shipborne surveying. A non-shipborne technology developed and used by CHS is airborne Light Detection And Ranging (LIDAR) charting with the LARSEN system [3]. The use of GPS for airborne LARSEN positioning was tested in 1983 [11]. The development of differential techniques is required for increased accuracy and quality assurance. In addition to reducing the effects of orbital errors, differential GPS reduces the effects of the ionosphere. This is important since large parts of Canada's Arctic are located in the auroral oval where long- and short-term ionospheric variabilities are particularly large [16].

For shipborne applications, the objective was to develop precise and highly reliable GPS positioning methodologies for optimal utilization of the full GPS constellation. The reliability aspect is of utmost importance for charting operations. In addition to standard statistical measures (e.g., 95 percent confidence level), CHS uses the concept of "accuracy assurance" to help better define its positioning requirements [4]. A near 100 percent assurance is required that errors will not exceed preset error bounds, without some degree of warning. Quality assurance is thus defined as outlier detection with a near 100 percent certainty. The concept of quality assurance is similar to that of integrity, used by other organizations such as the United States Coast Guard [17].

The vessel type of prime interest in the current project is a 10 m hydrographic survey launch. This is because in-shore hydrographic operations are mostly conducted with launches, and the dynamics of a launch are generally more turbulent than that of a larger vessel. The plan also calls for additional tests on a larger hydrographic vessel to confirm the results obtained with the launch. The first objective of the project was to develop and test a GPS technique for shipborne horizontal positioning which yields an accuracy of 5 m at the 95 percent confidence level. The 95 percent confidence region, which is formally defined as the error ellipse with semiaxes of 2.45 sigma (northing) and 2.45 sigma (easting), will be approximated herein by a 95 percent circle with radius obtained by multiplying the average sigma in northing and easting by a factor of 2.5. This approximation is acceptable under reasonable normality and orthogonality assumptions [1]. The second objective was to develop a 10 m accuracy assurance. A third objective was added immediately prior to the sea trials, namely the evaluation of a low cost attitude sensor for the measurement of ship's pitch and roll. Results pertaining to the first objective are reported herein. Development work pertaining to the second objective is currently in progress and will be reported at a later date. Preliminary results associated with the third objective are also presented.

As part of the first objectives, a differential positioning technique based on the combination of code (pseudorange) and phase data (carrier) was developed and tested extensively in controlled land kinematic mode to ascertain that better than 5 m accuracy could be achieved in a relatively benign environment. The accuracy objective was fully realized [12,13]. The next logical step was to evaluate the performance of the above positioning technique in the realistic shipborne environment where phenomena such as multipath caused by the ship's reflective steel structure and sea water, as well as ship's pitch, roll, and heave motion would truly test its effectiveness.

Many of the sea trial objectives described above are similar to those defined by the United States Coast Guard for the evaluation of GPS equipment onboard a 42 ft patrol boat [17].

SHIPBORNE TRIAL DESCRIPTION

The sea trials were conducted near Halifax, Nova Scotia, in November 1986, at latitude 44°40′N and longitude 63°35′W. Major features of the trials are given in Table 1. The trials were conducted on three specific sites, as shown in Figure 1. The first site, Bedford Basin, is sheltered and thus provides relatively calm conditions for the initial tests. Its proximity to the Bedford Institute of Oceanography, the headquarters of the campaign, and the excellent geometry

Table 1—GPS Shipborne Experiment Features

FEATURE	MAXWELL	HERON
Vessel Type	Hydrographic ship	Survey launch
Antenna Height	13.8 m (above water)	3.9 m
Trial Period	Nov 5–11, 1986	Nov 15–21, 1986
Vessel Speed	10 kn (5.1 m/s)	15 kn (7.7 m/s)
Temperature	2° to 7° C.	−5° to 9° C.
Wind Speed	5–22 kn	0–20 kn
Roll Motion (max.)	15° (3.5 m displacement)	30° (2 m displacement)
Pitch Motion (max.)	3°	15°
Roll and Pitch Period	7 s (average)	2.5 s
Heave Motion	1 m	Not measured

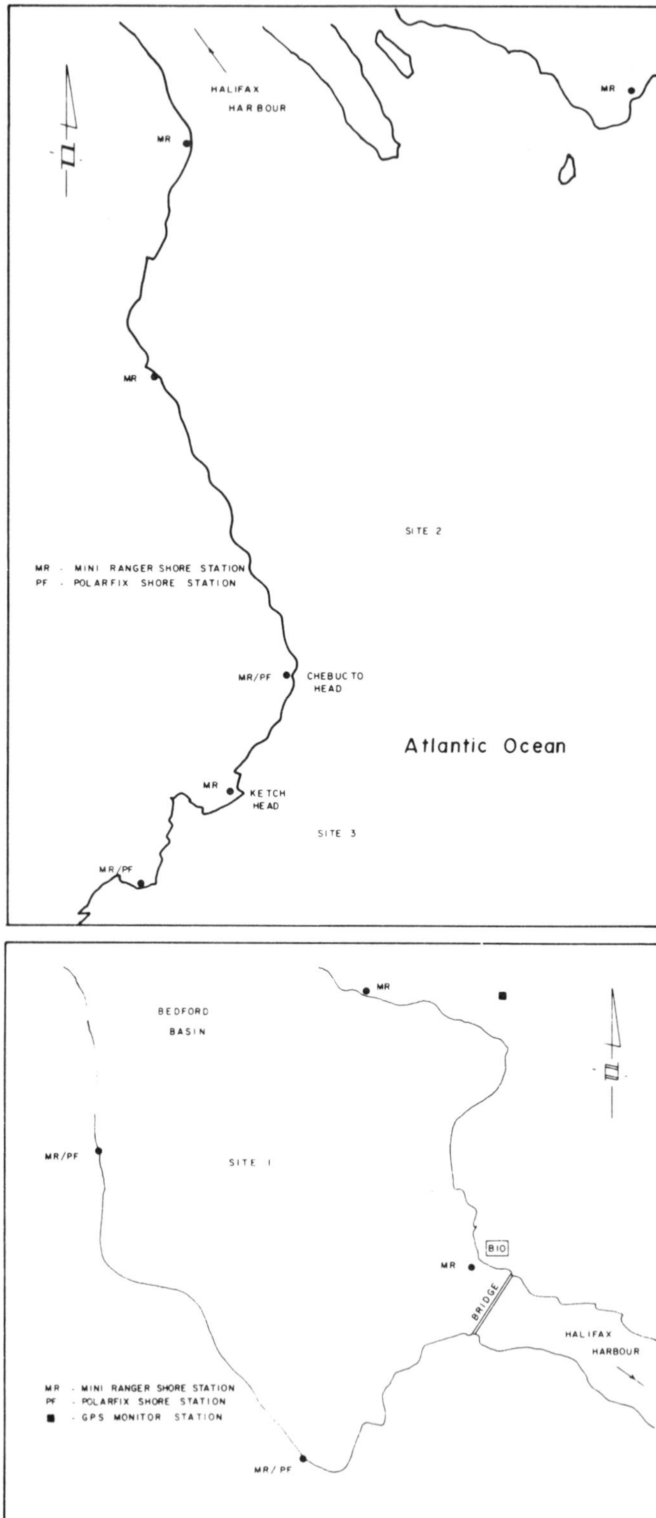

Fig. 1—Location of Shipborne Trials

of the shore-based Mini-Ranger system used to provide independent vessel positions, represented other important advantages. The second and third sites, located in the open sea outside Halifax Harbor, namely off Chebucto Head and Ketch Head, provide rougher sea states as required to evaluate thoroughly the performance of GPS. The two vessel types used, namely a 38 m hydrographic ship and a 10 m survey launch, are the most common used by CHS. GPS and shore-based radionavigation system data were collected simultaneously for comparison purposes. The four days during which simultaneous GPS, Mini-Ranger and Polarfix data were collected with the survey launch are given in Table 2.

The GPS equipment used for the trials consisted of two TI4100 GPS receivers on loan from the Geodetic Survey of Canada. Many of the advantages and disadvantages of this receiver for shipborne applications are discussed in [2]. Since the use of the differential method was required to reach the accuracy objective, a TI4100 unit was used throughout the project at a Monitor station located on the roof of Nortech Surveys' warehouse in Dartmouth. The second unit was successively mounted on the MAXWELL and HERON. The distance between the Monitor station and the vessel varied between 5 km when the sea trials were being conducted in Bedford Basin to 28 km when the vessel was operating off Ketch Head. On the vessels, the GPS antenna was mounted on the highest mast available to minimize shading effects (see Table 1).

With the TI4100, four User Dynamic (U.D.) modes are available, namely 0, 1, 2, and 3. These correspond to bandwidths of 0.7, 5, 8 and 16 Hz, respectively. The maximum accelerations allowed for each U.D., in order to avoid loss of phase lock, are 0, 6, 15, and 40 m/s^2. Phase noise increases with bandwidth. At the Monitor station, the TI4100 unit operated continuously in dual frequency P-code mode in U.D. 0. On the launch, data were collected in C/A code mode, and in U.D. 1, 2, and 3, as shown in Table 2. The P-code mode was selected for a 2 h observation period on Day 322 to assess its performance as compared to C/A code. The periods of time during which data were collected for each trial are also given in Table 2. The carrier phase measurements were collected during each trial. Both codes and various U.D. modes were also tested on the MAXWELL.

During November 1986, six healthy GPS satellites were available. These were PRN No. 3, 6, 9, 11, 12, and 13. The Geometric and Horizontal/Time Dilution Of Precision (GDOP and HTDOP) for these six satellites are shown in

Table 2—Description of Survey Launch Trials

DAY[1]	SITE[2]	CODE	U.D.[3]	OBS. PERIOD (h)
319	1	C/A	2	1.0
320	1	C/A	1	2.5
320	1	C/A	2	1.0
321	3	C/A	2	1.5
321	3	C/A	3	1.2
322	3	P	3	2.0
322	3	C/A	3	1.5

1 Julian Day—Day 319 is November 15
2 Bedford Basin-1, Chebucto Head-2, Ketch Head-3
3 User Dynamics Mode on TI4100 GPS Receiver

Figure 2. All trials were conducted between 12:00 and 18:00, GMT, i.e, between 8:00 and 14:00 AST (Atlantic Standard Time). The GDOP (latitude, longitude, height, and time unknown) varied from 4 to infinity during that period while the HTDOP (latitude, longitude, and time unknown, height fixed) ranged from 2 to 7. The satellite configuration was therefore adequate for the trials since the height was constrained when calculating the GPS horizontal positions of the vessels. Satellites at elevation angles less than 5 degrees were not used.

Up to three shore-based navigation systems were deployed during the trials to obtain the horizontal positions of the vessels independently from GPS, namely one Mini-Ranger Falcon 484 and two Polarfix systems. The Mini-Ranger system, which operates at C-band (5410 to 5600 MHz), is manufactured by Motorola Inc., USA, and has a quoted range accuracy of 2 m. Ten transmitters were deployed in such a way as to have access to four ranges at each of the three trial sites, as shown in Figure 1. Sufficient redundant observations were therefore available to monitor effectively the quality of ship's position accuracy. The system was calibrated with a Wild DI-20 EDM unit prior to and immediately following the trials. The Mini-Ranger antenna onboard the vessels was offset (laterally) by some 60 cm from the GPS antenna. The performance of the equipment was satisfactory throughout the trials.

Polarfix is a laser-based system manufactured by Krupp Atlas Elektronik Gmbh, FRG. A single shore station measures both range and direction to prism reflectors installed on the vessel. In this case, the reflectors were mounted on

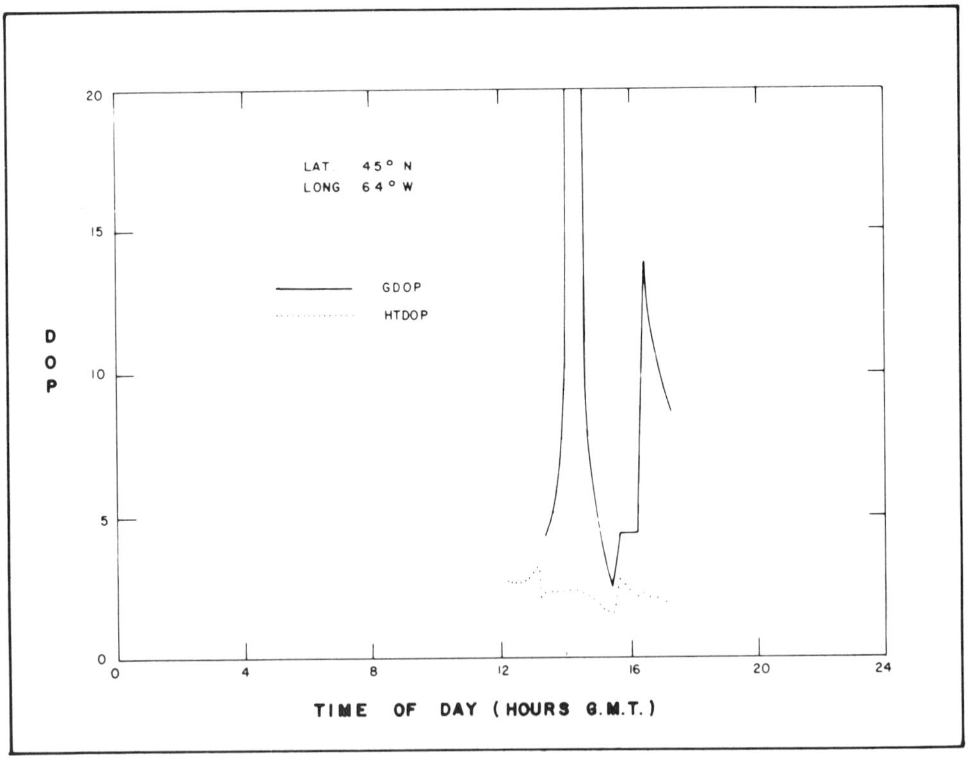

Fig. 2—GDOP and HTDOP of GPS Satellites During Shipborne Trials

the same vertical axis as the GPS antenna and no lateral offset was present. The position accuracy obtained during evaluation tests conducted by CHS was 4 to 6 m for ranges of 3 to 5 km [19]. During the launch trials, two Polarfix systems were used independently in addition to the Mini-Ranger system. On days during which the sea was relatively rough, the noise level on the Polarfix-derived launch positions increased sharply, presumably due to signal reflection on the waves and/or loss of prism lock. This was to be anticipated since the Polarfix was designed primarily for high accuracy barge positioning, in which case the system dynamics characteristics are lower than those encountered on a survey launch operating at 15 kn. Both Polarfix systems were calibrated prior to the experiments using a DI-20 EDM unit and a Wild T2 theodolite.

The WGS72 coordinates available initially for the Mini-Ranger and Polarfix shore station were third order. Selected shore stations were observed by GPS in differential static mode from the Monitor station. The carrier phase observations were processed using NOVAS, a NORTECH developed software package based on the double differencing technique [23]. This resulted in average corrections of 1.8 m (northing) and −1.1 m (easting) to the coordinates of the shore stations. This brought the shore station coordinates and the monitor station into a homogenous reference system. The contribution of shore station coordinate errors to the Mini-Ranger and Polarfix derived vessel position errors was expected to be approximately 50 cm.

A Telemetric Remote Inclination Monitor (TRIM) attitude sensor, manufactured by Measurement Devices Ltd., U.K., and made available for the trials by The University of New Brunswick, was used on both vessels to measure pitch and roll motion. These measurements were available at a rate of 5 Hz through an RS232 interface. The quoted accuracy for the unit is 0.05 degrees. This attitude sensor was used to assess its performance on hydrographic ships and survey launches, investigate the ability of GPS to detect pitch and/or roll motion, and study the feasibility of using such a low cost sensor to relate the swaying mast mounted GPS antenna positions of those of the ship's hydrographic sensors.

Measurements from both the shore-based systems and attitude sensor were time-tagged with GPS time using a Nortech-designed Computer Addressable Precision Timing System (CAPTS) interfaced to an HP9826 computer. Position errors due to the time interface limitations of all systems are expected to be well below 50 cm. All data were recorded on DC300XL magnetic tape cartridges using DIGI-DATA recording units. HP220 computers were used, at both the monitor station and on the vessels, to perform the real time processing of GPS code and carrier measurements. The differential corrections were applied post-mission. Single point GPS positions of the vessels were compared in real time to the Mini-Ranger and Polarfix positions for quality control purpose.

DATA PROCESSING, RESULTS AND ANALYSIS

The agreement between Mini-Ranger or Polarfix positions and GPS single point positions, based on combined code and phase measurements, was of the order of 15 m, which is satisfactory in view of the effects of orbital errors on single point GPS positioning. This real time result also confirmed that none of the systems was affected by large errors. The Mini-Ranger and Polarfix derived positions were intercompared to assess their external accuracy under prevailing conditions. The results for the four days of launch trials are summarized in Table 3. The reason for the lower sample number when comparing the two Polarfix systems is the frequent loss of prism reflector lock on either system, as discussed in the previous section. The agreement, in terms of rms difference, between any two of the three systems varies between 1.4 and 3.9 m in northing and between 0.7 and 2.4 m in easting. The mean differences vary from 0.0 to 1.8 m. A bias of about 1 m is apparent between Mini-Ranger and Polarfix derived northing coordinates. This is probably due to a combination of calibration and residual biases in the shore stations. Overall, the results are satisfactory in view of the estimated error components of the systems, as discussed in the previous section. The total rms differences were calculated as 2.1 m in northing and 1.8 m in easting. If we assume that the three systems are of comparable accuracy, the accuracy of a system is then 1.5 m in northing and 1.3 m in easting at the 1 sigma level. The corresponding two-dimensional accuracy of a system is therefore 3.5 m at the 95 percent confidence level.

Several data processing techniques now are available for differential kinematic GPS positioning using a combination of pseudorange and phase measurements [e.g., 2, 8, 17, 18, 21]. The technique used here, which is described in [12,13], was developed and implemented specifically for shipborne applications. The major objectives were to obtain a high level of reliability, while maintaining a 1 to 2 m positioning accuracy at the one sigma level in each of the two horizontal coordinates, with a mathematical formulation suitable for real time application on field portable computers. The combination of pseudorange and phase measurements is based on the method proposed in [6]. Pseudorange errors are filtered using phase measurements prior to their reduction to obtain position fixes. When a cycle slip is detected on a specific satellite, the filter is reset for this satellite. The respective weights assigned to a measured pseudorange and phase are a function of the time elapsed since the initial phase observation or since the latest cycle slip occurrence. A maximum weight

Table 3—Intercomparison of Shore-Based Positioning Systems

		PF1–PF2[3]			MR–PF1			MR–PF2		
Day	N/E[1]	n.s.[2]	mean	rms	n.s.	mean	rms	n.s.	mean	rms
319	dN	344	−0.9	1.4m	1221	−1.7	2.0m	2041	−1.3	1.8m
	dE		0.8	2.2		0.0	0.7		0.7	1.5
320	dN	144	−0.6	3.9	1041	−1.4	2.1	3874	−1.4	2.7
	dE		0.5	1.3		0.1	1.1		−0.8	1.4
321	dN	4771	0.1	2.0	6707	−0.9	1.5	5038	−1.3	2.6
	dE		−0.3	1.9		0.6	1.2		1.2	2.4
322	dN	2070	0.7	1.4	4307	−0.7	1.4	4696	−1.8	2.7
	dE		−0.4	1.5		0.5	1.6		1.1	2.3
All	dN	7329	−0.2	1.8	13276	−1.2	1.6	15699	−1.4	2.6
	dE		−0.2	1.8		0.2	1.3		0.2	2.1

1 N - northing, E - easting
2 n.s. - number of samples
3 PF1 - Polarfix 1, PF2 - Polarfix 2, MR - Mini-Ranger

of 0.999 on phase measurements is reached after a few hundred cycle slip-free phase observations. The corresponding weight on pseudorange thus ranges from 1.000 for an initial measurement, to 0.001, after a few hundred cycle slip free measurements. The height of the launch's GPS antenna was constrained to its known value with an assigned standard deviation of 1 m to account for heave motion.

The cycle slip detection method implemented in the processing technique described above is the phase velocity trend method described in [12]. The version used is specific to the TI4100 receiver. The accumulated phase difference measured over a code measurement interval (3.0 s in U.D. 0 and 1.0, and 1.2 s in U.D. 2 and 3) is compared with the phase difference predicted from phase velocities obtained at the two code measurement epochs. At a code measurement epoch, the phase velocity is derived from phase measurements available 80 ms before and after the code Fundamental Time Frame, respectively. If cycle slips occur over this 160 ms interval, they are easily detectable. The method is based on phase velocity linearity assumption between code measurement epochs (1.2 or 3.0 s). The cycle slip detection threshold is limited by this assumption and by phase measurement noise. For shipborne applications, the linearity assumption will be a function of the ship's attitude motion amplitude and period. If the number of cycle slips at a specific epoch is relatively low, say less than 5, it may go undetected. This will affect the accuracy of the corresponding range by up to 1 m. If a receiver other than the TI4100 is used and no phase measurements are available over short intervals but only every 1 s, the cycle slip detection threshold will go up. The use of a higher data output rate would evidently assist in lowering this threshold. The results presented in this section show that the detection method used here was effective to at least the 5 to 10 cycle slip level.

The GPS differential method utilized is well known, e.g., [7]. A differentially corrected position of the vessel was obtained by applying satellite phase smoothed range corrections derived at the monitor station to corresponding phase smoothed pseudorange observations on the vessels. The differential range corrections at the monitor station were obtained by differencing observed and calculated ranges. The differential pseudorange corrections used on the launch were based on 30 s averages at the monitor station. A software package developed by Nortech for CHS was utilized for the data processing. This software operates on Hewlett Packard Series 200 field portable computers.

The frequency of cycle slip occurrence was a function of several parameters, including the U.D. mode utlized, the roughness of the sea, and satellite elevation. The worst case in this respect was on Day 322 when trials took place off Ketch Head. The two previous days had been stormy and the sea was still relatively rough with a wind speed of 12 kn. The frequency of cycle slip occurrence for each satellite tracked on that day is shown in Figure 3, together with the corresponding satellite elevations. The interval between cycle slips on individual satellites varies from a few minutes to over 30 min. The frequency is significantly higher when satellites are at low elevations. This is due to the amplitude response pattern of the GPS antenna. Satellites were observed down to an elevation of 5 degrees above the horizon. Considering that the roll amplitude reached 20 degrees on Day 322, the antenna would have tracked satellites as low as 15 degrees below its horizon. The amplitude response of the TI4100

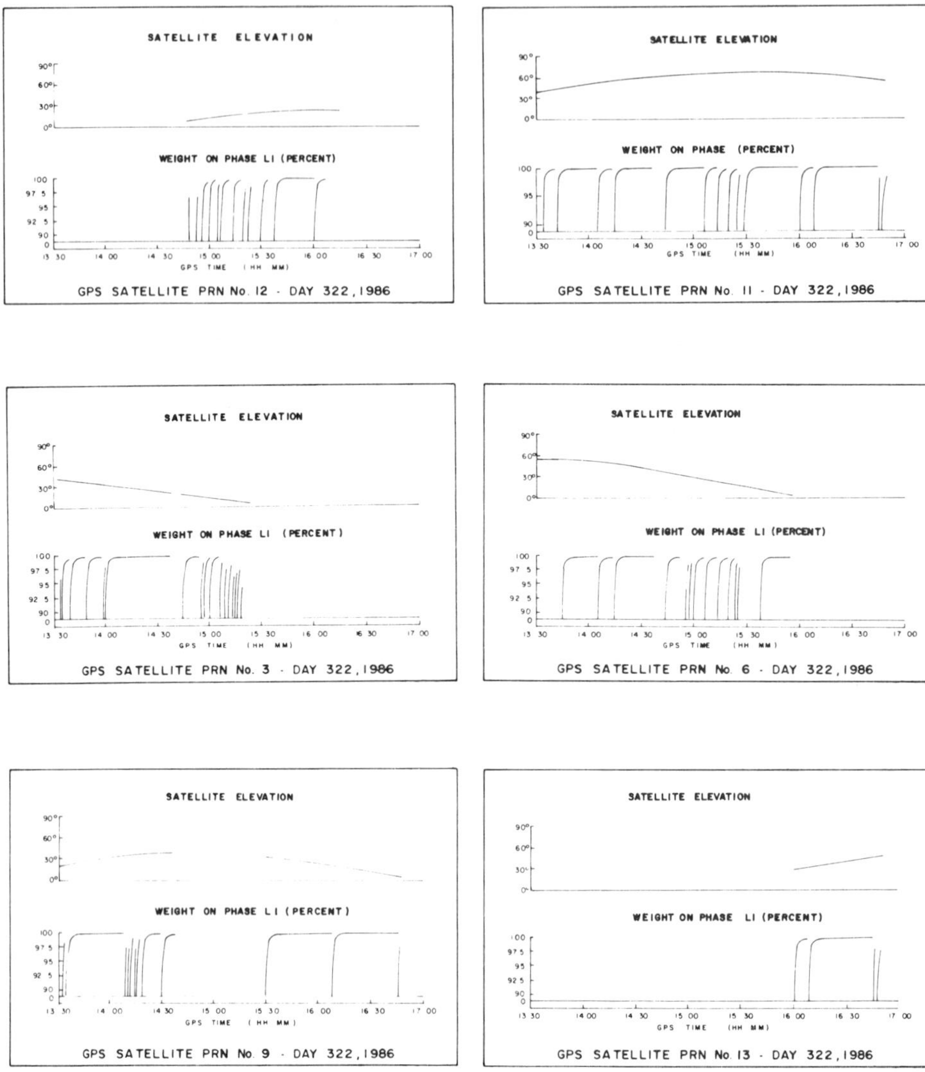

Fig. 3—Frequency of Cycle Slip Occurrence on Launch. The weight on phase data was reset to 0 percent when cycle slips occurred. (The P-code was used prior to 15:30, and the C/A code afterwards.)

antenna is down to −5 dB at −15 degrees as compared to 0 dB for elevations greater than 20 degrees [14]. The resulting signal loss contributed to the increased cycle slip occurrence. At 15:30, the TI4100 receiver on the launch was switched from P to C/A code mode. Cycle slips became less frequent. This is probably due to the 160 ms time interval required for the TI4100 to cycle through four satellites in P code (L1/L2) mode, as compared to 80 ms in C/A (L1) code mode, the nominal dwell time per satellite being 20 ms for each frequency. The use of a multi-channel receiver should result in a further decrease of cycle slips, due to continuous tracking of the satellites.

Simultaneous cycle slips on all satellites being tracked rarely occurred. More cycle slips would have occurred if a lower U.D. mode than 2 or 3 had been used. Unless the sea was relatively calm, as in Bedford Basin on Day 320 when there was practically no wind, the use of U.D. 2 or 3 generally resulted in fewer cycle slips. When cycle slips occurred, the filter for the affected satellite was reset as discussed earlier. The accuracy of differentially corrected positions obtained immediately after cycle slips on all satellites simultaneously would be the same as that obtained in pure code mode, namely 5 to 8 m (in each of the two horizontal coordinates) in C/A code and 3 to 5 m in P code [12]. If less than four satellites are affected, as it was generally the case during the trials, the accuracy will be higher than the above. At the monitor station, where the TI4100 operated in U.D. 0, very few cycle slips occurred throughout the trial period. The P code trial data of Day 322 was processed using L1 only since over such a short distance (approximately 25 km), the differential effect of the ionosphere is negligible when considering the accuracy level being sought.

Multipath effects can affect pseudoranges significantly. A major advantage of combining carrier phase measurements with pseudoranges is to dramatically decrease these effects, provided the frequency of cycle slips remains low. When cycle slips occur, the filter is reset and the relative weight of pseudoranges which could be contaminated by multipath increases. No significant multipath effects were detected on the launch. A cursory analysis of the pseudorange data at the monitor station revealed significant multipath effects. However, cycle slips were infrequent and carrier phase measurements were presumably successful in keeping pseudorange multipath effects on launch positions within the accuracy level being sought.

A comparison between GPS and shore-based system derived positions is presented in Table 4. No rejections were made of the positions obtained immediately following cycle slips. The mean differences reach 2 m and generally are below 1 m. The rms differences vary between 0.7 m and 3.8 m. The overall rms differences for all comparisons are 2.5 m and 2.1 m in northing and easting, respectively. These results are in agreement with those obtained in the Baltic Sea onboard a small surveying vessel using TI4100 receivers for GPS positioning and Polarfix as comparison [22].

Table 4—Comparison of GPS and Shore-Based Positioning Systems

		PF1–GPS			PF2–GPS			MR–GPS		
Day	N/E[1]	n.s.[2]	mean	rms	n.s.	mean	rms	n.s.	mean	rms
319	dN	593	0.9	1.5m	1068	0.1	0.7m	2093	−1.1	1.6m
	dE		0.0	3.8		0.9	1.6		0.1	0.8
320	dN	440	0.4	3.2	1807	−1.3	3.0	5105	−2.0	3.2
	dE		−0.7	2.8		0.1	1.2		−1.2	2.3
321	dN	3090	−1.1	3.8	1736	−0.1	2.6	4932	−1.3	2.1
	dE		0.4	3.3		0.3	2.0		1.0	1.7
322	dN	3401	−0.2	3.2	2817	−0.6	1.7	6446	−0.9	1.3
	dE		0.4	3.0		0.3	1.5		0.7	1.1
All	dN	7524	0.6	3.4	7428	−0.1	2.2	18576	−1.1	2.2
	dE		0.0	3.2		0.4	1.6		0.4	1.6

1 N - northing, E - easting
2 n.s. - number of samples

When the error contribution of the shore-based systems, namely 1.5 m and 1.3 m in northing and easting respectively, are taken into account, external accuracies of 2.0 m and 1.6 m in northing and easting, respectively, are obtained for differential GPS. This is considered satisfactory in view of the frequent cycle slips which occurred during some of the trials. The corresponding two-dimensional accuracy at the 95 percent confidence level is 4.5 m. The 5 m accuracy objective was therefore met.

The MAXWELL trials were conducted prior to the launch trials and minor problems occurred with the time tagging of the shore-based systems. However, the resulting effects could be removed during post-processing and an analysis of the results showed that the performance of the differential GPS positioning technique was similar to that obtained for the launch. Cycle slips were in fact less frequent than on the launch, due to the lower dynamics of the GPS antenna on this ship.

DETECTION OF ATTITUDE AND HEAVE MOTION

The TRIM attitude sensor unit was mounted successively on the MAXWELL and the HERON. The largest amplitudes were obtained on Days 315 and 322 for both vessels. On these two days, the trials were conducted on sites 2 and 3, and the wind speed was 6 and 12 kn, respectively. The peak-to-peak roll amplitude exceeded 15 degrees and 30 degrees for the MAXWELL and the HERON, respectively. The corresponding pitch values were 3 degrees and 15 degrees. On days when launch tests were conducted in a relatively rough open sea, the sensor roll and pitch data became noisy as compared to data obtained on other days, and overshooting seemed to affect the measurements. The period of the roll component was found to be approximately 7 s and 2.5 s for the MAXWELL and HERON, respectively. The difference is primarily due to vessel type and sea state. The period of the pitch varied significantly from day to day and is a function of the wave frequency. The amplitude variations in roll and pitch are mostly due to ship heading changes.

Since roll motion results in GPS antenna motion perpendicular to the ship's track, a comparison was made between differentially corrected horizontal GPS positions, derived at intervals of 1.2 s, with attitude sensor derived roll measurements. Sample results are shown in Figure 4 for the MAXWELL and HERON operating off Chebucto Head and in Bedford Basin on Days 311 and 325, respectively. The wind speed was 7 kn and the sea swell reached 4 m to 6 m on Day 311. The sea state in Bedford Basin on Day 325 was considered moderate despite a 20 kn wind. The non-periodic deviations of the GPS positions from the straight line are due to the nonlinearity of the ship's track. In the case of the MAXWELL, where the period of the roll motion was approximately 7 s, the successive 1.2 s interval GPS positions agreed fairly well with the sensor roll measurements made at a 5 Hz rate. In the case of the HERON, where the motion period was much shorter, namely 2.5 s, the GPS position sampling rate was insufficient to adequately model the sensor roll measurements. The use of GPS data to detect the ship's pitch motion has not been attempted herein. This would be a much more difficult task since the motion is along track. Overall, the preliminary results presented here demonstrate the potential of a low cost

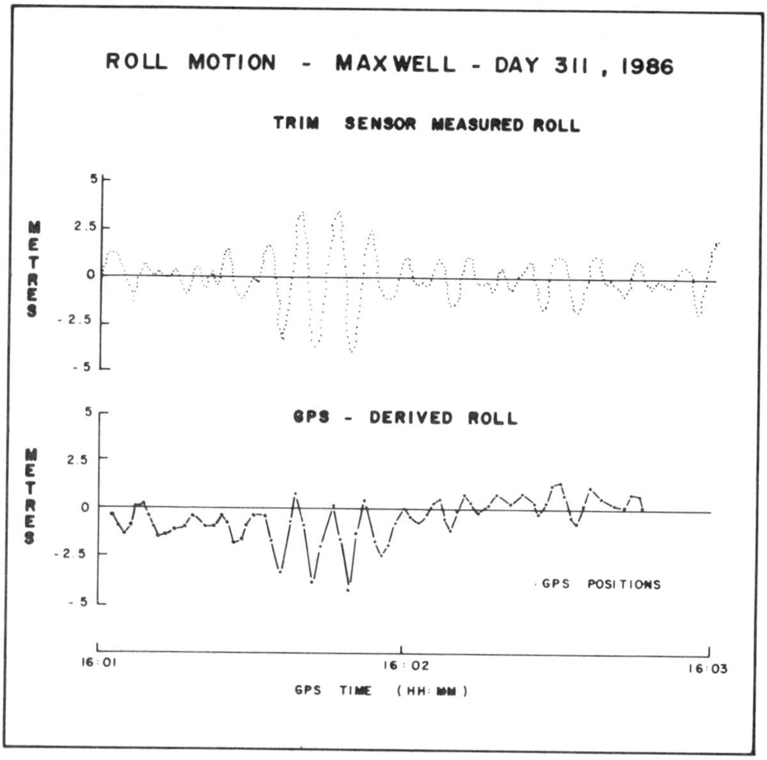

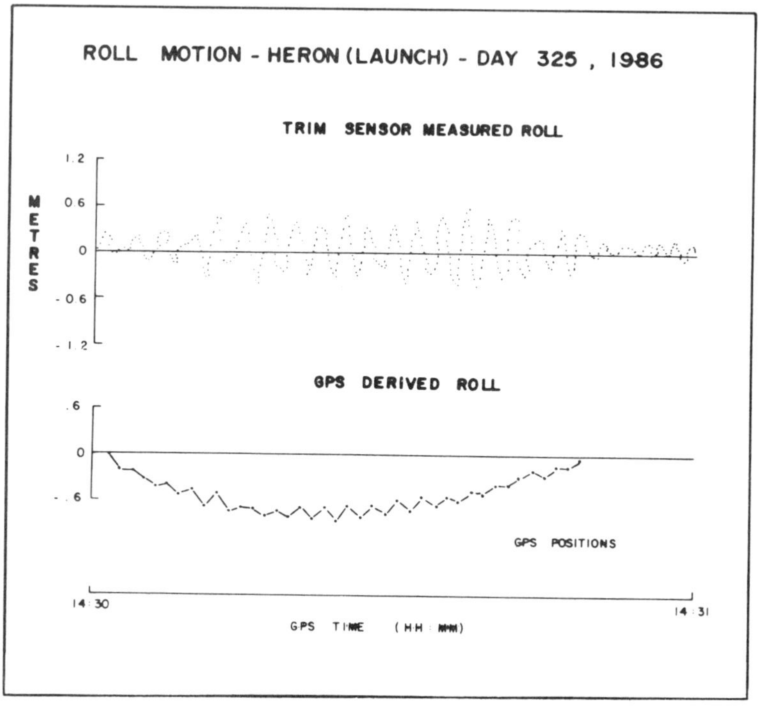

Fig. 4—Attitude Sensor and GPS-Derived Roll Motion

attitude sensor unit to measure ship's roll and pitch, provided that overshooting effects can be effectively controlled.

GPS also was tested to estimate ship's heave motion. Sample results obtained with the MAXWELL on Day 315 over a 1 min period are shown in Figure 5. No cycle slips occurred during this period and a four-dimensional differential solution was calculated. The antenna motion reached 1 m in amplitude. Similar results are shown in [15]. The results cannot be compared to independent measurements since no heave compensator was used during the trials. However, in the absence of cycle slips, the differential positioning technique used for the test depends mostly on carrier phase derived range differences and the short term relative positioning accuracy can be expected to be of the order of 5 to 10 cm. The variations shown in Figure 5 represent antenna motion. To obtain actual heave motion, the effect of roll and pitch would have to be accounted for. The combination of GPS and attitude sensor measurements would possibly be an effective solution to this problem.

CONCLUSIONS

The shipborne differential GPS tests described in this paper resulted in verifiable external accuracies of 2.0 m and 1.6 m (one-sigma level) in northing and easting, respectively. This corresponds to a two-dimensional accuracy of 4.5 m at the 95 percent confidence level, which meets the accuracy requirements of the project. The differential positioning technique used, which is used on the combination of carrier phase and pseudorange measurements, proved satisfactory for launch operations in the open sea, despite the frequent occurrence of cycle slips. The trials were conducted with TI4100 multiplex receivers. The use of dedicated channel receivers is likely to result in fewer cycle slips as a consequence of the improved signal-to-noise ratio, provided that high gain antennas at low elevation (down to −15 degrees) are used. The availability of receivers with high data output rate (> 1 Hz) would improve the cycle slip detection threshold. Since simultaneous cycle slips on several satellites is a

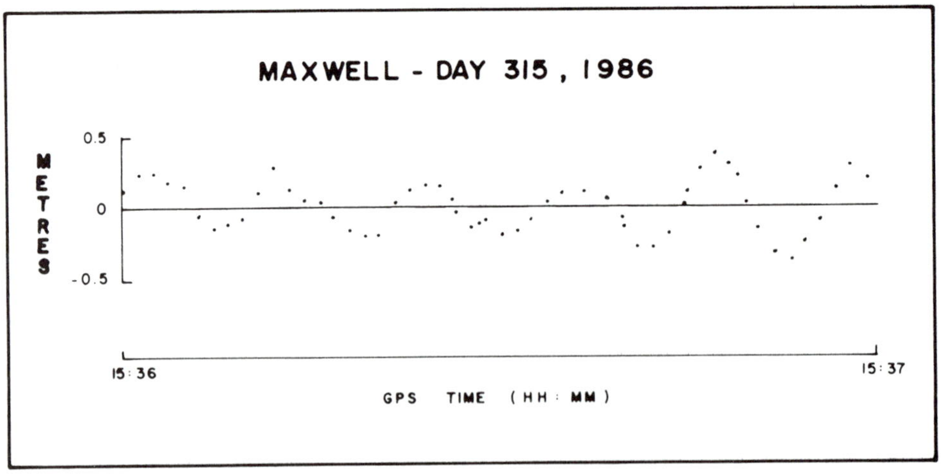

Fig. 5—GPS-Derived Heave Motion

rare occurrence, the use of high data output rate receivers with up to seven or eight dedicated channels would contribute to enhancing both accuracy and reliability when the full constellation is available, e.g., [14]. In such a case, the relative accuracy of differential GPS in the marine environment would be better than 1 m. Recent simulations and land kinematic tests performed by Nortech Surveys and other investigators, e.g., [20] have resulted in relative accuracies of 10 to 20 cm.

The preliminary attitude and heave motion results presented in the paper indicate that unaided differential GPS can be used to estimate ship's roll and heave motion. However, a low cost attitude sensor, such as the one tested during these trials, may be the most cost effective manner to measure attitude and heave components if combined with the output of a differential GPS system. Such a system would reduce, if not eliminate, the need for heave compensators for accurate in-shore hydrographic surveying.

ACKNOWLEDGEMENTS

The authors wish to acknowledge the efforts of Mr. R. M. Eaton, Head, Navigation Group, CHS/Bedford Institute of Oceanography, in his support of marine GPS development throughout the past seven years. Both his extensive expertise and unrelenting dedication have steered the program towards realistic, yet ambitious and challenging goals. The support and dynamism of The University of New Brunswick, in particular of Professor D. E. Wells, throughout the program, is also acknowledged. Work on the project reported in this paper was partially funded by the O.E.R.D. Program of the Department of Energy, Mines and Resources, and the Unsolicited Proposal Office of the Department of Supply and Services.

REFERENCES

1. Broughton, D. W., *Standardization within NATO on the Method of Expressing Navigation Accuracies,* NAVIGATION, Vol. 33, No. 2, 1986, pp. 83-89.
2. Cannon, M. E., *Kinematic Positioning Using GPS Pseudorange and Carrier Phase Observations,* Publ. No. 20019, Department of Surveying Engineering, The University of Calgary, 1987.
3. Casey, M. J., and Vosburgh, J., *Chartmaking with LARSEN,* The Canadian Surveyor, Vol. 40, No. 3, 1986.
4. Casey, M., Eaton, M., Kielland, P., MacDonald, G. D., and Eaton, G., *The Canadian Hydrographic Service GPS R&D Program.* Proceedings of Canadian Hydrographic Conference 1987 (Burlington, Ontario), Canadian Hydrographic Service/Canadian Hydrographic Association, Ottawa, 1987.
5. Hagglund, J., Lachapelle, G., and Eaton, R. M., *Loran-C Calibration off East Coast Using Navstar Differential Navigation,* Offshore Resources, Vol. 3, No. 1, 1985, pp. 30–34.
6. Hatch, R., *The Synergism of GPS Code and Carrier Measurements,* Proceedings of Third International Geodetic Symposium on Satellite Doppler Positioning, DMA/NOS, Washington, D.C.
7. Kalafus, R. M., Van Dierendonck, A. J., and Pealer, N. A., *Special Committee 104—Recommendations for Differential GPS Service,* NAVIGATION, Vol. 33, No. 1, 1986, Washington, D.C.

8. Kleusberg, A., *Kinematic Relative Positioning Using GPS Code and Carrier Beat Phase Observations,* Marine Geodesy, Vol. 10, 1986, pp. 256–261.
9. LaChapelle, G., Beck, N., Heroux, P. and Wells, D. E., *GPS Aided Navigation with a Slow Switching Receiver,* Presented at International Symposium on Point Positioning in Marine Geodesy, Maracaibo, February 23–27, 1983.
10. LaChapelle, G., Hagglund, J., Jones, H., and Eaton, M., *Differential GPS Marine Navigation,* Proceedings of Position Location and Navigation Symposium 1984, IEEE, 1984, pp. 245–255.
11. LaChapelle, G., Lethaby, J., and Casey, M., *Airborne Single Point and Differential Navigation for Hydrographic Bathymetry,* The Hydrographic Journal, No. 34, 1984.
12. LaChapelle, G., Hagglund, J., Falkenberg, W., Bellemare, P., Casey, M., and Eaton, M., *GPS Land Kinematic Positioning Experiments,* The Hydrographic Journal, No. 42, 1986.
13. LaChapelle, G., Falkenberg, W., and Casey, M., *Use of Phase Data for Accurate Differential GPS Kinematic Positioning,* PLANS '86 Symposium Proceedings, IEEE, New York. Bulletin Géodésique, Vol. 62, 1987.
14. LaChapelle, G., Casey, M., Eaton, M., Kleusberg, A., Wells, D. E.,and Tranquilla, J., *GPS Marine Kinematic Positioning—Accuracy and Reliability,* The Canadian Surveyor, Vol. 41, No. 2, 1987.
15. Nard, G., *Reliable High Accuracy Long Range Real Time Differential GPS Using A Lightweight High Frequency Data Link,* PLANS '86 Symposium Proceedings, IEEE, New York, 1986, pp. 377–388.
16. Nortech Surveys, *Study of Ionospheric Effects on Navstar/GPS Users in Canadian High Latitude Regions,* Report prepared for Bedford Institute of Oceanography under DSS Contract 10SC.FP901-4-R534, 1985.
17. Pietraszewski, D., Spalding, J., Viehweg, C. and Luft, L., *U.S. Coast Guard Differential GPS Navigation Field Test Findings,* NAVIGATION, Spring 1988.
18. Remondi, B. W., *Performing Centimeter-Level Surveys in Seconds with GPS Carrier Phase: Initial Results. Navigation,* NAVIGATION, Vol. 32, No. 4, 1985, Washington, D.C.
19. Sanfaçon, R., *Présentation et Évaluation de la Précision du Polarfix de Krupp-Atlas Lighthouse,* Journal of the Canadian Hydrographic Association, No. 34, 1986, pp. 11–17.
20. Schwarz, K. P., Cannon, M. E., and Wong, R. V. C., *The Use of GPS in Exploration Geophysics—A Comparison of Kinematic Models,* Presented at General Assembly of International Association of Geodesy, Vancouver, August 9–22, 1987.
21. Seeber, G., Schuchardt, A., and Wübbena, G., *Precise Positioning Results with TI4100 GPS Receivers on Moving Platforms,* Proceedings of Fourth International Geodetic Symposium on Satellite Positioning, Austin, Texas, 1986.
22. Seeber, G., Schuchardt, A., and Wübbena, G., *Precise GPS-Aided Marine Positioning Developments and Results,* Presented at International Symposium on Marine Positioning (INSMAP), Reston, Virginia, October, 1986.
23. Wanless, B., and LaChapelle, G., *NOVAS—An Automated Program for the Precise Reduction of GPS Static Carrier Phase Observations,* Accepted for publication in *Manuscripta Geodaetica,* November 1987.
24. Wells, D. E., and Grant, S. T., *An Adaptable Integrated Navigation System: BIONAV,* Proceedings of Colloquium III on Petroleum Mapping & Surveys in the 80s (Banff, Alberta), Canadian Petroleum Association, Calgary, 1981, pp. 195–207.
25. Wong, R. V. C., Schwarz, K. P., Hagglund, J., and LaChapelle, G., *Integration of Inertial and GPS-Satellite Techniques for Precise Marine Positioning.* Marine Geodesy, Vol. 9, No. 2, 1985, pp. 213–226.

NAVIGATION: *Journal of The Institute of Navigation*
Vol. 34, No. 1, Spring 1987
Printed in U.S.A.

Aircraft Positioning Using Global Positioning System Carrier Phase Data

WILLIAM B. KRABILL
NASA, Goddard Space Flight Center, Wallops Island, VA
CHRESTON F. MARTIN
EG&G Washington Analytical Service Center, Inc., Lanham, MD
Received March 1987

ABSTRACT

Results from a flight test of a GPS carrier phase tracking receiver operated on a NASA Orion P-3 aircraft are presented. The trajectory of the aircraft relative to a second, fixed receiver has been determined. The test was made over water, and the GPS vertical trajectory is compared to airborne LIDAR measured altitude data from the NASA Airborne Oceanographic LIDAR (AOL). The two data sets compare with a relative accuracy of 12 centimeters RMS, under poor GPS satellite geometry for vertical positioning. Analysis indicates that 1 to 2 centimeters relative vertical positioning is achievable with carrier phase tracking receivers and good GPS geometry.

INTRODUCTION

NASA has many reasons for interest in the Global Positioning System (GPS). From positioning of spacecraft and aircraft to making geodetic measurements, GPS has the promise for revolutionizing the way in which NASA conducts a significant portion of its work. The planned mode of precise positioning using GPS, in which a pseudo-random code (P-code) is tracked from 4 satelliltes on two L-band frequencies, has been tested by the Department of Defense (DoD) and found to meet accuracy specifications (~10 meters circular error).

For surveying applications, it has long been recognized that two GPS receivers could be used for relative positioning with accuracies considerably better than would be achievable with a single receiver. This is due to the fact that many systematic errors (satellite ephemeris, satellite clock, propagation) have very nearly the same effects on measurements made to receivers separated by distances of a few tens of kilometers or less. With surveying as the basic intended application, receivers have recently been developed which provide data with significantly better precision than P-code ranges. This precision is achieved via tracking the phase of the L-band carrier, with noise levels of approximately two degrees obtained. Ambiguous pseudoranges are thus measured with a noise level on the order of a millimeter for currently available receivers. With data sets taken over a period of a few hours, cycle ambiguities between receivers at the two ends of a baseline of a few tens of kilometers or less can be reliably determined along with the relative receiver coordinates. Accuracy for land based surveying applications with such systems has been demonstrated at the few centimeter level.[1,2]

Since NASA has great interest in precise positioning of moving vehicles, both aircraft and spacecraft, plans were made to test the operation of these new GPS receivers with one receiver on a moving platform (aircraft) and a stationary receiver located at a surveyed reference point. An obvious problem with such a test, which has as its primary objective to assess position measurements at the sub-ten-centimeter level, is "to what does one compare the data?" No other tracking system for an aircraft can compete at this level of precision, much less act as a standard. In order to assess the vertical positioning accuracy (and, by inference, the horizontal positioning accuracy) that is potentially achievable with GPS, the concept adopted was to operate an aircraft equipped with a precision laser ranging system over a relatively calm water surface. A comparison could then be made between the aircraft altitude determined by the laser system and the set of GPS-determined vertical coordinates.

Height accuracies produced with the calibrated laser system should be sub-ten-centimeter after smoothing and should possess noise levels which are on the order of 1–2 centimeters (or less). Assuming that GPS phase ambiguities can be resolved, the aircraft height accuracy relative to the reference station should be limited primarily by the noise on the phase measurements and the satellite geometry. Since the aircraft receiver must have a considerably greater bandwidth than a stationary receiver, a noise level of 1–2 centimeters rather that a millimeter could be expected. The Vertical Dilution of Precision (VDOP) for good GPS geometry is less than one, leading to a height noise level of 1–2 centimeters. However, during periods of poor geometry, the VDOP can exceed ten.

FLIGHT TESTS

The GPS flight tests, which were conducted on July 31, 1985 at the NASA Wallops Flight Facility (WFF), involved a cooperative effort between NASA and NOAA's National Geodetic Survey (NGS). NASA provided a P3-A aircraft (NASA 428) and the Airborne Oceanographic LIDAR (AOL) and associated personnel while NGS provided two geodetic quality GPS receivers (model TI4100, made by Texas Instruments), associated operators and survey personnel. One receiver was installed on the WFF P-3 aircraft with the antenna near the AOL. The second (fixed) receiver was deployed near the P-3 hangar. Nearby Chincoteague Bay was selected as the site for the experiment for the following reasons: (1) elimination of effects of large waves, (2) negligible geoid slopes in the area, and (3) proximity to WFF. NGS deployed tide gauges and a GPS receiver in the area to further refine the measurement surface relative to the reference ground receiver.

The basic objectives in the GPS tests were to:

- Determine the performance of the aircraft receiver in a dynamic environment, particularly in its ability to maintain lock on all four satellites;
- Determine the precision of the aircraft heights based on GPS phase data;

- Determine the existence of systematic errors in the aircraft heights, assuming that phase ambiguities are resolvable; and
- Develop guidelines for future operational usage for minimizing or avoiding data problems.

Development of the mission plan for the flight test was predicated on procedures used by NOAA (and others) for ambiguity estimation. Recognizing the possibility that the airborne receiver might lose lock on takeoff or landing, it was planned to collect a sufficient amount of data prior to takeoff, and also after landing, to perform independent ambiguity estimates. Data requirements for each period were considered to be one hour or more. With bias estimates from both periods, a loss of lock upon takeoff or landing could be tolerated and still leave the inflight data wth resolved ambiguities. Loss of lock inflight was expected to result in ambiguity problems that probably would not be resolvable.

It should be noted, however, that the TI4100 receivers track the phase of both L_1 and L_2 from each satellite, and that it is possible to lose lock at one frequency while retaining lock at the other frequency. Since L_1 phase changes and L_2 phase changes should differ only slightly due to the effects of ionospheric propagation, a brief loss of lock by one channel over 10–15 seconds should be a problem only in the operational sense. If L_1 data is continuous it can be used to bridge the gap in the L_2 data. Such problems did indeed arise during these tests, and L_1 data was used to correct for missing L_2 phase data and vice versa. This process cannot be made with 100% confidence, however, since there is always the possibility of cycle slips, and there is no inflight capability of detecting them in the TI4100 when there is track on only one frequency.

Figure 1 shows the basic elements in the tests: the aircraft with GPS receiver, a ground receiver, and four GPS satellites. The flight path flown for the mission is shown in Figure 2, including the aircraft path from just before takeoff to the parked position after landing. Shortly after takeoff, the aircraft was over the Chincoteague Bay and flew up and down the Bay twice, with AOL laser data taken during each pass up the Bay and each pass down the Bay for a total of four passes and approximately 25 minutes of data. AOL data was not taken during aircraft turns. The data gap evident in the middle of one of the passes is due to GPS data lost during switchover from one cassette to another in the ground receiver, but with no apparent signal loss.

Figure 3 shows a chronology of events during the mission, including times at which there were problems with the GPS data on one or more channels. These data problems are summarized in Table 1. Because of several problems around the time the aircraft left the parked position, it was decided not to attempt to estimate and use ambiguities from data taken before takeoff. In fact, due to an unfavorable GPS orbit configuration, it would have been very difficult to estimate ambiguities with pre-takeoff data.

Table 1 indicates that there was track on all channels on at least one frequency from before takeoff to several minutes after the aircraft had reached the parked position. In addition, the inflight problems occurred only for SV9, which was the GPS satellite with the lowest elevation angle (~20 degrees) and these problems occurred during an aircraft turn between passes. Since there was continuous track at L_1 or L_2, the satellite must have remained visible to the aircraft antenna. Possible causes for the loss of lock include a weak signal, or multipath from reflection off some part of the aircraft.

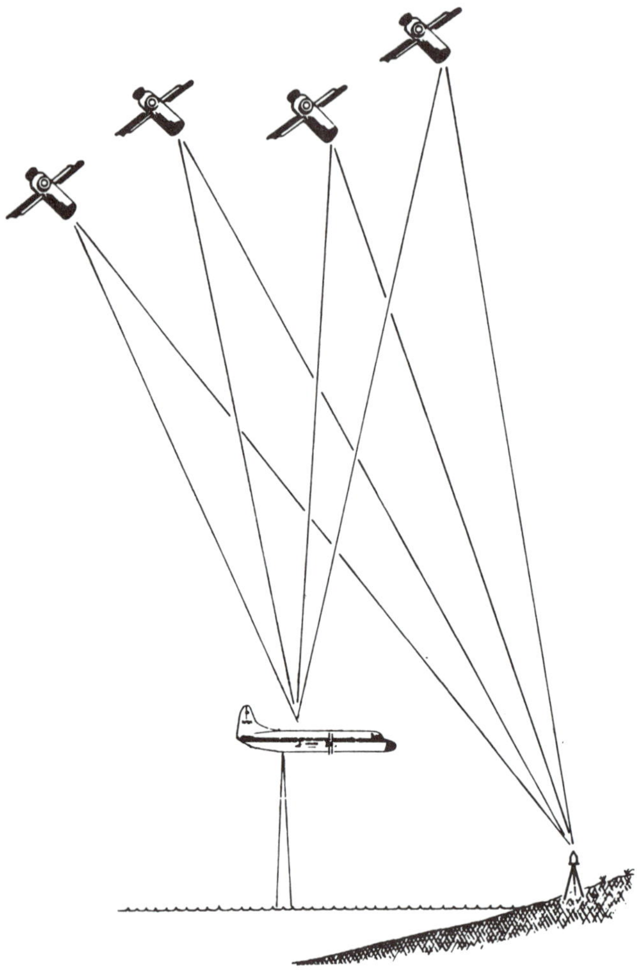

Fig. 1—Geometry of GPS Satellites, Aircraft and Ground Receivers used in GPS Tests.

The data segment taken with the aircraft in a parked position after landing, and before major data problems arose (< 8 minutes) is not sufficient to reliably estimate the phase ambiguity differences needed to convert the inflight phase data into unambiguous ranges. The good data span does not approach the 1+ hours of data desired and typically used for ground surveying. In addition, the aircraft data noise level is about a factor of ten higher than that from the ground receivers normally used by NOAA, because of the wider bandwidth used in the aircraft receiver. The procedure adopted for obtaining estimates of the ambiguities is discussed in the following section.

GPS Data Analysis and Aircraft Trajectory Estimation

In order to compute a GPS aircraft trajectory for the July 31 mission, it is necessary to resolve the inflight data problems, and to obtain estimates of the phase ambiguities. The data problems are considered first.

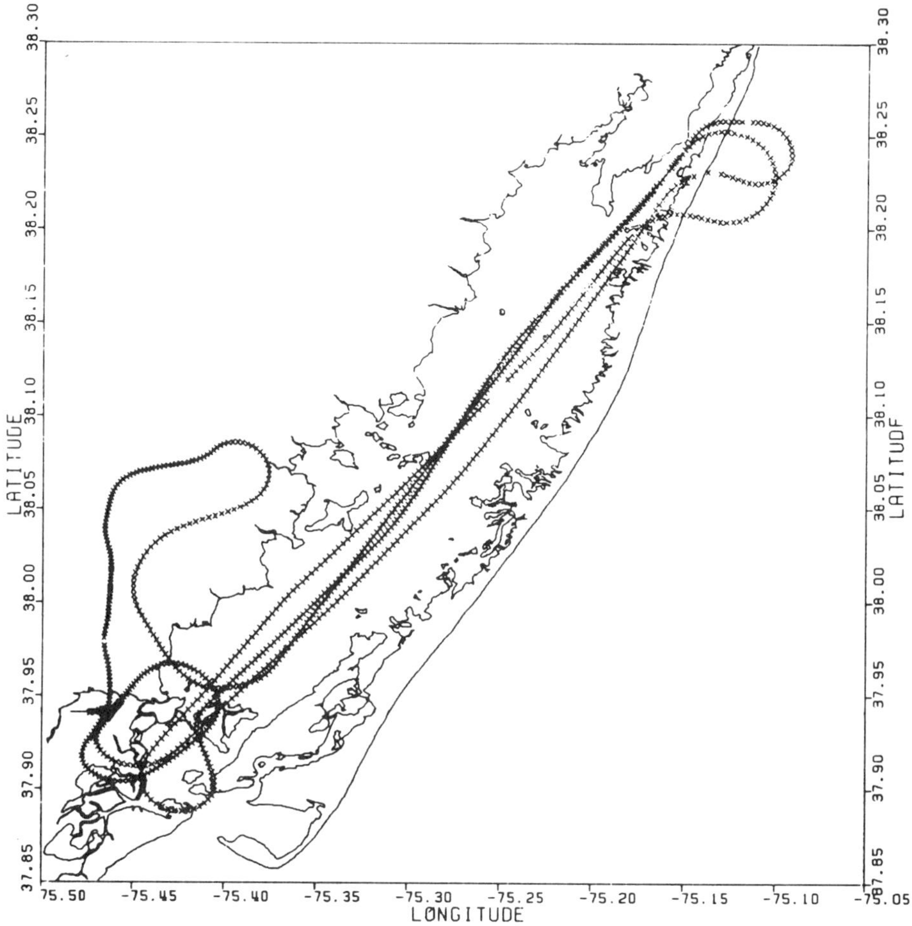

Fig. 2—Ground-Track Geometry Flown for GPS/AOL Flight Tests, July 31, 1985.

From Table 1, there are inflight problems for three 6–9 second segments for SV9. The first of these is evident at 79,826.74 seconds. Recorded pseudorange and phase data about this time point are shown in Table 2. In Table 3, the data in Table 2 has been differenced from the first time point listed, and the phase data has been converted into changes in range. The last column shows the differences in the L_1 and L_2 measurements of the change in range from that measured at 79,814.74 seconds, converted into L_1 cycles. Rounding to the closest integer cycle, this column indicates that 21,491 cycles have been dropped in the L_1 phase data between 79,823.74 seconds and 79,832.74 seconds and should be added to the data at the latter and subsequent time points. It should be noted that the changes in the L_1 and L_2 pseudoranges shown in Table 3 should be consistent with the L_2 phase changes. This is true to within a few meters, and confirms that the receiver has maintained lock on L_2. We cannot say, however, that there have been no cycle slips on L_2.

Table 4 shows data similar to Table 3 for the second inflight data segment with a problem. This table shows that the L_2 phase needs to be corrected for

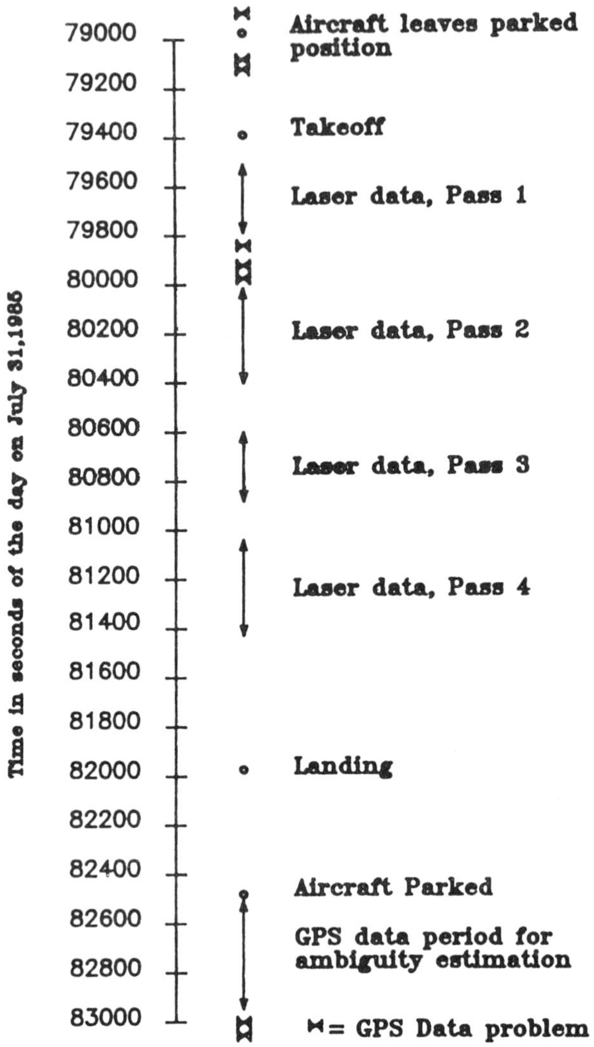

Fig. 3—Chronology of Major Events on GPS Mission of July 31, 1985.

79,913.74 seconds and subsequent time points by subtracting 59,720 cycles. Similarly, Table 5 for the final inflight problem data segment shows that 58,898 cycles should be subtracted from the L_2 phase for 79,940.74 seconds and subsequent time points. For all three data segments, the correction procedure only adds ambiguity corrections to presumably otherwise good data. The data is simply ignored at the frequency at which lock was lost. For the data segments shown in Tables 4 and 5, it is again possible only to say for L_1 that at most a few cycles could have been lost.

For inflight data other than at the problem times discussed above, comparisons of the changes in L_1 and L_2 phases between recorded data points indicate that there were no cycle slips on any channel with the possible exception of simultaneous slips of the same magnitude and direction in both L_1 and L_2. The

Table 1—Description of GPS Phase Tracking Problems on July 31, 1985, Affecting Capability for Aircraft Positioning.

Time (Seconds)		SV*	L_1 Frequency	L_2 Frequency
78944	6 sec	9	Loss of lock	Track with cycle slips
79088	12 sec	8	Loss of lock	Track with cycle slips
79112	9 sec	8	Loss of lock	Track with cycle slips
79826	9 sec	9	Loss of lock	None apparent
79907	9 sec	9	None apparent	Loss of lock
79934	6 sec	9	None apparent	Loss of lock
82979	201 sec	11	Loss of lock	Loss of lock
83003	Indef.	8,6,9,11	25 Hz freq. bias	25 Hz freq. bias

*Space Vehicle Number

Table 2—Data Recorded by Aircraft Receiver Tracking SV9 on July 31, 1985 Around Time of Data Problem.

Time (Seconds)	L_1 Pseudorange	L_1 Phase	L_2 Pseudorange	L_2 Phase
79814.74	22487700.9	−3864162.072	22487703.9	12330745.711
79817.74	22486150.3	−3874028.725	22486154.4	12359883.103
79820.74	22484638.8	−3884092.951	22484643.0	12388867.010
79823.74	22483168.4	−3894350.016	22483169.0	12417700.911
79826.74	bad	−3894349.516	22481734.8	12446388.099
79829.74	bad	−3894349.016	22480331.7	12474936.326
79832.74	22478913.3	−3904711.507	22478954.6	12503358.256
79835.74	22477604.7	−3915639.720	22477610.5	12531668.756
79838.74	22476282.4	−3926684.590	22476285.4	12559888.483

Table 3—Changes in Pseudorange and Phase Recorded by Aircraft Receiver Tracking SV9 on July 31, 1985 Around Time of Data Problem.

Time Past 79814.74	Change in L_1 Pseudorange	L_1 Phase Change	Change in L_2 Pseudorange	L_2 Phase Change	L_2 Phase-L_1 Phase (L_1 cycles)
0	0	0	0	0	0
3 sec	− 1550.6	− 1547.724	− 1549.5	− 1547.656	.36
6 sec	− 3062.1	− 3057.852	− 3060.9	− 3057.829	.12
9 sec	− 4532.5	− 4531.284	− 4534.9	− 4531.369	−.45
12 sec	—	− 7956.665	− 5969.1	− 5969.081	10444.83
15 sec	—	−11382.046	− 7372.2	− 7372.857	21068.43
18 sec	− 8787.6	−12835.416	− 8749.3	− 8745.789	21491.13
21 sec	−10096.2	−14181.132	−10093.4	−10091.510	21491.11
24 sec	−11418.5	−15504.649	−11418.5	−11415.062	21490.92

Table 4—Changes in Pseudorange and Phase Recorded by Aircraft Receiver Tracking SV9 on July 31, 1985 Around Time of Data Problem.

Time Past 79904.74	Change in L_1 Pseudorange	L_1 Phase Change	Change in L_2 Pseudorange	L_2 Phase Change	L_1 Phase-L_2 Phase (L_2 cycles)
0	0	0	0	0	0
3 sec	−1670.8	−1669.597	—	—	
6 sec	−3320.1	−3320.868	—	—	
9 sec	−4961.2	−4962.962	−4972.3	9621.271	−59719.998
12 sec	−6600.2	−6603.150	−6607.5	7981.924	−59720.034

Table 5—Changes in Pseudorange and Phase Recorded by Aircraft Receiver Tracking SV9 on July 31, 1985 Around Time of Data Problem.

Time Past 79931.74	Change in L_1 Pseudorange	L_1 Phase Change	Change in L_2 Pseudorange	L_2 Phase Change	L_1 Phase-L_2 Phase (L_2 cycles)
0	0	0	0	0	0
3	−1547.7	−1547.766	—	—	—
6	−3060.8	−3059.477	—	—	—
9	−4539.5	−4542.090	−4542.9	9841.493	−58898.366
12	−6003.8	−6000.485	−6003.0	8383.063	−58898.226

aircraft receiver noise level is not sufficiently low to rule out the latter possibility. For the ground receiver, which had a much lower noise level, the millimeter level agreement of L_1 and L_2 phases between time points very strongly indicates that there were no cycle slips.

We next consider estimation of the ambiguities in the phase measurements. As has been indicated, data problems occurred some eight minutes after the aircraft had reached the parked position after landing, and subsequent data is unusable. Accordingly, since P-code data was taken and relatively low noise pseudoranges (~1m) were available, it was decided to use the L_1 and L_2 pseudoranges for ambiguity or bias estimation. That the pseudoranges can be used to estimate the biases is clear. The only question is the accuracy which can be obtained. Melbourne[3] has concluded that when L_1 and L_2 pseudoranges and phases for a single receiver are processed, accounting for ionospheric effects, the estimation of phase ambiguities from pseudorange data is essentially impossible due to the amplification of noise from the ionospheric removal process. (On the other hand, the difference between L_1 and L_2 phase ambiguities is relatively easy to estimate.) However, when the reference station is sufficiently close that ionospheric effects nearly cancel, then L_1 and L_2 data can be separately processed, although the estimated ambiguities for aircraft and ground receivers must be differenced before obtaining useful results. In this procedure, we simply difference phase measurements (converted into ambiguous ranges) from pseudoranges and average over a number of points, thus reducing the effects of noise on the pseudoranges. If the pseudorange noise is uncorrelated, then the reduction is by a factor of $\sqrt{N}$, where N is the number of points averaged. For a twenty minute averaging period, a data rate of 1 point per 3 seconds (400 points), and a noise level of 1 meter, then the effect on the bias estimate is:

$$\frac{1 \text{ meter}}{\sqrt{400}} \bullet \sqrt{2} \cong 7 \text{ cm}$$

where the $\sqrt{2}$ is due to the use of both aircraft and receiver data. There is also some noise on the phase measurements (a few centimeters), but it is highly correlated and cancels to a very high degree.

Of the systematic errors, receiver clock errors cancel, as do tropospheric propagation effects. Ionospheric propagation effects do not cancel in the pseudorange-phase differencing, but double due to the opposite effects of the ionosphere on group and phase velocities. Ionospheric effects in the differences can, in fact, exceed ten meters. However, when the pseudorange-phase differences from the aircraft receiver are differenced from those from the ground receiver, the ionospheric effects should cancel to within ~0.1%, since the maximum separation of the aircraft from the ground station is about 30 kilometers and the minimum elevation angle of any satellite is about 20 degrees. The residual error should then not exceed a few centimeters—much less than a wavelength.

Tables 6 and 7 show how well this theory works in practice. The data is broken into three segments of about 20 minutes each. For the first segment, the data from SV9 is broken into segments before loss of lock and segments after loss of lock. For L_1, shown in Table 6, the bias estimates for SV6 for the three periods are reasonably consistent. A weighted mean of the biases would give a value of −19,036,370 wavelengths. The scatter is such that a value of −19,036,371 is also a possibility. For SV11, there is similar consistency, with −16,025,653 wavelengths the best estimate, but a value one wavelength less is also possible. For SV9, however, there is considerably more scatter. Values for the first two segments are consistent to within ~1.5 cycles but the third segment shows about a 4 cycle difference. Data from SV9 generally has a higher noise level due to the relatively low elevation angle, but this alone is not sufficient to explain the different bias in segment three. Cycle slips were not

Table 6—Estimated Differences between Ambiguities for Aircraft Receiver and Ground Receiver L_1 Phases, all Referenced to the SV8 Ambiguities, Based on Comparison of Measured Phases and L_1 Pseudoranges. The First Bias Listed for SV9 Has Assumed a Bias Correction of 21491 Cycles as Indicated in Table 3.

Time Segment	Bias for SV6 (cycles)	Bias for SV9 (cycles)	Bias for SV11 (cycles)
79118.74 sec to 80765.74 sec	−19036370.40	2867254.095* 2867252.909**	−16025653.64
80780.74 sec to 81878.74 sec	−19036369.82	2867254.649	−16025652.60
81887.74 sec to 82976.74 sec	−19036370.57	2867250.124	−16025653.74

*Beginning of segment to 79817.74 seconds.
**79838.74 seconds to end of segment.

Table 7—Estimated Differences between Ambiguities for Aircraft Receiver and Ground Receiver L_2 Phases, all Referenced to the SV8 Ambiguities, Based on Comparison of Measured Phases and L_2 Pseudoranges. The First Bias Listed for SV9 has Assumed a Bias Correction of 118, 618, Cycles Based on the Combined Cycle Losses Indicated in Tables 4 and 5.

Time Segment	Bias for SV6 (cycles)	Bias for SV9 (cycles)	Bias for SV11 (cycles)
79118.74 sec to 80765.74 sec	61302699.43	5925389.057* 5925388.025**	50249033.89
80780.74 sec to 81878.74 sec	61302700.95	5925388.418	50249035.12
81887.74 sec to 82976.74 sec	61302700.90	5925388.09	50249034.30

*Beginning of segment to 79901.74 seconds.
**79946.74 seconds to end of segment.

detected, but cannot be completely ruled out. Multipath effects are also a potential explanation, but no tests for multipath have been made with the aircraft, so they are difficult to quantify. Summarizing the bias estimates from Table 6, assuming that the problems are in estimating the biases and that the gap between loss of lock has been accurately bridged and that there are otherwise no whole cycle problems, we would have

SV6	$-19{,}036{,}370\ +0,\ -1$ cycles
SV9	$2{,}867{,}254\ +1,\ -4$ cycles
SV11	$-16{,}025{,}653\ +0,\ -1$ cycle

Looking now at Table 7 for the L_2 results, we find somewhat better consistency. Summarizing the results from this table in a manner similar to those for L_1, we have

SV6	$61{,}302{,}701\ +0,\ -2$ cycles
SV9	$5{,}925{,}388\ +1,\ -0$ cycles
SV11	$50{,}249{,}034\ +1,\ -0$ cycles

If we consider only data after the SV9 loss of lock periods, the L_2 biases look even more consistent. Since three of the four data passes occur during this period, we will adopt the above L_2 biases for the data reductions used in computing the aircraft trajectories for the mission. Because of the larger uncertainty in the L_1 biases and the small effect of the ionosphere on the phase differences between the aircraft and ground receivers, only the L_2 results will be presented. It should be noted that an uncertainty of 1–2 cycles remains in each bias.

AOL Data Processing

The AOL laser data provided the reference for comparing the GPS aircraft heights. The AOL system has been described in a number of papers. The most comprehensive description of the instrumentation as configured for the GPS investigation can be found in Reference 4. For present purposes, we can consider the AOL simply as a pulsed laser altimeter which measures the height of the aircraft above Chincoteague Bay during the four inflight passes. For this investigation the AOL was operated at a data rate of 400 ranges per second. Thus, at the nominal aircraft speed of approximately 100 meters/second, there is one height measurement per 25 centimeters of horizontal aircraft movement. The high sampling rate of the airborne laser system and the relatively calm water surface permit averaging of the laser ranging measurements prior to comparison with the GPS-determined heights. In this smoothing process advantage was taken of a previously developed technique to eliminate the effects of aircraft motion and to provide corrected height information. The technique, which is described in detail in Reference 5, involves the use of a vertical accelerometer mounted on the AOL optical table. Briefly, the instrument senses the vertical accelerations of the aircraft which can then be used in modeling the measured altitudes according to the relation:

$$H(t_i) = H_o + (t_i - t_o)V_o + \int_{t_o}^{t_i} \int_{t_o}^{t} a(t')dt'\,dt \qquad (1)$$

where

$H(t_i)$	is the aircraft altitude measured by the AOL laser at time t_i
t_o	is an initial time for a smoothing interval
$a(t')$	is the recorded acceleration (at 400 pps, the same as for the altitude data)
H_o, V_o	are constants to be solved for using the altitude data.

A data span of 15 seconds was used for the GPS test data. After solving for the constants H_o and V_o, the right hand side of (1) produces a very low noise vertical trajectory. Since the laser does not measure altitude directly, but rather a range to the surface in whichever direction the instrument is pointing, aircraft pitch and roll data from the aircraft inertial navigation system (INS) must be used to correct each measurement before it is used in (1).

For the GPS mission, there were two other important corrections which are not normally a part of AOL data processing. These corrections are necessary to properly reference the GPS antenna to the bay surface. First, a calibration of the laser ranging measurement was necessary. This calibration was performed in the WFF D-1 hangar used by the P-3 aircraft after the flight mission. Figure 4 shows the basic configuration used for this calibration. Seven retroreflectors were positioned around the hangar as shown in Figure 4 and were surveyed in a local coordinate system. A mirror, capable of rotating and tilting about a common point, was mounted underneath the aircraft and sequentially directed to each of the retroreflectors. A set of range measurements were made to each reflector. With this data set, a least squares solution was then made for the position of the mirror rotation/tilt point and the laser bias. In practice, the height of this point was weakly determined due to the near planar geometry of

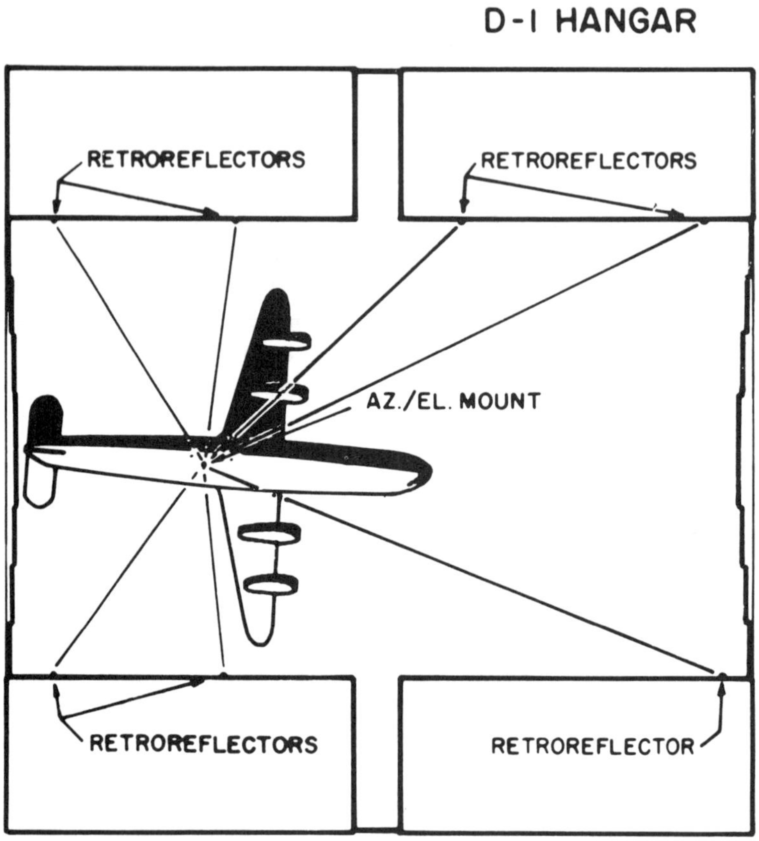

Fig. 4—Geometrical Configuration Used for In-Hangar Measurement of AOL Laser Bias.

the reflector/mirror system, and a measured height was used instead. The results of the calibration gave a bias estimate of

$$b_{AOL} = 8.35 \pm 0.10 \text{ m}$$

The AOL thus measured long by 8.35 meters, and this correction was applied in data processing.

It should be noted that the calibration point in the scheme described is to the rotation axes intersection of the mirror. In flight, this is simply a point in the air below the aircraft. The point does, however, have a definite geometrical relationship to the aircraft, and it is next necessary to measure/compute the vertical distance between this point and the GPS antenna. Figure 5 shows the aircraft in a parked position, along with the location of the GPS antenna and the laser calibration point. Relevant distances are also shown. From drawings, the aircraft height at the laser calibration point is 499.53 centimeters. The GPS antenna is forward of the vertical through the calibration point by 198.12 centimeters (measured along the aircraft surface) and the aircraft was pitched down by 0.52 degrees. The laser calibration point was measured to be 30.48 centimeters above ground. Based on these numbers, the GPS antenna was:

$$499.53 \text{ cm} - 198.12 \sin 0.52° = 497.73 \text{ cm}$$

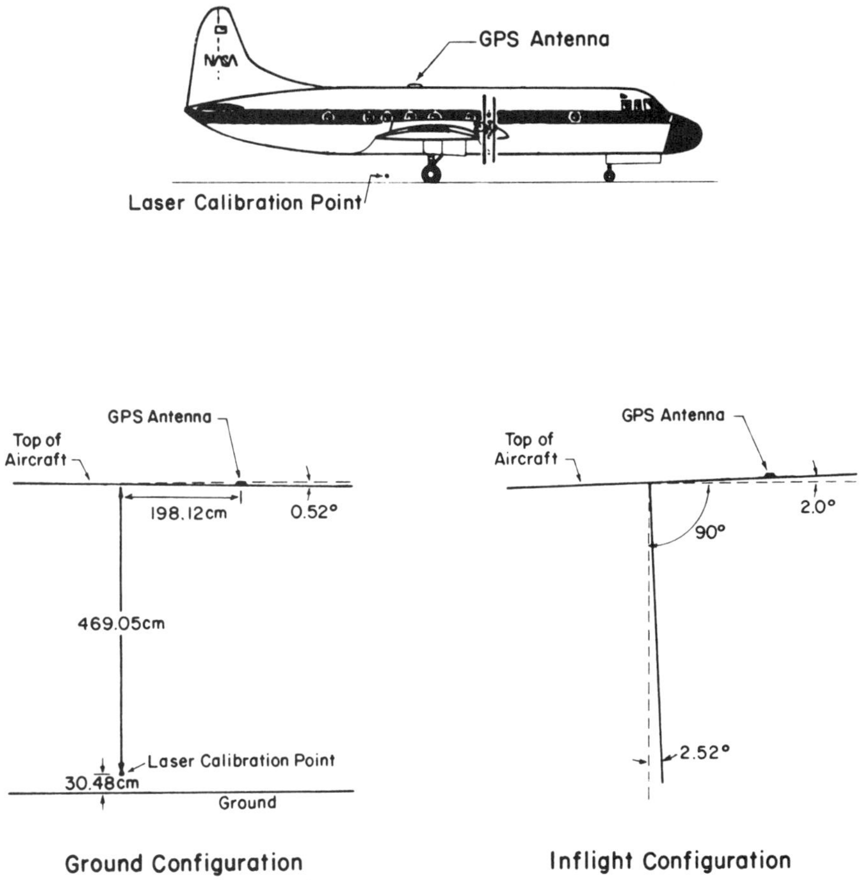

Fig. 5—Location of GPS Antenna and Laser Calibration Point.

above ground, while the laser calibration point was 30.48 centimeters above ground. In flight, the normal aircraft pitch is approximately 2 degrees up, so the configuration is rotated by 2.52 degrees. In this rotation, the GPS antenna moves up by

$$198.12 \text{ cm } (\sin 2^\circ + \sin 0.52^\circ) = 8.71 \text{ cm}$$

The calibration point moves up by

$$(499.53 \text{ cm} - 30.48 \text{ cm})\,(1 - \cos 2.52^\circ) = 0.45 \text{ cm}$$

Thus, for an aircraft pitch of 2 degrees, the vertical separation of the GPS antenna and the laser calibration point was

$$497.73 \text{ cm} + 8.71 \text{ cm} - (30.48 \text{ cm} + .45 \text{ cm}) = 4.76 \text{ m}$$

Measured laser altitudes thus need to have 4.76 meters added to them to put them on a common reference with heights estimated for the GPS antenna. This value is slightly pitch dependent (3 centimeters per degree of pitch), but a constant value was applied for all inflight data, since the pitch data varied much less than 0.5 degrees in this data set.

GPS Height Comparisons

There are a number of differencing techniques which have been used for ground position estimations using GPS phase data. For processing the aircraft data, however, only the so-called single difference technique is applicable. This technique has been described in the literature[1,6] and involves basically forming measurements which are the differences between the phase measurements made on the aircraft and on the ground. When these phase measurements are converted to ranges and the ambiguity corrections are made, the measurements are range differences (aircraft to satellite-ground to satellite). The differences between the aircraft and ground receiver clocks must, however, be estimated from the four range differences at each time point.

The height/altitude determinations made using the GPS phase data and the AOL laser data are depicted in Figure 6. From the GPS data, an estimate is obtained of the aircraft GPS receiver height above a reference ellipsoid. In the differencing procedure used, with a ground receiver in the vicinity, the ellipsoid height assumed for the ground receiver provides the reference height. The AOL laser measures the height of the aircraft above the local water surface. The procedure for referencing the AOL heights to the GPS receiver has been discussed above. The difference between the GPS heights and AOL heights then gives the ellipsoid height of the local surface.

First, considering the "raw" data (the GPS heights are directly calculated from the data; the AOL data has been smoothed as previously discussed), Figures 7 and 8 show the GPS and AOL heights on the second and third of the four passes, respectively. These figures show a general agreement in the height variations during the two passes. In Figure 9 and 10, the AOL data has been biased downward so as to approximately overlay the GPS heights. On the scale

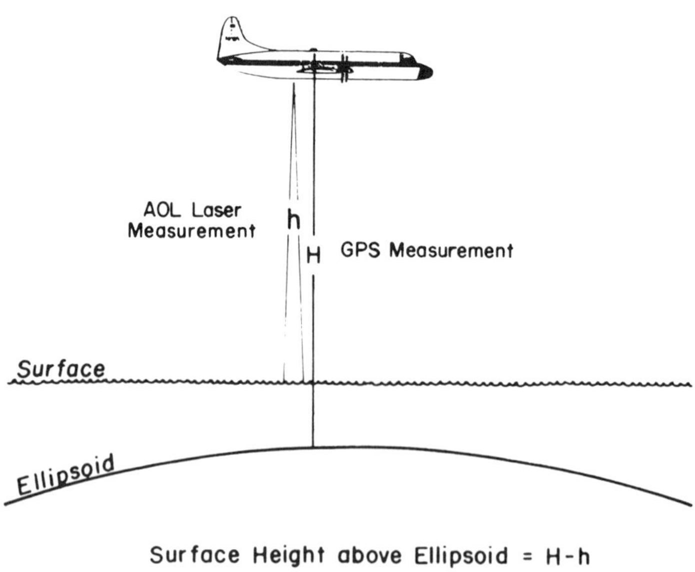

Fig. 6—Graphic Depiction of Measurements Made by GPS and the AOL laser. (Ellipsoid height is determined from GPS data. Altitude above local surface is measured by AOL laser.)

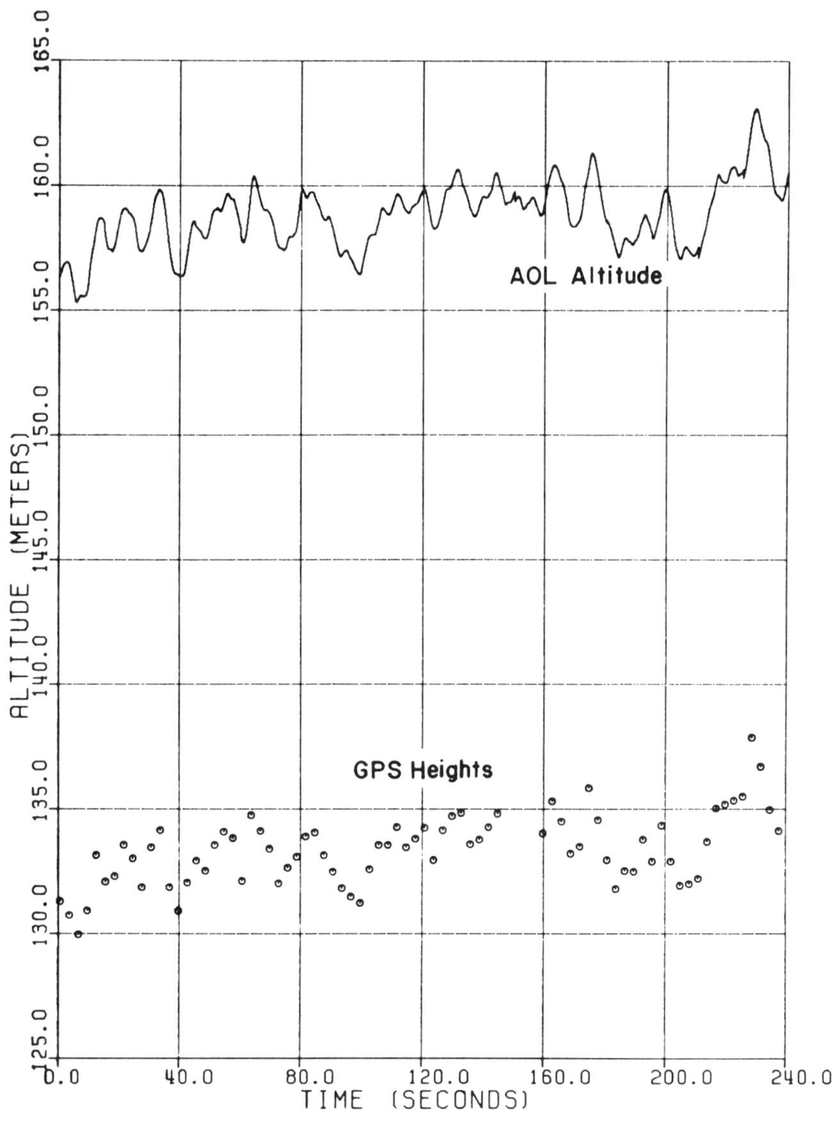

Fig. 7—GPS Ellipsoid Height Estimates and AOL Altitude Measurements, Pass 1/2 of GPS Tests.

of these plots, particularly in Figure 10, the agreement is nearly perfect. Figure 11 shows the first 60 seconds of Figure 10 on an expanded scale. The largest differences appear to be on the order of 10 centimeters.

Applying the calibration and offset corrections to the AOL data to reference it to the GPS antenna, and taking the GPS minus AOL height differences, we obtain the ellipsoid heights of the bay for the four passes. These results are shown in Figure 12. The heights have been plotted as a function of latitude rather than time. Since the longitude offsets between passes were very slight, the four passes should overlay one another except to the extent that the Bay tides vary between passes. This should be, at most, at the centimeter level. The

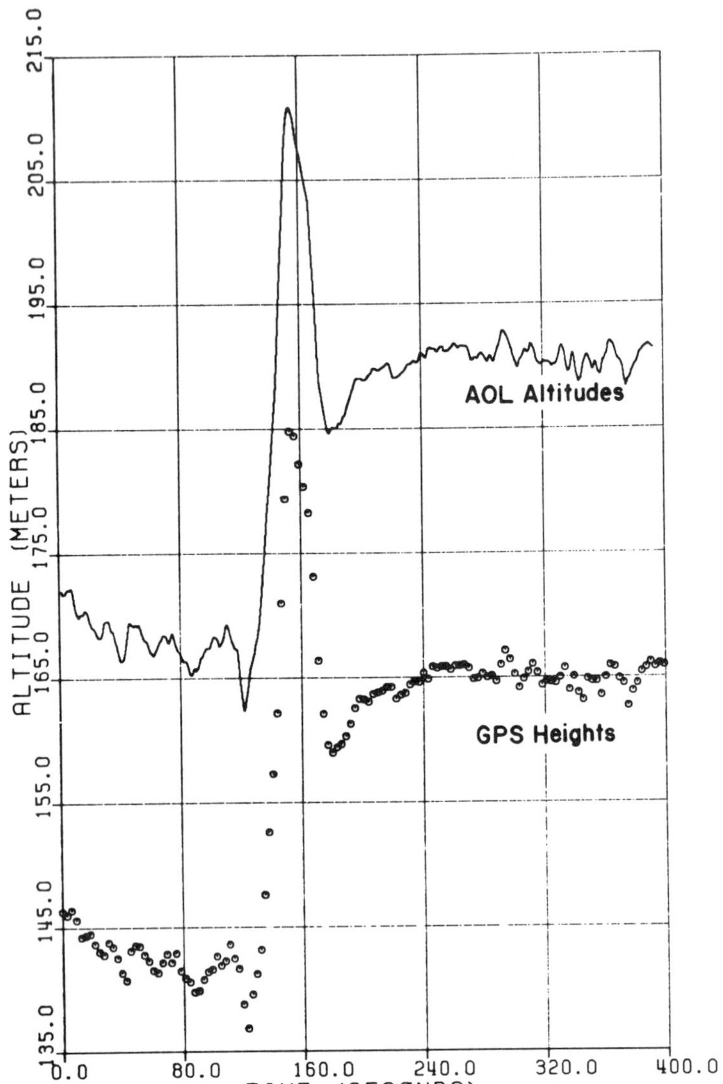

Fig. 8—GPS Ellipsoid Height Estimates and AOL Altitude Measurements, Pass 2/1 of GPS Tests.

along track ellipsoid heights should not be exactly constant but should have the slope of the local geoid. To first order, this slope should be constant. This assumption was made in order to obtain quantitative comparisons between the GPS and AOL data.

With a constant geoid slope along the Chincoteague Bay, the ellipsoid heights shown in Figures 12 should all be straight lines. Deviations about the line should be due to random errors in the GPS heights, since the rapid AOL data rate (plus the accelerometer data) allowed a high degree of smoothing in the AOL laser heights. Differences between the average ellipsoid heights for a pass and the surveyed ellipsoid height, as well as slope differences between those in

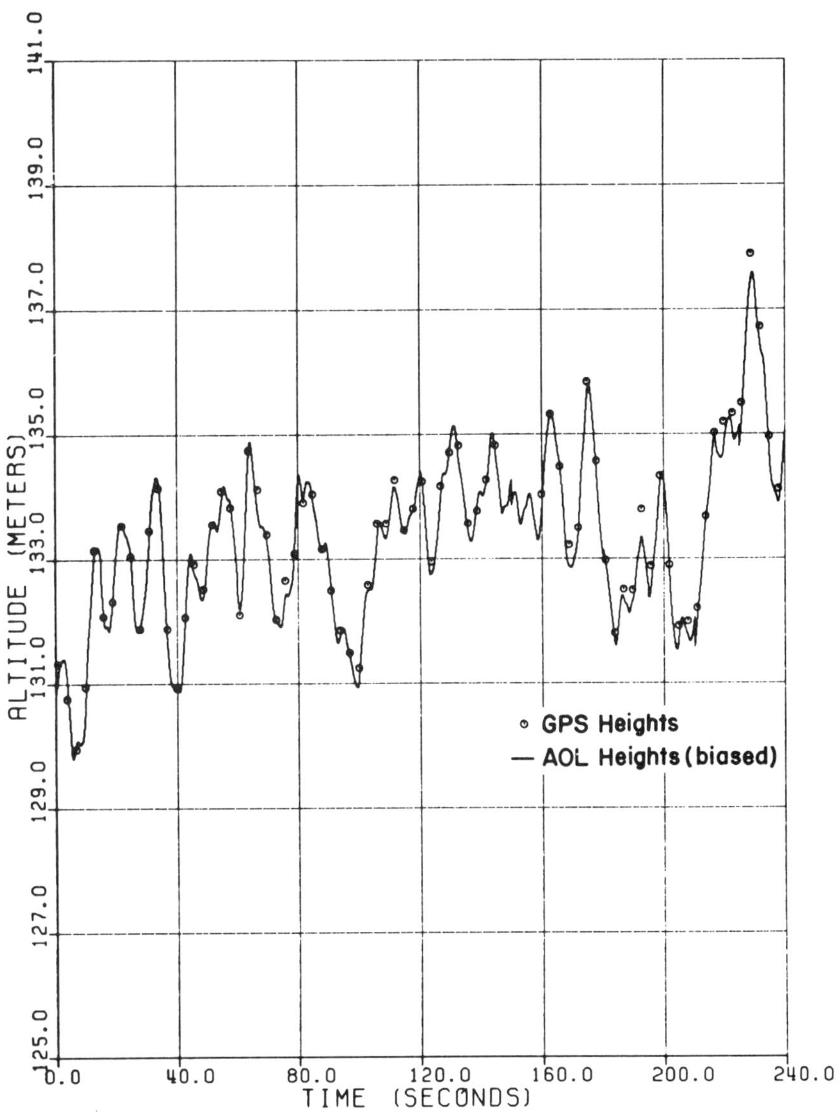

Fig. 9—GPS and AOL Heights for Pass 1/2 with AOL Heights Shifted Downward to Approximately Overlay GPS Heights.

Figure 12 and the actual geoid slope, are indicative of systematic errors in the GPS heights. Accordingly, each of the four passes in Figure 12 was fitted to a straight line. The average height, the slope of the fitted line, and the root-mean-square (RMS) of fit of the ellipsoid heights to this line were calculated. The RMS values are shown in Figure 12. The fit parameters, along with available "survey" data, are shown in Table 8. The average heights for the four passes show excellent agreement, and are well within ~1 sigma agreement with the "surveyed" ellipsoid height. No attempt was made during the tests to survey the ground receiver with respect to local survey markers, since the complete test included a survey by NOAA of the ground receiver relative to a

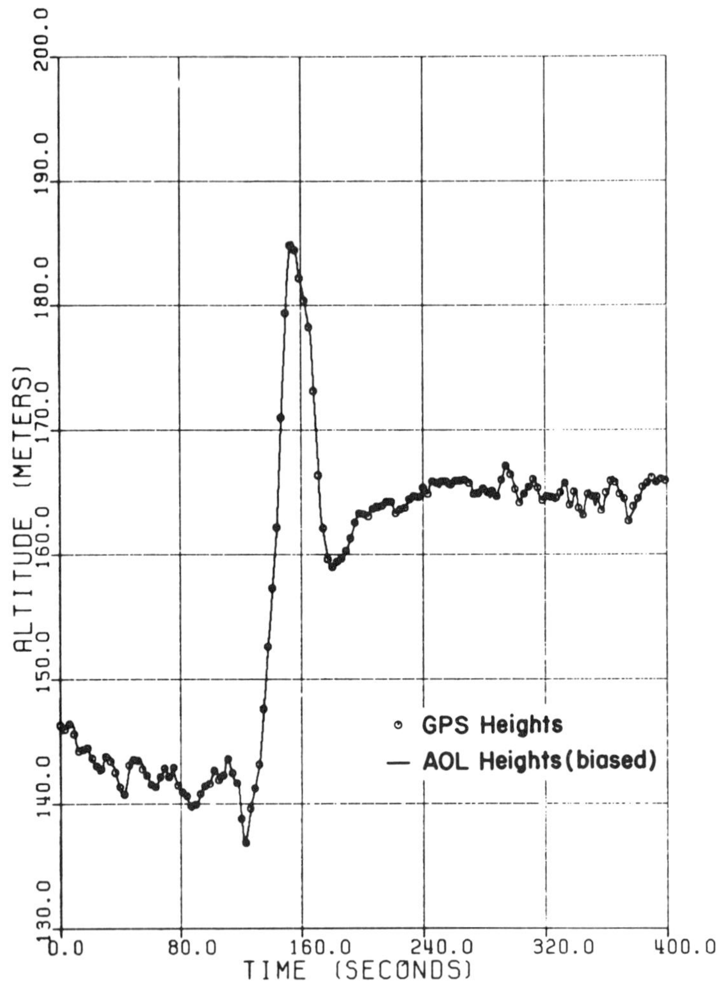

Fig. 10—GPS and AOL Heights for Pass 2/1 with AOL Heights Shifted Downward to Approximately Overlay GPS Heights.

receiver on the Chincoteague Bay shore, along with tide gauge data for the Bay. This survey data has not yet been received, so the ground receiver marker height relative to the nearest survey marker has been estimated, pending receipt of the GPS survey and tide gauge data.

All of the slopes in Figure 12 are consistent in sign with the geoid slope obtained from the geoid model. However, the variations between passes, as well as differences from the geoid model value, are sufficient to suggest that the ambiguities may not have been resolved exactly. In addition, Pass 2/2 shows an apparent change in slope in the middle of the pass, for which there is no ready explanation.

Assuming a straight line to be an appropriate model for the ellipsoid height during a pass,[7] the resultant RMS noise levels are interpreted as random errors in the GPS height estimates, and thus due to GPS phase noise. This is approximately the case for the first three passes. The first pass was expected to have

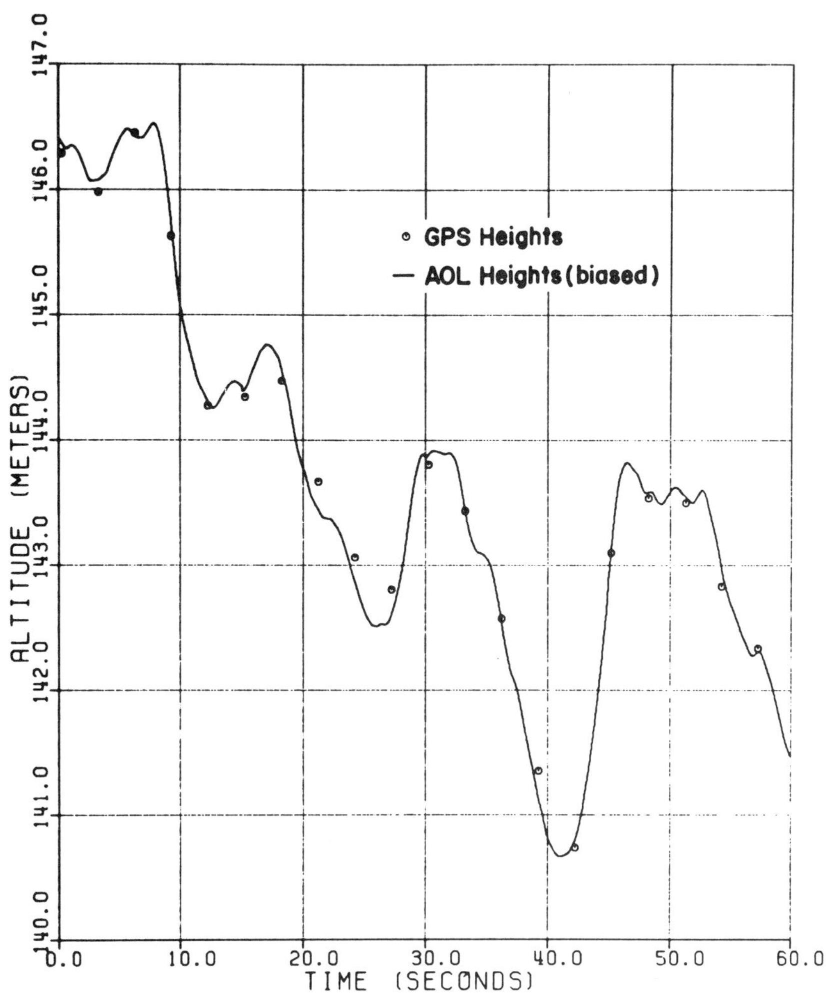

Fig. 11—Expansion of First Minute of Figure 10 Showing Details of Agreement Between GPS and AOL Heights on Page 2/1.

the highest noise level, since the GPS constellation passed through a near singular geometry shortly before aircraft takeoff. The geometry improved for each successive pass, so a corresponding reduction in noise level would have been expected. Also listed in Table 8 are the average VDOP values for each pass, along with the RMS/VDOP ratio. This ratio should be a rough indication of the height noise level to be expected with good GPS viewing geometry with a VDOP less than 1. This ratio varies from 1.4 to 1.8, consistent with the expected noise level of 1 to 2 centimeters. For the time segment after landing with the aircraft parked, the height noise level observed was 7.4 centimeters with a VDOP of 5.3. The ratio of 1.4 is consistent with the best inflight results.

Summary and Conclusions

The GPS tests demonstrated that:

1. A GPS phase tracking receiver in an aircraft could maintain lock during

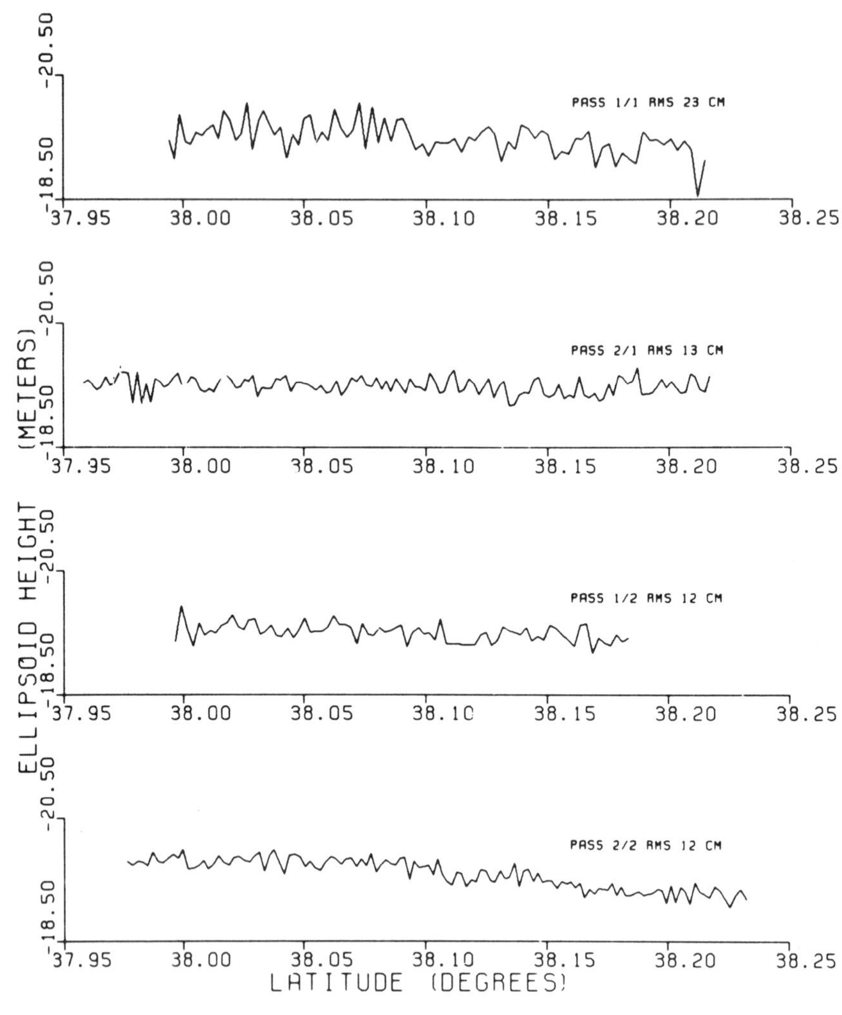

Fig. 12—Ellipsoid Heights Deduced from the GPS and AOL Data for the Four AOL Passes.

takeoff and landing, and during normal flight, without apparent problems and provide good data for an extended period of time. Lock was maintained for at least one frequency for a time period in excess of an hour on four satellites.

2. With receivers such as the TI4100, height noise levels on the order of 1 to 2 centimeters should be expected under good GPS viewing geometry.
3. Receiver operating experience is needed in order to determine the best procedure(s) for ambiguity removal.

The lessons from these tests for future missions, and for aircraft positioning using GPS phase measurements in general, are that positioning with phase measuring receivers works, and that experience is needed in order to determine the appropriate procedures for ambiguity estimation. The tests described in this paper, although using a 4 satellite receiver, benefited both from the P-code pseudoranges and from the dual-frequency phases, although such data was not

Table 8—Summary of GPS Flight Test Results.

Pass Number	1/1	2/1	1/2	2/2
Pass Duration	261 sec	394 sec	250 sec	384 sec
Average Ellipsoid Height*	−19.52 m	−19.48 m	−19.51 m	−19.59 m
Overall Slope	−1.45cm/km	−.22 cm/km	−.80 cm/km	−1.85cm/km
RMS Height Difference GPS-AOL	23 cm	13 cm	12 cm	12 cm
Average VDOP for pass	14.3	9.1	7.0	6.7
RMS/VDOP	1.6	1.4	1.7	1.8
Surveyed** Ellipsoid Height	−20 ± 1 m	−20 ± 1 m	−20 ± 1 m	−20 ± 1 m
Geoid Slope from Geoid Model (Marsh-Chang, 1978)	−.91 cm/km	−.91 cm/km	−.91 cm/km	−.91 cm/km

*Ellipsoid height using ground receiver ellipsoid height of −10.09 meters.
**Estimated using Wallops survey of marker in vicinity of ground receiver.

used for ionospheric corrections. On the other hand, had the aircraft receiver not lost lock, the ground ambiguity estimation could have worked. Thus, if a receiver is likely to have problems, then redundant channels and/or dual frequencies are desirable. If the receiver reliably maintains lock, then such features can be dispensed with. Experience with particular receivers is necessary in order to determine what to expect.

REFERENCES

1. Goad, C. C., Remondi, B. W. *Initial Relative Positioning Results using the Global Positioning System,* Bull. Geod., 58, 1984, pp 193–210.
2. Counselman, C. C., Ablott, R. I., Gourevitch, S. A., King, R. W., Paradis, A. R. *Centimeter-Level Relative Positioning with GPS,* Journal of Surveying Engineering, 109, 1983, pp 81–89.
3. Melbourne, W. G., *The Case for Ranging in GPS Based Geodetic Systems,* Proceedings, First International Symposium on Precise Positioning with the Global Positioning System, pp 373–386, Rockville, MD., April 15–19, 1985.
4. Hoge, F. E., Swift, R. N. and Frederick, E. B. *Water Depth Measurement Using an Airborne Pulsed Neon Laser System,* Applied Optics, 19, 1980, pp 871–883.
5. Krabill, W. B., and Martin, C. F. *Analysis of Aircraft Vertical Positioning Accuracy Using a Single Accelerometer,* NASA TM 84417, February 1983.
6. Remondi, B. W., *Modeling the GPS Carrier Phase for Geodetic Applications,* Proceedings of the First International Symposium on Precise Positioning with the Global Positioning System, NOAA, Rockville, Maryland, April 15–19, 1985.
7. Marsh, J. G., and Chang, E. S. *5′ Detailed Gravimetric Geoid in the Northwestern Atlantic Ocean,* Mar. Geod., 1, 1978, pp 253–261.

NAVIGATION: *Journal of The Institute of Navigation*
Vol. 38, No. 2, Summer 1991
Printed in U.S.A.

Design and Flight Test of a Differential GPS/Inertial Navigation System for Approach/Landing Guidance

LAWRENCE VALLOT, SCOTT SNYDER, BRIAN SCHIPPER

Honeywell Systems and Research Center,
Minneapolis, Minnesota

NIGEL PARKER

Honeywell Space and Strategic Systems Operation,
Clearwater, Florida

CARY SPITZER

NASA Langley Research Center, Hampton, Virginia
Received January 1990
Revised April 1991

ABSTRACT

A flight test project evaluating an integrated differential GPS (DGPS)/inertial navigation system as an approach/landing aid has recently been completed. The test objectives were to acquire a system performance database, and to demonstrate automatic landings using the integrated DGPS/inertial system augmented with barometric and radar altimeters.

The airborne and ground components of the DGPS/inertial system are built around standard transport aircraft avionics: a global positioning/inertial reference unit and two GPS sensor units. Flight tests were conducted using a specially modified Boeing 737-100. Performance of the DGPS/inertial system was evaluated in real time by comparison against Microwave Landing System (MLS) position. DGPS/inertial, MLS, and laser tracking system position data were recorded during the tests for postflight analysis. Over 120 landings were flown during the project, 36 of which were fully automatic DGPS/inertial landings.

The paper describes the objectives of the project, the system implementation, and the flight tests conducted. Laboratory test results and preliminary flight test results are also presented.

INTRODUCTION

A multiyear research and development effort at Honeywell has studied the potential of differential GPS (DGPS) as an approach/landing aid. The primary focus of the project has been to understand the capabilities and limitations of a DGPS/inertial navigation system as the primary navigation sensor supporting fully automatic landing of manned and unmanned space vehicles. Examples of manned vehicles of interest include the Assured Crew Return Vehicle (ACRV), Space Shuttle, and National Aerospace Plane. Studies conducted under this project in 1989 included analysis and simulation of a DGPS/inertial-based automatic landing system for ACRV [1].

Many other research activities have considered the DGPS-based automatic landing problem. Examples include the ongoing program at NASA-Ames, which has examined many of the design issues associated with a DGPS system to support terminal area operations [2, 3]. Work reported in [4] describes a DGPS-based automatic landing capability demonstrated in a Dornier DO-28 aircraft by researchers from the Technical University of Braunschweig, Germany. Approach guidance experiments using differential P-code GPS performed by the National Aerospace Laboratory in the Netherlands are addressed in [5].

In 1990, Honeywell and NASA-Langley Research Center conducted a joint flight test program to collect a DGPS/inertial system performance database and to demonstrate an automatic landing capability. The DGPS/inertial system equipment was built around standard air transport avionics: a global positioning inertial reference unit (GPIRU) and two GPS sensor units. The GPIRU is a new product based on an existing strapdown ring laser gyro (RLG) inertial reference system (IRS). The GPIRU differs from the RLG IRS by the addition of a GPS navigation processor, which mixes the GPS pseudorange measurements with inertial position and barometric altitude data.

The flight test experiments were performed in the NASA Transport Systems Research Vehicle (TSRV), a modified Boeing 737-100, at the Wallops Island Flight Test facility. Microwave Landing System (MLS) equipment and a laser tracking system at Wallops were used to provide the "truth" data against which the DGPS/inertial solution was compared.

After comparisons between the DGPS/inertial and MLS position data over a large number of approaches had shown good agreement, the DGPS automatic landing system was ready to be tested. On November 1, 1990, five DGPS/inertial automatic landings of the 737 TSRV were flown. A total of over 120 landings were flown during the project, 36 of which were fully automatic DGPS/inertial landings. We believe that these landings represent the first DGPS-based automatic landings in the United States, and the world's first DGPS/inertial automatic landings of a commercial jet transport class aircraft.

A notable aspect of the joint DGPS experiment is that the duration of the project, from the planning phase through DGPS/inertial system design, laboratory testing, and flight test, was less than 11 months. Future efforts will be directed at analysis of the extensive database of DGPS/inertial system performance relative to MLS and laser tracker data.

DGPS/INERTIAL SYSTEM CONFIGURATION

The basic DGPS concept is illustrated in Figure 1. A reference or ground station GPS receiver's antenna is installed at a surveyed site. Predicted pseudoranges to all satellites being tracked are computed based on the GPS satellite positions (as received in the GPS navigation message) and the known ground station antenna location. Differences between the predicted and observed pseudoranges are filtered to remove receiver-unique errors, and then transmitted

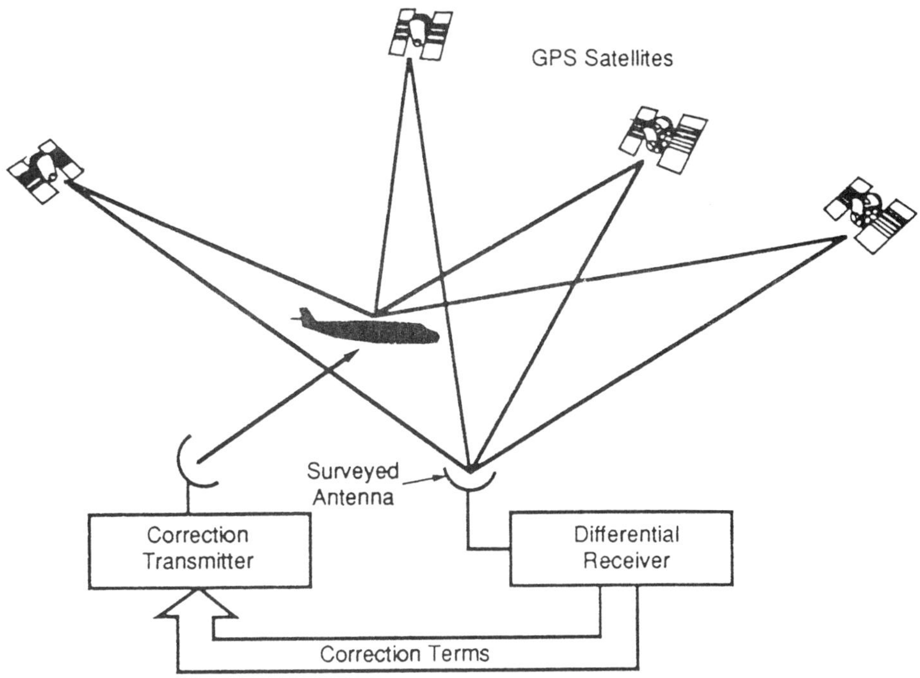

Fig. 1—Differential GPS Concept

over a differential correction data link to user vehicles in the vicinity of the ground station. The corrections are applied to the pseudoranges measured by the user vehicle's GPS before navigation calculations are performed.

The motivation for differential operation is that many of the largest GPS error sources, such as Selective Availability (SA) and ionospheric effects, are common to two or more receivers operating in spatial and temporal proximity. These common errors can be nearly eliminated in the differential mode, greatly enhancing the performance.

The DGPS/inertial system configuration used in the flight test project is shown in Figure 2. The equipment shown in the figure can be separated into four major blocks: the aircraft GPS/INS/altimeter equipment, the DGPS ground station, the data link, and the flight test instrumentation. A brief description of these four blocks is given below.

Aircraft GPS/Inertial Equipment

Primary aircraft equipment for the DGPS experiment consists of a GPS sensor unit (GPSSU) and a GPIRU. This equipment is currently in production, and is described in detail in [6]. A brief description of the equipment and the software modifications required for the DGPS/inertial experiment is provided here.

The GPSSU is a Honeywell model HG-2021 "all-in-view," C/A-code, two-channel sequential receiver. The receiver transmits satellite position, pseudorange, pseudorange rate, and time, as well as an autonomous GPS solution, over ARINC-429 data busses. A GPS pulse-per-second signal is also provided.

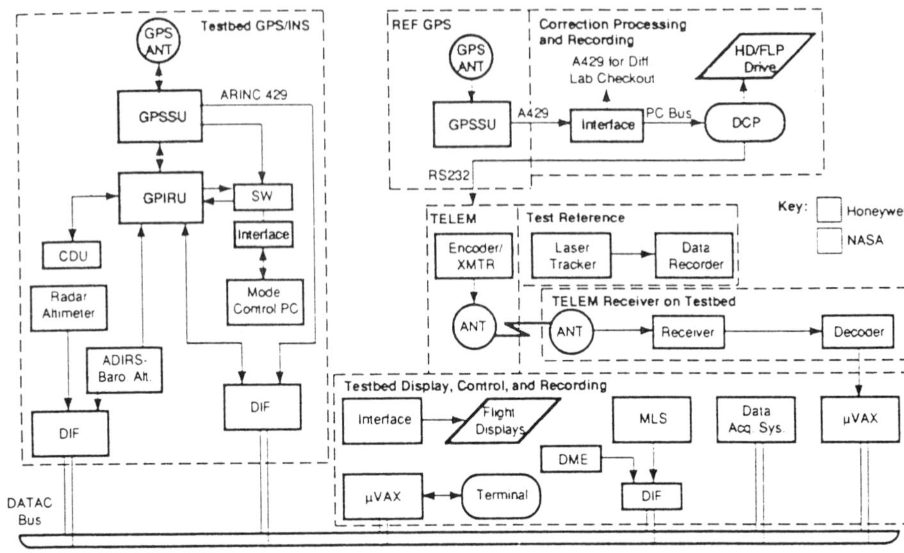

Fig. 2—TSRV Differential GPS/Inertial System Block Diagram

This particular receiver does not maintain continuous carrier track on each satellite, so integrated Doppler outputs are not available.

The autonomous GPS solution (position and velocity) is computed by the GPSSU once per second using a Kalman filter to process the pseudorange measurements. The Kalman filter models 12 states: 3 earth-centered, earth-fixed (ECEF) positions, velocities, and accelerations; pressure altitude bias error; user clock phase error; and user clock frequency error. The aircraft GPS antenna is a microstrip antenna mounted on the TSRV fuselage top centerline about 24 ft aft of the nosewheel.

A GPIRU provides conventional baro/inertial outputs and hybrid GPS/inertial outputs. As noted earlier, the GPIRU is an RLG-based INS with unaided inertial performance in the 2 nmi/h (95 percent radial error) class. The input/output interface follows the ARINC-704 IRS specification.

GPSSU-measured pseudorange, satellite position, and time data are blended with inertial data and barometric altitude to form the integrated GPS/inertial solution. This hybrid solution possesses the good short-term inertial system error characteristics, combined with the excellent long-term accuracy of GPS. An 18-state Kalman filter implemented in the GPIRU is used to integrate the GPS and inertial data. Figure 3 illustrates the processing performed by this filter. The 18 error states modeled in the filter are three positions, three velocities, three attitudes, three gyro biases, three accelerometer biases, pressure altitude, user clock phase, and clock frequency. GPS measurements are collected at an approximate 1 Hz data rate, prefiltered, and processed by the hybrid Kalman filter every 10 s.

Software in the GPIRU was modified to accommodate the differential correction message. The modifications consisted of I/O processing and application of the differential correction to the pseudorange measurement. A differential correction message consists of the following elements: satellite ID/health, time of correction, pseudorange correction, and pseudorange correction rate.

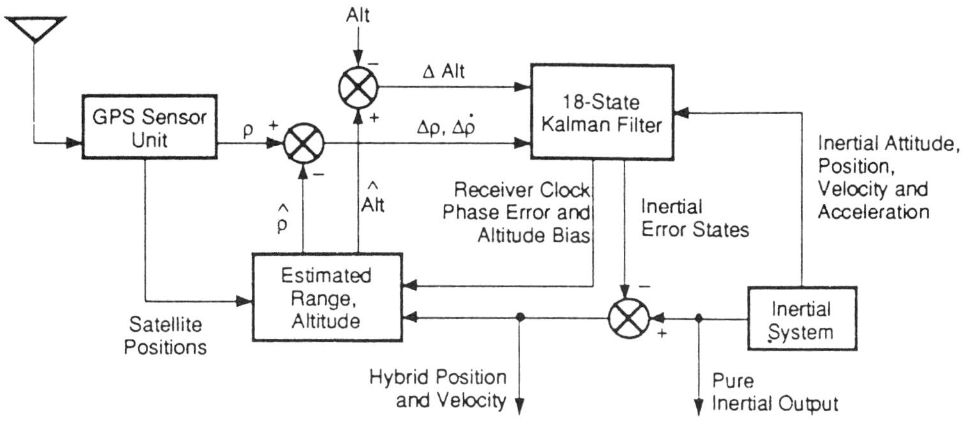

Fig. 3—Hybrid GPS/Inertial Kalman Filter

A DGPS mode word was added to the GPIRU software, allowing the system to be moded from AGPS (absolute GPS/inertial), the basic production system mode, to DGPS (differential GPS/inertial) mode, or vice versa. When the system is in DGPS mode, the hybrid Kalman filter will process only pseudorange measurements from space vehicles (SVs) for which a differential correction has been received. This reduces the complexity of the system and simplifies the postflight analysis task.

A 20 s differential correction time-out criterion was implemented in the software to prevent the processing of stale corrections. The system propagates the correction using the correction rate term until it receives another data link correction message or 20 s elapses.

With one exception, there were no parameter or structural changes made to the existing GPIRU Kalman filter software. The only change was that the filter position and clock phase error covariances were increased on a mode toggle to or from DGPS mode. This change was made to decrease the filter settling time after a mode change.

The GPS/inertial hybrid solution is transmitted on the GPIRU ARINC-429 output bus, along with the unaided inertial solution and the autonomous GPS solution received from the GPSSU. Position data is output in WGS-84 latitude, longitude, and altitude. Velocity data is provided in north, east, and up components.

The Mode Control Terminal (MCT) is an IBM PC clone with a custom ARINC-429 transmitter/receiver board. The MCT is used for GPIRU mode control (DGPS or AGPS), display of GPIRU and GPSSU status and data, and recording of data for postflight analysis.

The digital autonomous terminal access communications (DATAC) interface (DIF) serves as a gateway between the ARINC-429 bus and TSRV DATAC bus. The received and decoded differential correction message is transmitted on the ARINC-429 bus by the DIF. GPIRU, GPSSU, air data, and radar altitude data are transmitted onto the DATAC bus for recording and TSRV flight management/display processing.

Altitude Blending Filter

Because of expected DGPS/inertial performance limitations in the vertical channel, an altitude blending filter mechanization was designed to use radar altimeter and/or corrected barometric altimeter information. A schematic of this altitude blending approach is shown in Figure 4. The software associated with this mechanization is implemented in the TSRV flight management μVAX. Figure 4 shows two separate blending filters, one that uses the corrected barometric altimeter signal and another that uses the radar altimeter signal.

The barometric altitude blending filter was not flight tested because of schedule constraints. It was disabled by setting the low-pass filter gain to zero. However, the corrected baro data and DGPS altitude data were collected on data tape to assess corrected baro performance in postflight analysis. Radar altimeter blending with the DGPS altitude estimate is accomplished in the following manner. The GPS antenna-referenced DGPS/inertial/baro mean sea level altitude is received by the flight management computer (FMC) and transformed to an aircraft center of gravity position (CG) in runway coordinates. The radar altimeter measurement is converted from a ground-referenced altitude to a runway coordinate system-referenced altitude by correcting for the height of the runway altitude reference plane above the ground. This is done at 100 ft intervals. The difference between the DGPS altitude above the runway and corrected radar altimeter data provides a measurement to a two-state Kalman filter. This Kalman filter models a DGPS/inertial altitude bias error and a radar altimeter scale factor error. The resulting altitude bias error estimate is then subtracted from the DGPS Z-runway position estimate to provide the blended solution.

Terrain height is calculated from a two-dimensional terrain map in the X-runway and Z-runway axes. A plot of the terrain map implemented for the flight test is shown in Figure 5. This terrain profile was mapped using flight test data. The Figure 5 profile is the average difference between MLS Z-runway and CG-referenced radar altimeter from six different approaches. A zero terrain height is the runway altitude.

It should be noted that the radar altimeters are always used on the TSRV for autoland flare guidance/control below 42 ft indicated radar altitude, regardless of whether the source of navigation data is DGPS/inertial or MLS.

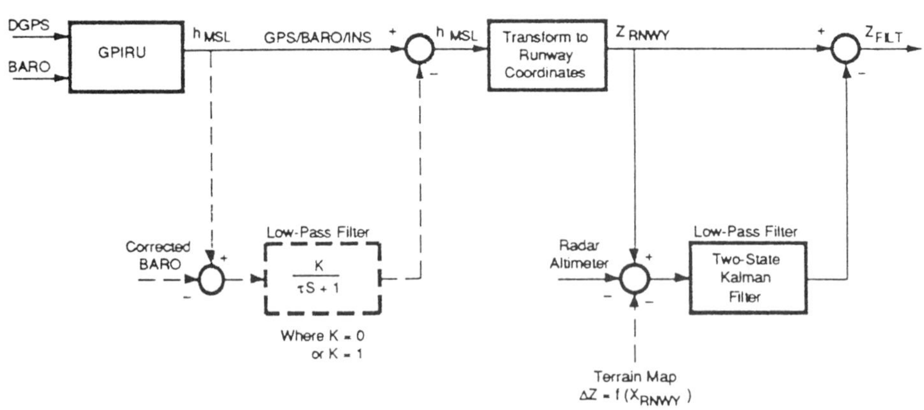

Fig. 4—DGPS/Inertial/Altimeter Blending Block Diagram

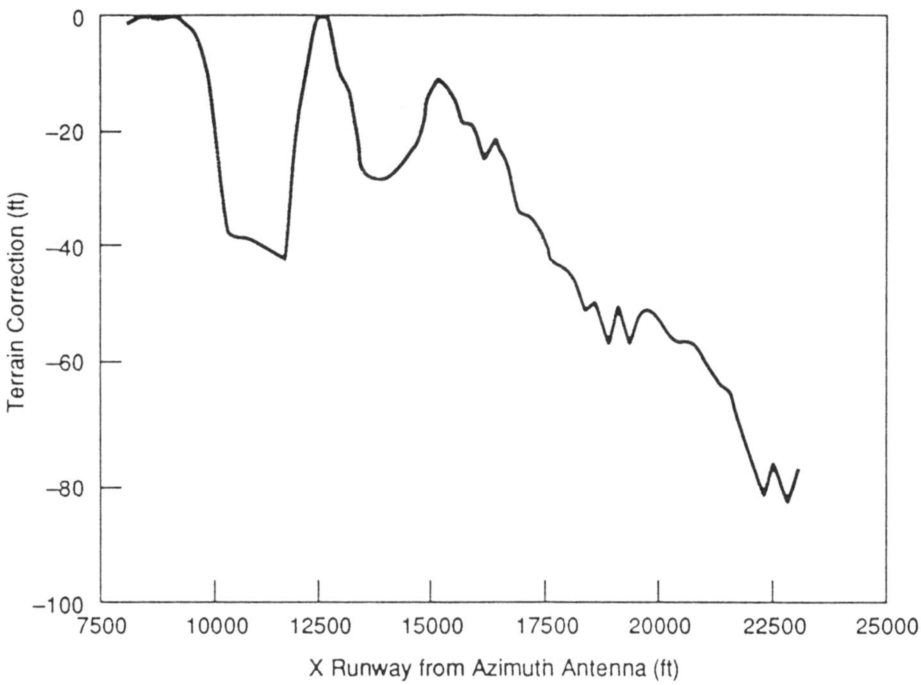

Fig. 5—Terrain Map for Wallops Runway 22 Approach

Differential GPS Ground Station

A DGPS ground station was installed at the NASA Wallops Flight Facility. The ground station uses an HG-2021 GPSSU that is identical to the unit in the TSRV aircraft. The GPSSU is operated with no modifications to its standard hardware or software.

The ground station's function is to calculate and transmit to the aircraft a differential correction message for each GPS SV that is considered valid. An SV is considered valid if it satisfies the dual criteria of having an elevation angle > 10 deg and a receiver-calculated carrier-to-noise ratio > 35 dB.

Ground station equipment includes the GPS sensor unit, an IBM PC clone, and radio equipment used to implement the VHF telemetry data link. The PC is subsequently referred to as the Differential Correction Processor, or DCP. A Kalman filter bank (one filter per SV) that estimates pseudorange errors is run in the DCP. The correction message outputs from the DCP are sent to the VHF data link over an RS-232 serial data bus.

The correction message for each SV contains the following quantities: SV ID number, time of correction, pseudorange correction, pseudorange correction rate, and message parity. The time of correction and the rate term are used to extrapolate linearly from the time of correction to the time of application in the aircraft. Corrections are computed in the DCP as shown in Figure 6. The general form of the ground station software is similar to that described in [7]. Within the DCP, pseudoranges are predicted using the surveyed position of the ground station antenna and the receiver's output of SV positions. The predicted pseudoranges are then differenced with the measured values. The pseudorange

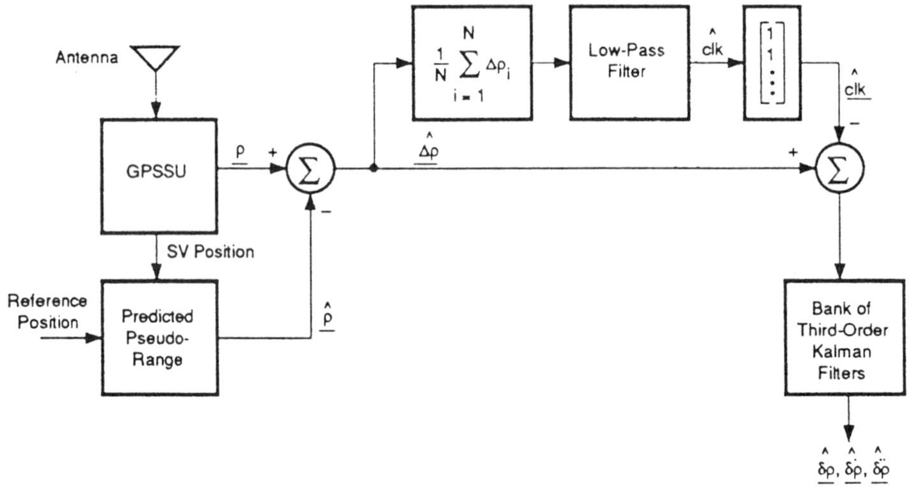

Fig. 6—Differential Correction Processor (DCP) Algorithm Structure

differences calculated for each SV contain a common error due to the receiver's clock error, which can be very large (range equivalent values on the order of 10^6 m or larger). The clock error is removed from each pseudorange difference by subtracting a heavily filtered average of all valid pseudorange differences from each individual difference. That is, the mean value of the pseudorange differences for all SVs tracked is interpreted as GPSSU clock phase error. Removal of the clock error from the pseudorange differences reduces the dynamic range of the pseudorange correction messages transmitted across the data link.

After removal of the clock error estimate, the individual clock-corrected pseudorange differences are Kalman filtered using a random acceleration third-order model. The filter design parameters (process and measurement noise variances) are chosen to filter receiver noise effects as much as possible without jeopardizing the ability of the filter to track variations due to SA errors. An estimate of the dynamics of SA errors derived from the data presented in [8] was used to establish frequency response requirements for the DCP range correction Kalman filters. (It should be noted that SA was apparently not active during the period of the DGPS flight tests.)

Differential Correction Data Link

A VHF telemetry data link was used to transmit the correction messages from the ground station to the TSRV aircraft. To make use of existing software, a decision was made to format the correction messages in ASCII. A more compact format for the correction messages could have been defined, but was not necessary for this project. The correction message for each SV is 25 ASCII characters in length, and is sent at 1200 baud to the data link. At this rate (allowing for transmission overhead), one correction message for a single SV is transmitted approximately every 2 s. With a typical operating condition of 5 SVs in track, the full set of differential corrections is updated every 10 s. The range capability of the data link was about 60 mi.

Test Instrumentation

Test instrumentation on the 737 TSRV supported recording an extensive set of GPS, inertial, and altimeter parameters during the testing for postflight analysis. The testbed display, control, and recording functions were served primarily by two μVAX computers. These two μVAXs are referred to as the FMC VAX and the Displays VAX. A DATAC serial data bus provides the link among the computers, all test instrumentation, and the Data Acquisition System (DAS). Input data from MLS, DGPS, and other aircraft sensors are received by the μVAXs off the DATAC bus. DIF units provide buffers for all sensor data before they are read by the FMC and the DAS.

The FMC VAX processes MLS and inertial data through a set of third-order complementary filters to provide an inertially smoothed MLS signal to the autopilot, which also resides in the FMC. The FMC VAX transforms the DGPS/inertial latitude, longitude, and altitude navigation solution to the aircraft CG location in runway coordinates for comparison with the MLS solution. The resulting difference data is output onto DATAC for data recording and display. Additional sources of hard-copy flight test information include 24 12-bit strip-chart channels, and snapshot data printed by a μVAX at the occurrence of preprogrammed events. The events chosen for the DGPS flight experiment were MLS altitudes of 900, 400, 200, 100, 50, and 30 ft, at the setting of a "land engage" discrete, and when the wheel spin discrete becomes true. The land engage discrete reflects capture of the 3 deg glideslope for a straight-in final approach, which typically occured between 400 and 900 ft of altitude. The wheel spin discrete is used as a touchdown indicator.

Two new displays were developed for the 737 TSRV that allowed real-time comparisons between the DGPS/inertial and MLS solutions. The primary engineering real-time display format is shown in Figure 7. This "bullseye" display

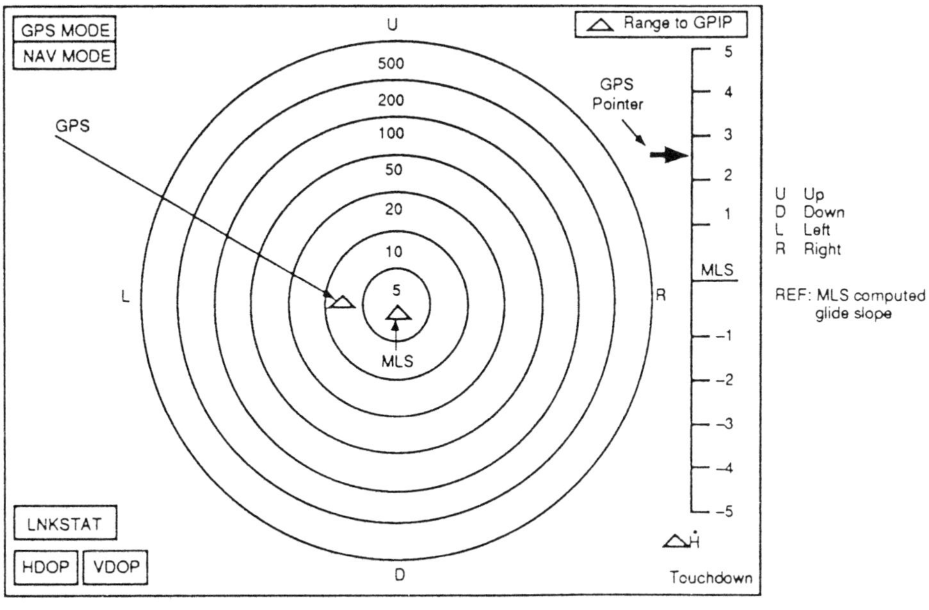

Fig. 7—GPS/MLS Difference "Bullseye" Display

showed the difference between DGPS/inertial and MLS position in all three axes, DGPS/inertial system and aircraft reference navigation system modes, difference in vertical velocities, GPS VDOP and HDOP, and the operational status of the telemetry link. The position differences were transformed to a rectangular coordinate system aligned with the glideslope. By aligning the differences to the glideslope, the component due to time synchronization error between MLS and DGPS is placed into the longitudinal axis, the least critical direction for the autoland application.

The primary flight display, which provides terminal guidance to the research pilot for manually flown approaches, was modified to show the GPS vertical and lateral deviations from the desired glideslope, along with the normally displayed MLS deviations. This display and the "bullseye" display were recorded for postflight analysis.

The Wallops Flight Facility includes a laser tracker, which was used to provide a third independent source of navigational data for postflight comparison with the DGPS/inertial and MLS solutions. The laser tracker provides aircraft position at 10 Hz and is believed to be accurate to within 3 ft at ranges up to 2 mi.

LABORATORY TEST RESULTS

Laboratory testing of the DGPS/inertial system fell into two categories. The first was designed to measure the level of receiver-unique errors, that is, those errors which cannot be removed by operation in a differential mode. These tests were performed to enable prediction of the expected performance during the flight test phase of the project. The second class of test operated the system on the ground in a configuration that matched (as closely as possible) the flight configuration. The purpose of these tests was to verify that all hardware and software interfaces in the DGPS/inertial system were performing properly.

The test setup for the first class of laboratory tests is shown in Figure 8. This test will be referred to as the dual differential correction processor test (or dual DCP test). In the dual DCP test, identical differential GPS ground station software operates on the outputs from two different GPS sensor units with their antennas essentially collocated. The known antenna position and the receiver outputs are operated on by a clock error estimator and a bank of Kalman filters (one for each SV) to produce estimates of the pseudorange errors for each SV.

Figure 9 shows a typical result of the dual DCP test. In this figure, differences between the estimated pseudorange errors for the two receivers are plotted. The figure shows a pseudorange correction difference for two SVs that were being tracked by the two receivers. Typical RMS values for the correction differences observed in the dual DCP tests were about 3.6 m, or nearly 12 ft.

Note that although the pseudorange correction differences in Figure 9 appear quite random on a time scale of several thousand seconds, they appear much more bias-like or slowly varying on the time scale of a single approach and landing. As a result, they can not be greatly reduced through filtering.

On the basis of the 12 ft RMS value for the receiver-unique pseudorange error, a flight test RMS lateral position error on the order of 10 to 15 ft was projected. For the vertical axis, where GDOP effects are larger, the predicted flight test RMS error was about 30 ft for the expected VDOP of about 2.5.

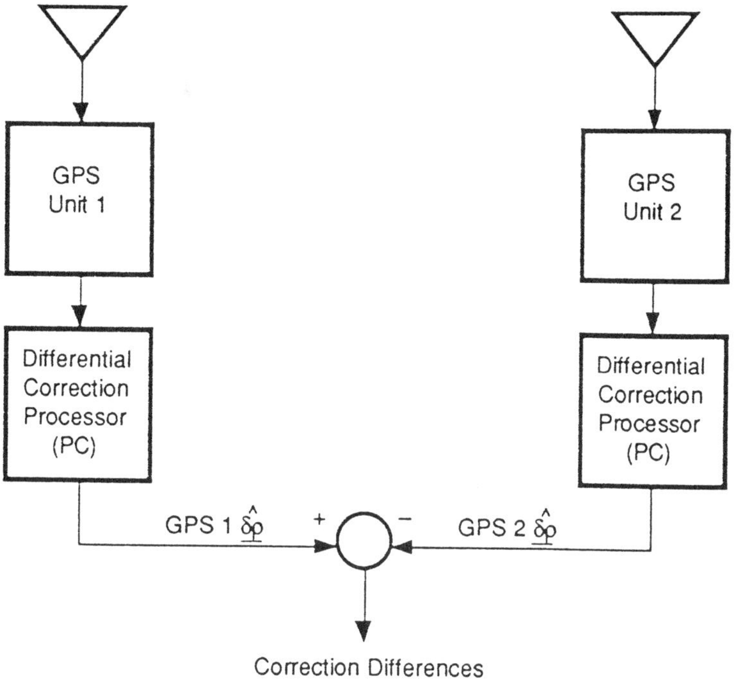

Fig. 8—Test Setup for Measurement of Receiver-Unique Errors

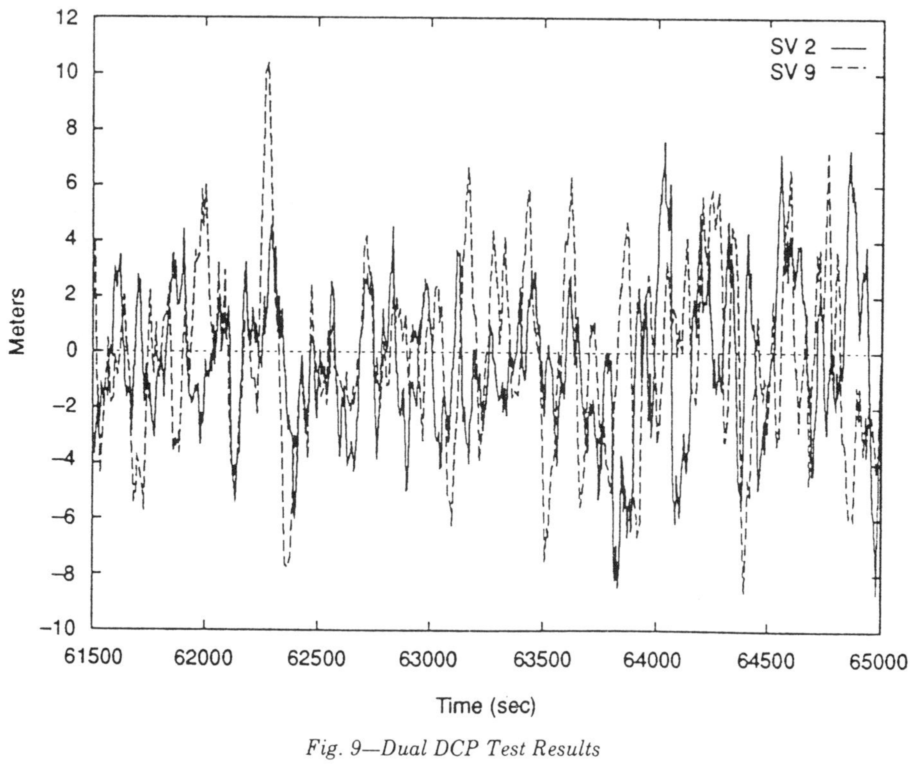

Fig. 9—Dual DCP Test Results

A second class of laboratory testing was performed to verify proper functioning of all system interfaces. Most of the testing was done using a single GPSSU to drive both the ground station and aircraft portions of the system. The single-GPS tests eliminated the effect of receiver-unique errors and allowed greater visibility of other system errors. These tests produced the desired results. They showed that, in the absence of receiver-unique error, the differentially corrected GPS/inertial solution would track the reference station position to an acceptable accuracy (typically within 1 m). The results verified that other errors in the DGPS/inertial system would not be significant relative to the receiver-unique errors measured in the dual DCP tests.

FLIGHT TEST DESCRIPTION

Flight Test Vehicle and Test Stages

The flight test vehicle was a Boeing 737-100, the TSRV, acquired by NASA Langley Research Center in 1974 for transport aircraft research. The TSRV is a unique aircraft in that it is outfitted with a second cockpit, the Research Flight Deck (RFD), located in what is normally the passenger cabin, as shown in Figure 10. While research is being conducted, the aircraft is normally flown from the RFD, which is coupled to the standard aircraft autopilot; however, the RFD only can command 50 percent of the autopilot control authority. Emergency disconnects are available in both the front cockpit and the RFD to allow

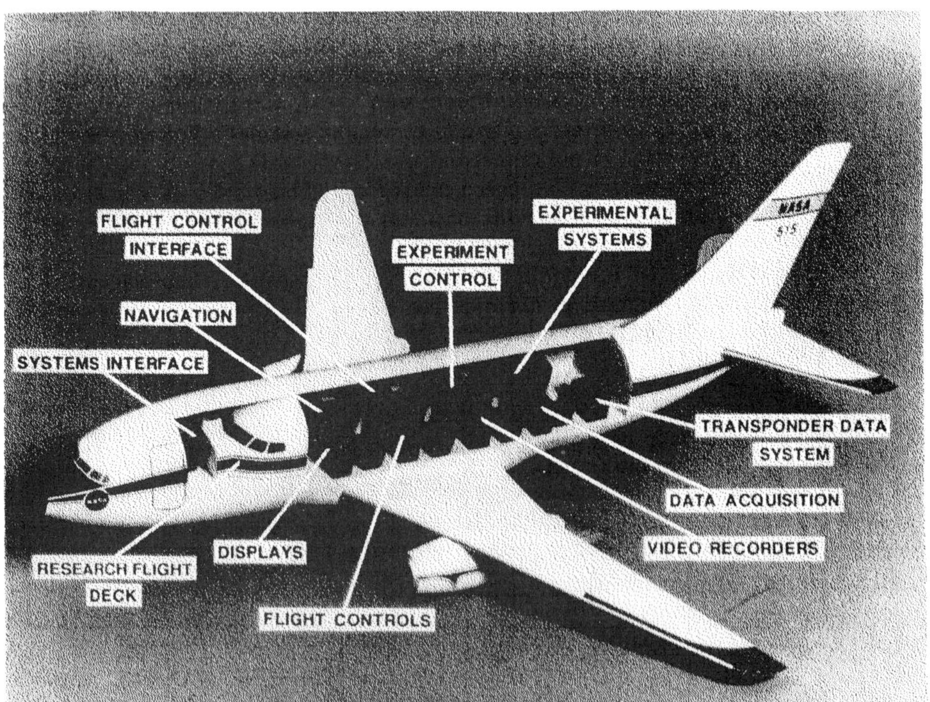

Fig. 10—TSRV Cutaway View

the front cockpit to take immediate control of the aircraft in case of a problem with the RFD. The front cockpit has a conventional 737-100 arrangement that was operated by two safety pilots.

The flight tests were conducted from October 2, 1990 to November 14, 1990. The MLS and laser tracking system at the NASA Wallops Island facility were used as references against which the DGPS/inertial system performance was compared.

The Wallops MLS uses a standard Distance Measuring Equipment (DME) rather than the precision DME normally used in an MLS installation. The standard DME has bias errors on the order of several hundred feet. A static calibration of the DME bias error was performed at the start of each flight day from a surveyed point on the runway. The result was a more accurate range measurement, which approximated the quality of a precision DME. The laser tracker was also calibrated over a surveyed test point each day.

The flight test consisted of four stages. During the checkout stage, DGPS/inertial system performance was evaluated in static testing and during manually flown landings. Flight management, telemetry, and data acquisition systems were verified during this stage.

In the second stage, data were gathered for various GPS/INS configurations while the aircraft was flown manually or coupled to MLS for several landings. During this stage, the accuracy of DGPS/inertial was confirmed for use in the autoland system. The approach paths flown during the project are shown in Figure 11. The solid line represents the pattern flown early in the test phase. To increase the number of approaches per flight hour, a tighter pattern (shown by the dashed line) was used for most of the testing.

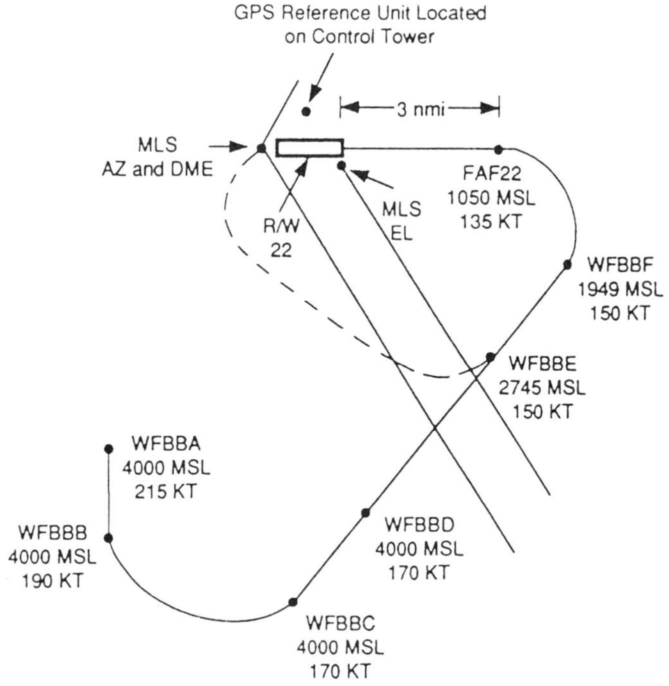

Fig. 11—Wallops Runway 22 Approach Patterns

During the third stage of testing, manually flown and automatic DGPS/inertial landings were performed. Manual approach/landings were flown by the research pilot using primary flight display guidance cues to ensure that GPS provided satisfactory glidepath guidance before GPS autolands were initiated. Then 27 DGPS/inertial autolands were performed under a variety of DOP conditions over three flight test days.

The fourth stage of the flight testing concentrated on radar altimer aiding to enhance the vertical accuracy of the DGPS/inertial solution. A total of 15 landings were conducted with the DGPS/inertial solution augmented using radar altimeters and a low-resolution terrain map.

Preliminary Flight Test Results

Over the seven flight test days, 109 DGPS/inertial approaches were flown. To reduce the 109 approaches to a set of known conditions quickly without selectively choosing individual approaches, the following criteria were imposed:

1) HDOP and VDOP < 5.0.
2) Removal of approaches immediately following moding to DGPS, to allow for settling time in the GPS/inertial Kalman filter.
3) Removal of approaches immediately preceding absolute (AGPS) moding. Moding to AGPS was typically due to loss of ground station corrections; therefore the preceding approach may not have been a DGPS approach through touchdown.
4) Removal of DGPS approaches with corrections available for only three or fewer SVs.
5) Removal of approaches made during the initial testing of a new ground station PC.

Application of the above criteria reduced the sample size from 109 to 70 DGPS approaches. All 109 approaches will be analyzed in more rigorous detail as the study progresses.

The set of 70 approaches considered in this section does not incorporate radar altitude blending. Vertical accuracy depends entirely on the DGPS/inertial/barometric altitude hybrid solution provided by the GPIRU. A subsequent section of this paper discusses the radar altitude blended results.

The results shown here are described as DGPS/inertial "errors." However, the results would be more correctly labeled DGPS/inertial versus MLS differences; that is, the results include both DGPS/inertial and MLS errors. It is assumed in the data reduction that the MLS errors are small compared with the DGPS/inertial errors. Planned postflight comparisons with laser tracker data will be used to examine/validate this assumption.

Using MLS as the reference, DGPS/inertial position errors in runway coordinates were examined for the 70 approaches. Results showed a 1-sigma lateral error (δY) of 8 ft about a mean of 3 ft (8.5 ft RMS) and a 1-sigma vertical error (δZ) of 14 ft about a mean of -6 ft (15.2 ft RMS). The longitudinal (δX) error, which includes the majority of the time synchronization error between DGPS/inertial and MLS, had a 1-sigma value of 30 ft about a mean of -6 ft.

Figure 12 is a plot of δZ vs δY, and Figure 13 is a plot of δX vs δY. These error data were computed from snapshot data of the MLS and GPS indicated positions at an MLS indicated altitude of 50 ft. The error was formed by subtracting the MLS indicated X, Y, or Z runway position from the DGPS indicated runway position. The mean and standard deviations of these 70 approaches represent the results presented in the preceding paragraph.

The error distribution in the Y and Z axes is reasonably clustered around a single "mean value." However, in the X axis, the distribution of the errors appears to be clustered around two distinct points. The suspected cause of this characteristic in the X axis is timing error in the synchronization of the DGPS/inertial and MLS solutions. The evidence is that the cluster of approaches in Figure 13 with an X error around 60 ft all occurred on one flight day immediately following a TSRV flight software reset. This software reset appears to have changed the timing involved in the comparison of the DGPS/inertial and MLS solutions.

The HDOP and VDOP, as determined by the aircraft GPSSU, were recorded for each of the approaches flown during the test program. The average HDOP for the set of 70 DGPS approaches was 1.4; average VDOP for the set of 70 was 2.5.

Flight results based on the above DOPs and the laboratory-measured 3.6 m (about 12 ft) receiver-unique RMS range error were predicted to be 17 ft horizontal and 30 ft vertical. However, a more detailed DOP analysis showed that most of the HDOP effect during the flight tests was in the longitudinal axis. The actual average value of the lateral or Y-axis DOP (YDOP) on the approach heading was about 0.8. This YDOP factor, together with the measured 12 ft

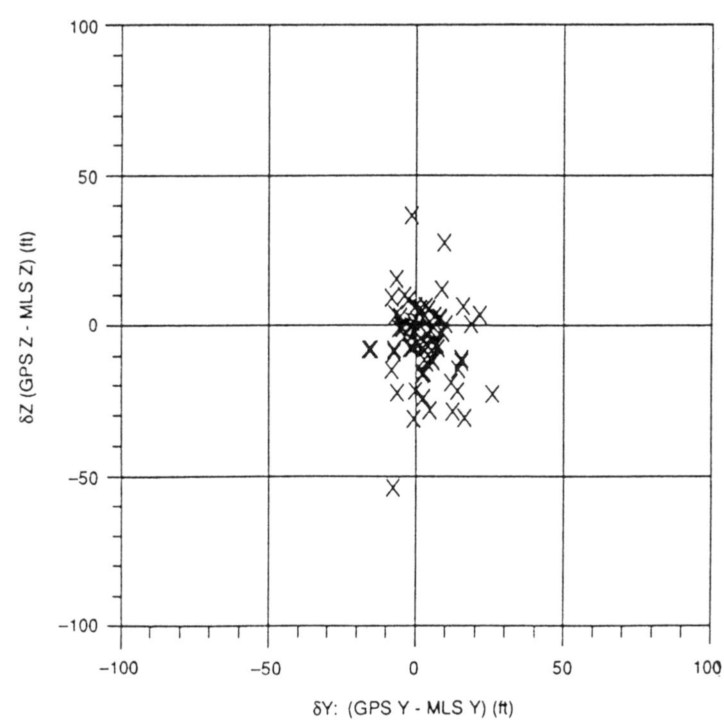

Fig. 12—DGPS/Inertial Lateral and Vertical Errors (relative to MLS as "truth reference")

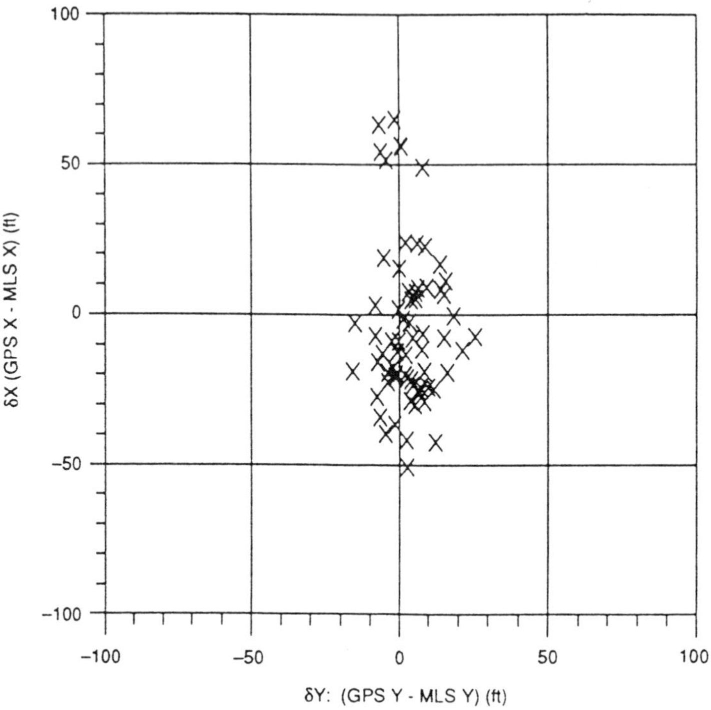

Fig. 13—DGPS/Inertial Lateral and Longitudinal Errors (relative to MLS as "truth reference")

range error, predicts flight test lateral RMS errors of about 9.6 ft, which agrees quite well with the actual 8.5 ft result.

Agreement in the vertical channel between the predicted and actual flight test results was not as good. While the predicted RMS vertical error was 30 ft, the actual 15.2 ft result was roughly a factor of 2 better. Additional analysis is required to determine why the vertical channel predictions based on laboratory results were so conservative.

As previously mentioned, the results presented above are taken from snapshot data at an MLS altitude of 50 ft. Snapshot data are also available at 400 ft, 200 ft, 100 ft, and touchdown. DGPS/inertial errors did not tend to change significantly over the short term. Errors remained essentially constant over the period of a single approach and landing. The 50 ft snapshot data are fairly representative of the DGPS errors for each approach from 400 ft altitude through touchdown.

Radar Altitude Blending Results

The radar altimeter blending filter, with terrain map, was flown on one flight test day for 15 approaches. Results from these cases showed a 1-sigma vertical dispersion of 4 ft about a mean of 3 ft (5 ft RMS). These results were calculated using snapshot data at an altitude of 400 ft. A vertical error histogram of the

15 approaches is shown in Figure 14. At touchdown, 1-sigma vertical dispersion was 1 ft about a 0.5 ft mean. The excellent touchdown results were expected because the filter had been processing radar altimeter measurements over the runway prior to touchdown.

Figure 15 shows a comparison of DGPS/inertial vertical error and DGPS/inertial/radar altimeter blended vertical error for a single landing. Radar altimeter measurements were masked out of the filter from approximately 200 ft altitude to 50 ft altitude. (This window was chosen to avoid processing measurements over a grove of 30 ft trees with roads through it, and through a creek bed prior to the runway. Small errors in X position within this window could create large vertical errors due to inaccurate terrain map data.)

During the period in which the radar altimeter measurements were ignored, the solution propagated like the DGPS/inertial solution, as indicated in Figure 15. After the radar altimeter was brought back into the solution at 50 ft altitude (over the runway), the error decreased to about 1 ft. These results demonstrate the feasibility of using a radar altimeter and a terrain map to reduce the DGPS vertical errors.

The results of the flight tests described above show promise that a DGPS/inertial/radar altimeter navigation system augmented with a low-resolution terrain map could support spacecraft automatic landing applications. Based on a preliminary analysis of the results, the DGPS/inertial/radar altimeter (terrain map-augmented) system appears capable of meeting the International Civil Aviation Organization Instrument Landing System (ILS) Category I (200 ft decision height) approach accuracy requirements as described in [9]. However, much work remains to be done prior to successful implementation of

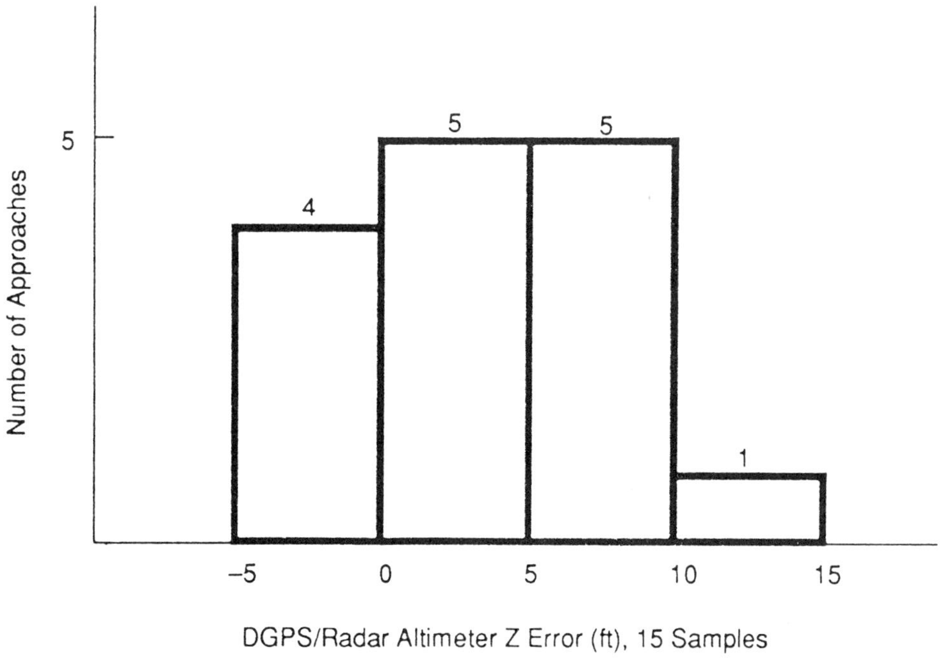

Fig. 14—Vertical Errors for Radar Altimeter-Aided Approaches

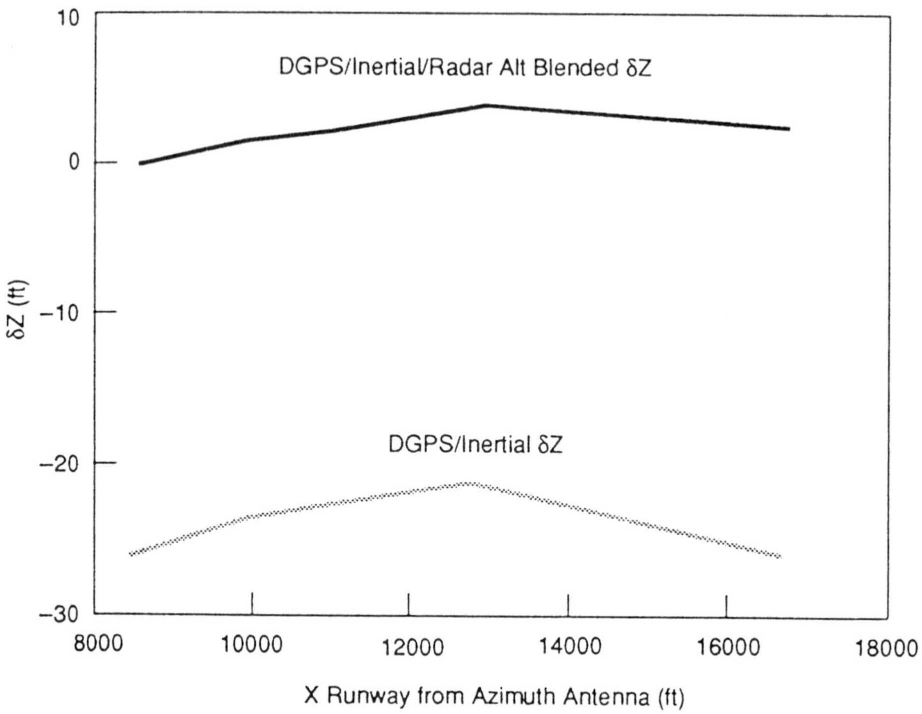

Fig. 15—Vertical Error With and Without Radar Altimeter

such a system, particularly with regard to system integrity and GPS coverage reliability.

SUMMARY

A differential GPS (DGPS)/inertial research project has been conducted to collect a system performance database and to demonstrate a DGPS-based automatic landing capability. The project has been successful with regard to both objectives. The highlight of the project was the successful execution of 36 differential GPS/inertial automatic landings of the 737 Transport Systems Research Vehicle (TSRV) test aircraft. In total, over 120 approaches were flown, for which an extensive set of DGPS/inertial, Microwave Landing System (MSL), and laser tracker position data was recorded. Further analysis of the data is in progress.

Preliminary analysis of the flight test results shows RMS lateral errors of 8.5 ft and RMS vertical errors of 15.2 ft in the DGPS/inertial mode. The DGPS/inertial vertical channel results are about a factor of 2 better than the predicted values based on the laboratory tests; the cause of this apparent discrepancy is currently being investigated.

With the addition of radar altimeter data and a low-resolution map of terrain height along the extended runway centerline, RMS vertical errors were reduced to about 5 ft prior to crossing the runway threshold.

The project was conducted under a very aggressive schedule that dictated the use of existing equipment and capabilities to the fullest extent possible. The entire project (with the exception of postflight data reduction) was performed in less than 11 months. One aspect of the demanding schedule is that the results presented in this paper could probably be improved upon with enhancements to both the hardware and software used in the system. For example, the results shown here were obtained using an off-the-shelf GPS receiver designed for use in commercial airliners, with no specific design requirements relating to use in the differential application. In addition, the schedule requirements of the project dictated that only minimal efforts could be directed at design refinements for the existing GPS/inertial integration filter to support DGPS operation.

Future efforts on this DGPS/inertial project will concentrate on analysis of the extensive flight test database. The results of the project will be used to evaluate the capabilities of DGPS/inertial systems for a wide range of applications. In particular, the results will be used to determine the suitability of a DGPS/inertial system for manned and unmanned spacecraft automatic landing systems, and to help define the requirements for such a system. Much additional effort will also be required in several important areas not addressed by the current work; for example, the integrity of a GPS-based landing system must be carefully evaluated.

ACKNOWLEDGMENTS

The success of the DGPS/inertial flight test project discussed in this paper depended on the contributions of nearly 100 individuals. We wish to thank the NASA Langley technical team, the NASA Advanced Transport Operating Systems Program Office, and the members of the Honeywell and NASA management teams who enthusiastically supported this project. We also wish to thank the Honeywell Air Transport Systems Division for providing the GPS and inertial equipment used during the project and technical support during the system design phase.

REFERENCES

1. Braden, K., Browning, C., Gelderloos, H., Smith, F., Marttila, C., and Vallot, L., *Integrated Inertial Navigation System/Global Positioning System (INS/GPS) for Manned Return Vehicle Autoland Application,* IEEE Position Location and Navigation Symposium, Las Vegas, NV, March 1990.
2. Edwards, F. and Loomis, P., *Civil Helicopter Flight Operations Using Differential GPS,* 41st Annual Meeting of The Institute of Navigation, Annapolis, MD, June 1985.
3. Edwards, F., Hegarty, D., Turner, R., van Grass, F., and Sharma, S., *Validating the Airborne and Ground Based Components of a Differential GPS System,* Proceedings of the National Technical Meeting of The Institute of Navigation, Santa Barbara, CA, January 1988.
4. Jacob, T. and Schaenzer, G., *Integrated System of Differential Global Positioning System and Inertial Measurement Unit: A Position Determination System for Automatic Landing,* AIAA/SFTE/DGLR/SETP Fifth Biannual Flight Test Conference, Ontario, Canada, May 1990.
5. van Driel, N. and Krijn, R., *Results of a Test Program for the Use of Differential GPS for Approach Guidance,* Proceedings of the Third International Meeting of the Satellite Division of The Institute of Navigation (ION GPS-90), Colorado Springs, CO, September 1990, pp. 621–30.

6. Hartman, R., *Integrated Laser Inertial/GPS Navigation (GPIRS)*, Paper presented at the Royal Institute of Navigation Satellite Navigation Conference, October 1989.
7. *Design Document for Differential GPS Ground Reference Station Pseudorange Correction Generation Algorithm*, Interim Report DOT-CG-N-03-86, Wright Aeronautical Laboratories, WPAFB, OH, December 1986.
8. Kremer, G., Kalafus, R., Loomis, P., and Reynolds, J., *The Effect of Selective Availability on Differential GPS Corrections*, NAVIGATION, Journal of The Institute of Navigation, Vol. 37, No. 1, Spring 1990, pp. 39–52.
9. Hogle, L., *Investigation of the Potential Application of GPS for Precision Approaches*, NAVIGATION, Journal of The Institute of Navigation, Vol. 35, No. 3, Fall 1988, pp. 317–34.

NAVIGATION: *Journal of The Institute of Navigation*
Vol. 40, No. 2, Summer 1993
Printed in U.S.A.

Flight Evaluation of a Basic C/A-Code Differential GPS Landing System for Category I Precision Approach

W. HUNDLEY, S. ROWSON, and G. COURTNEY
Wilcox Electric, Inc., Kansas City, Missouri

V. WULLSCHLEGER and R. VELEZ
FAA Technical Center, Atlantic City, New Jersey

P. O'DONNELL
The MITRE Corporation, McLean, Virginia
Received September 1992
Revised March 1993

ABSTRACT

Flight tests were conducted to evaluate the capability of local area C/A-code differential GPS (DGPS) to provide accuracy suitable for precision approach operations under Category I conditions.

An aircraft was equipped with DGPS receiving equipment and additional computing capability that derived GPS-based instrument approach paths, converted DGPS position information to lateral/vertical deviations from the approach path, and drove the appropriate aircraft flight instruments. A mobile ground station using identical DGPS receiving equipment and a VHF data transmitter was employed to send differential corrections to the airborne system. Other than the differential corrections, no other augmentation of the airborne data was employed.

Analysis of data collected from 82 hand-flown approaches using ground truth provided by theodolite or laser tracker shows that the system achieved vertical accuracies meeting Category I requirements with a substantial margin when narrow-correlator spacing GPS receivers were used. Lateral accuracies met Category III requirements.

INTRODUCTION

There is a broad-based effort taking place across the civil aviation community to capitalize on the benefits that can be obtained from the implementation of a Global Navigation Satellite System (GNSS). This effort has been sparked by the deliberations of the International Civil Aviation Organization (ICAO) Future Air Navigation Systems (FANS) Committee; the declaration by the U.S. and Russian governments of the guaranteed free use of GPS and GLONASS for 10 and 15 years, respectively, following operational availability; the establishment of a special FAA Satellite Operational Implementation Team (SOIT); and the recent impetus provided by the RTCA GNSS Task Force. One of the key early implementation goals defined by the task force is to achieve an initial operational capability for a Special Category I GNSS precision approach by 1994. This paper describes one of many efforts being conducted by industry, government, and the rest of the aviation community in order to reach that goal.

REQUIREMENTS FOR PRECISION APPROACH AND LANDING SYSTEMS

Precision approach and landing systems, as opposed to nonprecision approach systems, provide positive vertical guidance information in addition to lateral guidance information. Because of this, they are able to provide guidance to much lower minimums, including all the way to touchdown. Typically, the guidance for nonprecision approaches is provided by various standard navigation aids, such as VHF Omnidirectional Range (VOR), Distance Measuring Equipment (DME), or Nondirectional Beacon (NDB). For precision approach, the Instrument Landing System (ILS) has served the worldwide aviation community admirably for over 40 years. The Microwave Landing System (MLS) is the designated successor to ILS, although not implemented as yet.

The requirements for both these systems—ILS and MLS—have evolved over the years and are documented by ICAO in Annex 10, International Standards and Recommended Practices (SARPs) [1]. Many of the requirements are categorized in terms of operational capability, e.g., runway visual range (RVR) and decision height (DH) at which the runway approach end must be visually acquired or a missed approach procedure executed. This has resulted in the familiar Category I, II, and III classifications (see Table 1). The requirements become increasingly stringent as the operational capability increases. An exception to this occurs in the case of MLS accuracy in that all MLS ground installations are required to provide a signal-in-space with precision suitable for Category III operations.

Although many facets of ILS and MLS are specified in the SARPs, the key requirements are contained in the triad of accuracy, integrity, and continuity of service.

Most but not all of the work done to date on the suitability of DGPS for precision approach and landing has concentrated on accuracy. The same is also true of this paper. It is a logical approach since if a particular system concept

Table 1—Operational Categories for Precision Approach

Category	Decision Height (ft)	Runway Visual Range (ft)
I	200	2600*
II	100	1200
IIIA	**	700
IIIB	**	150
IIIC	**	0

*Can be 1800 ft for appropriately equipped runway.

**No decision height, although some authorities require 50 ft for Category IIIA.

cannot deliver the required accuracy, integrity and continuity of service become moot. However, even if a system satisfies all the accuracy requirements but fails to provide satisfactory integrity and continuity of service, it will not be suitable for precision approach. It is important to note that the integrity and continuity of service requirements for Category III operation are typically one to two orders of magnitude more stringent than those for Category I.

New ways of looking at the accuracy requirement that may eventually blur the distinctions among the traditional operational categories are presently being developed, most notably RNP, the required navigation performance. RNP is a measure of the navigation system performance in a defined airspace. In the case of accuracy, the RNP is defined by the total system accuracy (also known as system use accuracy), which is the root-sum-square of the navigation source error, airborne receiving error, and flight technical error. The combination of navigation source error and airborne receiving error leads to the position fixing error of the navigation system. For precision approach, the total system accuracy can be defined by a "tunnel in space" within which an aircraft must fly on its descent to the runway, and whose rectangular cross-section shrinks in a linear manner as the aircraft approaches touchdown. For a navigation system with a given position-fixing capability, application of the tunnel in space concept has the potential for rewarding the well-equipped aircraft (e.g., one with a high-quality flight control system) with operation to lower minimums than a lesser-equipped aircraft.

The position-fixing accuracy requirement for a precision approach system cannot be specified adequately by a single number. The modern approach for specifying landing system precision is to specify accuracy in terms of three parameters: (1) path-following error (PFE), (2) path-following noise (PFN), and (3) control motion noise (CMN). In specifying accuracy for MLS, ICAO Annex 10 defines PFE as that portion (spectral component) of the guidance system error which will result in an actual aircraft displacement from the desired flight path. It includes both a bias term and a noise term (PFN). CMN is defined as that portion of the guidance system error which during coupled flight (i.e., guidance system coupled to the autopilot) results in control surface, wheel, and column motion and possibly attitude angle change, but does not cause aircraft displacement from the desired flight path. Annex 10 also defines the guidance error spectral regions for PFE and CMN.

In the case of ILS, which is an older system, Annex 10 specifies the error in terms of course structure or bends in the flight path that can be considered equivalent to PFN, as well as bias errors. The combination of structure error and bias can be considered to be the PFE.

CMN is not specified for ILS; however, Attachment C to Annex 10 stresses the desirability of roll and pitch attitude changes of less than 2 deg in the final phases of the approach.

The question that arises, then, is the accuracy performance standard to which we should hold the Category I-capable DGPS system, and how it should be specified. Based on the ILS' outstanding safety record and long history of satisfactory performance, it is both logical and prudent to use the current ILS standards as the initial measuring stick. Employing ILS specifications for DGPS on an exact one-for-one basis does have some disadvantages. These arise not only because many of the principal error mechanisms differ, but also because ILS is basically an angular system and is specified as such, while GPS is a linear system. The angular specifications for ILS do not reflect the naturally superior accuracy of GPS farther out on the approach path, which could be turned to advantage for obstacle clearance and general airspace planning purposes. On the other hand, the fact that the linear accuracy of ILS improves as the runway is approached is advantageous for a landing system and is reflected in the ILS standards. In any event, meeting the current ILS standards down to the 200 ft decision height provides a proven benchmark. A system that meets these standards has the advantage of allowing a minimally equipped but IFR-capable pilot/aircraft to execute precision approaches to present Category I minimums, while not restricting more sophisticated aircraft from achieving lower minimums under the tunnel in space concept when that concept is implemented.

The ILS standards are found in ICAO Annex 10. Since CMN is not specified for ILS, we have used the value specified for MLS, which is a Category III requirement. Tables 2 and 3 show the required accuracies for a Category I capability assuming a 3 deg glideslope approach to a 10,000 ft runway with a 50 ft threshold crossing height (TCH).

SYSTEM APPROACH

The initial system development effort established a set of characteristics that were to be incorporated if at all possible in the experimental Wilcox DGPS precision approach system. Among the most important were the following:

1) Achieve an accuracy at least equivalent to present-day ILS Category I performance.
2) Achieve the required lateral and vertical guidance accuracy without requiring special equipment on the aircraft, such as inertial reference

Table 2—ICAO Annex 10 Category I Lateral Error Requirements (95%)
10,000 ft Runway

Height (ft)	Distance from Threshold (nmi)	Bias (m)	PFN (m)	PFE (m)	CMN* (m)
200	0.47	15.5	13.4	20.5	4.2
400	1.10	20.1	20.6	28.8	5.5
700	2.04	27.1	34.3	43.7	7.6
1000	2.98	34.0	51.6	61.8	9.8

*CMN is a Category III MLS requirement.

Table 3—ICAO Annex 10 Category I Vertical Error Requirements (95%)
3 deg Glide Path, 50 ft Threshold Crossing Height

Height (ft)	Distance from Threshold (nmi)	Bias (m)	PFN (m)	PFE (m)	CMN* (m)
200	0.47	3.4	2.9	4.5	1.2
400	1.10	6.8	5.9	9.0	2.5
700	2.04	11.9	10.2	15.7	4.5
1000	2.98	17.0	14.6	22.5	6.6

*CMN is a Category III MLS requirement.

units, radar altimeters, terrain contour maps, heads-up displays, and flight management systems.

3) Achieve required vertical accuracy with VDOP ≤ 4.

PROGRAM OBJECTIVES

The primary objective of the Wilcox development program to date has been to determine the precision approach performance, in terms of lateral and vertical deviations from a typical straight-in 3 deg glide path approach to ILS minimums, that could be achieved when all parameters available to the system designer are optimized. To this end, it was decided to employ the latest state-of-the-art receivers. In a field in which the technology is still evolving rapidly, this is important if one is to obtain a current benchmark. Some of the earlier test programs [2, 3] employed receivers which by today's accuracy standards are relatively crude.

The flight evaluations used the following receivers: (1) Ashtech Ranger and (2) Novatel Model 951R. Both of these receivers have features that are desirable for the precision approach application. In addition to being parallel multichannel (10 or 12 channel), both have higher position update rates than the conventional 1 Hz. The Ashtech has a 4 Hz update rate and the Novatel 5 Hz. At an update rate of 1 Hz, an aircraft on approach will typically travel about 200 ft and sink 10 ft between position fixes. Update rates of at least 5 Hz are necessary for good dynamic performance and smooth autopilot coupling. In addition, the Ashtech receiver achieves noise reduction through carrier smoothing of the code [4], while the Novatel receiver minimizes the effect of receiver noise through a novel narrow-correlator tracker [5], which also reduces multipath [6] effects. Another technique for reducing multipath that was evaluated was the use of a choke ring on the ground antenna. Since receiver noise and multipath are the two error contributors that cannot be corrected by differential techniques, reducing them to a minimum will result in a direct improvement in performance.

Another objective was to obtain data on what performance improvement, if any, might be obtained with a high-speed data link operating at 9600 bps and providing differential corrections at an update rate of 1 Hz, versus a lower-speed link at 1200 bps providing updates at 1/3 Hz. The required correction update rate is determined by the pseudorange accelerations imposed by Selective Availability (SA).

The flight tests were conducted in four phases. Phases A, B, and C were flown by Wilcox at Richards-Gebaur Airport in Kansas City using various combinations of GPS receivers and data links. Phase D was flown by Wilcox in cooperation with the FAA Technical Center at Atlantic City International Airport.

FLIGHT EVALUATION EQUIPMENT

The flight evaluation equipment consisted of several configurations of the local area DGPS (LADGPS) landing guidance system that was evaluated, plus a tracking system used as a reference against which the guidance system was compared. The guidance system consisted of a DGPS ground reference station that computed and transmitted differential pseudorange corrections over a

VHF data link, plus DGPS receiving, processing, and display equipment installed in a Beechcraft King Air 300. For the earlier phases of testing in Kansas City, the tracking system comprised a radio telemetering theodolite (RTT) on the ground, as well as equipment in the King Air that received and processed the theodolite data. The flight tests performed in Atlantic City used the FAA Technical Center's laser tracking system.

The ground reference station, depicted schematically in Figure 1, used an Ashtech Ranger 12 channel, standard-correlator-spacing GPS receiver with carrier-aided tracking for the first phase of testing, and a Novatel Model 951R 10 channel, narrow-correlator-spacing GPS receiver card installed in a laptop PC for the later phases. Differential corrections were input serially to a VHF transmitter/modem (136.2 MHz) that transmitted them to the aircraft at 1200 bps once every 3 s. This data link provided corrections at the aircraft receiver, with the age of the data ranging from 3.5 to 6.5 s. It was later replaced with a 9600 bps data link transmitting at 1 Hz, and reducing the correction ages to 0.5–1.5 s. Both of these data links had reliability problems, so a third data link was constructed that transmitted corrections at 1200 bps. This data link eliminated a buffering delay inherent in the other data links and implemented the RTCM SC-104 corrections format, the net result being a reduction in latency, and an update rate of approximately 1 Hz with ages of 1.0–2.0 s. At Richards-Gebaur Airport, the L-band antenna was mounted on a 7 ft pole located directly over a survey marker situated about 600 ft to the side of the stop end of runway 36. During narrow-correlator receiver tests, a choke ring was mounted under the L-band antenna to reduce multipath effects. At Atlantic City, the L-band antenna was laid a few centimeters above the ground (without choke ring), about 3500 ft to the side of the runway 13 midpoint.

The ground component of the tracking system initially consisted of an RTT, shown in Figure 1. An operator kept the theodolite cross hairs pointed at the nose of the aircraft at all times during approaches, and the angular deviation of the aircraft from the runway centerline or 3 deg glide path was transmitted to the airborne tracking equipment over a UHF data link. For lateral angle

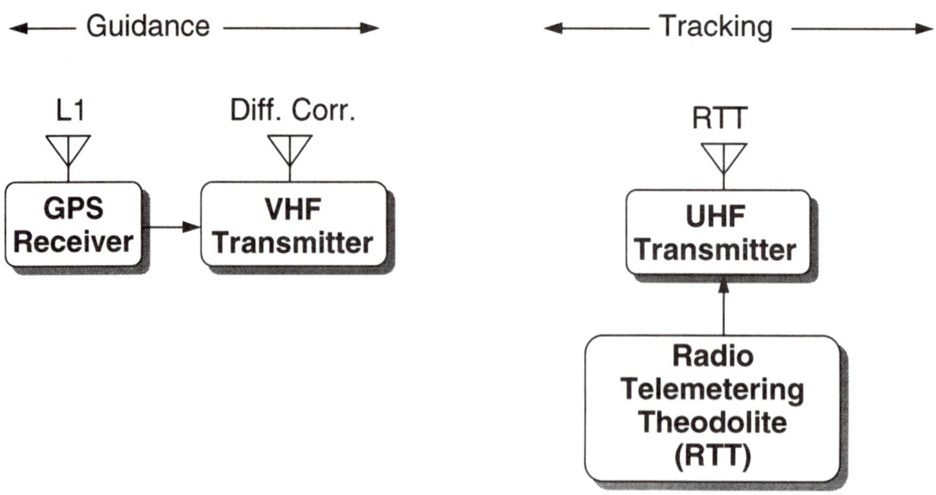

Fig. 1—Ground-Based Guidance and Tracking Equipment

measurements, the RTT was located on the extended runway centerline, 850 ft beyond the stop end. For vertical angle measurements, the RTT was located adjacent to the runway, 195 ft from the centerline and 1330 ft from the threshold. RTT accuracy was ± 0.02 deg, 95 percent, plus an operator pointing error of about 1 ft in smooth air. During periods of heavy turbulence, the operator cannot track sudden movements of the aircraft perfectly, and the pointing error increases. The laser tracker at Atlantic City was located about 2000 ft to the side of the runway midpoint. It locked onto a retroreflector mounted on the aircraft nose when the aircraft was approximately 6–8 mi from the runway threshold, and automatically tracked aircraft position in three dimensions throughout the approach. Position data were recorded on magnetic tape at 10 Hz with GPS-derived time tags. Tracker data were matched with time-tagged airborne GPS data during post-processing. Laser tracker angular accuracy was 20 arcsec, with a ranging accuracy of 1 ft.

The airborne guidance and tracking systems (RTTs) are shown in Figure 2. The airborne L-band antenna (Sensor Systems model S67-1575-39) was mounted in the top of the fuselage, about midway between the nose and tail. The airborne GPS receiver was initially an Ashtech Ranger and subsequently a Novatel Model 951R GPS receiver card. Differential corrections were input directly to the GPS receiver from a VHF receiver/modem.

The airborne GPS receiver computed three-dimensional position at a 2 Hz or 4 Hz (Ashtech), or a 5 Hz (Novatel) rate. The laptop PC converted position data in real time to angular deviation from the desired approach path, using a file of runway coordinates and elevations stored on disk. Horizontal and vertical deviations from the desired flight path were output in real time as analog voltages from the D/A converter that drove the course deviation indicator (CDI) located in the cockpit.

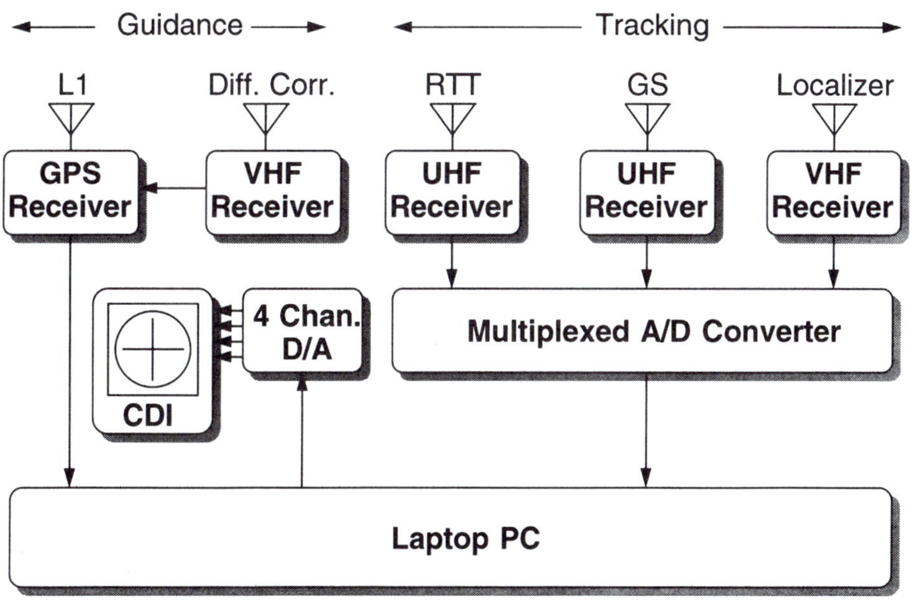

Fig. 2—Airborne Guidance and Tracking Equipment

The airborne equipment recorded time-tagged GPS position, ILS localizer angle, ILS glide slope angle, and RTT angles on the PC's hard disk for post-processing and plotting. The ILS data were recorded for use as a backup reference; however, ILS data were not used in the analysis reported in this paper. RTT, glide slope, and localizer data were received on three separate receivers (two UHF and one VHF), and provided as analog voltages to a multiplexed A/D converter mounted in the PC. RTT, glide slope, and localizer angle were sampled and recorded on disk at 2, 4, or 5 Hz, synchronized with GPS position measurements.

Post-processing software computed and plotted instantaneous angular deviation from approach path, PFE, and CMN. DGPS errors could be plotted and analyzed using the theodolite, laser tracker, or ILS signals as reference. Since the RTT angular measurements were not supplemented by independent distance measurements, one may question the accuracy of the conversion of angular errors (degrees) to cross-track errors (meters). Although this conversion was accomplished using distance from the RTT as measured by DGPS, the along-track DGPS distance errors are bounded, and they contributed negligibly to the calculated cross-track errors since the distance to the RTT was always large compared with the cross-track distance from the desired glide path.

DESCRIPTION OF TESTS

The first 47 approaches were flown at Richards-Gebaur Airport, a former Air Force base located on the southern edge of Kansas City. This facility was chosen because it has an operational ILS and light traffic. ILS data were monitored on the flight test panel in the rear of the aircraft during some approaches in order to verify in real time that the DGPS landing system was operating properly. All approaches were flown to runway 36, which provides the ILS approach. The next 35 approaches analyzed were flown to runway 13 at Atlantic City International Airport. A total of 43 approaches were flown at Atlantic City, of which 2 were aborted because of traffic, 1 was aborted when the system was inadvertently shut off in mid-approach, and 5 were not analyzed because of incomplete laser tracker data.

The pilot flew the approaches manually using the DGPS-driven CDI as the sole guidance instrument, although ILS indications, as well as visual cues, were also present during the Richards-Gebaur tests. During the Atlantic City tests, the pilot flew under the hood, and the cockpit ILS displays were turned off. This arrangement allowed the flyability of the system to be verified. In fact, the DGPS system flew indistinguishably from an ILS.

Since the RTT cannot measure lateral and vertical deviations simultaneously, separate lateral and vertical measurements were made at Richards-Gebaur. Each approach at Atlantic City provided both lateral and vertical data since the laser tracker provides three-dimensional aircraft position.

Equipment configurations for the four test phases are summarized in Tables 4 and 5. The receiver type (standard-correlator or narrow-correlator) was the same for the ground reference station and the airborne unit, depending on the phase (A, B, C, or D). However, the antenna configuration for the reference station changed somewhat, either by the use of a choke ring or by the type of mounting. During phase A, 15 approaches were flown with the standard correla-

Table 4—Cross-track Configuration and Errors ($|\mu| + 2\sigma$)

Parameter	ICAO Limit at DH (95%)	Phase A	B	C	D
Receiver type		Standard	Narrow	N/A	Narrow
Antenna		No choke ring	Choke ring	N/A	On ground
Position update rate		2 Hz	5 Hz	N/A	5 Hz
Baud rate		1200	1200	N/A	1200
Max. correction age		6.5 s	6.5 s	N/A	2.0 s
Tracker		RTT	RTT	N/A	Laser
SA		On	Off	N/A	On
Approaches		8	13	N/A	35
Average HDOP		1.9	1.2	N/A	1.4
Maximum HDOP		2.0	1.4	N/A	2.5
PFE at DH	20.5 m	2.6 m	3.9 m	N/A	1.0 m
PFE over final nmi		2.9 m	3.8 m	N/A	1.1 m
CMN* at DH	4.2 m	N/A	0.9 m	N/A	0.5 m
CMN* over final nmi		N/A	1.1 m	N/A	0.5 m

*CMN is a Category III MLS requirement.

Table 5—Vertical Configuration and Errors ($|\mu| + 2\sigma$)

Parameter	ICAO Limit at DH (95%)	Phase A	B	C	D
Receiver type		Standard	Narrow	Narrow	Narrow
Antenna		No choke ring	Choke ring	Choke ring	On ground
Position update rate		4 Hz	5 Hz	5 Hz	5 Hz
Baud rate		1200	1200	9600	1200
Max. correction age		6.5 s	6.5 s	1.5 s	2.0 s
Tracker		RTT	RTT	RTT	Laser
SA		On	On	Off	On
Approaches		7	7	12	35
Average VDOP		2.3	2.4	3.0	1.8
Maximum VDOP		4.0	2.9	3.6	3.6
PFE at DH	4.5 m	3.9 m	2.0 m	1.4 m	2.0 m
PFE over final nmi		4.0 m	2.3 m	2.0 m	2.0 m
CMN* at DH	1.2 m	1.1 m	0.9 m	0.8 m	0.7 m
CMN* over final nmi		0.8 m	0.8 m	0.8 m	0.7 m

*CMN is a Category III MLS requirement.

tor receivers and slower 1200 bps data link (6.5 s maximum correction age). For phase B, 20 approaches were flown; narrow-correlator receivers were substituted, and a choke ring was added to the ground station L-band antenna. For phase C, the data link was replaced by the 9600 bps unit (1.5 s maximum correction age), and 12 approaches were flown during which vertical data only were collected. It was discovered after the fact that SA was turned off during phase C and during lateral data collection for phase B. Because of the high correction update rates, especially for phase C, this is not believed to have affected the results materially. SA was in effect for all other tests. During phase D, the narrow-correlator receivers were tested with the faster 1200 bps data link (2.0 s maximum correction age). The purpose of phase D was to achieve greater statistical validity by flying a larger number of approaches

(35) and to use a more accurate three-dimensional position reference (laser tracker).

TEST RESULTS

Figures 3 through 11 show DGPS error data plotted for a representative approach flown at Atlantic City. All figures are for the same approach. They plot sensor (i.e., navigation) errors versus distance from threshold, and cover the range from the final approach fix at 4.3 nmi to the threshold. Figures 3, 4, and 5 show cross-track unfiltered guidance error, PFE, and CMN, respectively, using the laser tracker as ground truth. Figures 6, 7, and 8 show vertical errors versus the laser tracker. Figures 9, 10, and 11 show vertical errors using the RTT as ground truth. These figures illustrate the effects of PFE and CMN filtering. They also show the good agreement between the RTT and laser tracker near the runway threshold, and the progressive increase in RTT linear error at larger distances from the threshold.

Figures 12 and 13 show plots of cross-track and vertical unfiltered guidance error at 0.1 nmi increments for the final 2 nmi to the runway threshold for all approaches flown at Atlantic City with VDOP less than 4.

While these plots provide excellent visibility into the dynamic performance achieved on each individual approach, it is necessary to analyze the data on a statistical basis in order to quantify both the overall performance of the DGPS system and the effect on that performance of varying individual parameters of the system, such as type of receiver and correction data rate.

Two methods of data analysis were used. The first was to calculate the 95 percent errors at the 200 ft decision point for the various groups of

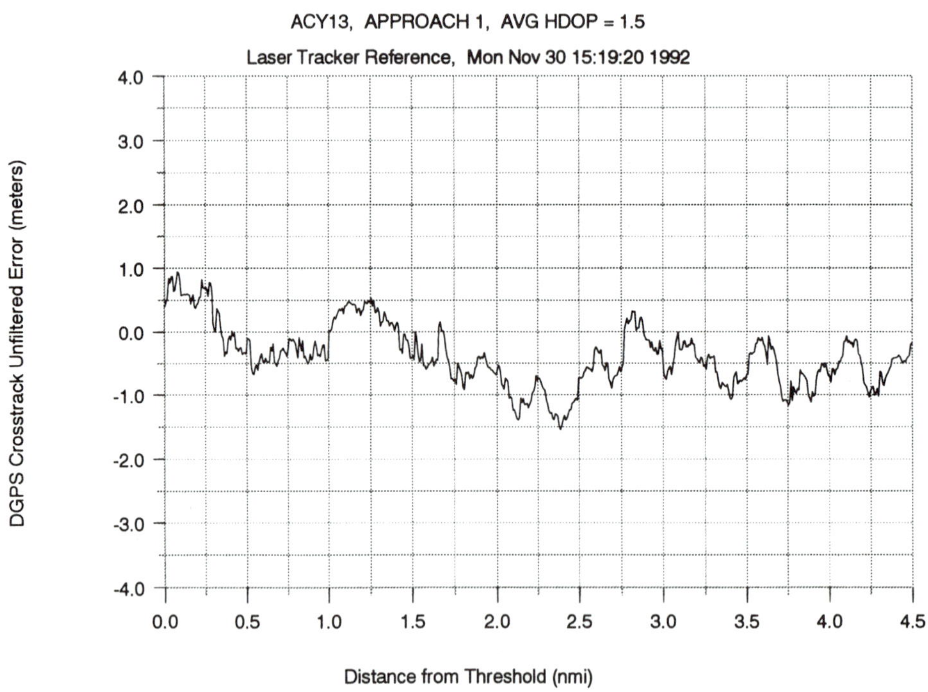

Fig. 3—Cross-track Unfiltered Error Using Laser Tracker Reference

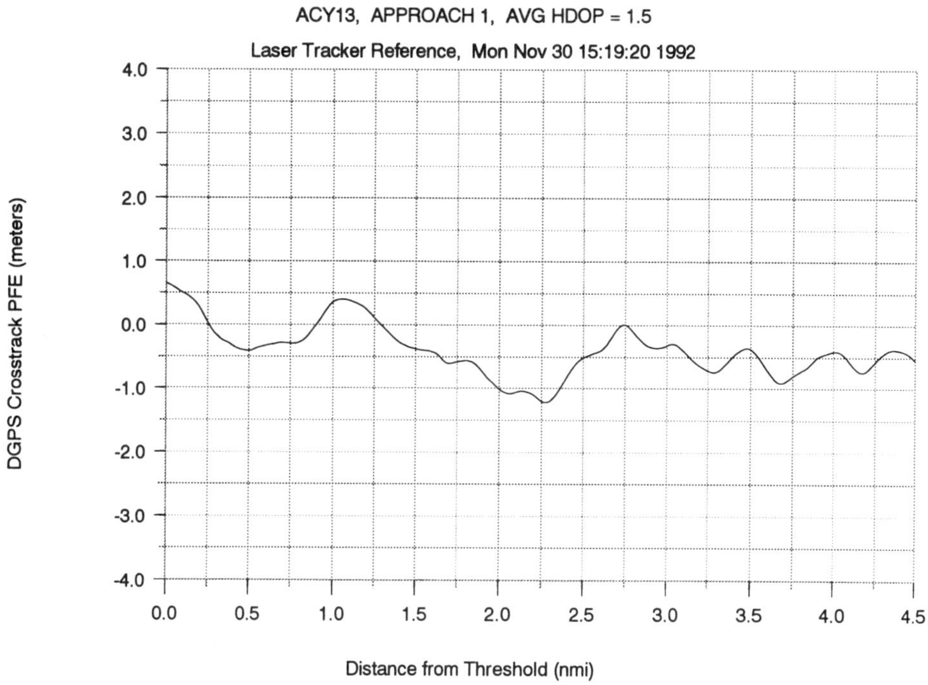

Fig. 4—Cross-track Path-Following Error Using Laser Tracker Reference

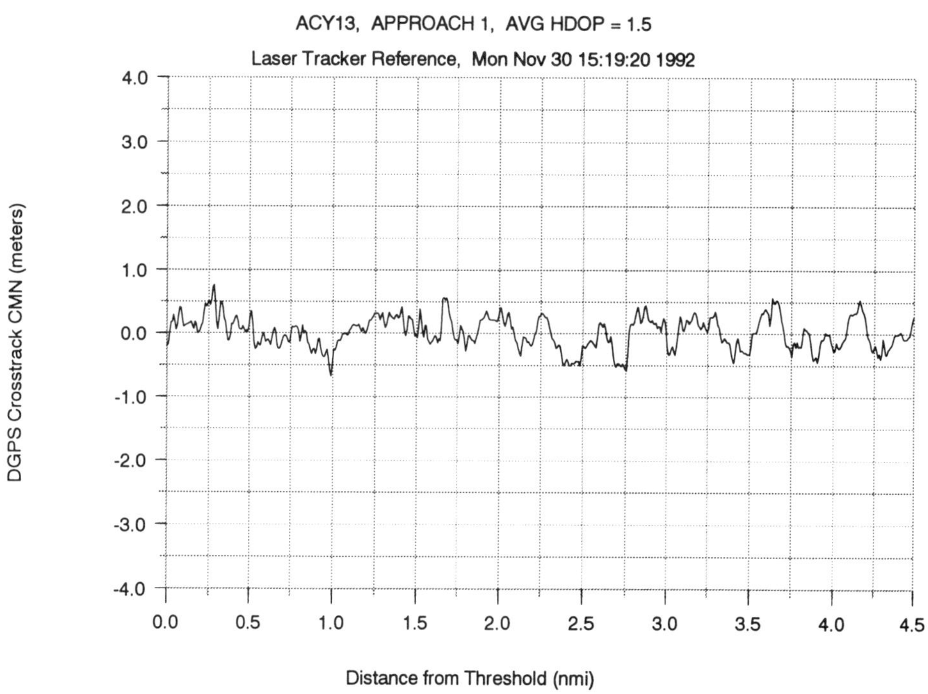

Fig. 5—Cross-track Control Motion Noise Using Laser Tracker Reference

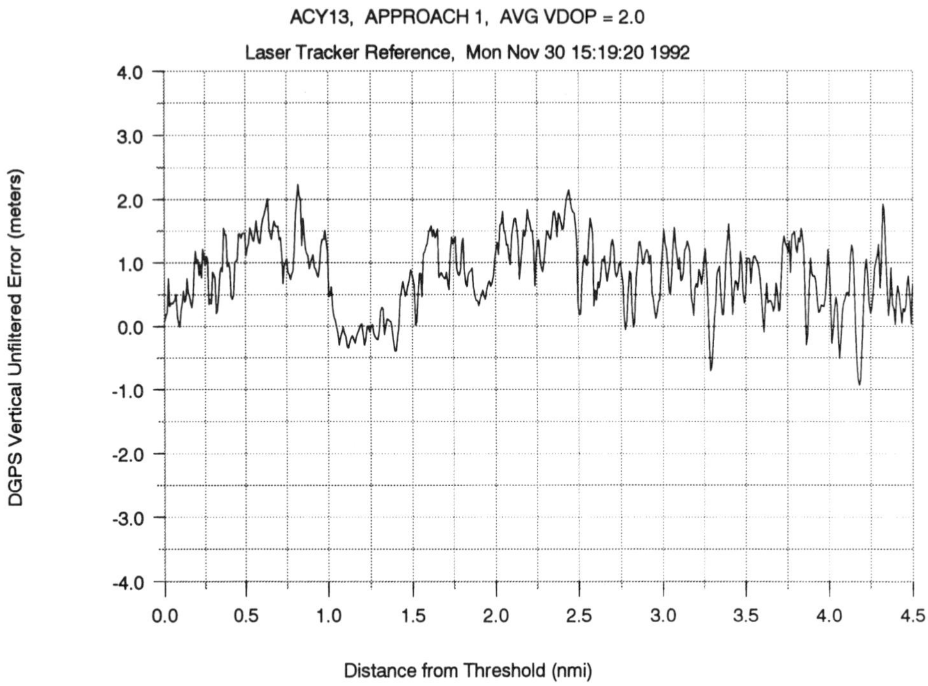

Fig. 6—Vertical Unfiltered Error Using Laser Tracker Reference

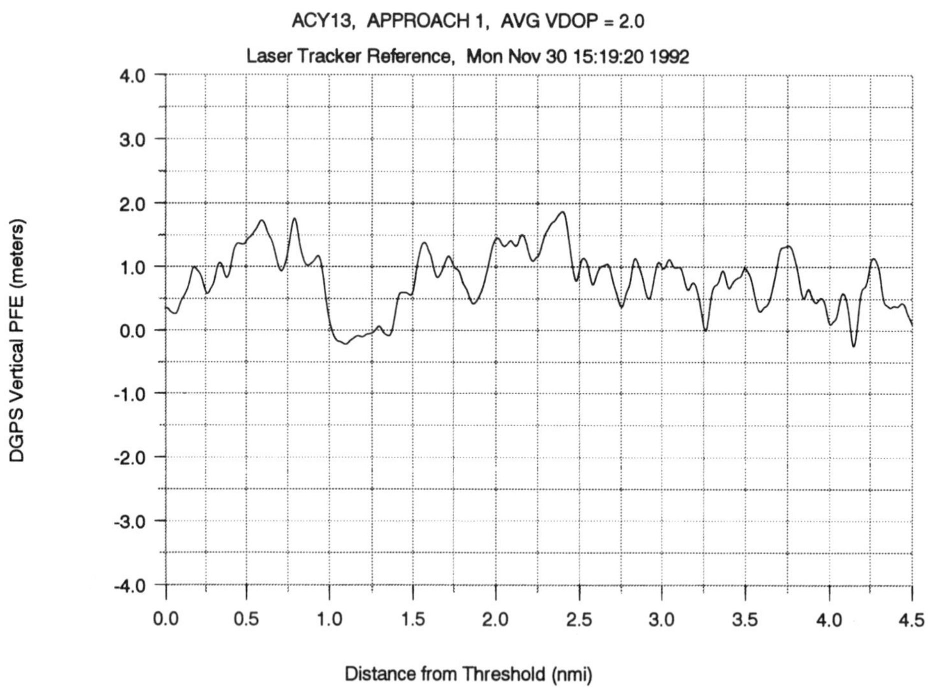

Fig. 7—Vertical Path-Following Error Using Laser Tracker Reference

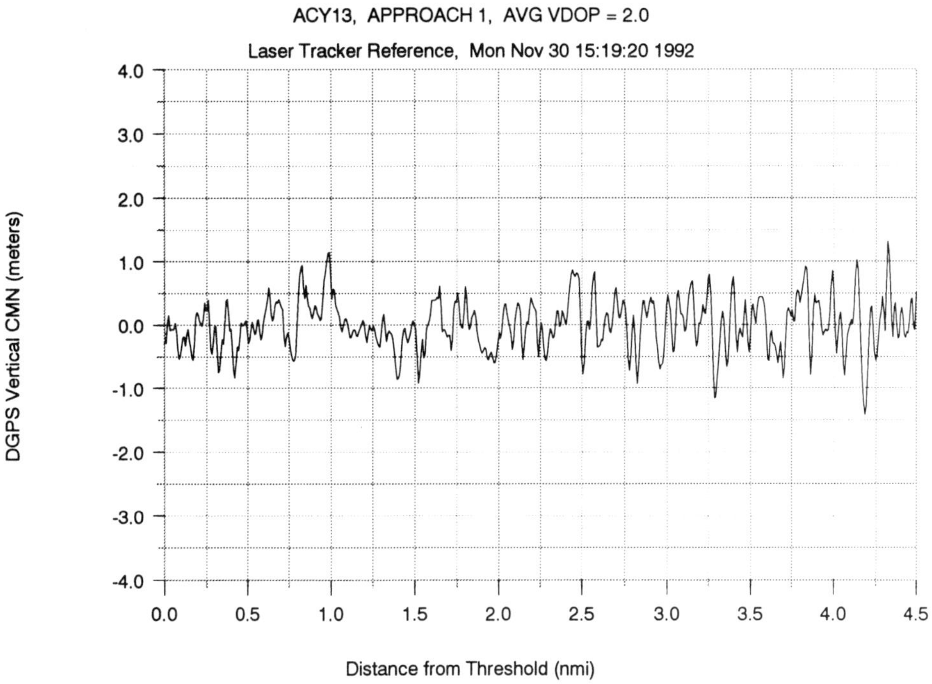

Fig. 8—Vertical Control Motion Noise Using Laser Tracker Reference

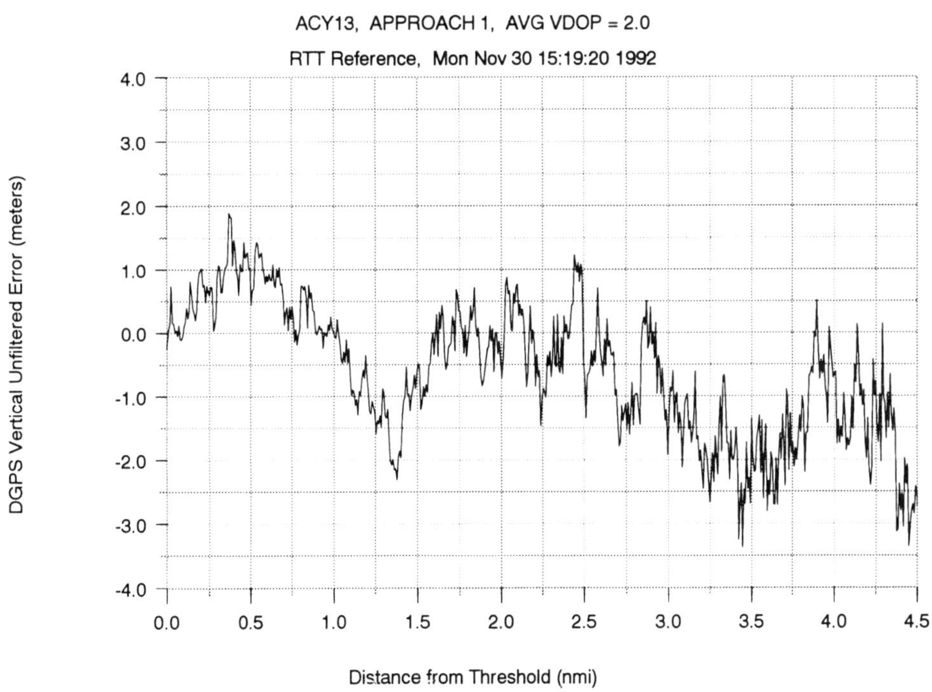

Fig. 9—Vertical Unfiltered Error Using Radio Telemetering Theodolite

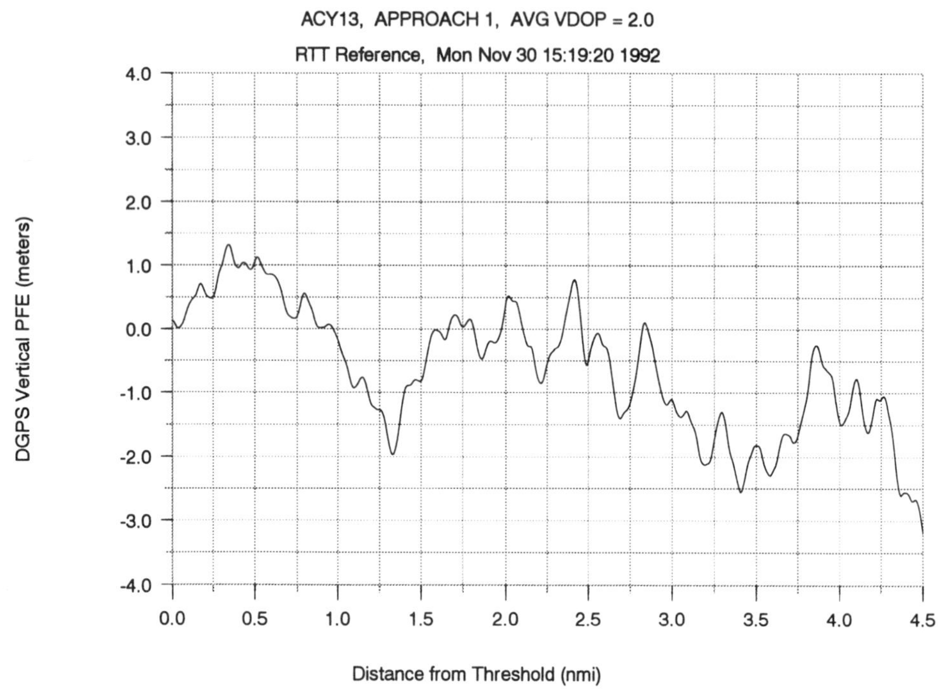

Fig. 10—Vertical Path-Following Error Using Radio Telemetering Theodolite

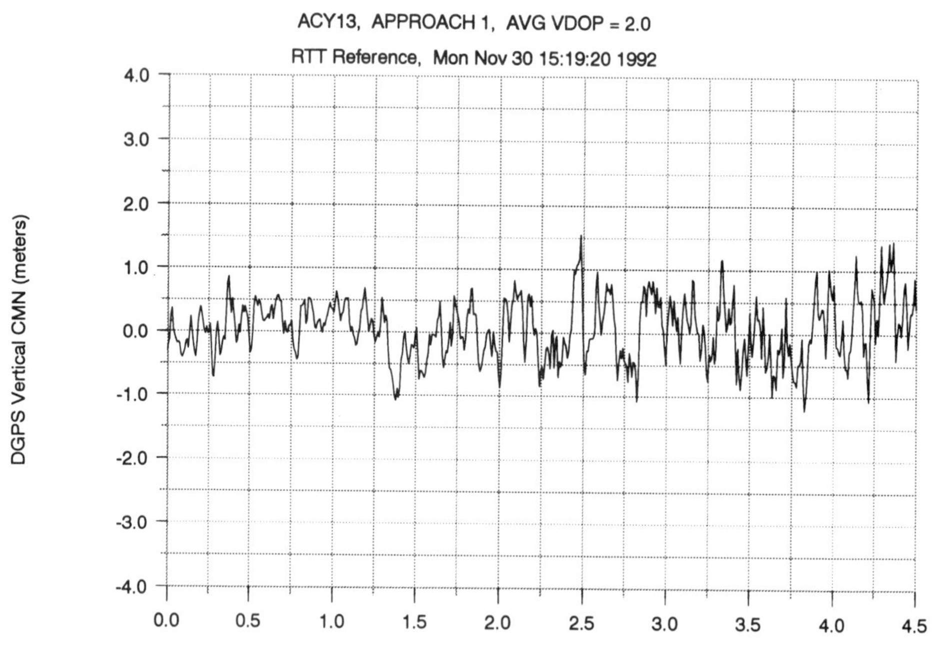

Fig. 11—Vertical Control Motion Noise Using Radio Telemetering Theodolite

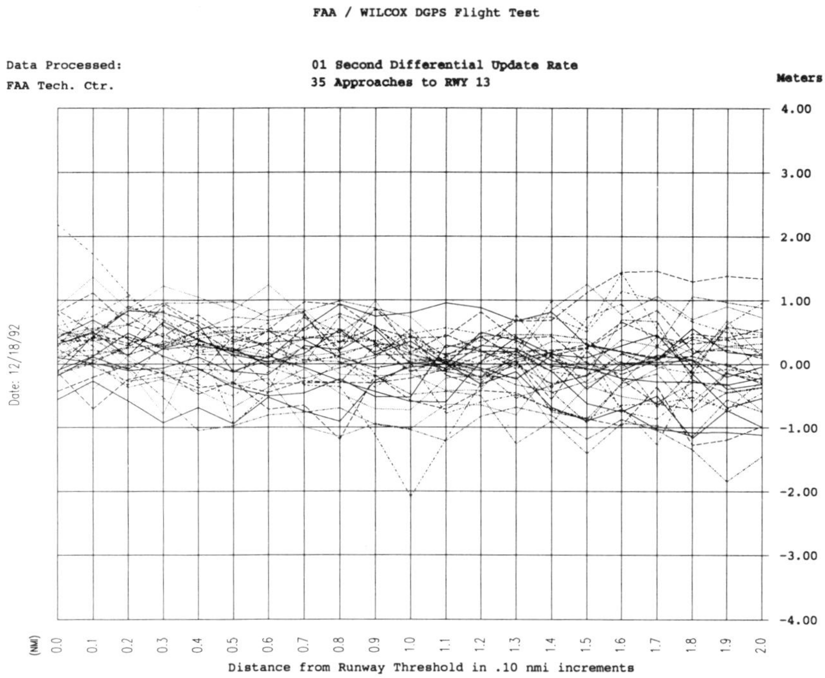

Fig. 12—Cross-track Unfiltered Error, 35 Approaches

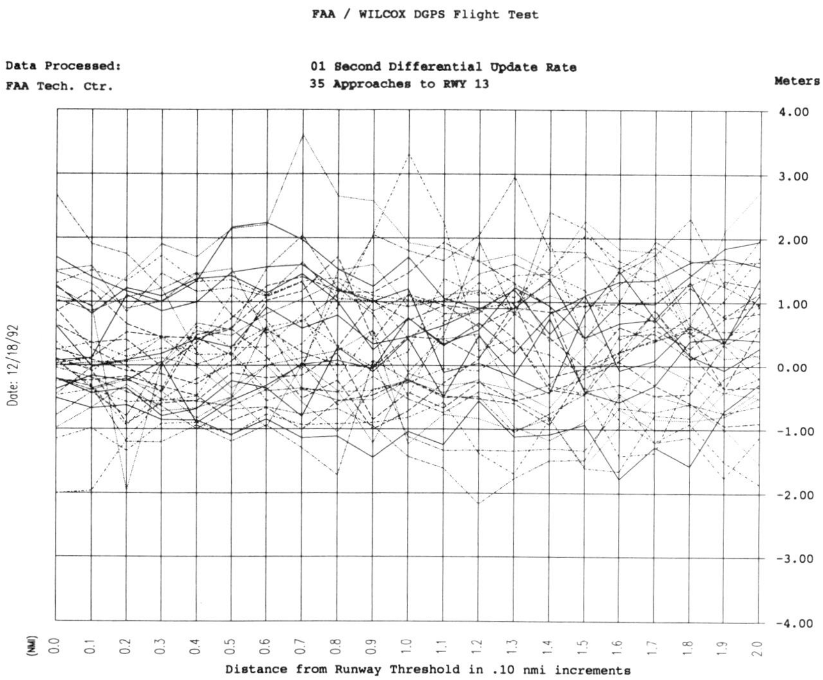

Fig. 13—Vertical Unfiltered Error, 35 Approaches

approaches. The second was to calculate the 95 percent error values for each group of approaches using data collected over the final 1 nmi to the threshold. This method provided many more samples with which to validate the statistical analysis. The 1 nmi section of approach was selected to optimize the RTT vertical tracking angle accuracy. Lateral RTT tracking is less than optimal in this region since a 0.02 deg RTT error contributes 4–5 ft of linear cross-track error. However, this magnitude of error is small compared with the ICAO lateral limits.

Tables 4 and 5 show the $|\mu| + 2\sigma$ (absolute value of mean plus 2 standard deviation) values of the PFE and CMN, at the 200 ft decision height and averaged over the final 1 nmi, for the cross-track and vertical errors, respectively. The ICAO limits are 95 percent specifications, and $|\mu| + 2\sigma$ provides a conservative estimate for 95 percent errors. The ICAO limits shown are those specified at the 200 ft decision height, which are the most stringent. In the case of cross-track deviations, Table 4 shows that all approaches and all configurations tested were well within Category I ILS accuracy tolerances. In fact, all approaches for all equipment configuration cases met Category III lateral accuracy tolerances.

Although the angular tolerances for lateral and vertical deviations are approximately the same, the linear requirement for vertical deviation is roughly 4.5 times more restrictive than the lateral requirement at the 200 ft decision height. Nevertheless, the 95 percent vertical dispersion for all configurations shown in Table 5 met the Category I PFE requirement, with the narrow-correlator-spacing Novatel receiver meeting the requirement by a wide margin. All equipment configurations also met the Category III CMN requirements with good margin. Although phases A and B differed in both receiver type and the use of a choke ring on the ground station GPS antenna, static ground tests have shown that only about 10 percent of the improvement in phase B vertical error is attributable to the choke ring.

Note in Table 4 that the cross-track errors for the narrow-correlator receiver versus the RTT (phase B) are unexpectedly large. They are greater than the vertical errors for the same equipment, cross-track errors for the standard-correlator receiver (phase A), and narrow-correlator cross-track errors based on the laser tracker (phase D). Because the RTT was located approximately 2 mi from the 200 ft decision point when making lateral measurements, the RTT linear error was large enough to mask differences in cross-track error between the two types of receivers. Also, the narrow-correlator receiver lateral data were collected under conditions of strong turbulence, which significantly increased RTT pointing error. The phase D data obtained with the laser tracker are a truer indication of the narrow-correlator receiver cross-track error.

Figure 14 shows the ensemble-cumulative 95 percent vertical guidance error for all 35 Atlantic City approaches, plotted at 0.1 nmi increments from the final approach fix (4.3 nmi) to the runway threshold. Note the slight increase in error at large distances from the threshold, caused by degradation of laser tracker accuracy at those distances.

CONCLUSIONS

For all cases tested, the local area differential GPS (DGPS) landing system provided the equivalent of ILS Category I accuracy for lateral guidance by a wide margin.

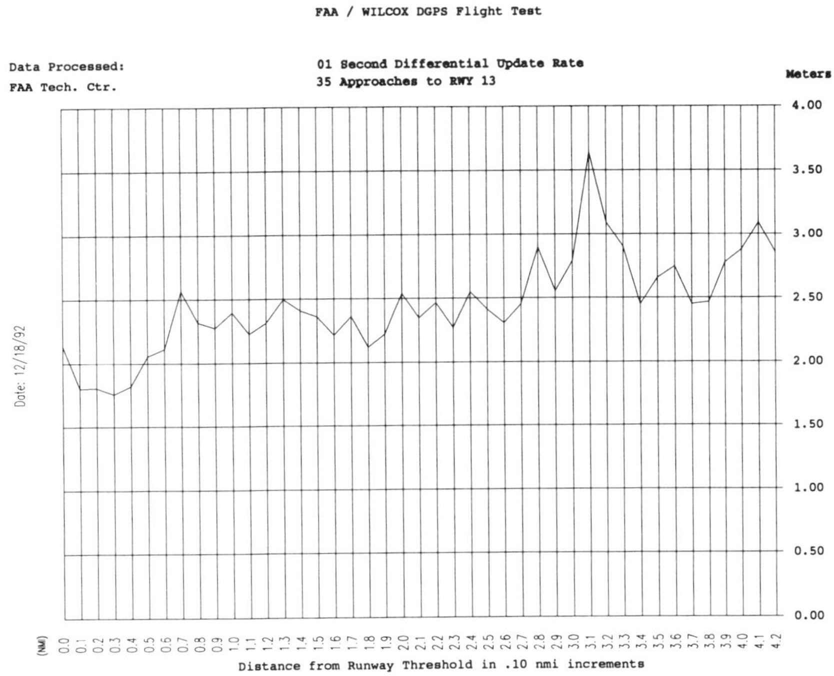

Fig. 14—Vertical Unfiltered Error, 35 approaches, $|\mu| + 2\sigma$

For vertical guidance, which has always been the difficult case for satellite-based systems, the most interesting result was the solid Category I performance provided by the narrow-spacing correlator design.

These results were obtained using guidance based solely on DGPS, without aiding from any other navigation sources.

ACKNOWLEDGMENTS

The authors would like to express their appreciation to all who participated in this program. At Wilcox: Larry Brown, chief pilot; Chuck DeAlbuquerque and Perry McLean, co-pilots; Kate Brown and Bob Ibach, theodolite operators. At the FAA and MITRE: Robert Loh, Deputy Satellite Navigation Program Manager; Robert Till, Senior Technical Program Manager; Ronald Braff, Principal Engineer.

Based partly on a paper presented at The Institute of Navigation Satellite Division GPS-92 meeting, Albuquerque, New Mexico, September 1992.

REFERENCES

1. *International Standards, Recommended Practices and Procedures for Air Navigation Services, Aeronautical Telecommunications,* Annex 10 to the Convention on International Civil Aviation.
2. Edwards, F. G. and Hegarty, D. M. *Flight Test Evaluation of Civil Helicopter Terminal Approach Operations Using Differential GPS,* AIAA Guidance, Navigation, and Control Conference, August 14–16, 1989.

3. Hueschen, Richard M. and Spitzer, Cary R., *Analysis of DGPS/INS and MLS/INS Final Approach Navigation Errors and Control Performance Data,* National Technical Meeting of The Institute of Navigation, January 27–29, 1992.
4. Ashjaee, J., Lorenz, R., Sutherland, R., Datilloy, J., Minazio, J. B., Ablahi, R., Eichner, J. M., Kosmalska, J., and Helky, R., *New GPS Developments and Ashtec M-XII,* ION GPS-89, September 25–29, 1989.
5. Fenton, P., Falkenberg, W., Ford, T., Ng, K., and Van Dierendonck, A. J., *Novatel's GPS Receiver, The High Performance OEM Sensor of the Future,* ION GPS-91, September 9–13, 1991.
6. Van Dierendonck, A. J., Fenton, P., and Ford, T., *Theory and Performance of Narrow Correlator Spacing in a GPS Receiver,* NAVIGATION, Journal of The Institute of Navigation, Vol. 39, No. 3, Fall 1992.

NAVIGATION: *Journal of The Institute of Navigation*
Vol. 38, No. 4, Winter 1991–92
Printed in U.S.A.

GPS Interferometric Attitude and Heading Determination: Initial Flight Test Results

FRANK VAN GRAAS and MICHAEL BRAASCH

Avionics Engineering Center, Ohio University, Athens, Ohio
Received June 1991
Revised November 1991

ABSTRACT

Attitude and heading determination using GPS interferometry is a well-understood concept. However, efforts have been concentrated mainly in the development of robust algorithms and applications for low-dynamic, rigid platforms (e.g., shipboard). This paper presents results of what is believed by the authors to be the first real-time flight test of a GPS attitude and heading determination system. Signals from four antennas are processed by a 24-channel GPS receiver. Data from the receiver are sent to a microcomputer for storage and further computations. Attitude and heading data are sent to a second computer for display on a software-generated artificial horizon. Demonstration of this technique proves its candidacy for augmentation of aircraft state estimation for flight control and navigation, as well as for numerous other applications.

INTRODUCTION

GPS will soon be capable of providing worldwide positioning and timing services to properly equipped users. In addition, interferometric techniques can be used to obtain real-time, highly accurate attitude and heading information from the GPS measurement data. GPS interferometry is a technique based on the measurement of phase differences of the GPS carrier signal observed at two or more antennas. Attitude and heading of a platform can be calculated from measurements with two linearly independent interferometers, where each interferometer consists of two separate antennas.

GPS interferometry has been used extensively in precise static and kinematic positioning. Experimental results of static relative positioning using GPS interferometry, first reported in 1982 [1], showed subcentimeter accuracies for baseline lengths between 8 and 52 m. Further stationary experiments were presented in 1983 [2]. After that time, implementation issues were emphasized, such as the development of fast initialization algorithms [3]. Real-time low-dynamic shipboard tests were performed in 1988 [4], with reported angular accuracies on the order of several degrees for a 60 cm baseline. The first non-real-time determination of aircraft pitch and heading using one GPS interferometer was reported in 1989 [5], revealing pitch and heading uncertainties below 1 mrad for a 23 m baseline. Today, submilliradian accuracies, even for

1 m baselines, are attainable for rigid platforms using off-the-shelf equipment [6]. In this paper, we present methods and results for real-time pitch, roll, and heading determination of a dynamic, flexible platform.

PRINCIPLES OF ATTITUDE DETERMINATION

In the formalism of rigid body kinematics, attitude is defined by the rotation transformation which relates a coordinate system fixed in space to a coordinate system fixed in the body [7]. For the purposes of aircraft navigation, the space coordinate system is typically defined to be locally level. The positive x-axis is pointed north, the positive y-axis is pointed east, and the positive z-axis is pointed down. Finally, since the transformation is concerned with rotation only (and not translation), the origin of the space coordinate system is coincident with the aircraft center of gravity. Similarly, the body (or aircraft) coordinate system also has its origin at the aircraft center of gravity. However, for the aircraft coordinate system, the positive x-axis is directed along the fuselage axis towards the nose, the positive y-axis points along the right wing (right according to the pilot), and the positive z-axis is directed towards the floor of the aircraft [8].

The two coordinate systems are related through the antenna locations. Under the rigid body assumption, the antenna locations remain fixed in the aircraft coordinate system. Their locations in this system are determined prior to the flight through standard GPS static surveying techniques. In flight, the antenna locations in the space coordinate system are determined using GPS interferometry. Thus at any given epoch, the antenna locations in the two coordinate systems are related as follows:

$$\mathbf{R}_S = \mathbf{T}\mathbf{R}_a \tag{1}$$

where $\mathbf{R}_s$ is the matrix containing the locations of the antennas in space coordinates, $\mathbf{R}_a$ is the matrix containing the locations of the antennas in aircraft coordinates, and $\mathbf{T}$ is the rotation transformation matrix.

The transformation is a 3×3 matrix containing the nine direction cosine elements. It is the product of three rotation matrices, each of which specifies a rotation about a certain aircraft coordinate system axis. Thus,

$$\mathbf{T} = \mathbf{T}_\phi \mathbf{T}_\theta \mathbf{T}_\psi \tag{2}$$

where $\mathbf{T}_\phi$ represents a rotation about the fuselage axis (roll), $\mathbf{T}_\theta$ represents a rotation about the wing axis (pitch), and $\mathbf{T}_\psi$ represents a rotation about the vertical axis (heading).

The transformation matrix can now be expressed in terms of the three rotations (also known as the Euler angles):

$$\mathbf{T} = \begin{bmatrix} \cos\psi\cos\theta & -\sin\psi\cos\phi + \cos\psi\sin\theta\sin\phi & \sin\psi\sin\phi + \cos\psi\sin\theta\cos\phi \\ \sin\psi\cos\theta & \cos\psi\cos\phi + \sin\psi\sin\theta\sin\phi & -\cos\psi\sin\phi + \sin\psi\sin\theta\cos\phi \\ -\sin\theta & \cos\theta\sin\phi & \cos\theta\cos\phi \end{bmatrix} \tag{3}$$

Determination of the instantaneous values of the Euler angles proceeds as follows. First, the matrix containing the antenna locations in aircraft coordinates is formed prior to the start of attitude determination. Second, real-time GPS interferometric measurements are used to form the matrix containing the antenna locations in the space coordinate system. Solving for the transformation matrix involves postmultiplying the space coordinate matrix by the generalized inverse of the aircraft coordinate matrix:

$$\mathbf{T} = \mathbf{R}_s\mathbf{R}_a^T(\mathbf{R}_a\mathbf{R}_a^T)^{-1} \tag{4}$$

Finally, the Euler angles can be calculated as follows:

$$\theta = \sin^{-1}(-T_{31}); \quad \phi = \sin^{-1}\left(\frac{T_{32}}{\cos\theta}\right); \quad \psi = \sin^{-1}\left(\frac{T_{21}}{\cos\theta}\right) \tag{5}$$

The calculations resulting in this Euler angle triad have been presented to provide insight into the attitude determination process. However, although this triad is probably the best understood means of attitude and heading expression, it is far from robust, and arguably should not be considered for use in a real-time integrated navigation system [9]. This lack of robustness can be seen most easily by examining equations (3) and (5) for the case of $\theta = \pm 90°$. Note that although θ can be recovered, ϕ and ψ are lost. As a result, the Euler angle triad is not an all-attitude parameterization. Although many parameterizations of **T** are possible, quaternions offer conciseness, efficiency, and stability not possible with other parameterizations [8, 9].

Further practical considerations involve the aircraft structure. The derivation given above rests on the assumption that the antenna locations are fixed in the aircraft coordinate system. Error mechanisms are thereby introduced by the nonrigid body behavior of the aircraft. The fuselage by itself is a good rigid body approximation, but inclusion of the wings (with their normal flexing) makes the overall aircraft far from being so. An additional consideration is more subtle. The derivation assumes that the location of the aircraft center of gravity is known relative to the antenna positions. Although a nominal location can be used, the true location will change over the course of the flight.

GPS INTERFEROMETRY

Consider a simplified GPS interferometer, as depicted in Figure 1, consisting of two GPS antennas and the electronics necessary to determine the instantaneous phase of a received GPS carrier signal. Usually, the distance between the antennas and a GPS satellite is much larger than the baseline length between the two antennas, such that the GPS carrier signal can be treated as a plane wave. Under the assumption that the satellite clock and the receiver clock are synchronized, the phase difference $\Delta\phi$ between simultaneous carrier phase measurements at the two antennas can be written as

$$\Delta\phi = \phi_1 - \phi_2 = \mathbf{b} \cdot \mathbf{e} + N\lambda \tag{6}$$

where ϕ_1 and ϕ_2 are the measured carrier phases in wavelengths for antennas 1 and 2, respectively; **b** is the three-dimensional baseline vector pointing from antenna 1 to antenna 2; **e** is the unit vector pointing from the interferometer to the satellite; and N is an integer number accounting for the ambiguity due to the limited length of the carrier wavelength λ. The phase difference between antennas is also referred to as a single difference [10]. The determination of the unit vector to the satellite requires the approximate coordinates of the

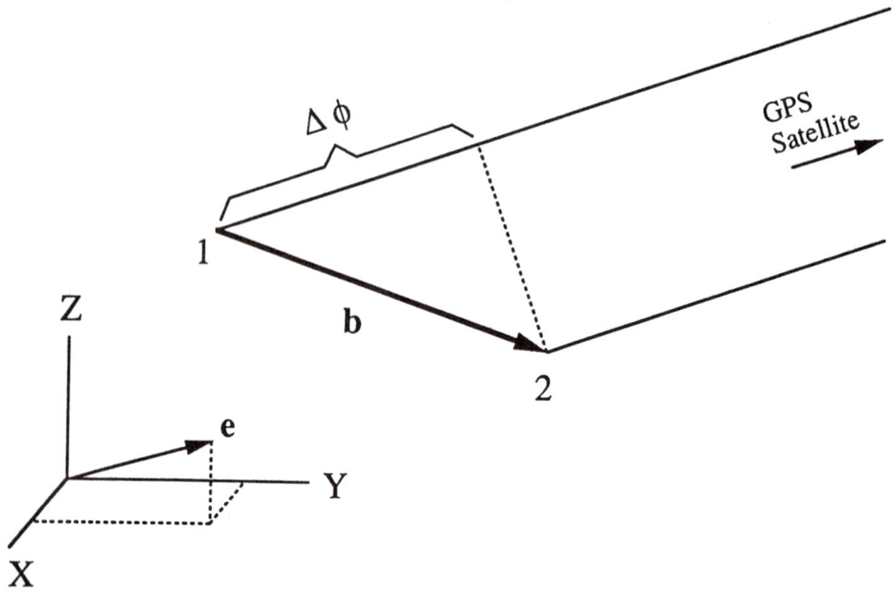

Fig. 1—Simplified GPS Interferometer

interferometer and the satellite coordinates. The latter are calculated from the navigation data transmitted by the satellites. A standard GPS navigation solution is used to determine the interferometer coordinates. The baseline solution is not very sensitive to the interferometer coordinates, as was shown by Jurgens [11], because of the large distance to the satellite with respect to the relatively short baseline length of the interferometer.

The unknown receiver clock offset can be eliminated by taking the difference between two independent single differences, also referred to as a double difference (DD):

$$DD = \Delta\phi_1 - \Delta\phi_2 = \mathbf{b} \cdot (\mathbf{e}_1 - \mathbf{e}_2) + (N_1 - N_2)\,\lambda \tag{7}$$

Thus a double difference is a linear combination of four phase measurements obtained from two satellites at two antennas, as shown in Figure 2. This can be extended to eliminate the unknown integer ambiguity number by differencing two double differences at two epochs to obtain a triple difference. The triple difference is useful for static positioning, but provides only the rate of change of the double difference for dynamic applications. Thus it does not provide an accurate instantaneous solution for the baseline vector.

Assuming that the integer ambiguities are known, then using equation (6), it follows that three independent double differences are required to solve for the baseline vector $\mathbf{b} = (b_1,b_2,b_3)^T$:

$$\begin{bmatrix} DD_1 \\ DD_2 \\ DD_3 \end{bmatrix} = \begin{bmatrix} (\mathbf{e}_1 - \mathbf{e}_2)^T \\ (\mathbf{e}_1 - \mathbf{e}_3)^T \\ (\mathbf{e}_1 - \mathbf{e}_4)^T \end{bmatrix} \begin{bmatrix} b_1 \\ b_2 \\ b_3 \end{bmatrix} \tag{8}$$

Furthermore, if the norm of the baseline vector, $|\mathbf{b}|$, is known a priori, then only two double differences are needed:

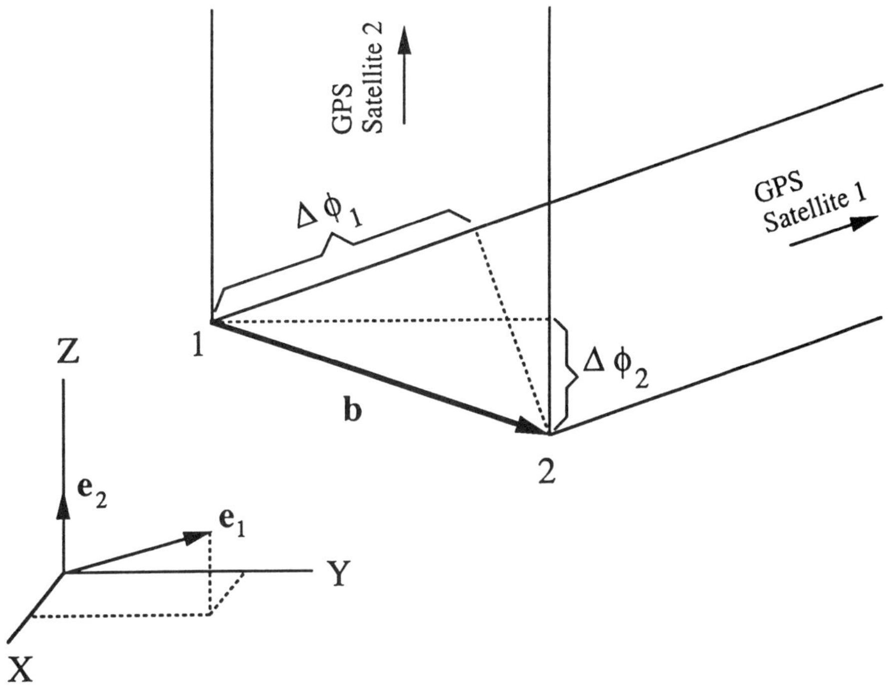

Fig. 2—Double Difference Observation

$$\begin{bmatrix} DD_1 \\ DD_2 \\ |\mathbf{b}| \end{bmatrix} = \begin{bmatrix} (\mathbf{e}_1 & - & \mathbf{e}_2)^T \\ (\mathbf{e}_1 & - & \mathbf{e}_3)^T \\ b_1 & b_2 & b_3 \end{bmatrix} \begin{bmatrix} b_1 \\ b_2 \\ b_3 \end{bmatrix} \tag{9}$$

Several techniques have been proposed for quickly determining the integer ambiguity for each of the double differences [12, 13]. In general, search methods are very effective in quickly resolving the integer ambiguities whenever redundant measurements are available. The basic idea is to evaluate the measurement residuals for potential baseline vector solutions. The techniques used are very similar to those applied for GPS fault detection (see [14] for a summary of these techniques). If, for instance, signals from five satellites are received by one GPS interferometer, four independent double differences can be formed. Equation (8) is rewritten as:

$$\mathbf{DD} = \mathbf{Hb} \tag{10}$$

where **DD** is a 4 × 1 double difference measurement vector, and **H** is a 4 × 3 data matrix containing the satellite unit vector differences. The residuals are the differences between the actual double difference measurements and those predicted based on the least-squares solution for **b**, the approximate interferometer coordinates, and the known satellite coordinates. The least-squares residual vector, $\mathbf{e}_r$, is obtained by projecting **DD** onto a space perpendicular to the estimation space spanned by the columns of **H**. Thus, $\mathbf{e}_r$ is orthogonal to the estimation space. An efficient method to obtain $\mathbf{e}_r$ is by decomposing the **H** matrix into the product of a real orthonormal matrix **Q** and an upper triangular matrix **R** using the QR factorization [15, 16]. Replacing **H** in equation (10) by the product **QR** and premultiplying both sides by $\mathbf{Q}^T$ yields ($\mathbf{Q}^T\mathbf{Q} = \mathbf{I}$)

$$\mathbf{Rb} = \mathbf{Q}^T\mathbf{DD} \tag{11}$$

The rank of **R** is equal to the rank of **H**; therefore, the bottom row of **R** consists of zeros only. Equation (11) can be divided into two equations:

$$\mathbf{Ub} = \mathbf{Q}_1\mathbf{DD} \text{ or } \mathbf{b} = \mathbf{U}^{-1}\mathbf{Q}_1\mathbf{DD}$$

$$\mathbf{0} = \mathbf{qDD} \tag{12}$$

where **U** consists of the first three rows of **R**, $\mathbf{Q}_1$ consists of the first three rows of $\mathbf{Q}^T$, and **q** is the transpose of the fourth row of $\mathbf{Q}^T$. The top expression of equation (12) relates the measurements to the baseline coordinates, forming the least-squares solution. The vector **q** and the double difference vector **DD** are orthogonal. The latter property is used to obtain the norm of the residual vector if measurement errors are present on the double differences:

$$|\mathbf{e}_r| = (q_1\ q_2\ q_3\ q_4)\begin{bmatrix} DD_1 \\ DD_2 \\ DD_3 \\ DD_4 \end{bmatrix} \tag{13}$$

The vector **q** has to be calculated only once for a search at one particular epoch. Thus, to calculate the residual for a potential solution, only four multiplications and three additions are required.

Some prior knowledge about the interferometer is required in order to reduce the potential number of solutions for **b**. Several methods can be used, including the following:

1) Limit the baseline length of the interferometer to, for instance, 1 m, such that the integer ambiguity number must lie in the interval [−5, 5] (one GPS wavelength, λ, is approximately 0.19 m).
2) Use a priori information about the orientation of the interferometer, such that the search can be limited to correspond to the approximate baseline orientation. Approximate attitude and heading information could, for instance, be obtained from a low-cost three-axis accelerometer and magnetometer.
3) Use double differences calculated from the GPS pseudorange measurements (code phase double differences) to obtain approximate values for the carrier phase double differences.

If the double difference uncertainty can be limited to ±5 λ, then even an exhaustive search will require only the calculation of $(11)^4 = 14{,}641$ residuals, which will require seven times that many, or 102,487 floating point calculations. This can easily be accomplished within 1 s using a microcomputer. In this paper, we estimate the initial double difference ambiguities by combining the relatively noisy code phase double differences with the carrier phase double differences using a complementary Kalman filter (see also [17]). In this filter, the carrier phase double differences are used to provide a reference for the code phase double differences. The performance of the complementary filter is very similar to that of the Hatch filter [18], but the computational and storage requirements are much lower. Although only one double difference is estimated, the complementary filter model can easily be extended to include the rate of change of the double difference as well. The equations for the (scalar) complementary Kalman filter are summarized below:

$$
\begin{aligned}
x_k^- &= x_{k-1}^+ + (DD_k - DD_{k-1}) \\
p_k^- &= p_{k-1}^+ + q \\
k_k &= p_k^- (p_k^- + r)^{-1} \\
x_k^+ &= x_k^- + k_k (z_k - x_k^-) \\
p_k^+ &= (1 - k_k)\, p_k^-
\end{aligned} \tag{14}
$$

where the superscript '−' indicates that the parameter is predicted, and the superscript '+' indicates that the parameter is estimated; x denotes the code phase double difference with variance q; r is the variance of the code phase double difference measurement z; and p is the variance of the estimation error. The first two lines in equation (14) propagate the estimated code phase double difference and the estimation error variance from time (k − 1) to time k, followed by the calculation of the Kalman gain (line 3), and the update of the estimated code phase double difference and the estimation error variance (lines 4 and 5). Figure 3 shows an example of both code phase and carrier phase double differences for a 20 min section of a flight test. The double differences were obtained from a GPS interferometer on the fuselage of a DC-3 research aircraft during roll maneuvers. The next section provides a detailed description

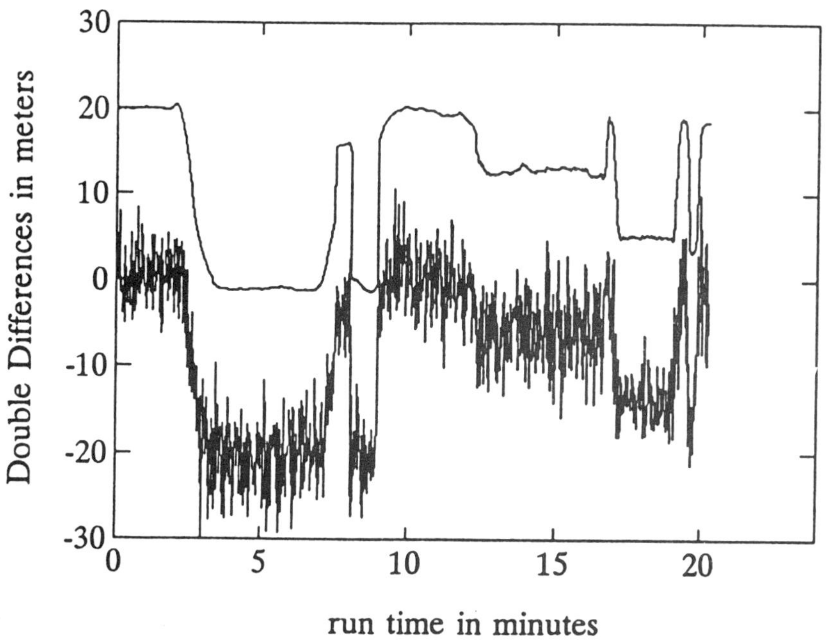

Fig. 3—Code Phase (bottom) and Carrier Phase (top) Double Differences

of the aircraft test bed. Note that the code phase double differences and the carrier phase double differences are separated by an unknown integer wavelength ambiguity. Figure 4 shows both the carrier phase double differences and the estimated code phase double differences. As expected, the estimated code phase double differences follow the dynamics very well. Finally, the difference between the carrier phase double differences and the estimated code phase double differences is shown in Figure 5. After the filter has been initialized with the first code phase double difference, it takes typically less than 20 s for the carrier phase double difference to be within $\pm 5\ \lambda$ with respect to the estimated code phase double difference. The remaining differences between the estimated code phase double differences and the carrier phase double differences are due to the code phase multipath from the aircraft structure.

The estimated code phase double difference is used to calculate an initial estimate for the absolute carrier phase double difference. Next, an exhaustive search is performed to find those combinations of integer ambiguities that will result in low residuals. Depending on the carrier phase measurement noise and the satellite measurement geometry, more than one potential set of ambiguities may be found. Further eliminations of potential sets are achieved by applying known baseline length constraints. For an interferometer mounted on a semirigid platform, the baseline length can be determined to within a few millimeters. The knowledge of the norm of the baseline vector is limited by differential antenna phase center migrations, differential multipath, and structural flexing of the platform. Next, additional redundant measurement data can be used to resolve the integer ambiguities, again by examining the residuals. Although statistically unlikely, it would still be possible to select the wrong set of ambiguities; however, by continuing to examine all measurement residuals throughout the experiment, the resulting inconsistencies in the data would be quickly detected.

Finally, after the integer ambiguities have been determined for each of the double differences, they can be used to maintain aircraft attitude and heading for long periods of time until loss of lock occurs. It is also important to add that once the integer ambiguities have been resolved, redundant measurements are no longer necessary to maintain attitude and heading. Assuming known baseline lengths, a rigid platform requires only continuous lock on the signals from three satellites at three antennas. In this case, to find the approximate interferometer coordinates, another sensor input, such as the baroaltimeter or a stable frequency reference source, is required. If the baselines are subjected to structural flexing, continuous lock on four satellites at three antennas is required.

DESCRIPTION OF THE TEST BED

The real-time attitude and heading determination test bed consists of three main components: (1) aircraft, antennas, and cabling; (2) GPS receiver; and (3) data storage, computation, and display units (see Figure 6). Four microstrip antennas have been installed on Ohio University's Douglas DC-3 research aircraft, N7AP. One antenna is located on each wingtip, and two antennas are located on top of the fuselage, with one near the nose and the other near the tail (Figure 7). RG-400 cable is used because of its high phase stability and

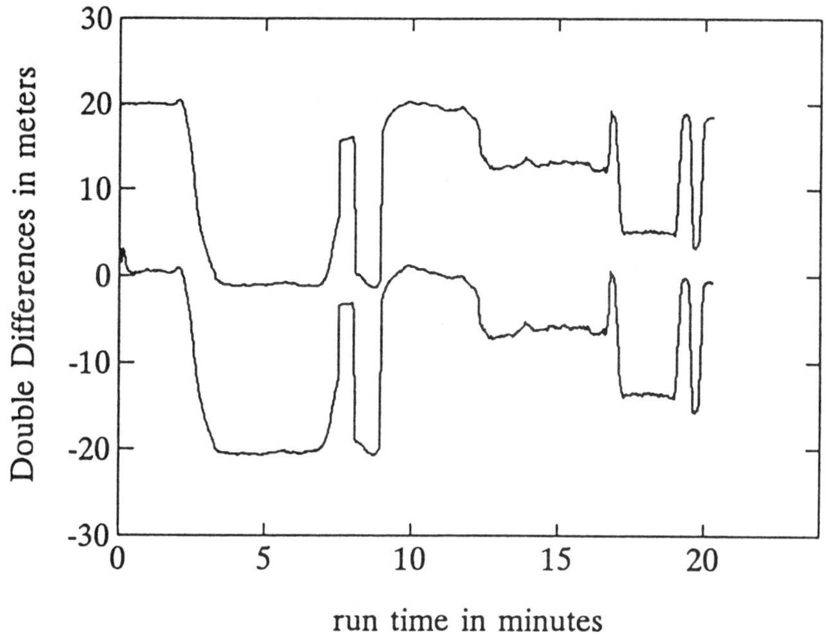

Fig. 4—Carrier Phase Double Differences (top) and Estimated Code Phase Double Differences (bottom)

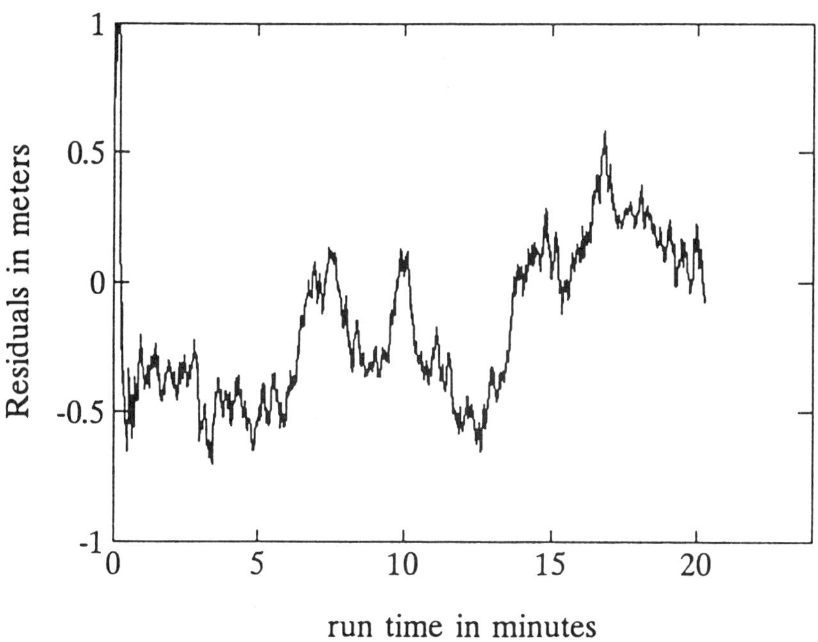

Fig. 5—Difference Between Carrier Phase Double Differences and Estimated Code Phase Double Differences

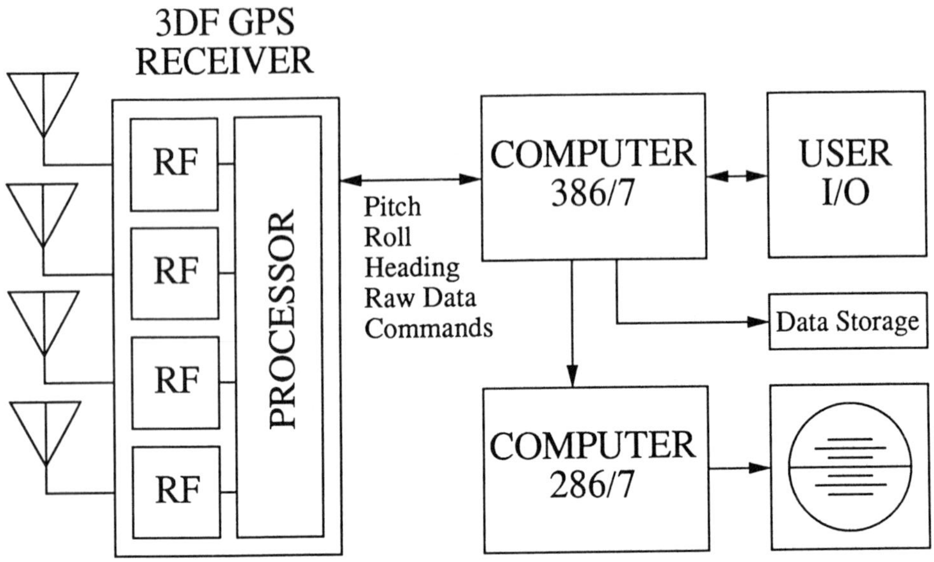

Fig. 6—Block Diagram of Real-time Attitude and Heading Determination Test Bed

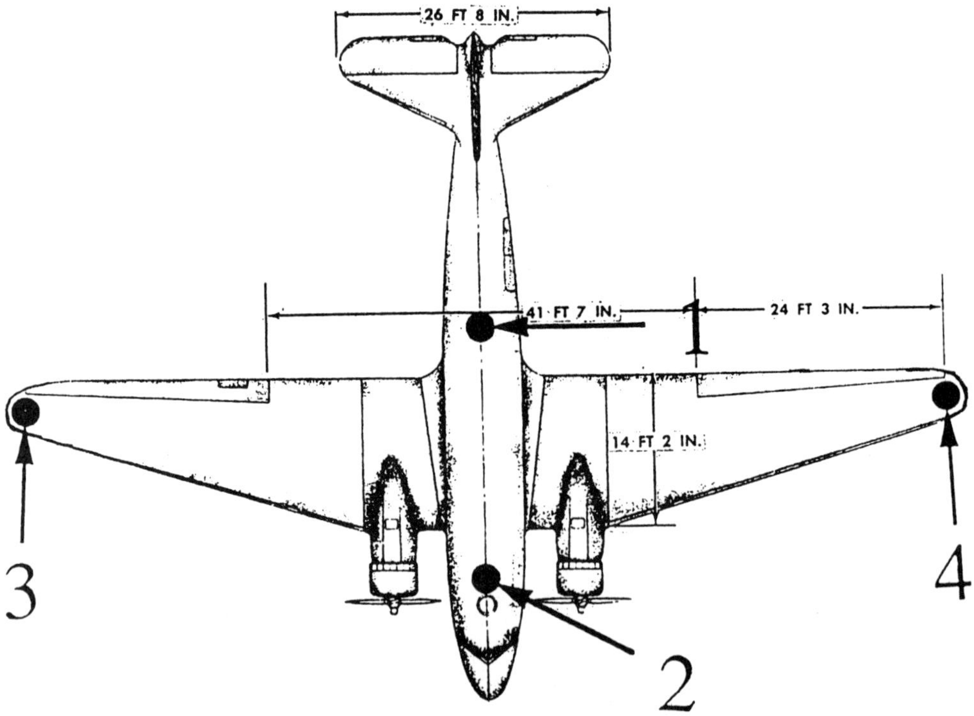

Fig. 7—Douglas DC-3 GPS Antenna Locations

small physical diameter. Phase stability is essential since each of the four cables is approximately 25 m in length. The cables feed the signal from the antenna preamplifiers to the GPS receiver, located in a standard equipment rack installed in the cabin midway between the nose and tail of the aircraft.

The GPS receiver being used is the Ashtech Three-dimensional Direction Finding (3DF) 24-channel receiver [6]. The 3DF is configured as four 6-channel sections, each receiving signals from a separate antenna. The receiver is packaged in a ruggedized 3⅝ in × 8½ in × 8 in housing, and requires a 12 W power source. Real-time measurement data at a selectable rate of up to 1 Hz (both raw and processed) are available over two serial ports at data rates up to 115,200 baud.

The storage, computation, and display units consist of two microcomputers. A 25 MHz 80386-based computer receives real-time data from the 3DF, performs storage and computation, and sends real-time attitude information to a second computer. The second computer is a 12 MHz 80286-based PC. It generates an artificial horizon and heading indicator, and drives this indicator with the attitude information supplied by the 80386 machine.

Baselines between the antennas were determined using standard GPS static surveying techniques. The lengths of these baselines are given in Table 1. These measurements were also verified using a steel tape measure and were found to agree to within 5 mm.

GROUND AND FLIGHT TESTING

Ground and flight testing of the GPS attitude and heading determination system started in April 1991. To date, three demonstration flights have been conducted (April 18, May 8, and May 22). Each lasted for approximately 1.5 to 2 h, and was preceded by a data collection session on the ramp at Ohio University airport (UNI) in Albany, Ohio. During the first two flights, only raw measurement data were collected, which were used for the continued development of real-time attitude and heading determination algorithms. During the third flight, long periods of real-time attitude and heading determination were achieved. Moreover, the 3DF GPS receiver maintained continuous lock on at least three satellites at three antennas during all maneuvers. This is an important result because it allows for continuous attitude and heading determination. We also found that whenever the receiver would lose lock on a particular satellite because of temporary shielding by the airframe, the receiver would typically reacquire the signal within a few seconds.

Table 1—Antenna Baseline Lengths for Douglas DC-3 Test Bed

Baseline	Length (m)
ant 1–ant 2	7.781
ant 1–ant 3	14.468
ant 1–ant 4	14.466
ant 2–ant 3	14.941
ant 2–ant 4	14.924
ant 3–ant 4	28.194

Aircraft flexing was found to be the main reason for discontinuities in the attitude and heading data. The carrier phase double difference residuals would increase, which caused the algorithm to reinitiate the integer ambiguity search. This could be avoided by allowing larger double difference residuals, in combination with using a higher baseline length uncertainty.

A representative section of the ramp test results collected from the 3DF receiver on May 22, 1991 is shown in Figures 8 through 10. It should be noted that the aircraft was not static during this data collection session; small changes in pitch, roll, and heading angles were introduced by wind gusts and by the flight crew's movement inside the aircraft. In addition, some multipath effects can be observed in the data. During this session, five healthy GPS satellites were visible (satellites 2, 6, 13, 14, and 15). The PDOP was approximately equal to 3 throughout the test. The noise level on the roll and heading angles was on the order of 0.1 mrad (0.006 deg). Pitch angle noise was approximately three times larger because of satellite geometry [19–21] and the shorter baseline used for the pitch angle determination. Absolute accuracy results are not available because of the lack of a reference for the attitude data, which must be aligned with the phase centers of the GPS antennas. However, previous comparisons between the 3DF and kinematic GPS results were found to agree to within 0.1 mrad [6].

Real-time data collected from the 3DF receiver during two sections of the third flight test conducted in the evening of May 22, 1991 are presented next. The noise levels on the pitch, roll, and heading angles are difficult to analyze directly because of the many attitude and heading variations with relatively large magnitudes due to normal flight. Instead, the emphasis of the results presented in this section is on the capability of GPS interferometry to provide real-time, in-flight attitude and heading information. Figure 11 shows the

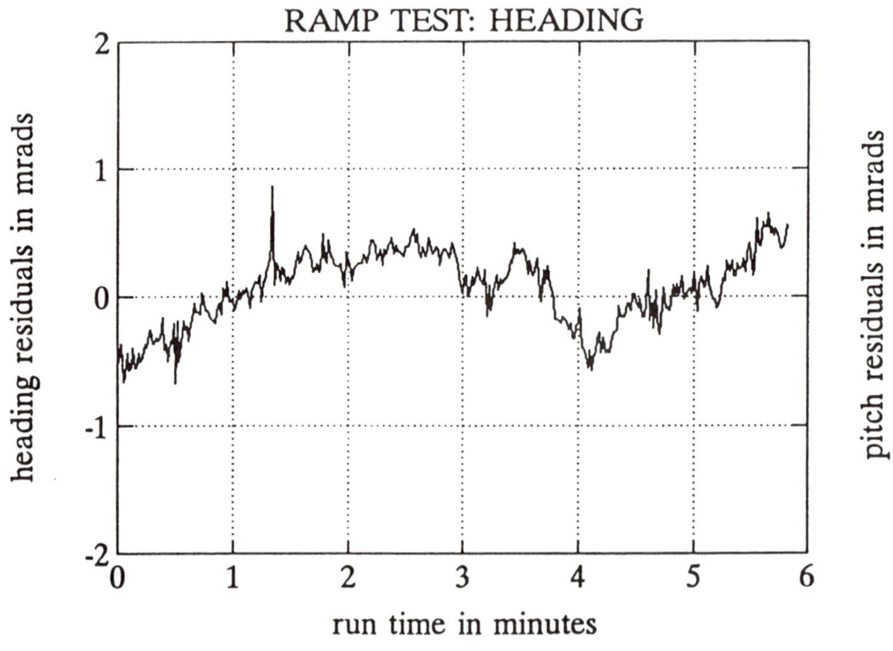

Fig. 8—Ramp Test Heading Residuals

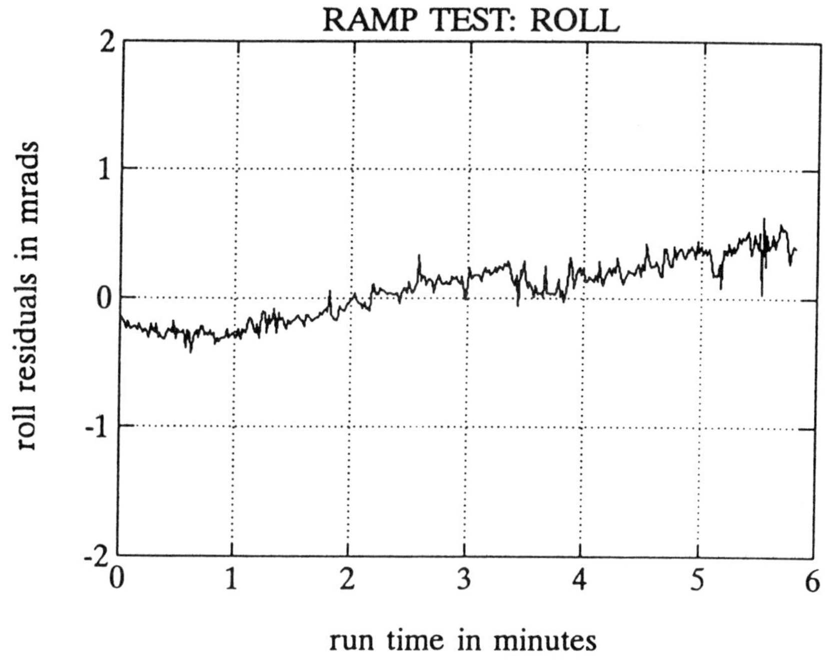

Fig. 9—Ramp Test Roll Residuals

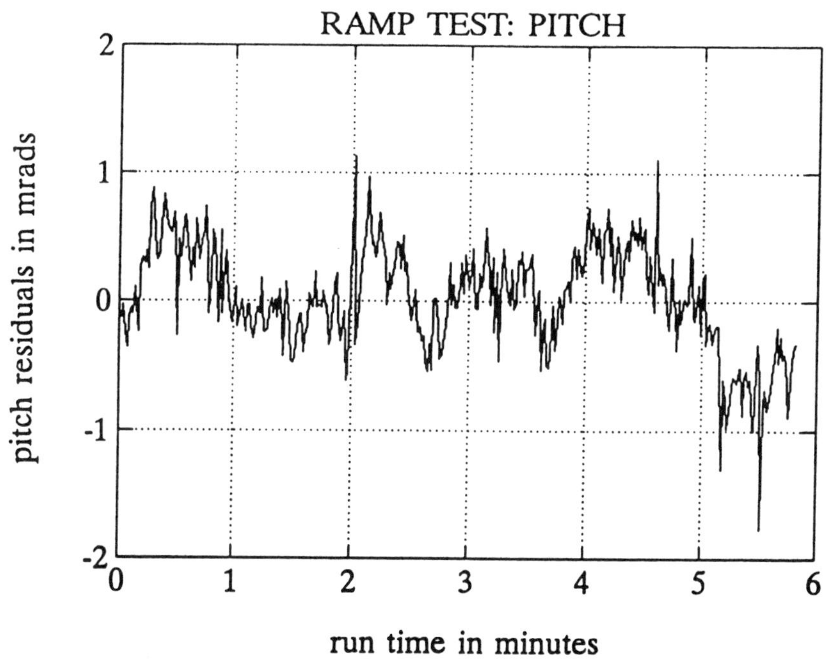

Fig. 10—Ramp Test Pitch Residuals

ground track of a 27 min section of the third flight, which was conducted in the vicinity of Ohio University. Figures 12 through 14 show the aircraft heading, roll, and pitch angles as a function of time. Note that the noise-like behavior of the angles is all due to actual aircraft dynamics (the period of the angle

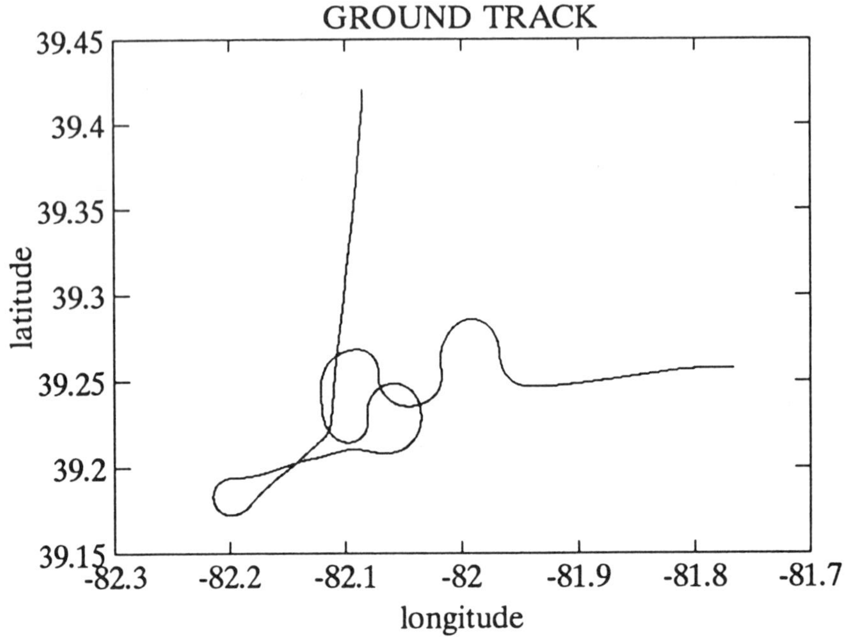

Fig. 11—Ground Track for a 27 min Section of the Third Flight Test

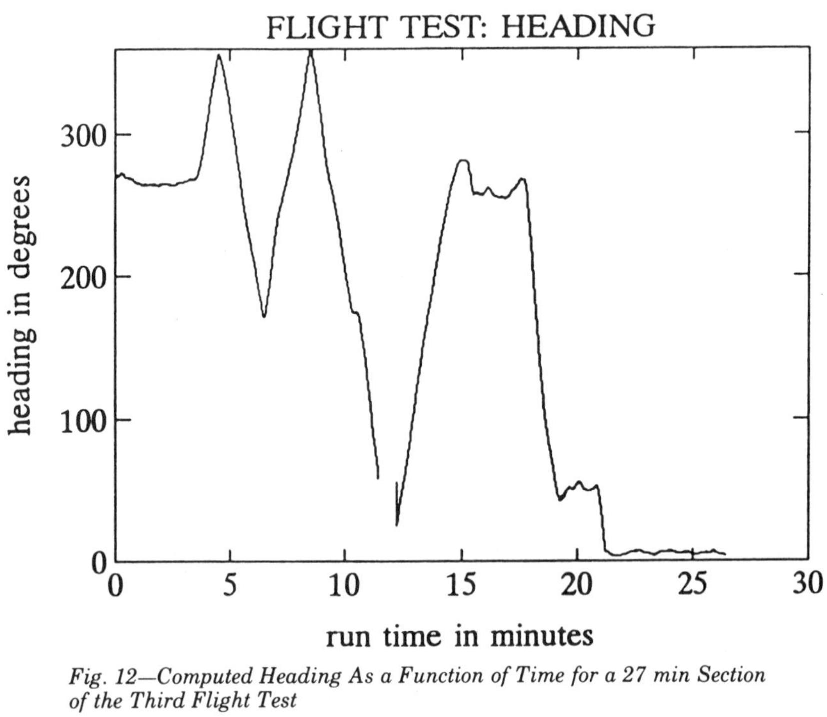

Fig. 12—Computed Heading As a Function of Time for a 27 min Section of the Third Flight Test

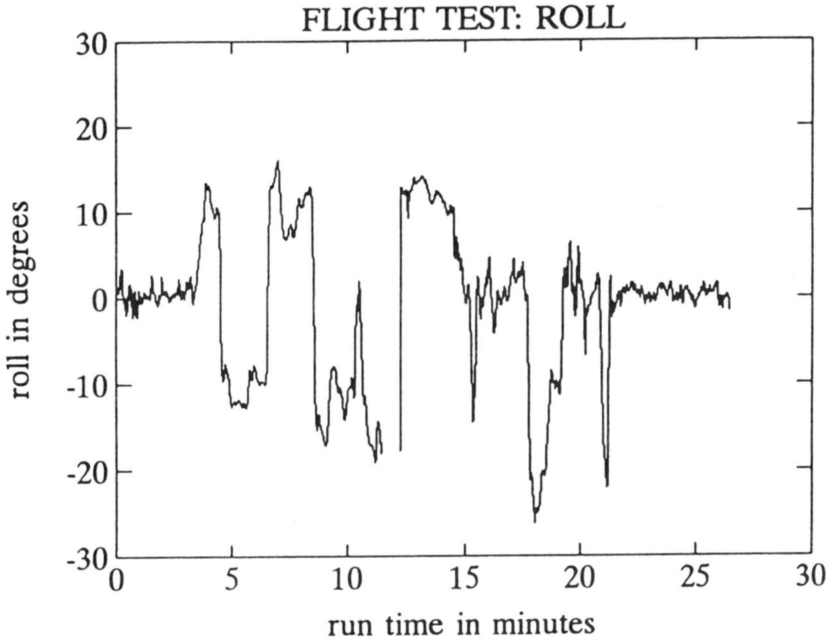

Fig. 13—Computed Roll As a Function of Time for a 27 min Section of the Third Flight Test

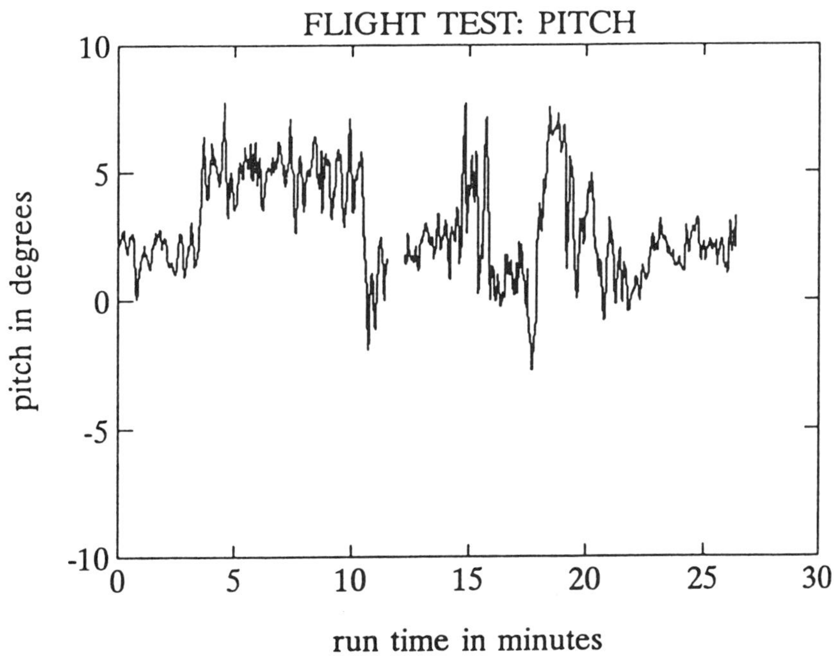

Fig. 14—Computed Pitch As a Function of Time for a 27 min Section of the Third Flight Test

variations is on the order of 10 to 20 s). Approximately 12 min after the start of the run, attitude and heading information was lost for approximately 30 s. As discussed earlier, this was caused by poor double difference residuals, which caused the attitude and heading algorithm to reinitiate the integer ambiguity search. Shown in Figure 15 is the lock history of the 3DF receiver; even in the presence of signal shielding due to airplane maneuvers, the receiver maintained lock on a minimum of four satellites at three antennas.

Figures 16 through 20 show the results for a 13 min section of the third flight test, during which a series of 10 and 30 deg roll maneuvers was performed. Continuous attitude and heading information was available using four or five GPS satellites. During the 30 deg rolls, only one of the five visible satellites was not observed continuously by at least three antennas.

SUMMARY, DEVELOPMENTS, AND APPLICATIONS

The feasibility of using real-time GPS interferometric attitude and heading determination for dynamic semirigid platforms has been demonstrated. During the course of this research, several issues were identified which require further study. Among these are continued development of robust fault detection and isolation algorithms for the detection and correction of cycle slips, characterization and mitigation of multipath, and integration with an inertial measurement unit.

GPS interferometry has numerous applications, including reliable aircraft/spacecraft state estimation, measurement of structural flexing of large space structures, synthetic aperture radar [5], aerial photogrammetry, pointing systems, and rapid alignment of inertial navigation systems for submarines and aircraft while maneuvering.

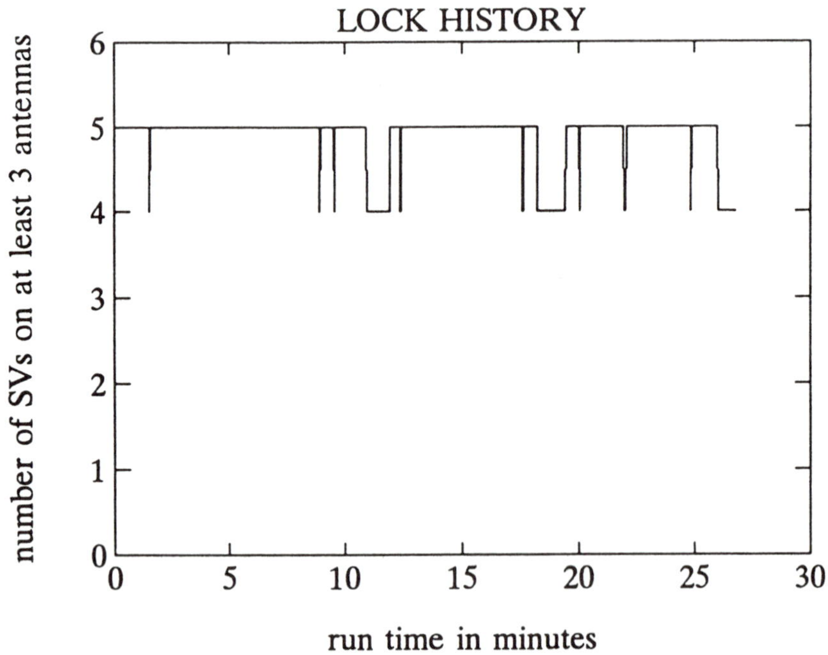

Fig. 15—Lock History for a 27 min Section of the Third Flight Test

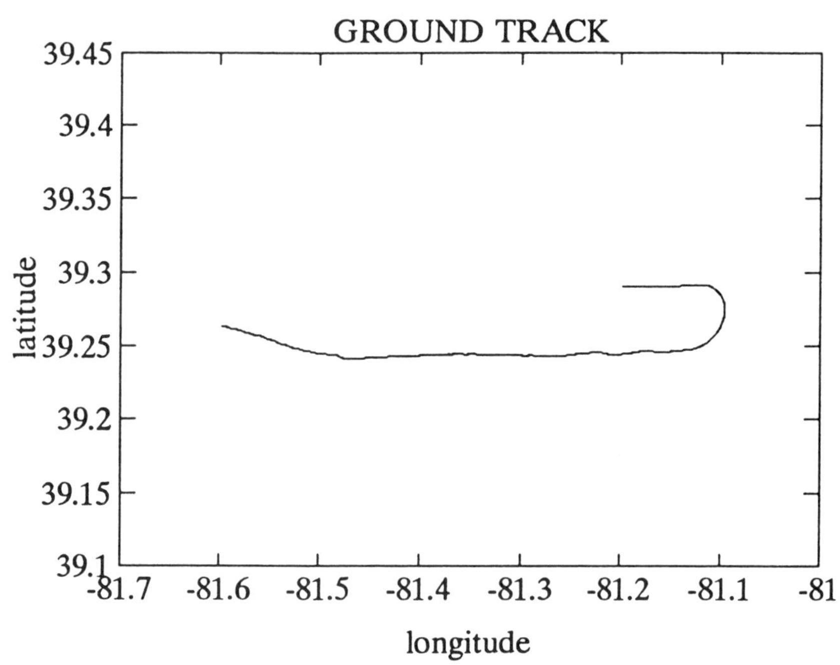

Fig. 16—Ground Track for a 13 min Section of the Third Flight Test

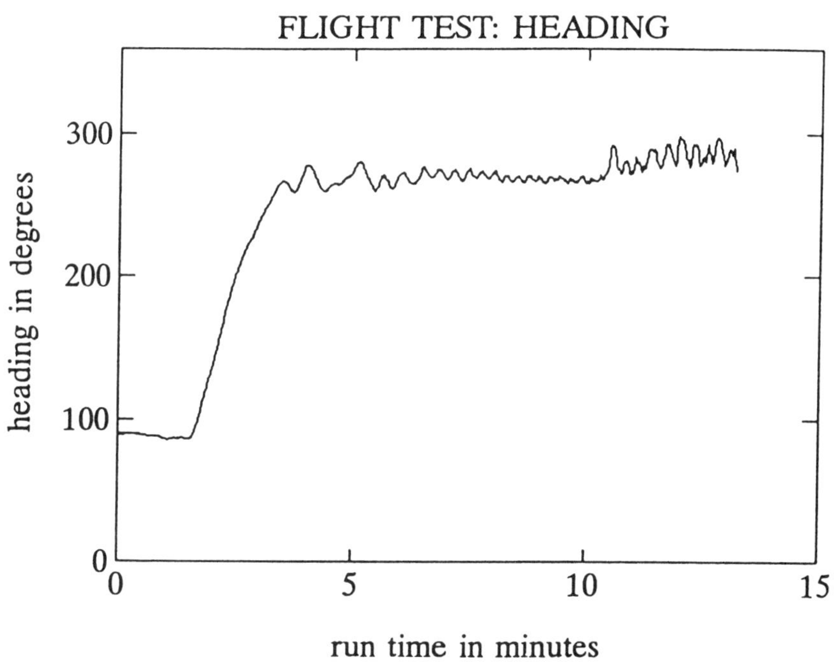

Fig. 17—Computed Heading As a Function of Time for a 13 min Section of the Third Flight Test

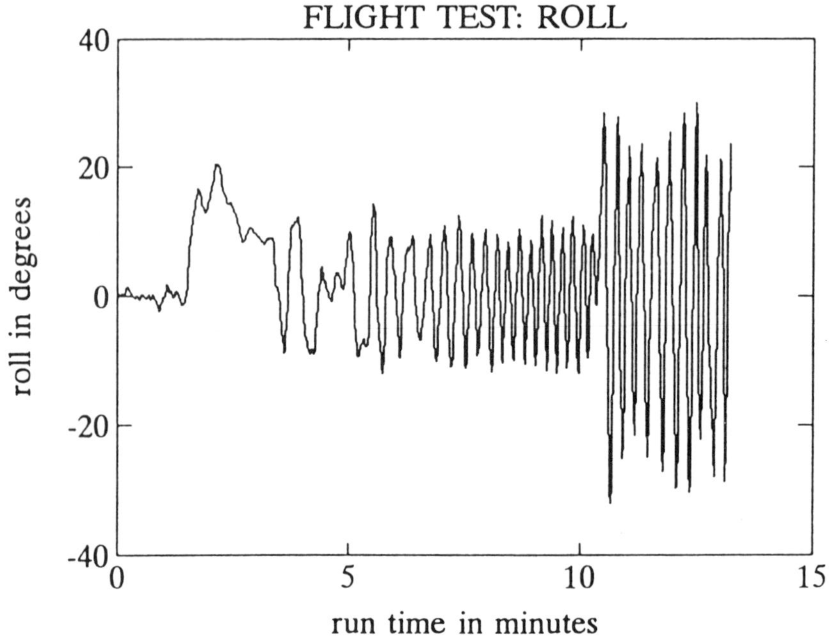

Fig. 18—Computed Roll As a Function of Time for a 13 min Section of the Third Flight Test

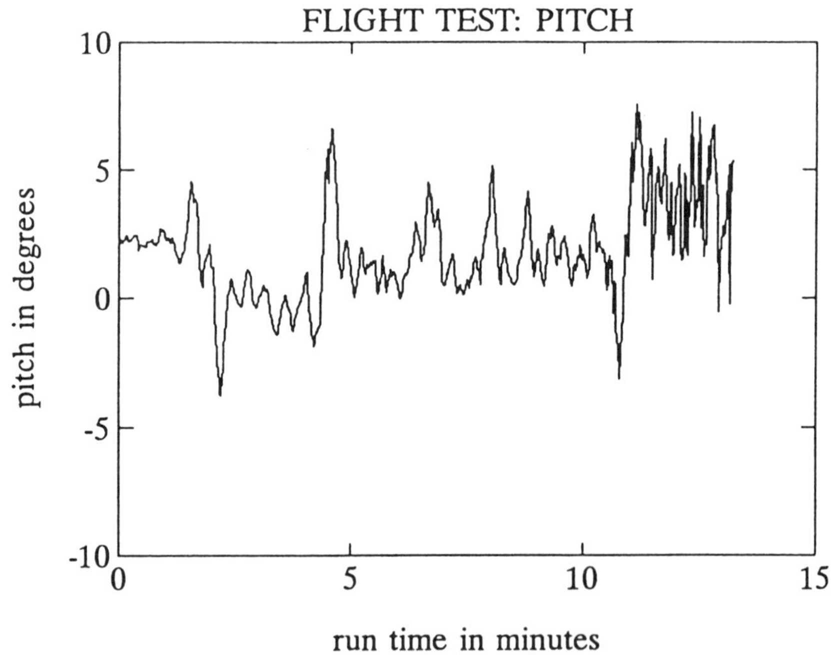

Fig. 19—Computed Pitch As a Function of Time for a 13 min Section of the Third Flight Test

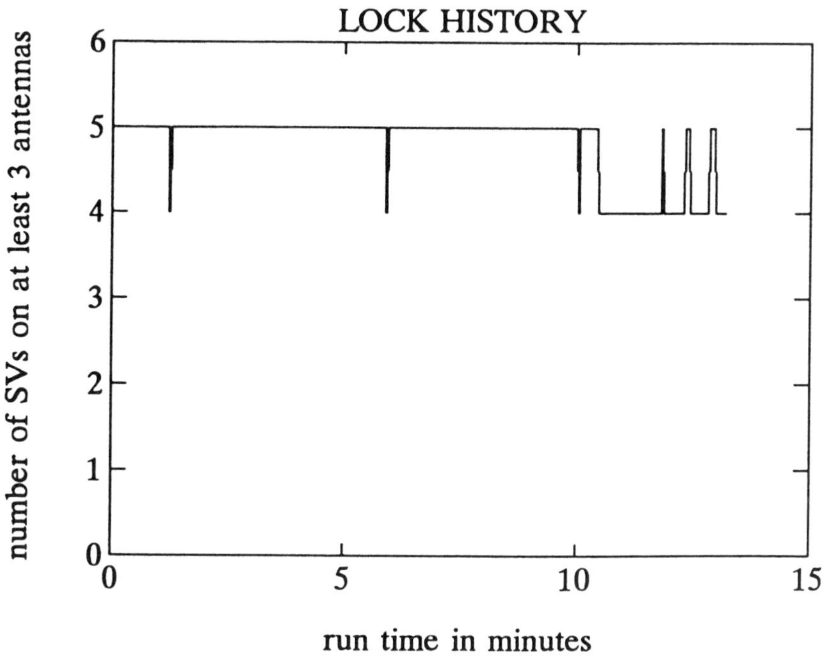

Fig. 20—Lock History for a 13 min Section of the Third Flight Test

ACKNOWLEDGMENTS

This work was supported by the FAA/NASA Joint University Program for Air Transportation Research (Grant NGR-36-009-17). The 3DF GPS receiver was provided by Ashtech, Inc. The authors are indebted to Dr. Javad Ashjaee and Mr. Mark Kuhl for their support, including many technical discussions on the topic of GPS interferometry, and their help in processing the raw measurement data. The authors would also like to thank the DC-3 flight crew, Dr. Richard H. McFarland, Mr. Dennis J. Atkeson, and Mr. Richard Zoulek, for their flexibility in dealing with the GPS satellite visibility windows.

Based on a paper presented at The Institute of Navigation Annual Meeting, Williamsburg, Virginia, June 1991.

REFERENCES

1. Greenspan, R. L., Ng, A. Y., Przyjemski, J. M., and Veale, J. D., *Accuracy of Relative Positioning by Interferometry with Reconstructed Carrier GPS: Experimental Results,* 3rd International Geodetic Symposium on Satellite Doppler Positioning, Las Cruces, NM, February 8–12, 1982.
2. Burgett, W. S., Roemerman, S. D., and Ward, P. W., *The Development and Applications of GPS-Determined Attitude,* Presented to the National Telesystems Conference, November 1983.
3. Remondi, B. W., *Performing Centimeter-Level Surveys in Seconds with GPS Carrier Phase: Initial Results,* NAVIGATION, Journal of The Institute of Navigation, Vol. 32, No. 4, Winter 1985–86.

4. Kruczynski, L. R., Li, P. C., Evans, A. G., and Hermann, B. R., *Using GPS to Determine Vehicle Attitude: USS Yorktown Test Results,* Proceedings of the 2nd International Technical Meeting of the Satellite Division of The Institute of Navigation, Colorado Springs, CO, September 27–29, 1989.
5. Purcell, Jr., G. H., Srinivasan, J. M., Young, L. E., DiNardo, S. J., Hushbeck, Jr., E. L., Meehan, T. K., Munson, T. N., and Yunck, T. P., *Measurement of Aircraft Position, Velocity, and Attitude Using Rogue GPS Receivers,* Proceedings of the 5th International Geodetic Symposium on Satellite Positioning, Las Cruces, NM, March 1989.
6. Ferguson, K., Kosmalska, J., Kuhl, M., Eichner, J. M., Kepski, K. and Abtahi, R., *Three-Dimensional Attitude Determination with the Ashtech 3DF 24-Channel GPS Measurement System,* Proceedings of the National Technical Meeting of The Institute of Navigation, Phoenix, AZ, January 1991.
7. Goldstein, H., *Classical Mechanics,* 2nd ed., Reading, MA: Addison-Wesley, 1980.
8. Farrell, J. L., *Integrated Aircraft Navigation,* San Diego, CA: Academic Press, 1976.
9. Farrell, J. L. and Van Graas, F., *That All-Important Interface,* Proceedings of the Third International Technical Meeting of the Satellite Division of The Institute of Navigation, Colorado Springs, CO, September 1990.
10. Remondi, B. W., *Using the Global Positioning System (GPS) Phase Observable for Relative Geodesy: Modeling, Processing, and Results,* Ph.D. Dissertation, The University of Texas at Austin, Austin, TX, 1984.
11. Jurgens, R. D., *Realtime GPS Azimuth Determining System,* Proceedings of the National Technical Meeting of The Institute of Navigation, San Diego, CA, January 23–25, 1990.
12. Hatch, R., *Ambiguity Resolution in the Fast Lane,* Proceedings of the 2nd International Technical Meeting of the Satellite Division of The Institute of Navigation, Colorado Springs, CO, September 27–29, 1989.
13. Hwang, P. Y. C., *Kinematic GPS for Differential Positioning: Resolving Integer Ambiguities on the Fly,* NAVIGATION, Journal of The Institute of Navigation, Vol. 38, No. 1, Spring 1991.
14. Brown, R. G., Chin, G. Y., and Kraemer, J. H., *RAIM: Will It Meet the RTCA GPS Minimum Operational Performance Standards?,* Proceedings of the National Technical Meeting of The Institute of Navigation, Phoenix, AZ, January 22–24, 1991.
15. Brenner, M., *Implementation of a RAIM Monitor in a GPS Receiver and an Integrated GPS/IRS,* Proceedings of the 3rd International Technical Meeting of the Satellite Division of The Institute of Navigation, Colorado Springs, CO, September 1990.
16. Golub, G. H. and Van Loan, C. F., *Matrix Computations,* 2nd ed., Baltimore, MD: The Johns Hopkins University Press, 1989.
17. Hwang, P. Y. C. and Brown, R. G., *GPS Navigation: Combining Pseudorange with Continuous Carrier Phase Using a Kalman Filter,* NAVIGATION, Journal of The Institute of Navigation, Vol. 37, No. 2, Summer 1990.
18. Hatch, R., *The Synergism of GPS Code and Carrier Measurements,* Proceedings of the 3rd International Geodetic Symposium on Satellite Doppler Positioning, Las Cruces, NM, February 1982.
19. Nesbo, I. and Canter P., *GPS Attitude Determination for Navigation,* GPS World, September/October 1990.
20. Brown, R. and Evans, A., *GPS Pointing System Performance,* ION GPS-90, Colorado Springs, CO, September 1990.
21. Martin-Neira, M., Lucas, R., and Martinez, M., *Attitude Determination with GPS: Experimental Results,* IEEE Aerospace and Electronics Systems Magazine, September 1990.